DATE DUE

OCT 23 02			

DEMCO 38-296

WETLAND INDICATORS

A Guide to Wetland Identification, Delineation, Classification, and Mapping

WETLAND INDICATORS

A Guide to Wetland
Identification, Delineation,
Classification, and Mapping

Ralph W. Tiner

LEWIS PUBLISHERS

Boca Raton London New York Washington, D.C.

Project Editor:	Maggie Mogck
Marketing Managers:	Barbara Glunn, Jane Lewis,
	Arline Massey, Jane Stark
Cover design:	Dawn Boyd

Library of Congress Cataloging-in-Publication Data

Tiner, Ralph W.
    Wetland indicators : a guide to wetland identification, delineation, classification, and mapping / Ralph W. Tiner.
        p. cm.
    Includes bibliographical references and index.
    ISBN 0-87371-892-5
    1. Wetlands--United States. 2. Wetland ecology--United States. 3. Plant indicators--United States. I. Title.
    GB624.T564 1999
    551.41--dc21

                                                                        98-51791
                                                                 CIP

Preface

The existence of wetlands has been and remains very important to people around the globe. Since the dawn of mankind, wetlands have provided a wealth of materials for food, fiber, and shelter that have supported many communities. In fact, the Tigris–Euphrates delta, long considered the "cradle of civilization," is an enormous wetland complex that has sustained hundreds of generations of families. The vital link between wetlands and water has created conditions that generate a host of ecosystem functions critical to the existence of many plants and animals and that also benefit mankind in many ways. Some wetlands are among the most diverse and productive natural systems on the planet, thereby serving as important places for the conservation of biodiversity. This productivity has also benefitted human civilizations. The fertile soils of many wetlands, especially those on floodplains, lured people to these areas who then put many wetlands into agricultural production. The staple crop for much of the world's population — rice — is a wetland plant. The location of many wetlands along rivers and other waterways was another important attraction as these waters served as transportation corridors facilitating trade, connecting one community to another. These uses led to the establishment of villages, towns, and cities near many wetlands. In other cases, they played a strategic role in military operations as roads through the larger wetland systems were critical arteries for supplying armies with supplies and equipment. These are but a few of the reasons people the world over have been interested in wetlands and their whereabouts.

As human cultures grew from small tribes reliant on natural subsistence to the agriculturally dependent and later the industrialized civilizations, human attitudes toward wetlands undoubtedly changed. With less dependence on natural products, the majority of people in these cultures likely viewed wetlands as wastelands whose best use could only be attained through conversion to agriculture or to "dry land" for development. Later, as the quality of the natural environment began to noticeably deteriorate (e.g., increasing water pollution, degrading air quality, more flooding of cities and towns, and accelerating wetland destruction), more Americans began to again appreciate the values of natural wetlands. In the U.S. today, wetlands are recognized as one of the nation's most valuable natural resources, providing valued functions such as water quality filtration, flood storage, erosion control, and shoreline stabilization. Wetlands also are essential habitats for many fish and wildlife species and areas where a variety of natural products (e.g., timber, wild rice, cranberries, blueberries, and peat moss) are produced. Due to these and other functions, wetlands are important enough for society to be concerned about their alteration (i.e., loss and degradation).

In the U.S., public recognition of the natural functions of wetlands and concern over accelerating wetland losses in the 1950s and 1960s led to the establishment of federal and state legislation to protect or at least attempt to minimize the adverse environmental impacts of development. The Federal Clean Water Act is the main federal law controlling the exploitation of wetlands in the U.S. Although this law is not a wetland protection law per se, it regulates wetland development due to the role of wetlands in maintaining water quality. Consequently, the federal government now regulates uses of wetlands that would adversely affect the quality of America's waters. Although the Clean Water Act has done much to improve the plight of wetlands over the past 2 decades, this law is not a wetland protection law, so many wetlands that are not important to water quality, but are vital to unique forms of plants and animals, are not currently protected by federal law. In contrast, numerous states have passed "wetland protection laws" specifically designed to protect the multiple-functions that wetlands provide, including fish and wildlife habitat. In the early to mid1960s, Massachusetts was the first state to pass wetland protection acts, first for coastal wetlands, and shortly thereafter for inland wetlands. Other coastal states followed this state's lead; today most states have laws that regulate uses of tidal wetlands. Several others have also passed similar laws for inland wetlands. Also, in more urban and expanding suburban areas, many local governments have established zoning ordinances to control development in wetlands. All of these laws and regulations now make it vitally important to know how to identify wetlands and their boundaries.

To meet their wetland protection mandates, federal and state governments had to develop techniques to aid in the identification and delineation of wetlands. These methods evolved over time with the earliest approaches using rather simple assessment techniques. As the science of wetland ecology grew in the 1970s and 1980s, more information on wetland plants, soils, and hydrology became available. This increased knowledge helped better articulate the concept of wetlands and necessitated improvements in wetland identification methodologies. Consequently, the extent of lands recognized as "wetlands" for regulatory purposes increased in geographic scope, with a rather obvious disapproving response from some segments of the regulated community.

During the past decade, there has been much public discourse and political rhetoric on what is a wetland (e.g., how wet should an area be in order to qualify as wetland) and how wetlands should be regulated. While the latter debate will undoubtedly continue, there appears to be widespread agreement among the scientific community on what constitutes a wetland. Since 1989, some politicians have been proposing different definitions of wetlands to curtail federal jurisdiction under the Clean Water Act. In 1993, Congress provided funds to the U.S. Environmental Protection Agency to commission a National Research Council (NRC) investigation of the wetland delineation issue. The NRC established the Committee on Characterization of Wetlands, a 17-member committee composed mostly of scientists (with expertise in wetlands, hydrology, soil science, and mapping) to review and evaluate the effects of alternative methods of wetland delineation and to summarize the scientific understanding of wetland functions with respect to wetland delineation. After 2 years of study, the NRC published its report, *Wetlands Characteristics and Boundaries*, in 1995. This document represents the view of the scientific community on wetlands. One of the report's major findings was that the federal government had, in general, been using appropriate scientific techniques for identifying wetlands. They recommended that a new federal wetland delineation manual be prepared because their study found more efficient ways of identifying wetlands in lieu of the federal government's current three-parameter/criteria procedures. The report made numerous other recommendations on how to improve the current system. In spite of a government proclamation in President Clinton's "No-Net-Loss" policy of using the best science in regulating wetlands, the federal government has yet to pursue development of a new manual with more efficient and effective methods based on current scientific knowledge.

In the past quarter century, wetlands have received more scientific study than in all the years before. More research articles have been published, several journals and newsletters are devoted to wetlands; and wetland maps have been produced by the federal government, many states, and some local communities. As a result of these efforts, wetlands are better understood, the concept of wetland is better explained, and the techniques for wetland identification and delineation are more refined and scientifically sound. Indicators used to identify wetlands have moved from a traditional dependence on plants and plant communities to an approach that considers a combination of factors including vegetation, soils, and signs of hydrology.

This book was written to aid readers in understanding the current concept of wetland and the use of various plant, soil, and other indicators for wetland identification in the U.S., and in learning how to identify, describe, classify, and delineate wetlands. The specific purposes of this book are: (1) to review the concept of wetland beginning with a review of wetland definitions (late 1800s to the present), (2) to examine the use of wetland indicators and provide background on the science behind them, (3) to review current wetland identification and delineation techniques, (4) to discuss problem situations for wetland delineation and offer technical guidance for their resolution, (5) to review wetland classification systems, (6) to introduce the diversity of wetland plant communities across the country, and (7) to address wetland photointerpretation and mapping.

In discussing wetland indicators, the book largely focuses on plants, soils, and other signs of wetland hydrology in the soil or on the surface of wetlands. The significance of learning to recognize certain landscape positions and landforms that facilitate wetland formation is also mentioned. The book does not emphasize wetland animals as possible indicators of wetlands, with few exceptions (e.g., aquatic invertebrate remains and crayfish burrows). Yet it must be recognized that the presence of many wetland-dependent animals (e.g., beaver, muskrat, wading birds, waterfowl, shorebirds, clams, mussels, and other aquatic invertebrates) also suggests the existence of wetland. Due to their mobility, however, many of these animals may be found in uplands or deepwater habitats, so their presence in a given area at a particular point in time does not make the area wetland. In addition, they tend to be associated with the more obvious, wetter wetlands where water is present much of the year, so their use in wetland identification is of limited added value.

Information presented in this book, coupled with the NRC's findings, should give readers a fairly thorough understanding of the current state-of-knowledge about the science behind wetland delineation and improved insight for evaluating the strengths and weaknesses of current delineation practices. Overall, it is hoped that this book will help individuals interested in wetland delineation strengthen their knowledge about wetlands and the use of various indicators and boost their confidence in applying professional judgement when making difficult wetland determinations.

The book should be required reading for anyone involved with wetland regulations and delineation. Both practicing wetland delineators and others seeking to become delineators should find that it provides critical insight into the development and significance of hydrophytic vegetation, hydric soils, and other wetland indicators and their use for wetland identification and delineation. Chapters on U.S. wetlands and wetland mapping and photointerpretation provide the reader with an understanding of the variety of plant communities associated with America's wetlands and some offsite tools to help identify them. The book may also serve as a supplemental textbook for upper-level undergraduate and graduate courses in wetland ecology, especially those emphasizing applied wetland science (i.e., wetland identification, classification, delineation, and mapping).

In writing this book, I became more concerned that certain wetlands are not being identified as "regulatory wetlands" because they fail to exhibit the suite of necessary indicators approved by some regulatory agencies. Consequently, these types are not protected by current laws. For some of them more information on their functions may be required to justify increased federal or state protection (e.g., marginally wet flatwoods), yet others have widely recognized values unique to these ecosystems. For example, West Coast vernal pools and small isolated vernal pools in eastern forests are vital areas for preserving biodiversity and critical breeding grounds for many amphibians, respectively. Nonwetland riparian habitats intricately linked to contiguous wetland and aquatic systems are also vital to local wildlife and migratory species. The above areas are rapidly disappearing and new legislation and/or government initiatives will be required to protect them by some combination of regulation, acquisition, and other means that promote private stewardship. Currently, wetlands are regulated nationally, largely under the Clean Water Act, due to their water quality maintenance function. Yet, recognizing the importance of preserving diverse ecosystems and the overall importance of wetlands (and associated riparian habitats) along rivers and streams and to preserving the integrity of these aquatic ecosystems and fish and wildlife dependent on these habitats, our nation's leaders need to muster the courage and support to establish a national wetlands and riparian habitat conservation act. This will help conserve, restore, and enhance wetlands and riparian habitat for their intrinsic values, not simply for their flood storage and water quality functions. The quality of the natural world around us is an important societal value that needs to be addressed in natural resource management.

Ralph Tiner
Wetland Ecologist

Author

Ralph Tiner is a nationally recognized authority on wetland identification and delineation. He is a wetland ecologist with nearly 30 years of experience in wetland identification, mapping, classification, and delineation and has advanced degrees from the University of Connecticut and Harvard University. He serves as an adjunct professor in the Department of Plant and Soil Sciences at the University of Massachusetts–Amherst. Since the mid-1980s, he has taught over 1000 students, mostly government personnel and consulting biologists, on how to identify and delineate wetlands using plants, soils, and signs of hydrology. In addition to an undergraduate/graduate course in wetland delineation at the university, he teaches short courses on this topic, plus wetland plant identification for environmental professionals at various institutions including Rutgers University and Ohio State University.

Professor Tiner has been identifying and mapping wetlands since 1970 when he helped conduct on-the-ground mapping of Connecticut's tidal wetlands for one of the nation's first state tidal wetland regulatory programs. Since then, he has been responsible for mapping wetlands from Maine through Virginia, and in South Carolina and other local areas of the country. He has written over 100 publications on wetland topics such as wetland status and trends, wetland identification, classification, mapping, restoration, hydric soils, and wetland plants. Among his more popular books are *Field Guide to Coastal Wetland Plants of the Northeastern United States* (1987), *Field Guide to Nontidal Wetland Identification* (1988), *Field Guide to Coastal Wetland Plants of the Southeastern United States* (1993), and *In Search of Swampland: A Wetland Sourcebook and Field Guide* (1998).

Acknowledgments

A book of this magnitude would not have been produced without the assistance of many individuals. First, I would like to recognize the following scientists whose writings and/or acquaintance have greatly influenced my understanding of wetlands over the past 28 years — Rob Brooks, Virginia Carter, Lew Cowardin, Frank Golet, Ted LaRoe, Mike Lefor, Maury Mausbach, Bill Niering, Eugene Odum, Blake Parker, John Rankin, Buck Reed, Charlie Rhodes, Bill Sipple, John Teal, Billy Teels, Russ Theriot, Peter Veneman, and Bill Wilen.

Several people contributed to the preparation of this book by reviewing various draft chapters of the manuscript — Wade Hurt (USDA Natural Resources Conservation Service), Bob Lichvar (U.S. Army Corps of Engineers), Ken Metzler (Connecticut Department of Environmental Protection), Sid Pilgrim (University of New Hampshire), Bill Sipple (U.S. Environmental Protection Agency), Steve Sprecher (U.S. Army Corps of Engineers), Peter Veneman (University of Massachusetts), and Mike Whited (USDA Natural Resources Conservation Service). Fred Weinmann (U.S. Environmental Protection Agency) and Kevin Bon (U.S. Fish and Wildlife Service) reviewed the tables of plant communities for western wetlands. Benoit Gauthier (Québec Ministère de l'Environnement et de la Faune) provided comments and clarification on Quebec's wetland identification and delineation methods. I appreciate their prompt and thoughtful responses.

Many individuals provided information and/or photographs that were used in this book — Dean Albro (Rhode Island Department of Environmental Protection), C.W.P.M. Blom (Catholic University of Nijmegen), Kevin Bon, Mark Brinson (East Carolina University), Rob Brooks (Pennsylvania State University), Mark Brown (U.S. Forest Service), Vic Carlisle (University of Florida), Ton Damman (Kansas State University), Chuck Elliott (U.S. Fish and Wildlife Service), Debbie Flanders (New England Wildflower Society), Benoit Gauthier, Art Gold (University of Rhode Island), Jon Hall (U.S. Fish and Wildlife Service), Tom Hruby (Washington Department of Ecology), Wade Hurt, Steve Faulkner (Louisiana State University), Mary Leck (Rider College), Bob Lichvar, Norm Mangrum (U.S. Fish and Wildlife Service), Irv Mendelssohn (Louisiana State University), Ken Metzler, Bill Patrick (Louisiana State University), Dennis Peters (U.S. Fish and Wildlife Service), Jimmy Richardson (North Dakota State University), Kim Santos (U.S. Fish and Wildlife Service), Bill Sipple, Dan Smith (U.S. Army Corps of Engineers), Steve Sprecher, Becky Stanley (U.S. Fish and Wildlife Service), Janice Stone (University of Massachusetts), Tom Tiner (Earthdata Int'l), Hank Tyler (Maine State Planning Office), Peter Veneman, Martin Wassen (University of Utrecht), Fred Weinmann, Mike Whited, Bill Wilen, and Robert Zampella (New Jersey Pinelands Commission). I'd also like to thank Pam Rooney and Katherine Sawyer for translating the article on Quebec's wetland delineation technique. The U.S. Fish and Wildlife Service, the Delaware Department of Natural Resources and Environmental Control, and the Massachusetts Department of Environmental Protection provided use of aerial photos to demonstrate the art of wetland photointerpretation.

The following persons and organizations provided photographs that appear as plates in this book (the number in parentheses is the plate number): John Carey (13), Vic Carlisle (14), Earthdata International (55), Phil Tant (25, 26, 40, 42), USDA Natural Resources Conservation Service (8, 15, 16, 20, 22, 23), U.S. Fish and Wildlife Service (51 to 54), Peter Veneman (7, 9, 12, 36, 39, 41), and Bill Zinni (46). A special thanks to them for their contribution. All other plates were photographs taken by the author.

Finally, I would like to thank my family for their indulgence and sacrificing opportunities for quality family time that allowed me to write this book. They well understand the seemingly infinite hours it takes to create this type of work and have seen the virtually endless piles of background material disappear from view in the final months of this book's preparation. A very special thank you is extended to my wife, Barbara, who came to my aid in the final 2 months by typing numerous tables and making sure that all the text citations were accounted for in the references of each chapter. She did a Herculian effort — many thanks, Barb!

I would also like to acknowledge the staff at Lewis Publishers for their assistance in bringing this book to print. I'd especially like to thank Jane Kinney for her patience and support during the final stages of manuscript preparation and Maggie Mogck for her help in getting this book to press.

Contents

Chapter 1   Wetland Definitions

Introduction 1
Shaler's Wetland Definition (1890) 3
Recent Scientific Definitions 3
    1950s and 1960s Fish and Wildlife Service Wetland Definition 4
    1970s to Present Fish and Wildlife Service Wetland Definition 4
    National Research Council Definition 6
    Canadian National Definition 7
    Other International Definitions 8
        Ramsar Wetland Definition...8
        Australian Wetland Definition 8
        Zambian Wetland Definition..9
Regulatory Definitions 9
    Federal Clean Water Act Definition 9
    Food Security Act Definition...10
    Selected State Definitions 11
Field Application of Wetland Definitions..13
References 14

Chapter 2   Wetland Concepts for Identification and Delineation

Introduction 17
How Wet is a Wetland?..18
    Flooding or Waterlogging Duration...19
    Frequency of Prolonged Wetness 20
    Critical Depth of Saturation...20
    Timing of Wetness 20
    Wetland Hydrology Defined 21
The Significance of "Growing Season" for Defining Wetland Hydrology 21
    Federal Regulatory Definition of Growing Season...21
        Shortcomings of the Regulatory Concept 22
    Ecological Considerations in Defining the Critical Period for Wetland
    Hydrology...24
        Botanical Evidence 24
        Faunal Evidence..28
        Wetland Functions Evidence 29
        Concluding Remarks...29
Are all Wetlands at least Periodically Anaerobic, Reduced Environments? 30
What are Common Types of Wetland Hydrology?...31
Are Wetlands Ecotones? 34
    Ecotonal Wetlands...35
    Non-ecotonal Wetlands 37
Do Wetlands Really Become Dryland?..39
Are Some Wetlands Ephemeral Cyclical Features on the Landscape? 42
References 43

Chapter 3 Plant Indicators of Wetlands and their Characteristics

Introduction 51
Environmental Changes Due to Flooding and Waterlogging 51
Plant Adaptations to Flooding and Waterlogging 53
    Morphological Adaptations...55
        Hypertrophied Stems..56
        Fluted Trunks 57
        Hollow Stems 58
        Shallow Root Systems 59
        Adventitious Roots..60
        Soil Water Roots 62
        Pneumatophores 62
        Aerenchyma 62
        Hypertrophied Lenticels...65
    Oxidized Rhizospheres 66
    Life-Form or Habit Changes...67
    Germination and Seedling Survival 68
    Accelerated Stem Growth 69
Wetland Plants — Hydrophytes 69
    Evolving Concept of a Hydrophyte...70
    Plant Specificity 71
    Wetland Ecotypes...71
Using Plants to Identify Wetlands 76
    Wetland Plant Lists 76
    Field Indicators of Hydrophytic Vegetation 78
    Problems Interpreting FAC– and FACU Species as Hydrophytes..........78
    Human Effects on Wetland Plant Distribution 80
Using Plants to Predict Environmental Conditions in Wetlands 81
    Plants as Hydrology Indicators...81
    Plants as Indicators of Water Sources 81
    Plants as Water Chemistry Indicators 83
    Plants as Soil Type Indicators..86
    Plants as Indicators of Nativeness 87
Useful Guides for Identifying Wetland Plants 87
References 89

Chapter 4 Vegetation Sampling and Analysis for Wetlands

Introduction 101
Stand Identification 102
Locating Sample Areas 103
Timing of Sampling 103
Sampling Strata..103
Sampling Variables...104
    Areal Cover 104
    Basal Area 106
    Height Classes...110
    Density 110
    Frequency 110

(Color Insert — Plates 1 to 55)

Dominance 111
Weighted Averages 111
Basic Vegetation Sampling Techniques 114
Plot or Quadrat Sampling 114
Sample and Plot Sizes 114
Number of Plots 116
Shape of Plots 117
Point Intercept Sampling 118
Other Sampling Methods 119
Exercises in Determining the Presence of Hydrophytic Vegetation 122
References 124

Chapter 5   Soil Indicators of Wetlands

Introduction 127
Basic Soil Properties...128
Soil Definition 128
Soil Composition..129
Soil Color 130
Waterlogging Effects on Soil Properties 133
Other Effects of Flooding on Soil Development 138
Factors Affecting the Color of Soil 139
Wetland Soils 140
The Evolving Hydric Soil Definition and Technical Criteria 141
General Concept of Hydric Soils 146
Lists of Hydric Soils and Hydric Soil Map Units 148
Major Categories of Hydric Soils...149
Organic Soils...149
Mineral Hydric Soil 151
Lists of U.S. Hydric Soil Indicators...152
Hydric Soil Field Indicators in the U.S. 160
Histic Materials..160
Organic Soils...160
Histic Materials in Mineral Soils 165
Thick, Dark Surface Layers..168
Redox Dark Surface Layers...168
Gleyed Matrix 168
Depleted Matrix 169
Mottled Depleted Matrix 170
Sulfidic Materials 170
Depressional Indicators...170
Other Hydric Soil Indicators 171
Indicators of Flooded or Ponded Soils..171
Internationally Recognized Hydric Soil Properties..174
Measuring Reduction or Soil Redox Potential..175
Interpreting Soil Taxonomy to Identify Potential Hydric Soils..176
References 178

Chapter 6   Wetland Identification and Boundary Delineation Methods

Introduction 187
Evolution and Use of Wetland Identification and Delineation Methods 188

Differing Wetland Identification and Boundary Delineation Protocols 191
    Corps Manual 192
    EPA Manual 194
    Food Security Act Manual 195
    Federal Interagency Manual 196
    Primary Indicators Method 198
    Connecticut Method 200
    Florida Manual 200
    Massachusetts Manual 202
    New York Manual 203
    Rhode Island Method 204
    Wisconsin Method 204
    Quebec Method 205
Wetland Delineation Methods 205
    Corps Manual 205
    EPA Manual 208
    Federal Interagency Manual 209
National Research Council Findings and Recommendations 214
Wetland Delineation Tips 215
    Preparation for the Field 215
    Field Tips 216
Professional Judgment 218
References 218

Chapter 7   Problem Wetlands and Field Situations
Introduction 221
Problematic Wetland Plant Communities 222
    Wetlands Dominated by Often Dry-Site Species 222
    Seasonally Variable Wetland Plant Communities 225
    Cyclical Wetlands in Arid to Subhumid Regions 226
    Cyclical Wetlands in Permafrost Regions 229
    Tropical Wetlands 229
Problematic Soils 230
    Sandy Hydric Entisols 231
    Nonsandy Hydric Entisols 233
    Hydric Spodosols 234
    Hydric Mollisols 235
    Hydric Vertisols 236
    Wet Aridisols 237
    Red Parent Material Soils 238
    Saturated Soils Not Forming Hydric Properties 238
    Dark-Colored Parent Material Soils 239
    Cryic Soils 240
    Newly Formed Soils 240
        Anthraquic Soils 241
    Relict Hydric Soil 242
    Drained Hydric Soils (Partly Drained Wetlands) 243
        Farmed Wetlands 246
Hydrologically Problematic Wetlands 247
    Groundwater-Driven Wetlands 247

Drier-End Wetlands 248
    Flatwoods 248
    Floodplain Wetlands 248
    Western Riparian Habitat 248
Problematic Field Conditions 249
    Complex Landscapes 249
    Rocky Landscapes..250
    Irrigated Lands 250
National Research Council Recommendations 250
Other Recommendations...250
References 251

Chapter 8   Wetland Classification

Introduction 257
Features Used to Classify Wetlands 259
    Vegetation...259
    Hydrology...260
    Water Chemistry...260
    Origin of Water 262
    Soil Types...262
    Landscape Position and Landform 262
    Wetland Origin 262
    Wetland Size...262
    Ecosystem Form/Energy Sources 263
Early Ecological Classifications 263
Wetland Classification in the U.S. 264
    1950s to 1960s Fish and Wildlife Service Classification 265
    Current Fish and Wildlife Service's Wetland Classification System.....267
    Hydrogeomorphic (HGM) Classification 272
    FWS/HGM-Type Classification Systems 274
Wetland Classification in Canada 276
Other Wetland Classification Systems...279
    Ramsar: A Multinational Classification..279
    Peatland Classifications..280
    Australia 282
    Western Australia 284
References 285

Chapter 9   Wetlands of the U.S.: An Introduction, with Emphasis on their Plant Communities

Introduction 291
    Wetland Distribution 291
Palustrine Wetlands 294
    Emergent Wetlands 296
    Scrub-Shrub Wetlands..301
    Forested Wetlands 303
Estuarine Wetlands..306
    Estuarine Emergent Wetlands 306
    Estuarine Scrub-Shrub Wetlands 308

       Estuarine Intertidal Shores..309
   References 332

Chapter 10   Wetland Mapping and Photointerpretation
  Introduction 347
  Maps to Aid in Wetland Identification 347
     Map Limitations...348
     National Wetlands Inventory Maps 350
        Mapping Procedure 350
        Features Mapped...351
        NWI Map Strengths and Weaknesses...355
        Uses of NWI Maps 356
     County Soil Surveys 358
        Mapping Procedures 359
        Features Mapped...359
        Soil Survey Strengths and Weaknesses 361
        Uses of Soil Surveys...362
     Comparing NWI Maps with Soil Survey Maps..362
     Use of Maps for Regulatory Determinations 365
     Future Maps 365
  Wetland Photointerpretation..366
     Photointerpretation Concerns..366
     Interpreting Estuarine Wetlands..369
     Interpreting Palustrine Wetlands..371
     Interpreting Submerged Aquatic Vegetation..375
  References 376

Index 381

Dedication

*to all professionals who seek to use
the best of science
to identify, delineate, evaluate,
and conserve wetlands*

1 Wetland Definitions

INTRODUCTION

Wetland is a generic term used to define the universe of wet habitats including marshes, swamps, bogs, fens, and similar areas. Wetlands are environments subject to permanent or periodic inundation or prolonged soil saturation sufficient for the establishment of hydrophytes* and/or the development of hydric soils or substrates** unless environmental conditions are such that they prevent them from forming. They are places where a recurrent excess of water imposes controlling influences on all biota (plants, animals, and microbes). Given regional differences in hydrologic regimes, climate, soil-forming processes, and geomorphologic settings, a vast assemblage of wetland plant communities and hydric soil types have evolved worldwide. Numerous terms have been applied to individual wetlands because of these differences. Some common wetland types in North America include salt marsh, freshwater marsh, tidal marsh, alkali marsh, fen, wet meadow, wet prairie, alkali meadow, shrub swamp, wooded swamp, bog, muskeg, wet tundra, pocosin, mire, pothole, playa, salina, salt flat, tidal flat, vernal pool, bottomland hardwood swamp, river bottom, lowland, mangrove forest, and floodplain swamp.

Within regions, wetlands naturally form in places on the landscape where surface water periodically collects for some time and/or where groundwater discharges, at least seasonally, sufficient to create waterlogged soils. Common wetland landforms include (1) depressions surrounded by upland and with or without a drainage stream, (2) relatively flat low-lying areas (floodplains) along major waterbodies (e.g., rivers, lakes, and estuaries) usually with fluctuating water levels, (3) in shallow water of protected (low-energy) embayments and slow-flowing channels, (4) broad flat areas lacking drainage outlets (e.g., interstream divides), (5) vast expanses of arctic and subarctic lowlands where a permafrost layer occurs near the surface (muskegs), (6) sloping terrain below sites of ground water discharge (e.g., springs, seeps, toes of slopes, and drainageways), (7) slopes below melting snowbeds and glaciers, and (8) flat or sloping areas adjacent to bogs and subject to paludification processes in regions with cold, wet climates (Figure 1.1). While most wetlands tend to develop in the above areas, some wetlands form on steep slopes. In regions of high rainfall, steep slopes with clayey soils may support extensive wetlands, as in Puerto Rico where 200 in. (500 cm) of rainfall creates such conditions (Lugo et al., 1990).

Some people have said that wetland is a euphemism for "swamp" that has evoked negative feelings due to its usage in literature and everyday speech (National Research Council, 1995). If this is true, it is likely an unintentional euphemism, since it is clear that the term "wetland" is simply a combination of two words, "wet" and "land," that simply describes the overall condition of the land. Such land has water present at or near the surface for significant periods that affect land use. It may be wet too long to grow most crops without some artificial drainage or long enough to pose significant constraints on construction activities (e.g., can't build without filling and/or draining). The term "wetland" has been in use for some time. Shaler (1890) referred to "the wet

---

* Hydrophytes are individual plants growing in water or on substrates (e.g., soils) that are periodically subject to anaerobic conditions due to excessive wetness.
** Hydric soils are soils that are saturated at or near the surface (by flooding or high ground water tables) usually frequent and long enough to promote the development of anaerobic reducing conditions that affect plant growth and promote the establishment of erect (self-supporting) hydrophytes. Hydric substrates are permanently or nearly permanently inundated or saturated substrates lacking erect hydrophytes; they are mostly nonvegetated areas (e.g., the bottoms of permanent waterbodies), but may be vegetated with submergent species or floating-leaved plants.

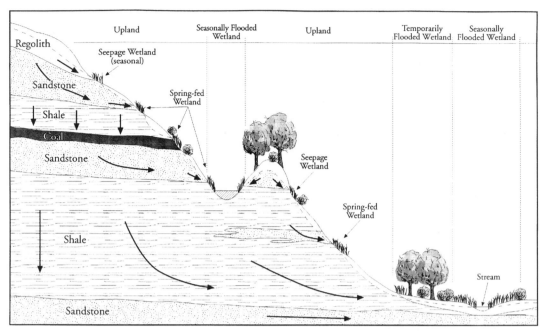

Vertical scale greatly exaggerated.

FIGURE 1.1   Wetlands form on the landscape where there is an abundance of water such as seeps and springs, and along waterbodies (e.g., rivers, streams, lakes, ponds, and estuaries). Water flow paths are shown by the dark arrows. (From Tiner, R.W. and D.G. Burke, 1995.)

lands of the Old World" when comparing the significance of sphagnum growth in the U.S. to Europe. The common usage of "wetland" is further witnessed by Dachnowski's complaint that the usage of too many generic terms, including wetland, overflowed lands, swamp land, and muck, were hindering scientific studies of peat (Dachnowski, 1920).

Wetland does not imply that the land has to be permanently flooded. If this was the case, such lands would simply be called "submerged lands" or "flooded lands", rather than wet land. Instead, wetland suggests that land that is wet for various periods — land that is periodically wet and when not, may be dry or at least, exposed to air for some time. Wetlands include both periodically and permanently flooded lands (see Chapter 2), yet the latter are typically restricted to shallow water areas that often support either free-standing plants like reeds, shrubs, or trees, or rooted floating-leaved aquatic plants like water lilies, unless wave energy and other factors prevent their colonization.

When considering periodically wet lands, the frequency and duration of flooding and/or soil saturation (near the surface) needs to be taken into account. It is the recurring prolonged wetness that exerts significant stress on vegetation to exclude most plants from growing in wetlands. The fundamental questions for defining wetlands, therefore, include (1) how long and at what frequency must an area be wet, (2) must wetness occur during the growing season or is nongrowing season wetness also important to consider, and (3) should wetness occur to the ground surface or, if not, what soil depth is critical? These questions should be kept in mind when reviewing existing wetland definitions (see Chapter 2 for discussion). The answers to these questions are vital to developing criteria and a set of reliable indicators (e.g., plant, animals, and soils) to validate the occurrence of wetland.

Numerous wetland definitions have been developed for various purposes. The earliest wetland definitions were created for scientific studies or management purposes, with the more widely used ones created to serve as the foundation for wetland mapping projects. Since the 1960s, other definitions have been formulated to establish limits on what "wet" lands should be regulated through various wetland laws and government wetland regulations.

SHALER'S WETLAND DEFINITION (1890)

One of the earliest reports on the nation's wetlands was Nathaniel Shaler's *General Account of the Freshwater Morasses of the United States* published in 1890. This U.S. Geological Survey report focused on inundated lands and included what is perhaps the nation's first national wetland classification system (see Chapter 8). In preparing a table with the estimated acreage of inundated lands east of the Rockies, Shaler included the following definition of "swamp".

"all areas…in which the natural declivity is insufficient, when the forest cover is removed, to reduce the soil to the measure of dryness necessary for agriculture. Wherever any form of engineering is necessary to secure this desiccation, the area is classified as swamp."

In calculating the total wetland acreage, he also included "areas of alluvial lands subject to inundations in the tillage season to such an extent that agriculture is unprofitable until the land is drained or diked." Consequently, Shaler's concept of wetland (swamp) is quite broad and includes consideration of the effect of soil wetness on land use, namely farming.

RECENT SCIENTIFIC DEFINITIONS

Depending on the field of study, wetland definitions may focus on particular attributes as noted by Lefor and Kennard (1977) in their review of inland wetland definitions. A hydrologist's definition of wetlands would focus on the fluctuations in the water table and on the frequency and duration of flooding. A soil scientist's definition might center on the presence of certain soils, mainly poorly drained and very poorly drained soils, plus soils that are frequently inundated for long or very long duration, that affect crop production (e.g., soils that are too wet to grow corn and other crops without artificial drainage). A botanist's definition would emphasize the occurrence of certain plant species and certain plant communities and the wetness conditions promoting their colonization. A fish and wildlife biologist's definition might stress the characteristics associated with fish spawning and nursery grounds and with wetland-dependent wildlife like waterfowl, shorebirds, wading birds, beaver, muskrat, frogs, salamanders, turtles, and alligators. A civil engineer's definition of wetland would likely highlight the wetness conditions in the soils that affect the ability of the soil to support construction of roads, bridges, buildings, and similar structures. Thus, a plethora of wetland definitions could be developed based on different areas of expertise or interest.

Current wetland definitions are largely biologically based, since professionals in wildlife biology and botany were among the first to recognize the values that wetlands contribute to society in their unaltered state. Since vegetation patterns are readily observed in most cases, wetland definitions tend to emphasize certain plant communities associated with waterbodies or waterlogged soils. Such communities provide vital areas for fish spawning and nurseries and support unique forms of wildlife and plant species, and depending on their landscape position and wetness properties, provide other valued functions such as flood storage, nutrient recycling, water quality renovation, and shoreline stabilization.

The most widely used of the scientific definitions in the U.S. were developed by the U.S. Fish and Wildlife Service (FWS) which has a long history of interest in wetland conservation given their significance as critical habitats for many of the nation's fish and wildlife species. In order to provide more effective management of these habitats, the FWS has been monitoring the status of wetlands in the country since the 1950s. Two wetland definitions were developed to aid in performing inventories of U.S. wetlands. More recently, another scientific wetland definition was formulated by the National Research Council's Committee on Characterization of Wetlands as the reference for conducting a review of the federal government's wetland delineation practices. Canada and other countries have developed wetland definitions for their own inventories. These definitions are briefly discussed below.

1950s AND 1960s FISH AND WILDLIFE SERVICE WETLAND DEFINITION

During the 1950s and 1960s, the FWS initiated a nationwide inventory of wetlands important to waterfowl and followup surveys of key areas (e.g., coastal wetlands). In conducting these inventories, the FWS used the Martin et al. (1953) wetland classification system based on the following definition.

> Wetlands are "lowlands covered with shallow and sometimes temporary or intermittent waters. They are referred to by such names as marshes, swamps, bogs, wet meadows, potholes, sloughs, and river-overflow lands. Shallow lakes and ponds, usually with emergent vegetation as a conspicuous feature, are included in the definition, but the permanent waters of streams, reservoirs, and deep lakes are not included. Neither are water areas that are so temporary as to have little or no effect on the development of moist-soil vegetation. Usually these temporary areas are of no appreciable value to the species of wildlife considered in this report." (Shaw and Fredine, 1956)

While the focus on habitat for certain wildlife (mainly waterfowl) and vegetation is clear, the definition recognizes the variability of the hydrology (shallow to temporary or intermittent waters) in wetland habitats. Most of the wetland types listed are largely vegetated areas. Shallow waterbodies are included in the definition, whereas deeper waterbodies are omitted. This definition was also adopted for use by the USDA Soil Conservation Service for determining when it was appropriate for providing technical assistance for wetland drainage on farms. The Martin et al. system was criticized for inconsistent application, ignoring ecologically significant differences such as the distinction between fresh and subsaline wetlands, and lumping dissimilar wetland types like boreal spruce bogs with southern cypress-gum swamps (e.g., Leitch, 1966; Stewart and Kantrud, 1971; Cowardin et al., 1979).

1970s TO PRESENT FISH AND WILDLIFE SERVICE WETLAND DEFINITION

In 1974, the FWS established the National Wetlands Inventory Project to map the nation's wetlands for resource conservation purposes. For this survey, wetland maps were to be prepared to inform the public on the location of these significant natural resources. To accurately map these resources, the FWS had to determine where along the natural soil moisture continuum wetland ends and upland begins. An ecologically based definition was constructed by the FWS to help ensure accurate and consistent wetland determinations. The FWS did not attempt to legally define wetland, since it was recognized that each state or federal regulatory agency may define wetland somewhat differently to suit its administrative purposes. A science-based definition would serve more uses and users outside the FWS.

In January 1975, the FWS brought together wetland scientists from across the country to review the Martin et al. system in light of 20 years of new information on wetlands and experiences applying the system. They concluded that a new classification, should be developed (Sather, 1976). In July 1975, the FWS sponsored a wetland classification workshop in College Park, Maryland. During this meeting, wetland scientists from federal and state agencies, the Canadian government, and academia discussed wetland programs, concepts, and classification and reviewed a draft classification by Lewis Cowardin of the FWS and Virginia Carter of the U.S. Geological Survey. Participants indicated that many people regarded lakes and ponds as wetlands and that the seaward limit of wetland might be considered to be the edge of the continent shelf. There was widespread agreement that wetlands should be defined by plants, soils, and hydrology (water regimes) and that the definition should be general (not specifying individual plants or soil types), concise, easily understood, and scientifically based. Agricultural lands that had wetland water regimes and soil types should be included in the definition. Also, the presence of existing vegetation (e.g., agricultural crops or certain trees or herbaceous species not typical of wetlands) should not prevent classification of such areas as wetlands if their water regime and soils were typically wetland in nature. Western scientists felt that riparian vegetation should be included in the wetland definition. Questions also were raised on whether the concept of wetland should include the entire 100-year floodplain as wetlands or at least in the wetlands inventory, recognizing the public interest in these resources (Sather, 1976). After this meeting, a four-person team (Cowardin, Carter, Frank Golet, and Ted LaRoe) developed the FWS's interim wetland classification system (Cowardin et al., 1976) that included the following wetland definition for review.

"Wetland is land where an excess of water is the dominant factor determining the nature of soil development and the types of plant and animal communities living at the soil surface. It spans a continuum of environments where terrestrial and aquatic systems intergrade. For purposes of this classification system, *wetland* is defined more specifically as land where the water table is at, near, or above the land surface long enough each year to promote the formation of hydric soils and to support the growth of hydrophytes, as long as other environmental conditions are favorable… In certain wetland types, vegetation is absent and soils are poorly developed or absent as a result of frequent and drastic fluctuations of surface-water levels, wave action, water flow, turbidity, or extremely high concentrations of salts or other substances in the water or substrate. Wetlands lacking vegetation and hydric soils can be recognized by the presence of surfacewater at some time during the year and their location within, or adjacent to, vegetated wetlands or aquatic habitats."

An April 1977 version contained a similar definition with only minor revisions (Cowardin et al., 1977a). This version was made available for public review and comment. In October 1977, the FWS produced an operational draft version of its classification system for initiating wetland mapping (Cowardin et al., 1977b). The definition is essentially the same as the 1976 version except that the first two sentences were eliminated from the definition and moved to the concepts discussion preceding the definition.

"Wetland is defined as land where the water table is at, or near or above, the land surface long enough to promote the formation of hydric soils or to support the growth of hydrophytes. In certain types of wetlands, vegetation is lacking and soils are poorly developed or absent as the result of frequent and drastic fluctuations of surface-water levels, wave action, water flow, turbidity, or high concentrations of salts or other substances in the water or substrate. Such wetlands can be recognized by the presence of surface water or saturated substrate at some time during each year and their location within, or adjacent to, vegetated wetlands or deep-water habitats."

After using the operational draft for a couple of years and receiving public input, the FWS amended its classification and revised its wetland definition (Cowardin et al., 1979). The following definition has served as the FWS's official wetland definition for nearly 2 decades and continues to be used for wetland mapping.

"Wetlands are lands transitional between terrestrial and aquatic systems where the water table is usually at or near the surface or the land is covered by shallow water. For purposes of this classification, wetlands must have one or more of the following three attributes: (1) at least periodically, the land supports predominantly hydrophytes; (2) the substrate is predominantly undrained hydric soil; and (3) the substrate is nonsoil and is saturated with water or covered by shallow water at some time during the growing season of each year."

In defining wetlands from an ecological standpoint, the FWS emphasizes three key attributes of wetlands: (1) hydrology — the degree of flooding or soil saturation, (2) wetland vegetation (hydrophytes), and (3) hydric soils. Although not mentioned specifically in the definition, the document's discussion of concepts and definitions also makes reference to animal communities stressed by prolonged saturation. The concept of wetland is a broad-based ecological one recognizing both plants and animals adapted to these conditions.* All areas considered wetlands must have enough water at some time during the year to stress plants and animals not adapted for life in water or saturated soils. Most wetlands have hydrophytes and hydric soils present, yet many are nonvegetated (e.g., tidal mudflats). The FWS prepared a list of plants occurring in the nation's wetlands and has revised the list periodically (Reed, 1988; Reed, 1997). Likewise, the USDA Natural Resources Conservation Service (formerly the Soil Conservation Service) developed a national list of hydric soils which also has been updated periodically (USDA Soil Conservation Service, 1991). These lists have greatly contributed to a better understanding of wetlands and their identification.

According to the FWS definition, wetlands typically fall within one of the following four categories: (1) areas with both hydrophytes and hydric soils (e.g., marshes, swamps, and bogs); (2) areas without hydrophytes, but with undrained hydric soils (e.g., farmed wetlands); (3) periodically inundated or permanently flooded areas without soils but with hydrophytes (e.g., seaweed-covered rocky intertidal shores and aquatic beds on hydric substrates in shallow water); and (4) periodically flooded areas without soil and without hydrophytes (e.g., gravel bars and tidal mudflats). All wetlands must be periodically saturated or covered by shallow water during the year (especially the growing season), whether or not hydrophytes or hydric soils are present.

Various altered wetlands also are contained within the FWS wetland definition. Farmed wetlands are wetlands where "the soil surface has been mechanically or physically altered for the production of crops, but hydrophytes will become reestablished if farming is discontinued." Such areas typically possess hydric soils and wetland hydrology. Partly drained wetlands still retain some degree of wetland hydrology, sufficient to support the continued growth of hydrophytes. Artificial wetlands are wetlands created either purposefully or accidentally. They include vegetated wetlands built in tidal areas by deposition of dredged spoil or constructed in impoundments or excavated basins in nontidal areas and nonvegetated wetlands such as rocky shores (e.g., jetties and groins) that provide artificial habitats for marine life. Other altered wetlands are excavated, impounded, or diked wetlands.

Drainage activities in wetlands can modify the hydrology to the point where the area is no longer functioning as a wetland. Effectively drained hydric soils that are no longer capable of supporting hydrophytes due to a major change in hydrology are not considered wetland. Areas with effectively drained hydric soils are, however, good indicators of historic wetlands, which may be suitable for restoration.

The FWS definition includes shallow waters as wetland, but does not include permanently flooded deep water areas. These deeper waterbodies are defined as deepwater habitats, since water and not air is the principal medium in which dominant organisms live (Cowardin et al., 1979). Along the coast in tidal areas, the break between wetland and deepwater habitat begins at the extreme spring low tide level. In nontidal freshwater areas, this split starts at a depth of 6.6 ft (2 m) because shallow water areas are often vegetated with emergent wetland plants (e.g., bur-reeds) or floating-leaved rooted vascular plants (e.g., water lilies) (Figure 1.2).

NATIONAL RESEARCH COUNCIL DEFINITION

During the early 1990s, there was considerable interest and controversy in wetland delineation because the federal government significantly strengthened its regulation of wetlands under the Clean Water Act, thereby increasing the geographic scope of public and private lands subject to such regulation. Due to significant opposition from the development and the agricultural communities, political pressures mounted to the extent that certain members of Congress and the Bush Administration attempted to change the definition of wetland to reduce the scope of federal jurisidiction. During this time, environmental groups lobbied Congress and recommended that they fund a study

---

* In fact, some wetland types are formed by animal communities, e.g., coral, mollusk, and worm reefs and numerous animals are listed as dominance types for unconsolidated shores (a type of nonvegetated wetland).

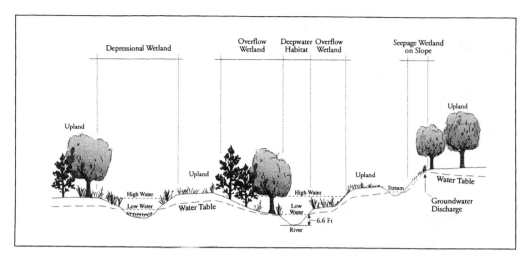

FIGURE 1.2 Schematic diagram showing wetlands, deepwater habitats, and uplands on the landscape. Note that in freshwater systems, wetlands include shallow water areas less than 6.6 ft (2 m) at mean low water (according to the Cowardin et al. system). In tidal systems, wetlands extend seaward to the extreme low spring tide mark, thereby including all areas that are periodically exposed by the tides but not permanently inundated areas. (From Tiner, R.W. and D.G. Burke, 1995.)

of wetland delineation. The study would focus on the scientific foundation for wetland delineation and review existing methods used by the federal government to determine their scientific fitness. In 1993, Congress provided funding to the Environmental Protection Agency for the National Research Council (NRC) to conduct such a study. A 17-member committee (representing the academic community, industry, the environmental community, and the legal profession) — the Committee on Characterization of Wetlands — was assembled to undertake this study. The results of their 2-year study are presented in *Wetlands: Characteristics and Boundaries* (National Research Council, 1995).

In order to initiate its study, the Committee needed a definition to provide the framework for its investigation. They drafted the following reference definition.

"A wetland is an ecosystem that depends on constant or recurrent, shallow inundation, or saturation at or near the surface of the substrate. The minimum essential characteristics of a wetland are recurrent, sustained inundation, or saturation at or near the surface and the presence of physical, chemical, and biological features reflective of recurrent, sustained inundation or saturation. Common diagnostic features of wetlands are hydric soils and hydrophytic vegetation. These features will be present except where specific physiochemical, biotic, or anthropogenic factors have removed them or prevented their development."

This definition is consistent with the official FWS definition. Both definitions recognize that hydrology, soils/substrates, and vegetation are key elements, with added attention given to biota in general (including animals) in the NRC definition. They both speak to the variability of hydrology ranging from permanent water (constant inundation or soil saturation) to periodic wetness (recurrent shallow inundation or saturation). Both emphasize that when it comes to soil saturation, it is saturation "at or near the surface" that determines wetlands.

CANADIAN NATIONAL DEFINITION

The National Wetlands Working Group of the Canada Committee on Ecological Land Classification defined wetlands in its national wetlands classification for Canada. The group was represented by technical wetland experts from across the country.

Wetland is defined as "land that is saturated with water long enough to promote wetland or aquatic processes as indicated by poorly drained soils, hydrophytic vegetation, and various kinds of biological activity which are adapted to a wet environment" (National Wetlands Working Group, 1987).

Two basic types of wetlands are recognized — organic wetlands on peat and muck soils and mineral wetlands (including mineral soils beneath shallow water and others) "which are tilled and planted but which, if allowed to revert to their original state, become saturated for long periods and are then associated with wet soils (e.g. Gleysols) and hydrophytic vegetation." This definition also is consistent with U.S. wetland definitions of the FWS and NRC, and like the latter also stresses biological activity resulting from prolonged wetness as an indicator of wetlands.

OTHER INTERNATIONAL DEFINITIONS

Wetlands have been defined in other countries largely by scientists and natural resource agencies interested in the ecological significance of these areas as habitats for waterbirds and other animals and potential land uses.

Ramsar Wetland Definition

The Ramsar convention, an international government body of more than 90 countries, interested in worldwide wetland conservation, developed the following defintion.

Wetlands are "areas of marsh, fen, peatland, or water, whether natural or artificial, permanent or temporary, with water that is static or flowing, fresh, brackish, or salt, including areas of marine water the depth of which at low tide does not exceed 6 m." Wetlands "may incorporate riparian and coastal zone adjacent to wetlands, and islands or bodies of marine water deeper than 6 m at low tide lying within the wetlands" (Ramsar Information Bureau, 1998).

This definition is quite broad, extending the waterward depth of wetlands to 6 m, thereby including many "deepwater habitats" of the U.S. definition in the concept of wetland. It also recognizes that wetland ecosystems may contain other habitats (e.g., riparian habitats and deep open water areas) as vital components that are virtually unseparable from the wetland itself. Although not explicit in the definition, it also includes "human-made wetlands" (Ramsar Information Bureau, 1998). Fish and shrimp ponds, farm ponds, irrigated agricultural land (e.g., rice paddies), salt pans, reservoirs, gravel pits, sewage farms, and canals are within the scope of this wetland definition.

Australian Wetland Definition

A review of Australian wetlands by Paijmans et al. (1985) contained the following definition.

Wetlands are "land permanently or temporarily under water or waterlogged. Temporary wetlands must have surface water or waterlogging of sufficient frequency and/or duration to affect the biota. Thus, the occurrence, at least sometimes, of hydrophytic vegetation or use by waterbirds are necessary attributes. This wide definition includes some areas whose wetland nature is arguable, notably land subject to inundation but having little or no hydrophytic vegetation, and bare 'dry lakes' in the arid interior."

This definition emphasizes hydrology and wetland biota, specifically hydrophytic vegetation and waterbirds. It also, however, recognizes that wetlands include both vegetated and nonvegetated periodically wet areas. Moreover, the latter types are expected due to the dry climate of the country's interior. The section on wetland definition also gives Lake Eyre as an example of one of the questionable wetlands since it is filled only a few times in 90 years. Yet, given that, it "may occasionally be significant for waterbirds and should count as a wetland." This type of area is a good example of a cyclical wetland (see Chapter 7). Such wetlands are common in arid regions of the world.

Another Australian wetland definition was developed by the Wetlands Advisory Committee (1977). This definition has been adopted for inventorying wetlands in western Australia.

> Wetlands are "areas of seasonally, intermittently, or permanently waterlogged soils or inundated land, whether natural or artificial, fresh or saline, e.g., waterlogged soils, ponds, billabongs, lakes, swamps, tidal flats, estuaries, rivers and their tributaries."

Wetlands can be identified by the presence of water or waterlogged soils that support vegetation typical of wet conditions and/or hydric soils, with the limits determined by dampness conditions, characteristic vegetation, or hydric soils (Semeniuk Research Group, 1992). This wetland definition includes all water bodies in addition to vegetated and nonvegetated periodically waterlogged lands.

Zambian Wetland Definition

Perera (1982) generally described wetlands as "either an area of land in which periods of flooding and emergence of ground alternates or as areas that are permanently flooded with a water layer not exceeding several meters in depth." This definition contains about 20% of the country's land area.

REGULATORY DEFINITIONS

Environmental laws have been passed to protect wetlands or to control certain activities that may jeopardize public resources or cause damage to public and private property. Legally, a wetland is whatever the law says it is. Regulatory definitions are largely grounded on scientific concepts, but often include other considerations specifying lands of particular concern. For example, some definitions exclude certain areas due to tidal vs. nontidal influence, or interest in land use (e.g., agriculture in Alaska), or include deepwater areas (e.g., lakes and rivers) and other areas (e.g., banks and buffers) as wetlands for the purposes of regulating certain activities in these places.

FEDERAL CLEAN WATER ACT DEFINITION

The federal government regulates wetlands under two laws — the Rivers and Harbors Act and the Clean Water Act. The former is focused on navigable waters and deals more with disposal of dredged material and construction of potential hazards to navigation, while the latter is far more expansive in geographic scope and is mostly concerned with the deposition of fill in waters of the U.S. Consequently, it is the Clean Water Act, particularly Section 404, that gets the most attention (for a review of federal wetland regulations, see Want, 1992; Dennison and Berry, 1993; Environmental Law Institute, 1993).

The Section 404 program is jointly administered by the U.S. Army Corps of Engineers (Corps) and the U.S. Environmental Protection Agency (EPA), with the Corps being largely responsible for issuing permits and the EPA providing program oversight. Wetlands are considered part of the waters of the U.S. since such waters ebb and flow over and/or through such lands and they are vital resources that help maintain the biological, chemical, and physical integrity of the nation's waters.

A 1975 U.S. district court decision (Natural Resources Defense Council v. Callaway; 392 F. Supp. 685, 5 ELR 20285) forced the Corps to expand its role in wetland regulation. To comply with the court's decision, the Corps proposed the following definition (40 Federal Register 31328, July 25, 1975).

> "Wetlands are those land and water areas subject to regular inundation by tidal, riverine, or lacustrine flowage. Generally included are inland and coastal shallows, marshes, mudflats, estuaries, swamps, and similar areas in coastal and inland navigable waters."

After receiving significant public comment about this definition, the final regulatory definition was published (42 Federal Register 37125-26, 37128-29, July 19, 1977).

> Wetlands are "those areas that are inundated or saturated by surface or ground water at a frequency and duration sufficient to support, and that under normal circumstances do support, a prevalence of vegetation typically adapted for life in saturated soils. Wetlands generally include swamps, marshes, bogs, and similar areas."

This definition requires that certain vegetation be present or capable of growing in the area to be classified as wetland. Such vegetation has been termed *hydrophytic vegetation* in manuals developed to apply this definition on the ground (Chapter 6). It is evident that this definition is limited to vegetated wetlands. It does not include mudflats, sandbars, beaches, rocky shores, and similar nonvegetated areas that may be viewed as wetlands from the scientific standpoint. It also does not include aquatic beds, such as water lily beds, since this vegetation is not growing on soil, but on a submerged substrate. The Corps does, however, recognize nonvegetated wetlands and aquatic beds as "special aquatic sites" and part of "other waters of the U.S." for regulatory purposes, but they are not regulated wetlands.

The Section 404 definition includes the term "under normal circumstances" in regard to the presence of vegetation. This term was required to prevent people from seeking to remove a wetland area from regulation by simply eliminating the vegetation. Prior to 1989, the Corps' interpretation of normal circumstances was typically determined by the existing land use — determined on the basis of an area's characteristics and use, at present and in the recent past (Offringa, 1986). If the area was farmed for some time, it was viewed as not supporting a prevalence of hydrophytic vegetation and was not wetland, even recognizing that "if left unattended for a sufficient period of time, [it would] revert to wetlands solely through the devices of nature" (Offringa, 1986). Consequently, farmed wetlands were excluded from the wetland definition. In 1989, the Corps and EPA adopted the concept of normal circumstances used in the Food Security Act which defined normal circumstances in the absence of vegetation, on the basis of existing hydrology and the presence of hydric soils. This was a significant reinterpretation of the term, with major implications on the extent of regulated wetlands. Since 1989, farmed wetlands have been included in the scope of the federal regulatory definition.

FOOD SECURITY ACT DEFINITION

In 1985, Congress passed the Food Security Act (FSA) (P.L. 99-198, 99 Stat. 1504) that contained provisions to discourage wetland drainage for crop production — the Swampbuster provision. This provision is quasi-regulatory, in that the federal government would deny federal subsidies (agricultural loans, crop support payments, and other benefits) to farmers who drained wetlands to produce a commodity crop after December 23, 1985. For administering this provision, wetlands were defined as:

> ...areas that have a predominance of hydric soils and that are inundated or saturated by surface or ground water at a frequency and duration sufficient to support, and under normal circumstances do support, a prevalence of hydrophytic vegetation typically adapted for life in saturated soil conditions, except lands in Alaska identified as having a high potential for agricultural development and a predominance of permafrost soils" (USDA Soil Conservation Service, 1988, 1996).

Wetlands are determined by three features (hydric soil, hydrology, and hydrophytic vegetation), so it is a definition of vegetated wetlands. It contains a geographic exclusion for wetlands on certain agricultural lands in Alaska, since Congress did not want to prevent conversion of such wetlands (see Chapter 7 for discussion of this wetland type). This definition is actually the first wetland definition published in federal public law (16 U.S.C. Sect. 801 (a) (1)). It also has been used to define wetlands in the Emergency Wetlands Resources Act of 1986, but without the geographic exclusion for Alaska.

The FSA definition uses the term "normal circumstances" to refer to "soil and hydrologic conditions that are normally present, without regard to whether the vegetation has been removed" (USDA Soil Conservation Service, 1996). Consequently, it includes many farmed wetlands. For purposes of the Act, farmed wetlands are wetlands that were drained, dredged, filled, leveled, or otherwise manipulated before December 23, 1985 with the intent to grow an agricultural commodity crop, but still retain some degree of wetland hydrology. In practice, farmed wetlands must have a 50% chance of being inundated for at least 15 consecutive days during the growing season or 10% of the growing season, whichever is less. Exceptions to this general rule are wetlands that are potholes, playas, or pocosins which when farmed still qualify as wetland if either flooded or ponded for at least 1 week or saturated for at least 2 consecutive weeks during the growing season. Thus, identifying a farmed wetland as one of the latter three types is critical to the determination of whether it is a FSA wetland. Wet pastures and hayfields also are included in the FSA definition provided they were used as such prior to December 23, 1985 and they are inundated for at least 1 week or saturated for at least 2 weeks during the growing season.

Excluded from the FSA definition are farmed wetlands that do not meet the hydrology requirements and "prior converted cropland." The latter category is a former or existing wetland that meets certain criteria: (1) has been used to produce a commodity crop at least once before December 23, 1985, (2) as of this date did not support woody plants, and (3) has the same hydrology as "farmed wetlands" mentioned in the preceding paragraph. Basically these are farmed wetlands that are not affected by FSA because they were cultivated prior to the establishment of the Act. Wetlands cultivated or altered with the intent of growing a commodity crop after the Act are considered "converted wetlands" for administrative purposes.

SELECTED STATE DEFINITIONS

Since the 1960s, many states have recognized the significance of wetlands as natural resources and have passed laws to protect them from unwise development. State wetland definitions are quite variable, but most have a common thread based on vegetation and hydrology. Soils may be referred to explicitly or may be inferred from the hydrologic conditions stated in the definitions. More recent state laws use the federal definition or something quite similar for inland wetlands (e.g., Maryland, New Jersey, and Wisconsin). Table 1.1 lists examples of some state wetland definitions, including some of the early definitions and more recent ones.

State wetland definitions also may include areas beyond ecological wetlands that the state considers important for maintaining the functions to be protected by state law. For example, both Massachusetts and Rhode Island include river banks and the 100-year floodplain in their definitions. Rhode Island also includes a 50-foot buffer in its designated wetland (Rhode Island Department of Environmental Management, 1994). New Jersey's Pineland Commission (1980) definition of wetlands includes submerged lands (lakes, ponds, rivers, and streams).

The interpretation of the definitions may lead to significant differences, even though the definitions appear similar. For example, Wisconsin's definition is quite similar to the federal regulatory definition. The state clearly points out that the definition closely follows the 1989 manual's concept, but recognizes that there may be situations where differences occur: (1) in "somewhat poorly drained" soils capable of supporting hydrophytic vegetation and (2) where hydrology is altered but the site is still capable of supporting wetland plants (Wisconsin Coastal Management Program, 1995). These sites may be considered regulated wetlands by the state.

TABLE 1.1
Examples of Wetland Definitions Used by State Regulatory Programs in the U.S.

State	Definition
Connecticut	"Wetlands are those areas which border on or lie beneath tidal waters, such as, but not limited to banks, bogs, salt marshes, swamps, meadows, flats, or other low lands subject to tidal action, including those areas now or formerly connected to tidal waters, and whose surface is at or below an elevation of 1 ft above local extreme high water." (State's tidal wetland definition; the definition also includes a list of indicator plants.)
Connecticut	"Wetlands mean land, including submerged land, which consists of any of the soil types designated as poorly drained, very poorly drained, alluvial, and floodplain by the National Cooperative Soils Survey, as may be amended from time to time, of the Soil Conservation Service of the United States Department of Agriculture." (State's inland wetland regulatory definition; it focuses on soil types.)
Florida	"Wetlands" means those areas that are inundated or saturated by surfacewater or groundwater at a frequency and a duration sufficient to support, and under normal circumstances do support, a prevalence of vegetation typically adapted for life in saturated soils. Soils present in wetlands generally are classified as hydric or alluvial, or possess characteristics that are associated with reducing soils conditions. The prevalent vegetation in wetlands generally consists of facultative or obligate hydrophytic macrophytes that are typically adapted to areas having soil conditions described above. These species, due to morphological, physiological, or reproductive adaptations, have the ability to grow, reproduce, or persist in aquatic environments or anaerobic soil conditions. Florida wetlands generally include swamps, marshes, bayheads, bogs, cypress domes and strands, sloughs, wet prairies, riverine swamps and marshes, mangrove swamps, and other similar areas. Florida wetlands generally do not include longleaf or slash pine flatwoods with an understory dominated by saw palmetto." (Statewide definition includes tidal and nontidal wetlands.)
Maryland	Tidal wetlands are "all state and private tidal wetlands, marshes, submerged aquatic vegetation, lands, and open water affected by the daily and periodic rise and fall of the tide within the Chesapeake Bay and its tributaries, the coastal bays adjacent to Maryland's coastal barrier islands, and the Atlantic Ocean to a distance of 3 m offshore of the low water mark." (State's tidal wetland definition; it includes deepwater areas.)
Massachusetts	"Salt Marsh means a coastal wetland that extends landward up to the highest high tide line, that is, the highest spring tide of the year, and is characterized by plants that are well adapted or prefer living in saline soils. Dominant plants within salt marshes are salt meadow cord grass (*Spartina patens*) and/or salt marsh cord grass (*Spartina alterniflora*). A salt marsh may contain tidal creeks, ditches, and pools." (State's coastal wetland definition; it was published in the first state wetland law in the nation in 1962.)
Massachusetts	"Bordering Vegetated Wetlands are freshwater wetlands which border on creeks, rivers, streams, ponds, and lakes. The types of freshwater wetlands are wet meadows, marshes, swamps, and bogs. They are areas where the topography is low and flat, and where the soils are annually saturated. The ground and surface water regime and the vegetational community which occur in each type of freshwater wetland are specified in the Act." (State's inland wetland definition; it only pertains to wetlands that border waterbodies.)
New Jersey	Coastal wetlands are "any bank, marsh, swamp, meadow, flat, or other lowland subject to tidal action in the Delaware Bay and Delaware River, Raritan Bay, Sandy Hook Bay, Shrewsbury River, including Navesink River, Shark River, and the coastal inland waterways extending southerly from Manasquan Inlet to Cape May Harbor, or any inlet, estuary or those areas now or formerly connected to tidal waters whose surface is at or below an elevation of 1 ft above local extreme high water, and upon which may grow or is capable of growing some, but not necessarily all, of the following:" (19 plants listed). Coastal wetlands exclude "any land or real property subject to the jurisdiction of the Hackensack Meadowlands Development Commission…" (State's tidal wetland definition; it contains a geographic exclusion.)
New Jersey	"Freshwater wetland means an area that is inundated or saturated by surface water or groundwater at a frequency and duration sufficient to support, and that under normal circumstances does support, a prevalence of vegetation typically adapted for life in saturated soil conditions, commonly known as hydrophytic vegetation; provided, however, that the department, in designating a wetland, shall use the three-parameter approach … developed by the U.S. Environmental Protection Agency, and any subsequent amendments thereto…" (State's inland or freshwater wetland definition.)

State	Definition
New York	"Freshwater wetlands means lands and waters of the state as shown on the freshwater wetlands map which contain any or all of the following: (a) lands and submerged lands commonly called marshes, swamps, sloughs, bogs, and flats supporting aquatic or semi-aquatic vegetation of the following types: (lists indicator trees, shrubs, herbs, and aquatic species); (b) lands and submerged lands containing remnants of any vegetation that is not aquatic or semi-aquatic that has died because of wet conditions over a sufficiently long period … provided further that such conditions can be expected to persist indefinitely, barring human intervention; (c) lands and waters substantially enclosed by aquatic or semi-aquatic vegetation … the regulation of which is necessary to protect and preserve the aquatic and semi-aquatic vegetation; and (d) the waters overlying the areas set forth in (a) and (b) and the lands underlying (c)." (State's inland wetland definition; the definition includes lists of indicator species for each wetland type; the state generally regulates wetlands 12.5 acres or larger.)
Rhode Island	"Coastal wetlands include salt marshes and freshwater or brackish wetlands contiguous to salt marshes. Areas of open water within coastal wetlands are considered a part of the wetland. Salt marshes are areas regularly inundated by salt water through either natural or artificial water courses and where one or more of the following species predominate: (eight plants listed). Contiguous and associated freshwater or brackish marshes are those where one or more of the following species predominate: (nine plants listed)." (State's coastal wetland definition; the definition includes lists of indicator species.)
Rhode Island	Fresh water wetlands are defined to include, "but not be limited to marshes, swamps, bogs, ponds, river and stream flood plains and banks; areas subject to flooding or storm flowage; emergent and submergent plant communities in any body of fresh water including rivers and streams and that area of land within fifty feet (50') of the edge of any bog, marsh, swamp, or pond." (State's inland wetland definition; various wetland types are further defined based on hydrology and indicator plants.)
Vermont	"Wetlands means those areas of the state that are inundated by surface or groundwater with a frequency sufficient to support significant vegetation or aquatic life that depend on saturated or seasonally saturated soil conditions for growth and reproduction. Such areas include but are not limited to marshes, swamps, sloughs, potholes, fens, river and lake overflows, mud flats, bogs and ponds, but excluding such areas as grow food or crops in connection with farming activities."
Wisconsin	"An area where water is at, near, or above the land surface long enough to be capable of supporting aquatic or hydrophytic vegetation and which has soils indicative of wet conditions."

Note: Some states use the federal regulatory definition for inland wetland programs (e.g., Maryland, Oregon, and Pennsylvania).

FIELD APPLICATION OF WETLAND DEFINITIONS

Applying any of the wetland definitions on-the-ground requires knowledge of several disciplines (especially botany, soil science, and hydrology, with an emphasis on the first two). Consequently, plant, soil, and hydology indicators must be established and the more reliable ones used to identify wetlands. Scientific definitions have been interpreted largely by biologists, applying their knowledge of ecology (plant–soil–hydrology relationships) and geomorphology (e.g., landscape position and landform). In contrast, federal agencies have developed manuals for interpreting the regulatory definition in an attempt to achieve consistency and repeatability in application among regulators and the regulated public. Since these definitions serve to identify lands subject to government regulations, precise and accurate delineations need to be performed in the name of good government. The manuals use certain biologic or physical features, namely plants, soils, and signs of hydrology, to establish criteria for verifying the presence of potentially regulated wetlands (see Chapter 6 for discussion of methods). It is important to recognize that a criterion can be validated by any evidence that has bearing, so for example, a hydric soil property (such as undrained organic soil) could be used to verify wetland hydrology (National Research Council, 1995).

A nationally applied definition needs to consider three levels for interpretation — criteria, indicators, and regional variation (National Research Council, 1995). Wetland identification criteria can include biological criteria (organisms) and physical criteria (wet substrates and soils, hydrology) as well as political considerations for regulatory definitions. Biological indicators are certain plants and animals found in water or wetlands. Wet substrates/soil indicators are hydric soil properties plus submerged substrates and periodically flooded rocks or bedrock. Hydrology indicators can be any of the above indicators as well as indirect and direct evidence of flooding or waterlogging. Regional variation can be addressed by listing typical plants and/or plant communities and soil types, for example, used to indicate or validate a criterion. Regulatory considerations can include interpretations of what constitutes a significant "resource area" based on matters of size, location in the watershed, wetland type, and regional abundance of wetlands, for example. These policy factors may be explicit or implicit.

The key to accurate identification of wetlands is to have a well-conceived, science-based definition and fairly explicit guidance on the appropriate use of various wetland indicators to verify the presence of wetlands on-the-ground. Such guidance should be able to handle the majority of wetlands without difficulty and provide a list of known exceptions with specific instructions on how to evaluate them. This type of approach would promote accurate, precise, and consistent wetland identification and delineation. Once identified as wetland, regulators could then superimpose their value system to determine which wetlands should be regulated and to what degree. The distinction between ecological wetlands and the subset that warrant regulatory considerations should be made clear by the regulatory community.

REFERENCES

Cowardin, L.M., V. Carter, F.C. Golet, and E.T. LaRoe. 1976. Interim Classification of Wetlands and Aquatic Habitats of the United States. U.S. Fish and Wildlife Service, Office of Biological Services, Washington, D.C., March 1, 1976.

Cowardin, L.M., V. Carter, F.C. Golet, and E.T. LaRoe. 1977a. Classification of Wetlands and Aquatic Habitats of the United States. U.S. Fish and Wildlife Service, Washington, D.C., April 1977.

Cowardin, L.M., V. Carter, F.C. Golet, and E.T. LaRoe. 1977b. Classification of Wetlands and Deepwater Habitats of the United States (an operational draft). U.S. Fish and Wildlife Service, Washington, D.C., October 1977.

Cowardin, L.M., V. Carter, F.C. Golet, and E.T. LaRoe. 1979. Classification of Wetlands and Deepwater Habitats of the United States. Publ. No. FWS/OBS-79/31. U.S. Fish and Wildlife Service, Washington, D.C.

Dachnowski, A.P. 1920. Peat deposits in the United States and their classification. *Soil Sci.*, 10: 453-465.

Dennison, M.S. and J.F. Berry (Eds.). 1993. *Wetlands: Guide to Science, Law, and Technology.* Noyes Publications, Park Ridge, NJ.

Environmental Law Institute. 1993. *Wetlands Deskbook.* The Environmental Law Reporter, Washington, D.C.

Lefor, M.W. and W.C. Kennard. 1977. *Inland Wetland Definitions.* Report No. 28. Institute of Water Resources, University of Connecticut, Storrs, CT.

Leitch, W.G. 1966. Historical and ecological factors in wetland inventory. *Trans. N. Am. Wildl. Nat. Resour. Conf.*, 31: 88-96.

Lugo, A.E., S. Brown, and M.M. Brinson. 1990. Concepts in wetland ecology, chap. 4. In, *Forested Wetlands.* A.E. Lugo, M.M. Brinson, and S. Brown (Eds.). Elsevier Publishers, Amsterdam. 53-85.

Martin, A.C., N. Hotchkiss, F.M. Uhler, and W.S. Bourn. 1953. Classification of Wetlands of the United States. Spec. Rep. Wildlife No. 20. U.S. Fish and Wildlife Service, Washington, D.C.

National Research Council (NRC). 1995. *Wetlands: Characteristics and Boundaries.* National Academy Press, Washington, D.C.

National Wetlands Working Group. 1987. The Canadian Wetland Classification System. Land Conservation Branch, Canada Committee on Ecological Land Classification, Canadian Wildlife Service, Environment Canada, Ottawa, Ontario. Ecological Land Classification Series No. 21.

New Jersey Pinelands Commission. 1980. *Comprehensive Management Plan for the New Jersey Pinelands National Reserve and Pinelands Area.* New Lisbon, NJ.

Offringa, P.J. 1986. Clarification of "normal circumstances" in the wetland definition (33 CFR 323.2 (c)). Regulatory Guidance Letter 86-9 (August 27, 1986). U.S. Army Corps of Engineers, Office of Chief of Engineers, Washington, D.C.

Paijmans, K., R.W. Galloway, D.P. Faith, P.M. Fleming, H.A. Haantjens, P.C. Heyligers, J.D. Kalma, and E. Löffler. 1985. *Aspects of Australian Wetlands.* Division of Water and Land Resources Technical Paper No. 44. Commonwealth Scientific and Industrial Research Organization, Melbourne, Australia.

Perera, N.P. 1982. Ecological considerations in the management of the wetlands of Zambia. In, *Wetlands Ecology and Management.* B. Gopal, R.E. Turner, R.G. Wetzel, and D.F. Whigham (Eds.). Proceedings of the First International Wetlands Conference, New Delhi, India. National Institute of Ecology and International Scientific Publications, Jaipur, India, 21-30.

Ramsar Information Bureau. 1998. What are Wetlands? Ramsar Information Paper No. 1. Gland, Switzerland.

Reed, P.B., Jr. 1988. National List of Plant Species that Occur in Wetlands: 1988 National Summary. Biol. Rpt. 88(24). U.S. Fish and Wildlife Service, Washington, D.C.

Reed, P.B. Jr. (compiler). 1997. Revision of the National List of Plant Species that Occur in Wetlands. In cooperation with the national and regional interagency review panels. Department of Interior, U.S. Fish and Wildlife Service, Washington, D.C.

Rhode Island Department of Environmental Management. 1994. Rules and Regulations Governing the Administration and Enforcement of the Freshwater Wetlands Act. Providence, RI.

Sather, J.H. (Ed.). 1976. Proceedings of the National Wetland Classification and Inventory Workshop. FWS/OBS-76/09. U.S. Fish and Wildlife Service, Washington, D.C.

Semeniuk Research Group, V. & C. 1992. Classification and Mapping of Wetlands in the Moore River to Perth Area. Report to the Water Authority of Western Australia.

Shaler, N.S. 1890. General account of the freshwater morasses of the United States, with a description of the Dismal Swamp District of Virginia and North Carolina. 10th Annual Report 1888-1889. U.S. Geological Survey, Washington, D.C. 255-339.

Shaw, S.P. and C.G. Fredine. 1956. Wetlands of the United States. Circular 39. U.S. Fish and Wildlife Service, Washington, D.C.

Stewart, R.E. and H.A. Kantrud. 1971. *Classification of Natural Ponds and Lakes in the Glaciated Prairie Region.* Resource Publ. 92. U.S. Fish and Wildlife Service, Washington, D.C.

Tiner, R.W. and D.G. Burke. 1995. Wetlands of Maryland. U.S. Fish and Wildlife Service, Region 5, Hadley, MA and the Maryland Department of Natural Resources, Annapolis, MD. Cooperative publication.

USDA Soil Conservation Service. 1988. National Food Security Act Manual. U.S. Department of Agriculture, Washington, D.C.

USDA Soil Conservation Service. 1991. National List of Hydric Soils. In cooperation with the National Technical Committee for Hydric Soils. Washington, D.C.

USDA Soil Conservation Service. 1996. National Food Security Act Manual, 3rd ed. U.S. Department of Agriculture, Washington, D.C.

Want, W.L. 1992. *Law of Wetlands Regulation.* Clark Boardman Callaghan, Deerfield, IL.

Wetland Advisory Committee. 1977. The Status of Wetland Reserves in System Six. Report of the Wetland Advisory Committee to the Environmental Protection Authority. Australia.

Wisconsin Coastal Management Program. 1995. Basic Guide to Wisconsin's Wetlands and their Boundaries. Wisconsin Department of Administration, Madison, WI.

2 Wetland Concepts for Identification and Delineation

INTRODUCTION

Wetland identification and delineation require an understanding of certain fundamental concepts, since this work involves more than simply following guidance in a government manual. The use of best professional judgment is commonly needed to interpret key provisions, especially for problematic situations. Wetland delineators therefore need to be familiar with plant–soil–water relationships and general wetland ecology. Mitsch and Gosselink's (1993) textbook, *Wetlands*, is an excellent starting point. There is universal recognition that the forcing function of wetlands is hydrology, that is an excess of water sufficient to exert a controlling influence on plant or animal life or soil development (National Research Council, 1995). Yet to really understand this dominant influence, there are many issues pertaining to wetland hydrology that are significant for wetland identification and delineation. Some issues are at the root of the controversy over what habitats actually constitute wetlands, at least from the regulatory point of view. Ecological concepts relating to wetland formation and development also have influenced today's concept of wetland.

Seven key questions reveal the underlying concepts upon which wetland definitions are based or expose some shortcomings of current definitions.

1. How wet is a wetland?
2. What is the significance of "growing season" in defining wetland hydrology?
3. Are all wetlands at least periodically anaerobic environments?
4. What are the more common types of wetland hydrology?
5. Are wetlands simply ecotones between land and water?
6. Do wetlands eventually become uplands in a natural order of terrestrialization?
7. Are some wetlands ephemeral features on the landscape and cyclical in occurrence?

There may be other relevant questions, but these seven should include the most important ones. Most of these questions relate to hydrology which is the master variable of wetlands and a topic that has received the least study in terms of wetlands. Only recently has more attention been given to this crucial topic.

The following discussion of these questions is not intended to be an exhaustive treatise, but should be sufficient to help readers better understand the current concept of wetland and some key issues underlying much of the current debate about wetland definitions. Noticeably absent from this list is a question dealing with wetland functions, although some discussion of this is provided in the response to question No. 2. Some scientists have proposed that defining wetlands by functions would help solve debate raised by question No. 1. Yet, this position appears naive as a significance test would still be required, e.g., the minimum frequency, duration, and intensity (level) of the performance of a given function, especially since many valued "wetland" functions, such as flood storage, shoreline stabilization, and groundwater recharge, are not unique to wetlands. Once wetlands are defined, other significant questions are worth considering in the regulatory context (e.g., regarding wetland evaluation — is wetter better, should created wetlands be regulated the same as "natural" wetlands, and are restored or created wetlands acceptable functional replacements for

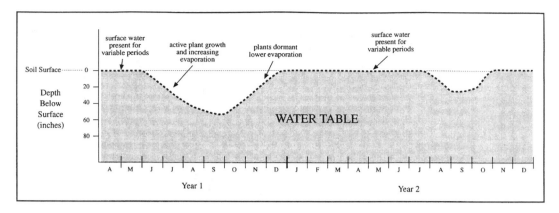

FIGURE 2.1   Water tables fluctuate in most, if not all, wetlands throughout the year and from year to year. The hydrology is not exactly the same in any 2 years. Year 1 in the figure is a typical year, whereas Year 2 experienced an extremely wet spring with obvious effects on the water table. (From Tiner, R.W. and D.G. Burke, 1995.)

existing wetlands?). But they are not the subject of this book (see Environmental Defense Fund and World Wildlife Fund, 1992; Leidy et al., 1992 for further discussion).

HOW WET IS A WETLAND?

This is the definitive question for wetland identification and if the answer was simple, much academic and political debate could be eliminated. The name *wetland* implies land having significant wetness (see Chapter 1 for definitions). Flooding virtually eliminates gas exchange between the soil and the atmosphere and the supply of dissolved oxygen in the soil is soon exhausted by soil microbial respiration in only hours or days, causing the soil to become anaerobic (Evans and Scott, 1955; Takai et al., 1956; Turner and Patrick, 1968; Monhanty and Patnaik, 1975; Jackson and Drew, 1984; Wakely et al., 1996). Significant wetness, as one can see in reviewing the wetland definitions in Chapter 1, has traditionally been defined as wet enough to create certain plant communities and/or hydric soils. Yet the minimum hydrology required to support such vegetation and to form such soils has not been the object of scientific study. Thus one has to examine information gleaned from existing studies of vegetation and soils to develop a best approximation. This is a challenging assignment due to the diversity of wetland types and associated environmental conditions.

Hydrology is by its very nature dynamic, varying annually, seasonally, and daily from wetland to wetland (no two are exactly alike), and from region to region (Figure 2.1; see subsection on common wetland hydrologies later in this chapter). Consequently, hydrologic assessments require long-term studies to document the fluctuations in surface water levels and in the position of the water table. While considerable scientific study has been devoted to assessing the hydrology of rivers and streams and forests, relatively few studies have examined the hydrology of wetlands. Scientific research has not focused on examining these long-term hydrologic relationships in wetlands, especially along their upper limits, for several reasons: (1) the recent interest in this topic due to regulatory needs, (2) wetland identification by plants and/or soils was widely accepted as a practical approach and using specific hydrologic conditions to determine wetland limits is a relatively new approach, and (3) the long-term commitment of resources (dollars and time) required to undertake such a task (Tiner, 1993a). Groundwater-flow systems are particularly difficult and expensive to monitor due to the need for instrumented well fields and often discontinuous confining beds (Woo and Winter, 1993).

The lack of specific hydrologic data to establish the limits of wetland coupled with the high variability in hydrology among wetlands are undoubtedly the primary reasons why definitions of "wetland" avoid specificity and merely state that the area is wet enough to support hydrophytic vegetation and/or to form hydric soils (see wetland definitions in Chapter 1). Reliance on certain plants and soils as indicators of wetland has been and will undoubtedly continue to be the main criteria used to identify and delineate wetlands (Chapters 3 and 4), since they are more readily observed than the presence of water at a given site, especially during a single site visit. Furthermore, the presence of water at a given point in time does little to indicate the presence of wetland, given the temporal nature of water in most wetlands and uplands. Observation of water in the soil on a given day does not give any indication of how long it has been there, how long it will persist, and how frequently such events occur. Plants and soil properties at a site are, in large part, the expressions or manifestations of site wetness and are the indicators of how wet the site really is, provided it has not been drained.

Vegetation itself has a significant effect on the hydrology of a site. For example, the USDA Forest Service's loblolly pine management guide cautions that "a mature, well-stocked stand of trees often appear to be better drained than they really are. These sites may become exceedingly wet after transpiration losses are eliminated by harvesting the trees. As the new stand develops, moisture loss through transpiration will supplement drainage through the soil, drying the site" (McKee, 1989). This document correctly points out that increased transpiration from tree cover often has a significant effect on the local water table and seemingly dry areas are wetter than their superficial appearance. Measurements of the water table in forests may not reflect site wetness without tree cover, e.g., marginally wet flatwoods are wetter than they appear.

Although an interesting theoretical topic, the real need to establish a lower threshold for wetland hydrology is for regulatory determinations. In evaluating altered sites where drainage has resulted in a drying out of the wetland, wetland regulators and the regulated public need a threshold to determine which wetlands are now effectively drained and which are not. Yet, given the paucity of long-term hydrologic studies for the variety of wetland types across the country, it must be recognized that only an approximation of the minimum threshold for wetland hydrology can be established. And while it may be practical and necessary for assessing the status of hydrologically disturbed sites for regulatory purposes, it must be recognized that some unaltered natural wetlands may not satisfy this condition for reasons yet to be determined. Consequently, caution should be exercised when applying such threshold to unaltered wetlands.

In defining wetland hydrology, four main factors to consider are (1) duration of wetness, (2) frequency of wetness, (3) depth of saturation, and (4) timing or seasonality of wetness. These four factors working together influence plant communities and soil formation as well as animal life and wetland functions.

FLOODING OR WATERLOGGING DURATION

Many plants are intolerant of flooding and their seedlings perish after only hours of inundation (Kozlowski, 1984). The rate of oxygen depletion depends on several factors, including soil temperature, organic matter content, and chemical oxygen demand from ferrous iron and other reduced elements (Gambrell and Patrick, 1978). Flooding a soil for a day has been shown to create anaerobic conditions under certain situations (Turner and Patrick, 1968; as reported by Mitsch and Gosselink, 1993) and if it occurs during a period of active plant growth, such flooding can have a limiting effect under some circumstances. Plant morphological responses to flooding and saturation are important plant adaptations (see Chapter 3) and can develop in a relatively short time. For example, Gill (1975) observed formation of adventitious roots in experimentally flooded alders just 2 to 5 days after bud-break. Topa and McLeod (1986b) noted that after 30 days of anaerobic conditions, there was an abundance of adventitious roots and hypertrophied lenticels in wet-site loblolly and pond pine seedlings. In another paper, they reported that "only 15 days of anaerobic conditions were necessary to increase internal root porosities of loblolly and pond pine seedlings"; evidence of aerenchyma tissue, an important adaptation for life in waterlogged soils (Topa and McLeod, 1986a). McKevlin et al. (1987) also found that after 3 weeks of flooding, loblolly pine seedlings developed hypertrophied lenticels and stems. After reviewing the scientific literature, the National Research Council (1995) concluded that flooding for 14 consecutive days was sufficient to create a wetland environment.

FREQUENCY OF PROLONGED WETNESS

In much of the coterminous U.S., especially east of the Mississippi Valley, precipitation is in excess of evaporation, so water is readily available throughout the year in most years. This has created favorable conditions for wetland formation and, as a result, most of the wetlands in the lower 48 states are located here. Precipitation patterns in this region are more or less predictable, recognizing seasonal and annual variations, so considering wetland hydrology in terms of an "average year" or conditions that prevail in most years has merit and utility. Consequently, extended wetness occurring in most years (roughly every other year on average) has been the standard for wetland delineation in the U.S. following government manuals (Environmental Laboratory, 1987; Sipple, 1988; Federal Interagency Committee for Wetland Delineation, 1989).

Precipitation patterns are much different in arid and semiarid regions, being characterized by annual water-deficits and by frequent long-term droughts. "Average" wetness conditions are simply mathematical calculations with no ecological meaning in evaluating wetland hydrology for these regions. During wet periods in the normal hydrologic cycle, wetlands may form in normally dry or moist depressions on the landscape. These ephemeral, cyclical wetlands are not addressed in the current federal regulatory wetland definition. In studying soils in semiarid Australia, Coventry and Williams (1984) found that hydromorphic properties may develop in soils saturated for less than 5 weeks at a frequency of once every 3 years. Perhaps conditions that prevail more than 33 out of 100 years should be the metric for assessing wetland hydrology in dry climates. Interestingly, the Australian concept of wetland also includes cyclical wetlands that may be flooded by episodic events once every 100 years because of their importance to waterbirds (Paijmans et al., 1985). Is a moist soil in the desert, wet enough to be a wetland? The ecological significance of cyclical wetlands in the U.S. needs further evaluation (see discussion on cyclical wetlands at the end of this chapter). Also, moist riparian habitats along rivers and streams in the arid West also are not considered wetland because they lack the required frequency and duration of wet conditions.

CRITICAL DEPTH OF SATURATION

Saturation in the root zone is fundamental for defining wetland hydrology. The bulk of the roots in wetland plants is generally restricted to the upper, partly aerated zone of the soil (Costello, 1936; Boggie, 1972; Whigham and Simpson, 1978; Montague and Day, 1980; Lieffers and Rothwell, 1987; Sjors, 1991). Although there are reports of rooting to 2 ft or more in some wetland species, the majority of the roots tend to occur within 1 ft (30 cm) of the surface (Day and Dabel, 1978; Powell and Day, 1991; NRC, 1995). Rice, the most widely cultivated wetland species in the world, has an extensive root system of hundreds of adventitious roots and thousands of lateral roots and most of these occur within the top foot (30 cm) of soil (Morita 1993). In studying root distribution in wet meadows of the Nebraska Sandhills, Moore and Rhoades (1966) found that one-half to two-thirds of the roots were located within the upper 2 in. of soil, although some roots went as deep as 4 ft. After reviewing the literature and combining this with his personal observations, Sipple (1992) concluded that most of the roots in wetland herbs occurred within the upper 6 in. and in woody plants within the upper 6 to 18 in.

TIMING OF WETNESS

Excessive wetness during the growing season is limiting to most plants. Most plants cannot tolerate a couple of weeks of flooding during the height of the growing season. Certain stages of plants (e.g., seedlings) are more susceptible than others to early season wetness (see Chapter 3). While plant growth is inhibited by prolonged waterlogging or flooding during the growing season, similar conditions during the so-called "non-growing season" also may adversely affect plants. Certain plants may have a competitive advantage for life in wetlands if they can continue root growth, germinate, or initiate shoot growth prior to the growing season for other plants (see discussion on growing season later in this chapter).

WETLAND HYDROLOGY DEFINED

In their wetland characterization report, the National Research Council (NRC, 1995) concluded that "wetland hydrology should be considered to be saturation within 1 ft of the soil surface for 2 weeks or more during the growing season in most years (about every other year on average)." The upper foot contains most of the plant roots that would be adversely affected by anaerobic conditions resulting from prolonged saturation. The NRC acknowledged that there may be regional variations due to climate, vegetation, soils, and geologic differences, but found that no such data currently exist. Until information to the contrary is produced, this threshold should be considered the minimum time necessary to create conditions that support the growth of hydrophytic vegetation and sufficient to define wetlands hydrologically.

The definition of wetland hydrology used for regulations and federal wetland policies seems to fall short of the technical standard defined by the NRC. For example, in identifying wetlands subject to the Federal Clean Water Act, the Corps of Engineers uses a period between 5 to 12.5% of the growing season (in most years) as the standard according to their wetland delineation manual (Environmental Laboratory, 1987) and a March 6, 1992 guidance memorandum (Williams, 1992), although their wetland delineation training materials, refer to 5% as the wetland hydrology threshold. This causes much confusion in establishing a minimum threshold for consistent application across the country. In addition, there is considerable discretion by Corps districts for interpreting the Corps manual, especially for defining growing season (see section on growing season below) which can lead to significant inconsistencies. In applying the Swampbuster provision of the Food Security Act of 1985 (and amendments), the National Resources Conservation Service requires that farmed wetlands be flooded for at least 2 consecutive weeks during the growing season or 10% of the growing season in most years, whichever is less unless it is a pothole, playa, or pocosin which is inundated for at least 1 week or saturated for 2 weeks or more during the growing season (USDA NRCS, 1996). Both of these definitions of wetland hydrology were developed before the NRC reached its conclusions and were established presumably with an intent to have federal regulations and policies cover what these agencies believed to be "significant" wetlands for their respective programs. No effort has been made to revise their guidelines regarding NRC's conclusions and recommendations.

THE SIGNIFICANCE OF "GROWING SEASON" FOR DEFINING WETLAND HYDROLOGY

Extended flooding or prolonged saturation during the *growing season* exerts considerable stress on plants. Many species cannot tolerate even a few days of flooding during the peak of the growing season (Chapter 3). Given this limiting effect on plant activity, most wetland definitions emphasize wetness during the growing season, yet wetness during the so-called "nongrowing" season may have a beneficial effect on some plants. The definition of growing season has become critically important in setting the minimum threshold for wetland hydrology, especially for regulatory purposes.

FEDERAL REGULATORY DEFINITION OF GROWING SEASON

Growing season has many definitions. In its common usage, it is an agricultural term referring to the period for growing cultivated crops — for germination (shoot emergence) of planted annual crops. Traditionally, the frost-free period has been used to define this growing season since it defines a period of no risk of crop damage due to killing frosts. It has little relevance to native plants as many are already growing well before this period. The Corps wetland delineation manual suggests using the "frost-free" period to identify the growing season for assessing wetland hydrology, while the period was actually defined by the concept of "biological zero" used by soil scientists* (Environmental Laboratory, 1987). This recommendation seems based on an incomplete understanding of wetland plant growth and/or an attempt to use the traditional concept of growing season in defining wetlands.

In a March 1992 guidance, the Corps clarified its interpretation of growing season (Williams, 1992). Using information in local soil survey reports, the growing season can be defined for a given geographic area by determining the period when air temperatures are 28°C or more in more than 5 years in 10, except "in the South" where air temperatures 32°C or more may be used at the discretion of the local Corps district. The exception allows Corps districts in the South some flexibility in establishing the limits of wetlands, since a higher base temperature requirement for growing season would translate into less area to regulate in a region with an abundance of wetlands. Table 2.1 gives some examples of growing seasons across the country using both the 28°C and 32°C thresholds.

Shortcomings of the Regulatory Concept

Although this approach is better than a strictly frost-free concept to defining "growing season" for wetland delineations, it is still not an accurate means of assessing when plants are growing or when soil processes are occurring that help develop hydric soil properties. It remains a conservative estimate that results in a significant decrease in the amount of wetlands that meet regulatory criteria. Soil temperatures typically lag behind air temperatures in the fall, so the end of the growing season estimates should be underestimated based on air temperature considerations (Sprecher et al., 1996). Woo and Winter (1993) reported soil temperatures above 5°C through December for a semipermanently flooded marsh in North Dakota, while air temperatures remained below this temperature from early November. In northern Minnesota, the current estimates of growing season based on 28° and 32°F "would shorten the growing season from 1 to 3 weeks in the spring and 2 to 3 weeks in the fall" (Bell et al., 1996). In western Oregon, the growing season following the 28°F threshold would be about 3 months shorter than one based on soil temperature of 41°F measured in the upper part of the soil (10 in.) (Huddleston and Austin, 1996; Sprecher et al., 1996). Also while defining growing season based on the concept of biological zero is better than using the frost-free period, it still is not technically correct as soil microbial activity occurs below these temperatures and certain animals are actively using some types of wetlands at temperatures below biological zero (Flanagan and Bunnell, 1980; Grishkan and Berman, 1993; Zimov et al., 1993; Clark and Ping, 1997).

Microflora of northern and southern soils develop under temperatures ranging between 3 to 35°C with certain microbes active at temperatures close to the freezing point of water (Volobueu, 1964). Recent observations in Massachusetts floodplain wetland soils suggest that microbial action takes place essentially year-round (Peter Veneman, University of Massachusetts, personal communication). Special cloth used for microbial studies was placed in floodplain wetland soils in November and re-examined in early February and some microbial action was witnessed. In February, however, microbial activity increased rapidly. This is about 1 month before budburst of silver maple (*Acer saccharinum*), the predominant wetland tree in this wetland. In studying a Wyoming subalpine meadow, Sommerfeld et al. (1993) documented microbial oxidation at temperatures below biologic zero. Studies of hydric soil hydrology and morphology in Oregon found that

---

* Biological zero is defined by a temperature of 5°C (or 41°F) measured at approximately 20 in. (50 cm) below the surface; it is supposed to represent the temperature at which biological activity (including microbial action) ceases and reduction does not occur in the soil.

TABLE 2.1
Examples of Growing Seasons Across the U.S. Using Either
28°F or 32°F at a Frequency of More than 5 Years in 10

	28°F or Lower		32°F or Lower	
Location	**Last Freeze in Spring[a]**	**First Freeze in Fall[b]**	**Last Freeze in Spring[a]**	**First Freeze in Fall[b]**
Orono, ME	Apr 25	Oct 12	May 9	Sep 29
Boston, MA	Mar 28	Nov 20	Apr 8	Nov 7
Burlington, VT	Apr 29	Oct 16	May 12	Oct 2
Buffalo, NY	Apr 14	Nov 1	Apr 27	Oct 18
Philadelphia, PA	Mar 30	Nov 12	Apr 10	Oct 29
Baltimore, MD	Apr 4	Nov 8	Apr 12	Oct 26
Norfolk, VA	Mar 8	Dec 3	Mar 23	Nov 19
Raleigh, NC	Mar 26	Nov 11	Apr 13	Oct 26
Charleston, SC	Feb 23	Dec 4	Mar 16	Nov 18
Atlanta, GA	Mar 10	Nov 26	Mar 26	Nov 12
Orlando, FL	Jan 20	Jan 12	Jan 31	Jan 8
Birmingham, AL	Mar 14	Nov 15	Mar 30	Nov 6
New Orleans, LA	Feb 3	Dec 21	Feb 21	Dec 7
Little Rock, AR	Mar 7	Nov 20	Mar 22	Nov 9
Memphis, TN	Mar 4	Nov 22	Mar 23	Nov 11
Cincinnati, OH	Apr 3	Nov 2	Apr 19	Oct 20
Chicago, IL	Apr 6	Nov 3	Apr 21	Oct 25
Minneapolis, MN	Apr 19	Oct 14	May 1	Oct 4
Fargo, ND	May 3	Oct 3	May 13	Sep 25
Lincoln, NE	Apr 14	Oct 16	Apr 28	Oct 6
St. Louis, MO	Apr 3	Nov 1	Apr 12	Oct 18
Tulsa, OK	Mar 14	Nov 18	Mar 26	Nov 8
Dallas, TX	Feb 22	Dec 11	Mar 7	Nov 26
Houston, TX	Feb 2	Dec 28	Feb 15	Dec 12
Denver, CO	Apr 21	Oct 16	May 1	Oct 5
Boise, ID	Apr 28	Oct 15	May 10	Oct 5
Albuquerque, NM	Apr 6	Nov 6	Apr 18	Oct 27
Sacramento, CA	Jan 6	Dec 30	Jan 25	Dec 17
Portland, OR	Feb 24	Dec 2	Mar 27	Nov 10

[a] Probability of occurring 5 years in 10 later than this date.

[b] Probability of occurring 5 years in 10 earlier than this date.

Source: USDA Natural Resources Conservation Service, unpublished data for 1961-
1990.) For wetland delineations, the former should be used, although Corps districts
have discretion to use the latter in southern states (Williams, 1992).

microbes are active year-round and that the growing season for soil microbes could not be predicted
from either 28°F or 32°F air temperatures as currently done for federally regulated wetlands
(Huddleston and Austin, 1996). In studying microbial activity in southern bottomland hardwood
forests from South Carolina to Louisiana, Megonigal et al. (1993) found sufficient soil respiration
and oxygen consumption to create anoxia in saturated soils in winter. The soil temperatures,
however, never fell below 5°C at 50 cm. Based on these findings, they recommended a 12-month
microbial activity season in the South for purposes of wetland hydrology assessment. Other studies
documenting microbial activity during the winter, include Pickering and Veneman (1984) and
Groffman et al. (1992) for New England and Megonigal et al. (1996) for the Southeast.

Using the 5°C soil temperature to define the beginning of the growing season would eliminate any "growing season" for arctic and subarctic regions with shallow permafrost soils, despite the presence of tundra and taiga vegetation. Ping et al. (1992) acknowledged this problem, and the NRC (1995) noted that it also may not be appropriate to use this concept for the growing season in some temperate communities such as Minnesota bogs (Dise, 1992). Ping et al. found that microbial activity was more related to soil water content than to soil temperature, as they observed reduction when temperatures were below 2°C in Alaska. Likewise, Bell et al. (1996) recorded similar findings in Minnesota mollisols, reporting reduction at 2.6°C. Clark and Ping (1997) recommended lowering the temperature to 0°C based on their observations of soil reducing conditions below 5°C. There are serious shortcomings to the current concept of biological zero as applied to soils. (See Chapter 4 for more discussion.)

ECOLOGICAL CONSIDERATIONS IN DEFINING THE CRITICAL PERIOD FOR WETLAND HYDROLOGY

The current regulatory definition of growing season for wetland assessments is not scientifically sound for several reasons. First, as mentioned above, the concept of biological zero is really not accurate at least when considering soil microbial activity in cold regions. Second, there is ample evidence of wetland plant activity occurring beyond the period currently defined by the federal government's procedure for estimating it. Third, the entire concept of growing season does not consider the requirements of wetland-dependent animals that need water outside of this period. Fourth, the emphasis on growing season for defining wetland hydrology also ignores the fact that many important wetland functions occur during the non-growing season (e.g., flood storage and winter wildlife habitat). From an ecological standpoint, the significant "season" or period for wetland hydrology should be based on the requirements of wetland organisms (both plants and animals, including microbes). Defining the critical hydrology period by these considerations would be superior and more scientifically sound than one based on either the frost-free period or biological zero. Such a period might be better termed the "ecologically critical hydrologic period" or the "functionally critical hydrologic period" rather than "growing season" which emphasizes plants and ignores animal life in wetlands. The following provides a rationale for basing the critical period for defining wetland hydrology on ecological and functional grounds.

Botanical Evidence

It is well established that many native plant species begin growing well in advance of this frost-free period, especially spring-bloomers, evergreen trees and shrubs, and herbs with evergreen leaves and stems (e.g., *Juncus effusus*). Growing season is plant-specific and should include growth of any kind including both shoot and root elongation. It is different for each type of plant and for plants of the same species in different locations within a state and even on one side of the mountain vs. the other. For the most part, at least some native species are actively growing roots, moving nutrients and water through their cells, and even flowering before the beginning of the frost-free period. Most people know that the sap flows in sugar maples well before buds break. In New England, this event takes place from late February into May. Skunk cabbage (*Symplocarpus foetidus*), the dominant herb in most seasonally flooded red maple swamps throughout the northeastern U.S., begins flowering about this time and does so even when snow is still on the ground. The heat emitted from cellular respiration associated with its rapid growth melts the snow or ice around it (Niering, 1985). The fact that skunk cabbage flowers and leafs out before other species emerge undoubtedly gives it a significant edge over would-be competitors.

Early-spring blooming species provide visual evidence of above-ground plant activity. Besides skunk cabbage, these species include some others that grow in wetlands, such as pussy willow (*Salix discolor*), red maple (*Acer rubrum*), silver maple (*Acer saccharinum*), marsh marigold (*Caltha palustris*), leatherleaf (*Chamaedaphne calyculata*), spring beauty (*Claytonia virginica*), trout lily (*Erythronium americanum*), speckled alder (*Alnus rugosa*), eastern red cedar (*Juniperus virginiana*), golden club (*Orontium aquaticum*), and Atlantic white cedar (*Chamaecyparis thyoides*). These and other plants typically bloom before the trees around them are fully leafed out. Many shrubs and trees also produce flowers before their leaves appear, including highbush blueberry (*Vaccinium corymbosum*), alders (*Alnus* spp.), American elm (*Ulmus americana*), ashes (*Fraxinus* spp.), oaks (*Quercus* spp.), and sweet gum (*Liquidambar styraciflua*). Table 2.2 presents examples of early blooming wetland species in various parts of the country (compare these dates with those in Table 2.1). Following an ecological approach to growing season, the beginning of this season might be defined by the average time when the buds first begin to swell for the earliest blooming wetland species. Yet, evergreen trees and shrubs and herbs with overwintering green parts (e.g., shoots of sedges and grasses) undoubtedly commence growth before budswell is noticed in deciduous woody species. Moreover, root growth prior to flower and leaf emergence and shoot development also should be included in an ecological definition of growing season. Bachelard and Wightman (1974) studying balsam poplar (*Populus balsamifera*) reported three phases of root activity prior to budburst on May 13 and such activity initiated around March 17, about 2 months before budburst. Similar processes undoubtedly occur in other species, so budburst is probably 1 to 2 months after initial root activity in northern climes.

As far as the ecological end of the growing season for wetland species, fall flowering, seed development, and root growth activity are relevant factors. Many typical late summer–fall blooming species, like goldenrods (*Solidago* and *Euthamia*), asters (*Aster*), and beggar-ticks (*Bidens*), possess flowers well into October in the Northeast and undoubtedly are forming seeds for some time thereafter. Also, witch hazel (*Hamamelis virginiana*) flowers in the fall, with flowers observed in early December. Autumn is a prime time for root and rhizome growth as many plants continue to send nutrients to these organs for winter storage vital to next year's shoot growth. Prentki et al. (1978) reported growth of shoot bases and rhizomes in broad-leaved cattail (*Typha latifolia*) from mid-summer until December in a lakeside Wisconsin marsh. In studying sedges in New York, Bernard and Gorham (1978) noticed that most shoots of beaked sedge (*Carex rostrata*) had appeared in October, but some new ones emerged in early December.

The average citizen in the eastern U.S. also realizes that the fall is an excellent time to reseed a lawn as the absence of a tree canopy (no shading) and relatively high soil moisture (due to lower air temperatures and reduced evapotranspiration) create favorable conditions for germination and root growth of cool-season grasses. Many woody plants exhibit significant root growth at this time. Lopushinsky and Max (1990) observed very small amounts of new root growth on some seedlings at a soil temperature of 0.5°C and 5°C — 53% of the Pacific fir seedlings (*Abies amabilis*) and 37% of Noble fir seedlings (*Abies procera*) developed new roots longer than 1 cm at these soil temperatures. They concluded that the ability of these firs to produce new root growth at low soil temperatures may partially explain their occurrence at relatively high elevations in the Pacific Northwest. Lyr and Hoffman (1967) reported that minimum soil temperatures for root growth range from slightly above 0° to 7°C, while optimum temperatures are from 10° to 25°C. DeWald and Feret (1987) studying root growth of loblolly pine (*Pinus taeda*) in Virginia found that root growth continued upon cessation of shoot growth (late October) through most of the winter and, in late February, increased photosynthesis supported elevated root metabolic activity. Root activity in apple, plum, and oak trees is most active during periods of inactive shoot growth and in temperate trees there is considerable root growth during autumn and winter (Head, 1967; Reich et al., 1980). Mature trees of white oak (*Quercus alba*) in Missouri exhibited root growth from fall into winter until soil temperatures reached 2° to 5°C and root growth was observed for 63.5% of the 145-day period between leaf drop (in the fall) and budburst (in spring) (Reich et al.). Several other references that document plant growth at low temperatures include Lawrence and Oechel (1983) for root growth of arctic deciduous trees, Vézina and Grandtner (1965) for northern hardwood forest spring-herb leaf development, and Chapin and Shaver (1985) for tundra plant photosynthesis (see NRC 1995 for additional references).

TABLE 2.2
Phenological Data on Early-Blooming Wetland and Nonwetland Species in Different Parts of the Northern Conterminous U.S.

Location (Source)	Species	Date of First Flower
Eastern Massachusetts	*Acer rubrum*	April 8–14
(Debbie Flanders, personal communication, 1998)	*Alnus rugosa*	April 1–7
	Lindera benzoin	April 15–21
	Salix candida	March 25–31
	Salix discolor	April 15–21
	Symplocarpus foetidus	April 8–14
Washington, D.C.	*Acer rubrum*	March 11[a]
(Shetler and Wiser, 1987)	*Acer saccharinum*	February 22[a]
	Alnus serrulata	March 10[a]
	Cardamine hirsuta	February 27[a]
	Corylus americana	March 8[a]
	Lindera benzoin	March 27[a]
	Populus grandidentata	March 29[a]
	Salix discolor	March 15[a]
	S. sericea	March 31[a]
	Stellaria media	January 28[a]
	Symplocarpus foetidus	February 9[a]
	Taraxacum officinale	February 1[a]
	Ulmus americana	March 2[a]
Blue Ridge Mountains, NC	*Acer rubrum*	March 11–16
(Day and Mark, 1977)	*Cornus florida*	April 15–22
	Quercus prinus	April 15–22
Northeastern Minnesota	*Picea glauca*	April 29–May 28
(Ahlgren, 1957)	*Abies balsamea*	April 30–May 8
	Larix laricina	April 25–May 9
	Thuja occidentalis	April 10–May 6
	Ulmus americana	April 25–May 8
	Betula papyrifera	April 2–23
	B. allegheniensis	April 2–May 16
	Acer saccharinum	April 18–30
	A. rubrum	April 18–30
Kansas	*Acer saccharinum*	January 1–March 23
(Hulbert, 1963)	*Ulmus pumila*	February 7–March 26
	Ulmus rubra	February 11–April 12
	Taraxacum officinale	January 1–April 12
	Lamium amplexicaule	January 22–April 25
	Vinca minor	February 3–April 30
Great Basin, NV	*Hilaria janesii*	March 24–April 7[b]
(Everett et al., 1980)	*Atriplex confertifolia*	March 28[b]
	Oryzopsis hymenoides	March 24–April 2[b]
	Chrysothamus viscidiflorus	March 18–24[b]

[a] Mean dates
[b] Break in dormancy

Note: In southern parts of the U.S., flowering occur, year-round, with some species in bloom during winter.

A most interesting example of plant growth in winter in New England (central New Hampshire) was reported by Muller (1978). In studying the spring ephemeral herb, trout-lily, Muller described winter growth as "growth leading to the early spring development of photosynthetic tissue begins with fall root growth and continues through a long winter phase during which the shoot elongates from the perennating organ, through the soil and into the snowpack. Following snowmelt, the shoots begin rapid unfurling and maturation of the photosynthetic tissue." Vegetative growth for this species begins in mid-October with root apices elongating from the base of the bulb (corm) and the shoot elongates from mid-November to mid-December and continues through winter.

Wetzel (1990) reported that periphyton associated with aquatic macrophytes typically develop most prolifically during the autumn and winter, possibly in response to significant release of nutrients from the macrophytes. Interestingly, this stimulated microfloral growth may actually conserve nutrients within the community by preventing their release to open water, thereby recycling these vital elements and reducing possible eutrophication.

Photosynthesis may occur throughout the year in evergreen species. In North Carolina, loblolly pine and white pine (*Pinus strobus*) seedlings respired and photosynthesized year-round although rates were lowest in late January (Davis McGregor and Kramer, 1963). Lopushinsky and Kaufmann (1984) found that Douglas fir (*Pseudotsuga menziesii*) transpired at 1.3°C at a rate that was 18.8% of its rate at 20.2°C. Kramer (1942) reported that when the soil was near freezing, pine seedlings of eastern species took up water at 14 to 38% of the rate that they did at 25°C. Teskey (1982), studying Pacific silver fir, found that stomata closure occurred when temperatures fell below 1.5°C. Cranberry vines (*Vaccinium macrocarpon*) in Wisconsin are "alive and respiring during the dormant period" (Roper et al., 1991). Forsyth and Hall (1967) observed that cranberry leaves respire, fix CO_2 and release O_2 at a higher rate than that of respiratory oxygen consumption at low temperatures under winter flood conditions. Roper et al. further stated that "the selection of 41°F as biological zero…for an evergreen perennial vine appears arbitrary and, in all likelihood, is incorrect. This temperature does not define the lower limit of the growing season for this non-deciduous perennial plant that continues to respire and grow below 41°F." In addition, they reported that a layer of ice over the vines in winter (result of winter hydrology) is important to prevent dessication of cranberry vines exposed to cold temperatures and wind. Ice or water also may prevent dessication of other wetland species in winter.

Seeds of certain plants can germinate below 5°C. In a classic study, DeCandolle (1865) found that mustard (Cruciferae; *Sinapis alba*) germinated in 11 to 17 days at nearly 0°C, while pepper-grass (*Lepidium sativum*) and flax (*Linum usitatissimum*) germinated in 30 and 34 days, respectively, at 1.8°C.

Another important consideration is that winter wetness is critically important to some plant species. Certain evergreens are highly susceptible to the effects of winter drying or drought due to evaporation from cold, dry winds. Evergreen rhododendron and evergreen bog plants (including cranberries and other members of the heath family – Ericaceae) all might significantly benefit from winter wetness. Winter flooding of cranberry bogs is a standard management practice. Loblolly pine (*Pinus taeda*) is well adapted to prolonged anaerobic conditions found in flatwoods that are wet from winter through early spring. In a Louisiana loblolly pine stands, Hu and Linnartz (1972) found that the wettest site had the highest site index values (most productive sites). Winter wetness increased productivity of both loblolly pine and slash pine (*P. elliottii*) (Haywood et al., 1990). The availability of water early in the growing season may be more important for plant growth than an adequate oxygen level in the soil. In studying the effect of waterlogging on floodplain herbs in The Netherlands, Van der Sman et al. (1988) noted winter inundations among several significant factors affecting plant distribution.

Wetland plants capable of performing vital life functions and beginning reproduction during cold season wetness may have a competitive advantage over other species in terms of colonizing wetlands. The fact that evergreen species, cool-season grasses, and certain sedges, grow year-round or virtually so in northern climes (Bernard and Gorham, 1978; Prentki et al., 1978; Wisconsin State

Cranberry Growers Association, 1991; David Cooper, Colorado School of Mines, personal communication, 1991) makes it likely that any green plant in winter is probably active at some time (although at reduced levels). Many submerged aquatic species overwinter in an "evergreen" condition and continue basal metabolism under ice when temperatures are below 5 to 10°C, despite little or no net growth (Wetzel, 1990). The significance of winter wetness to many wetland species cannot be overlooked, nor can year-round soil microbial activity. For arctic species, thawing of soil is an important event since nutrients in the frozen soil then becomes available for plant uptake and shoot development (Prentki et al., 1978).

Faunal Evidence

Many wetlands serve as valuable wildlife habitats during the nongrowing season — critical to the survival of many animals. Some examples follow. Evergreen forested wetlands (cedar swamps) serve as winter yards for deer and moose, northeastern coastal marshes as overwintering grounds for black ducks, and bottomland hardwood swamps as overwintering grounds for other waterfowl. In the Upper Midwest, migrating waterfowl begin arriving in pothole wetlands well before the frost-free period as temporary and seasonal potholes become inundated as snow melts during the last week of March and first week of April (Cowardin and Kantrud, 1991). In the Northeast, male red-winged blackbirds are among the first migratory birds coming back to freshwater marshes. They begin establishing their territories in early March, a month before the beginning of "growing season" defined by federal wetland regulations. Floodplain forested wetlands inundated in winter and early spring are important fish habitats. Some fish species spawn during these periods, especially in the spring, while others use the flooded wetlands as nursery grounds (Leitman et al., 1991; Hoover and Killgore, 1998).

Some wetland-dependent amphibians begin breeding in vernal pool wetlands in the northeastern U.S. before such ponds are ice-free. The Jefferson salamander is the first amphibian to use vernal pools for breeding in the new year. Welsch et al. (1995) state that this salamander "migrates over the snow on rainy nights in late winter to slip into the pond (vernal pool) through cracks in the ice" (see photo on p. 42 of this reference for picture of Jefferson salamander migrating over snow). Klemens (1993) had to chip holes in the ice to collect specimens from the cold pond waters of southern New England.* The spotted salamander is the next salamander to use these icy ponds, entering along the ice-free edges of their natal ponds. Klemens noted that its migration occurred when soil temperatures at 30 cm were at least 4.5°C and when the soil was warmer at the surface than it was at 30 cm (i.e., when soil temperature profile reversed). In southern New England, this turnover probably occurs from early to mid-March. Wood frogs are the earliest frogs to begin breeding in the Northeast. They begin their courtship choruses in similar ponds while ice is still present.

There also is significant breeding activity by some species after the "growing season." Marbled salamanders migrate to dry vernal pools in the fall (late August to October) in the Northeast (Klemens). Males leave sperm sacks on the dry pool bottom and females later arrive to fertilize the eggs with these sacks, leaving the fertized eggs beneath leaf litter or rocks in the pool. As fall rains fill the pools, the eggs hatch and larvae feed on aquatic invertebrates (Welsch et al., 1995). Thus, late fall through winter inundation is vital to the hatching, larval development, and ultimate survival of this species. Another example of this type of life history requirement is found in the pitcher-plant bog crayfish (*Fallicambarus gordoni*). It is active from late fall through late spring in Mississippi pitcher-plant bogs when flooded and aestivates during summer drawdown (Johnston and Figiel, 1997). Mating occurs in late spring, with females retaining the fertilized eggs until fall or winter when the eggs are laid in the flooded bogs. The above species are just a few of the animals that depend on winter wetness for survival.

---

* Aquatic life is active through winter in the waters below ice-covered lakes and ponds. Invertebrates including rotifers and fairy shrimp thrive at such times and fish also are active as evidenced by ice fishing.

FIGURE 2.2   Temporary storage of flood waters by wetlands at any time of the year helps reduce the risk of flood damage.

Wetland Functions Evidence

Besides providing fish and wildlife habitat, wetlands perform other important wetland functions outside the "growing season." Winter and early spring (pre-growing season) flood storage is a common function of many floodplain wetlands. Significant groundwater recharge may take place during the dormant season. Woo and Winter (1993) reported significant ground-water recharge from North Dakota pothole wetlands in early spring just after seasonal frost disappears.

Many wetland functions are carried out in all seasons and are independent of plant activity (Tiner, 1991b). Flood storage at any time during the year should be a highly valued function to people living downstream (Figure 2.2). Other wetland functions, such as shoreline stabilization, erosion control, sediment retention, and water quality renovation, continue year-round (e.g., Simmons et al., 1992). Holland (1996) notes that "it must be remembered that wetland functions are a product of all components of the wetland ecosystem (not just vascular plants), that the wetland functions year-round (not just when vascular plants are actively growing), and that critical functions (such as flood protection) will only occur at irregular intervals." A Rhode Island study of nitrate dynamics reported >80% nitrate removal from streamside wetlands in both the growing season and the dormant season (Simmons et al., 1992). Groundwater temperatures were between 6.5 to 8.0°C throughout the latter season, so microbial activity was not limited by temperature. Nelson et al. (1995) observed the highest nitrate removal rate in November when water tables were high and the lowest rate in June when the deepest water tables occurred. This study suggests that microbial processes (immobilization and denitrification) were more responsible for groundwater nitrate removal than the wetland plants. The hydrology of wetlands must clearly be viewed in the year-round context to be most meaningful, both ecologically and functionally. Megonial et al. (1996) and Bedford et al. (1992) also supported this view.

Concluding Remarks

After considering the above facts, one might reasonably ask the question, when is hydrology not important to wetland life? Given the difficulty of establishing the "growing season," plus considering ecological and functional significance of "nongrowing season" flooding and soil saturation, wetland hydrology should perhaps best be considered in a year-round context or at least whenever the soil is not frozen (Tiner, 1991b). The NRC (1995) reached similar conclusions and recommended that the growing season concept be replaced with another approach that better addresses the hydrologic, biotic, physical, and climatic differences in wetlands across the country. The NRC further indicated that it may be more appropriate to use a time–temperature concept. This concept would establish the duration of wetland hydrology on a regional basis related to the effect of temperature on plants, animals, soil microbes, and soil formation.

Until regional studies define a different critical period for wetland hydrology, hydrologic conditions during the entire year (except for areas experiencing frozen soils for extended periods) should probably be considered as important for defining wetland hydrology for the following reasons:

1. Evergreen plants and persistent grasses and other grasslike plants (e.g., sedges) continue to grow during the "dormant period" for nonevergreens, and saturation at this time should affect these plants and competing species.
2. Water conditions during the dormant period (winter) have a profound influence on the hydrologic conditions during the early part of the "growing season" and probably help prevent winter dessication of some wetland plants.
3. Hydric soil properties have developed under conditions that extend beyond the "growing season".
4. Critical activities of some animals require nongrowing season flooding or soil saturation.
5. Aquatic animals (e.g., fish and red-spotted newts) depend on water during winter as well as the rest of the year.
6. The functions of wetlands do not cease with the "growing season".
7. Wetness limitations during the "dormant period" also affect the potential uses of the land (Tiner, 1991b).

Besides the scientific merits of doing so, an important advantage of defining the period of critical wetland hydrology based on ecological and functional grounds is that it would be easier to explain the significance of this period to the general public. Currently, largely due to the narrower definition of growing season, the minimum wetness for wetland hydrology is established at 7 days of inundation or 14 days of saturation at or near the surface. The general public may think that this is a relatively short period of time for an area to be wetland, as they have undoubtedly seen mudpuddles with water for a week or more in the spring. Yet if an ecological and functionally based concept of wetland hydrology is adopted, the period of soil saturation to qualify as wetland would likely be expanded, depending on the wetland type and climate. For example, most wetlands in the Northeast are wet at or near the surface throughout the winter and into spring and not just for 2 weeks, although many are only wet for 2 weeks during the growing season as currently defined. In studying the hydrology of headwater streams in Massachusetts forests, Patric and Lyford (1980) observed that from winter through early spring, the poorly drained soils in the pits of an area with pit and mound relief were filled with about 6 to 10 in. of water. By summer, the water tables dropped to 3 ft or more, with heavy rains again saturating the soil to the surface for varying periods. Evans (1996) also found that saturation was most pronounced in wetland (hydric) soils from late fall through early spring in New Hampshire. Even the wet flatwoods of the Southeast, locally called "winter wet woods" are wettest during the so-called "nongrowing season" and only briefly wet during the traditional "growing season."

ARE ALL WETLANDS AT LEAST PERIODICALLY ANAEROBIC, REDUCED ENVIRONMENTS?

Prolonged flooding and waterlogging create anaerobic conditions in most soils. The recurrent nature of such events promotes the formation of wetlands and has a profound effect on plant and animal life. Most wetlands experience significant reduction and anaerobiosis due to frequent extended wetness. Yet, waterlogging does not have to result in anaerobic (oxygen-deficient) substrates or promote reduction. Patrick and Mahapatra (1968) reported that the redox potential of waterlogged soils ranges from 700mV to −300mV, including both aerobic and anaerobic environments. A few studies have identified soils saturated for extended periods before reduction commenced or without any detectable reduction. Daniels et al. (1973) detected no reduction in some Aqualfs that were saturated up to 5 months. The upper 0.5 m of soil had sufficient oxygen to preclude reduction. Vepraskas and Wilding (1983) reported mixed signals from several Texas soils — some were reduced longer than they were saturated, while others were saturated longer than reduced.

Flowing water has higher oxygen content than stagnant water and highly oxygenated water flowing rapidly through wetlands may prevent the establishment of anaerobic conditions. For example, sandy soils with extremely rapid flow-through with highly oxygenated water (e.g., coldwater streams) may be aerobic environments, yet most soils that are inundated for significant periods are anaerobic (Robert Wetzel, University of Alabama, personal communication, 1991). Wet aerobic environments also may occur when through-flow flooding or soil saturation occurs in winter or early spring when temperatures are low and there is a lack of organic matter for bacteria. Some examples of wetlands that may not experience any or significant anaerobiosis are (1) cobble substrates associated with mountain streams, including willow-dominated bars in these streams; (2) seasonal seeps which flow only during late winter and early spring (these areas may be fed with well-oxygenated water arising from underground aquifers); (3) tidally flooded, algae-covered rocky shores (barnacles and blue mussels also may colonize these wetlands); and (4) other wetlands on rocky substrates, such as beds of aquatic mosses. Bell et al. (1996) observed in studying Minnesota Mollisols that stratified and highly permeable sands and gravels could have oxiaquic (wet, but aerobic) conditions following heavy rains. An Oregon study of hydric soils found that a poorly drained soil was continuously saturated for months, but did not have low redox potentials sufficient for iron reduction (Huddleston and Austin, 1996). The researchers thought that groundwater flowing through these soils must have had sufficient oxygen to prevent reduction during this observation period.

For reduction to occur in saturated soils, there must be a source of organic matter, the temperature must be high enough to support microbial activity, and free oxygen must be absent (Diers and Anderson, 1984). There are certain situations where reduction does not occur despite the soil being saturated repeatedly for long periods. Moormann and van de Wetering (1985) listed four conditions where saturated soils may not be reduced.

1. Cold climates with average temperature less than 1°C.
2. Very saline waterlogged soils of deserts where high salinity restricts growth of reducing microbes.
3. Areas with little or no organic matter and moderate to high amounts of calcium carbonate that limit reduction in arid and semiarid regions (e.g., irrigated rice basins in northwest India lack a low chroma matrix).
4. Groundwater discharge areas where considerable dissolved oxygen is present (e.g., areas of moderate relief and lateral water movement and soils on the edges of valleys).

WHAT ARE COMMON TYPES OF WETLAND HYDROLOGY?

Factors influencing wetland hydrology include landscape position, soils, underlying geology, precipitation patterns, groundwater relations, surface water runoff, and tidal action. Winter and Woo (1990) summarize information on the hydrology of lakes and wetlands. Two general categories of wetland hydrology exist — tidal and nontidal. Tidal wetland hydrology is associated with estuaries and coastal rivers where tides exert a major influence on site wetness. Nontidal wetland hydrology is the most widespread covering all types of wetlands that are beyond the reach of the tides. Each of these hydrologies can be subdivided into other categories. The following discussion uses terminology largely derived from Cowardin et al. (1979) which has been used to classify wetlands across the country.

Tidal wetland hydrology is associated with salt and brackish marshes, mangrove swamps, and tidal freshwater wetlands along all ocean coastlines and tidal rivers. The frequency and duration of tidal flooding in local areas is greatly affected by elevation. Lower elevations that extend from mean sea level to mean low water are flooded at least once daily by the tides. These areas are considered regularly flooded. The lowest elevations of periodically flooded lands occur below the mean low water mark. These areas are flooded most of the time and exposed for short periods, in other words, intermittently exposed. Above the regularly flooded zone, tidal wetlands are flooded less often than daily — irregularly flooded. Here flooding may occur weekly, monthly, or just a few times a year. When not flooded, however, the soils remain saturated continuously throughout the year. Note that permanently flooded (subtidal) areas in tidal waters are not considered wetlands for federal regulatory purposes or by the federal wetlands inventory.

Beyond the penetration of salt-laden tides, tidal action affects the level of strictly freshwater wetlands. Here water levels may fluctuate in sync with the tides during certain seasons. At maximum discharge periods, tidal influence is usually eliminated from many of these areas, but at low flows, maximum tidal influence can be experienced. Consequently, there are seasonal tidal effects. Some portions of these freshwater wetlands experience daily tidal flooding (regularly flooded), while most are either semipermanently flooded-tidal (flooded for the entire growing season in most years), seasonally flooded-tidal (flooded for extended periods during the growing season), or temporarily flooded-tidal (inundated for brief periods, usually early in the growing season, with the adjacent river or stream undergoing tidal fluctuations at other times).

Given the diversity of nontidal wetlands, their hydrologies are quite varied (Figure 2.3). At the wettest end are permanently flooded areas — wetlands occurring in shallow water, like cattail and tule marshes or water lily beds. Next are the semipermanently flooded wetlands which are inundated throughout the growing season in most years, followed by seasonally flooded wetlands where surface water is present for extended periods during the growing season, but is usually absent by late summer. Temporarily flooded wetlands are inundated for brief periods during the growing season often 2 weeks or less.* Saturated wetlands have little evidence of standing water, but have high water tables at or near the land surface for most of the growing season. Seasonally saturated wetlands typically lack signs of surface water ponding (except in micro-depressions) and have high water tables at or near the soil surface during the winter and early spring, but water tables are low during most of the growing season due to high evapotranspiration rates. (Note: This term is not described in Cowardin et al., 1979, but is an outgrowth of wetland mapping on the coastal plain where many such wetlands, also known as "flatwoods," abound.) Temporarily flooded and seasonally saturated types are the focus of much of the debate over what areas should qualify as regulated wetlands.

Other terms may be used to define the flow of water associated with wetlands (Brinson, 1993; Tiner, 1997, among others). Outflow wetlands are wetlands that are sources of water flow for streams or subsurface flows; water flows out of these wetlands. Throughflow wetlands have water entering and leaving them from either surface water or subsurface sources; water flows through these wetlands. Inflow wetlands are sinks for water flow; water flows into them but does not exit via a stream or subsurface.

Soil scientists have developed several terms to describe inundation and soil saturation. Flooding and ponding represent two inundated conditions derived from different processes. Flooding is caused by waters overflowing streambanks, tidal flowage, or runoff from adjacent uplands (or some combination of them), whereas ponding is the result of water collecting in a closed depresssion. Four categories of frequency of flooding are recognized.

1. Frequent (more than 50 times in 100 years)
2. Occasional (5 to 50 times in 100 years)
3. Rare (1 to 5 times)
4. None

* Many wetlands are temporarily flooded, especially along floodplains and the Gulf–Atlantic coastal plain. For example, over half of Maryland's wetlands were classified as temporarily flooded by the U.S. Fish and Wildlife Service's National Wetlands Inventory (Tiner and Burke, 1995). Many of these are better described as seasonally saturated.

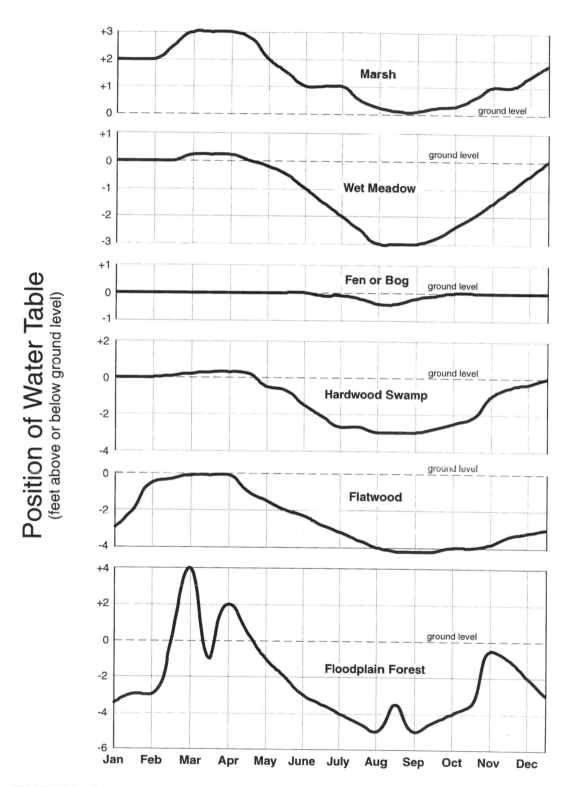

FIGURE 2.3 Generalized hydroperiods of some common wetlands in the country based on eastern U.S. examples. Note that there is considerable variability within types and that the hydrographs depicted are presented for illustration purposes.

Duration of flooding or ponding may be one of five categories.

1. Very long (≥1 month)
2. Long (1 week to 1 month)
3. Brief (2 to 7 days)
4. Very brief (4 to 48 h)
5. Extremely brief (<4 h) (Soil Survey Division Staff, 1993)

Soil saturation may be one of three types.

1. Endosaturation (wet throughout the top 2 m)
2. Episaturation (perched water tables; wet from above, an underlying layer is unsaturated within 2 m)
3. Anthric saturation (controlled by people as in cranberry and rice cultivation) (Vepraskas, 1996).

These terms are applied to soils to describe their wetness characteristics.

The *National Water Summary on Wetland Resources* (Fretwell et al., 1996) contains numerous state wetland chapters that provide overviews of wetland hydrology. Some other background references on wetland hydrology include: Carter et al. (1979), Novitzki (1982, 1989), Ingram (1983), Siegel (1983, 1992), Carter (1986), Routlet and Woo (1986), Ford and Bedford (1987), Siegel and Glaser (1987), Winter and Woo (1990), and Gilman (1994).

ARE WETLANDS ECOTONES?

Wetlands have been considered to be ecotones or ecotonal habitats between land and water by many scientists (Mitsch and Gosselink, 1993; among others). While certain wetland plant communities or portions of wetlands represent such ecotones and do serve as excellent examples, the concept of ecotones has been greatly misapplied to and overused for wetlands (Tiner, 1993b). Ecotones are no more attributed to wetlands than they are to other habitats (upland forests, fields, and thickets).

An ecotone has been defined as a transition zone of tension between two or more communities (Clark, 1954; Odum, 1959). This zone emphasizes the struggle between adjacent communities brought about by physical conditions (e.g., soil, drainage, and pH) and/or biological competition. Ecotones may vary in scale from broad transitions of vegetation (e.g., sometimes more than 100 km in width) between major vegetation communities to narrow belts of transition between adjacent plant communities, such as the edge of the forest next to a grassy field, a streambank beside a meadow, or the transition from a soft bottom to a hard bottom marine community. Clark's concept of ecotone requires that there be tension between the communities and he admits that this is difficult to demonstrate for large geographic areas. Odum suggests that the ecotone is a tension belt of considerable linear dimension that is narrower than the communities on either side. More recently, Holland (1988) has defined ecotones as "a zone of transition between adjacent ecological systems having a set of characteristics uniquely defined by space and time scales and by the strength of the interactions between adjacent ecological systems."

Physical conditions in the ecotone are different from those in either adjacent community as it is influenced by the adjacent communities. Ecotones may be preferred habitats for some species, so certain plants and animals may only occur in these areas and may be absent or virtually absent from the bordering communities. Clark gives the margin of a pond where willows and cattails predominate as an example of an ecotone between land and water.

An important, yet often overlooked, requisite of an ecotone is that "the ecotone inhabitants owe their existence to the presence of the particular conditions on each side and that the ecotone assemblage would disappear or be considerably modified, if the bordering communities or conditions were removed or seriously changed," according to Clark. The "edge-effect" is typified by ecotones, leading to increased variety and diversity of organisms (Odum). This would seem to restrict the concept of ecotones mostly to areas relatively narrow in width but possibly quite long in length (the edges) between adjacent communities as suggested by Odum.

The difference between an edge and an ecotone should be emphasized. An edge is any recognizable boundary between two adjacent communities. It is usually defined by an observed change in the plant community. This change may be abrupt as the edge between a lake and a rock cliff or a floating bog mat and open water, or it may be more gradual and transitional in nature. An edge may be an ecotone when there is a zone of tension between the two communities. While there may be a zone of tension between any two communities, the concept of an ecotone implies a tension zone more than just a few centimeters in width — it should be wide enough to be of ecological significance. Evidence of this tension is usually reflected by vegetation patterns on land, while substrate differences have been used to identify ecotones in submerged marine systems (Odum). Areas of low topographic relief, such as the Gulf–Atlantic coastal plain, may have relatively broad ecotones where environmental conditions change gradually over considerable distances creating transitional communities where plant species from adjacent habitats intermix. This situation applies equally to wetlands and nonwetlands (aquatic and terrestrial communities).

ECOTONAL WETLANDS

Wetlands have been generally considered "ecotonal communities" presumably because (1) many wetlands occur between dryland and permanent waterbodies and, therefore, appear as the first sign of "land" (a substrate or soil at least periodically exposed to air) and are easily recognized as ecological tension zones;* (2) many wetland plant communities appear transitional in their composition, containing species unique to wetlands mixed with true aquatic species or mesic terrestrial species;** and (3) a belief that the vegetation of wetlands would eventually succeed into an upland plant community. The term "terrestrialization" has sometimes been used to define the latter process — hydrarch succession (Kangas, 1990).

Wetlands occurring in the shallow water zone of lakes and rivers are classic examples of ecotonal communities (Figure 2.4). These ecotones are the "littoral ecotones" of Hillbricht-Ilkowska (1993) or the "wetland/open water ecotones" of Holland (1996). These wetlands are permanently inundated or nearly so throughout the growing season in most years, or at least dominated by flows from the open water body. They tend to be dominated by aquatic species in deeper areas, by a mixture of aquatic and obligate emergent*** hydrophytes in shallower water, and by the latter in sites flooded for long periods but exposed for some time. Emergent species usually cannot survive near permanent inundation deeper than 1.5 to 2.0 m as suggested by Sculthorpe (1985) and Cowardin et al. (1979), while aquatic species typically cannot survive long periods of exposure to air. The shallow water marsh appears to be an ecotone between the adjacent aquatic bed and seasonally flooded marsh. One would expect that the plant composition in this area would change depending on water conditions. In wet years, aquatic species may be expected to predominate the site, while in dry years pronounced summer drawdown could result in an extension of the seasonally flooded marsh community. Consequently, there is noticeable tension in this zone. This situation commonly occurs

---

* The water's edge and the obvious interaction between land and water in these wetlands make it easy for people to visualize the concept of ecotones, hence they do serve as useful examples of the concept.

** For purposes of this discussion, *aquatic* species are those plants that live exclusively in open water habitats; they may tolerate only brief periods of exposure to air. *Terrestrial* species are defined as plants that typically occur on dryland (upland) sites. *Wetland* species are those plants that occur in wetlands (e.g., marshes, swamps, and bogs) with a higher frequency than in uplands.

*** *Emergent* refers to plants that are free-standing, self-supporting, and therefore includes both herbaceous (nonwoody) and woody plants.

FIGURE 2.4   Ecotonal wetland between land and water.

in wetlands with variable annual or cyclical water levels. North American examples include Great Lakes coastal marshes (except Lake Ontario with its stabilized water levels), prairie pothole wetlands, and playa wetlands. Kantrud et al. (1989) and van der Valk and Davis (1976) describe vegetative patterns and dynamics in these wetlands. Littoral ecotones also provide important functions such as buffering adjacent waters from upland impacts (e.g., nutrient and sediment inputs) (Holland, 1996).

The upper edges of many wetlands are transitional or ecotonal plant communities. Where the wetlands border streams, rivers, or lakes, the upper zone may be called the *riparian ecotone* as classified by Hillbricht-Ilkowska, yet the upper edges of isolated depressional wetlands cannot be considered riparian. In either situation, wetland plant species are intermixed with mesic terrestrial species to form what some have called the "transition zone" (Environmental Laboratory, 1987) or the "upland/wetland ecotone" (Holland, 1996). This very condition has made it virtually impossible to use vegetation alone for defining the upper limits of wetlands (Sipple, 1985, 1988; Tiner, 1988; Federal Interagency Committee for Wetland Delineation, 1989). This same problem has been recognized by many scientists well before wetland delineation became a common practice (Tiner, 1991a). For example, in studying vegetation of the Pocomoke Swamp on Maryland's Eastern Shore, Beaven and Oosting (1939) identified an "upland border" zone that was "transition from swamp to upland and not the upland itself." They observed a co-mingling of species especially where small short streams diverged and "lost their identity," and that raised islands in the swamp, although not flooded, had a "constantly high water table." Rather than being ecotones between land and water, these upper communities are ecotones between wetland and dryland (upland). Obligate hydrophytes, especially aquatic bed species, are conspicuously absent from these communities. The limits of these wetlands can only be established by assessing soil properties, since the plant community is dominated by facultative-type* hydrophytic species that occur in both wetlands and nonwetlands to varying degrees.

Many macrophytic species (obligate hydrophytes) occur only in wetlands, while many others (facultative wetland plants) are more common in wetlands than in nonwetlands. The wettest wetlands (permanently or semipermanently flooded) often have both wetland species and aquatic species, but lack terrestrial plants. Conversely, the driest wetlands (seasonally saturated in the root zone or temporarily flooded) typically have terrestrial and wetland species, with aquatic species

---

* In the U.S., species occurring in wetlands have been rated in five wetland indicator categories based on their frequency of occurrence in wetlands. (1) Obligate Wetland (>99% frequency of occurrence in wetlands), (2) Facultative Wetland (67 to 99% frequency of occurrence in wetlands), (3) Facultative (34 to 66%), (4) Facultative Upland (1 to 33%), and (5) Upland (<1% frequency of occurrence in wetlands) (Reed, 1988). See Chapter 3 for details.

FIGURE 2.5 Non-ecotonal wetland, an isolated swamp with seasonal ponding of surface water and no contiguous deepwater habitat.

absent. Terrestrial species found in these areas have adapted in some way to living in periodically anaerobic soils, with shallow root systems being perhaps the most common adaptation (Tiner, 1991a; see Chapter 3). These species exemplify the tension and transitional nature of this zone. The wettest and the driest wetlands, therefore, also may be considered ecotones, but not simply as ecotones between water and dryland, rather as wetland/open water and wetland/upland ecotones, respectively. Holland (1996) offers some insights into the functional significance of these ecotonal wetlands in both tidal and nontidal environments.

Non-ecotonal Wetlands

The majority of wetlands in the conterminous U.S. are not found between land and large waterbodies, but are surrounded by upland (Figure 2.5). Some may be drained by a shallow, narrow stream. These wetlands do not represent an ecotone between dryland and deep water. Their plant communities are usually distinctive, but do not typically contain mixtures of aquatic and terrestrial species living together. They are generally characterized by obligate and/or facultative wetland plant species, while many are dominated by facultative species with no affinity for wetland or upland habitats. Ecotones may occur within these wetlands.

Besides changes in macrophytic vegetation, significant differences occur in the soil of wetlands and nonwetlands. Soil types are distinctly different, with most wetlands typified by at least periodically anaerobic soils (e.g., organic soils or gleyed mineral soils with organic-enriched surface layers). On the other hand, adjacent upland soils are better aerated mineral soils and adjacent deepwater habitat substrates of sands, silts, and clays are often anoxic, but may lack comparable accumulations of organic matter. The differences in soil properties are so significant that they may be used to separate wetlands from nonwetlands in the U.S. (Tiner and Veneman, 1987; Federal Interagency Committee for Wetland Delineation, 1989; see Chapter 4). Also for nonvegetated communities in deep water, substrate differences are used to define ecotones and differences in benthic organisms can be found. This rationale could be equally applied to vegetated areas with differences in soils and soil organisms.

Odum (1959) recognized that communities may not have ecotones and gave as a generic example, a transition community with characteristics of its own. Many wetlands fit this type of categorization, although it may be preferable to consider them as "intermediate" communities between dryland and water, rather than "transitional" communities due to their location along the soil moisture gradient. Included among these wetlands in North America are the interior portions of large wetland complexes, the Florida Everglades, arctic and subarctic muskegs, many boreal bogs and fens, the majority of prairie pothole marshes, wet meadows, playas, vernal pools, seepage wetlands, forested wetlands in interstream divides along the Atlantic–Gulf coastal plain (including pocosins, swamp-on-a-hill), and Carolina bay wetlands. These wetlands and similar ones do not possess vegetation in distinct zones that gradually change from open water to dryland. Different vegetation patterns and soil properties may reflect microsite variations in the frequency and duration of flooding and/or soil saturation, but not a water–land zonation. The gradient is based on variable soil moisture resulting from a complex set of environmental factors including elevation, soil type, geomorphology, and local hydrology. The change in vegetation patterns in wetlands within a given climatic region largely reflects differences in these important factors plus human intervention and the activity of other animals, such as beaver, muskrat, nutria, snow geese, and other herbivores.

Despite their juxtaposition between land and water, coastal salt marshes may not be valid examples of an ecotone between land and water. Some form in areas of increased sedimentation, while others develop on submerged uplands. In New England, these marshes develop peaty substrates unlike the mud or sand bottom substrate of adjacent waters. This difference is as abrupt as the macrophytic vegetation. The salt marsh community does not contain any significant vascular species of adjacent habitats, except along the edges. For example, at the seaward edge in New England salt marshes, marine algae — rockweeds (*Ascophyllum nodosum* and *Fucus vesiculosus*) — may be found in the low marsh intermixed with smooth cordgrass (*Spartina alterniflora*). At the landward edge, switchgrass (*Panicum virgatum*) is a common border plant that grows in both wetlands and upland fields (Tiner, 1987). The salt marsh–upland border zone has the highest plant diversity, while the marsh itself generally has low species diversity (Tiner, 1987; 1993c). This border zone has ecotonal qualities of increased diversity and intermixing of macrophytic species from adjacent communities. The salt marsh is characterized by plants that are mostly unique to these marshes, halophytes (salt-tolerant) specially adapted for life in saline soils. Salinity is an extremely effective limiting factor for plant growth and the relatively few species that are adapted for life in saline soils have extensive areas available for colonization both in coastal areas and in inland arid and semiarid regions where salts also concentrate in the soil. Most of these species do not grow in adjacent upland communities (outcompeted by terrestrial plants) nor in adjacent waters (cannot withstand permanent inundation or extreme exposure to tidal currents), so salt marsh plant communities have little in common with adjacent communities and are not ecotones. Chabreck (1988) recognized two types of ecotones in coastal marshes: (1) the broad intertidal region representing different marsh types varying with salinity, and (2) the zones between different marsh types (e.g., brackish marsh to intermediate or oligohaline marsh).

The greater change in hydrology (the frequency and duration of flooding and/or soil saturation) between adjacent communities, the more distinctive the difference in their plant composition. Where slopes adjacent to any wetland are steep, vegetation changes are abrupt and no transitional plant community develops. This commonly occurs in recently glaciated regions where many depressional wetlands exist in former kettleholes. Another example is an artificially created pond (excluding ponds excavated within wetlands). These dugout ponds often have a margin of cattails (*Typha* spp.) or other obligate wetland emergent plants in the shallow water zone, yet they have no transition zone on the landward side, the adjacent plant community is cropland or pasture with vegetation completely different from the marsh. The extreme difference in the hydrology between the adjacent communities is reflected in the vegetation patterns. There is no zone of tension or transition between upland and the pond and the cattail marsh would still exist in all probability if a parking lot replaced the cropland or pasture community. Depending on the size of the marsh and the permanence of water, the lower part of the wetland may be ecotonal, but it would be an ecotone between aquatic communities (e.g., aquatic beds) and wetland communities and not simply between water and dry land.

While the lower and upper edges of many wetlands represent ecotones between the wetland and an adjacent waterbody or upland, the majority of wetlands, including many distinct wetland types, are not ecotonal between dryland and water (aquatic communities). These wetlands do not occur along a waterbody and cannot be considered transitional between water and land in the strict sense. Most wetlands possess distinct assemblages of plant species and diagnostic soil properties that clearly separate them from adjacent dryland communities. The ecotone concept has been too broadly applied to wetlands in a way that oversimplifies the issue by ignoring the high variability among wetlands, the lack of a permanent waterbody by many wetlands, and the distinctiveness of wetlands vs. other habitats in terms of plants and/or soils. The variety of wetlands in the world resulting from differences in climate, hydrogeomorphology, soils, and other factors, plus the wealth of upland plant communities provide ample evidence that wetland plant communities no more exemplify ecotones than terrestrial communities. The concept of ecotones seems best applied to individual plant communities, allowing for the recognition of ecotones within wetland and upland complexes, between wetlands and nonwetland communities, and between adjacent upland communities (e.g., oak–hickory forest and oak forest) (Tiner, 1993b).

DO WETLANDS REALLY BECOME DRYLAND?

One of the reasons for the traditional notion of considering wetlands as ecotones and transitional habitats could be a belief that all wetlands eventually become uplands. If wetlands were part of the evolutionary pathway from open water to uplands, there would be good reason to consider them as possible ecotones or transitional habitats in a generic sense. Gates' models of hydrarch succession for the Douglas Lake region of Michigan showed a progression of communities from aquatic bed communities to marsh–fen communities to shrub wetlands to forested wetlands to the climax — "dry land" (Gates, 1926). The "dry land" state is, in all likelihood, a wetland forest that is not as obviously wet as the earlier successional stages. This notion of wetlands evolving into uplands is still widespread, at least among the public.

Plant communities may be extremely dynamic within some wetlands especially those with fluctuating water levels like prairie pothole marshes, coastal wetlands along the Great Lakes (except Ontario), or those subjected to disturbances brought on by fire or sediment deposition. Others may experience changes due to anthropogenic actions like drainage, timber harvest, peat mining, grazing, burning, mowing, and altered hydrology. Despite the changing vegetation pattern in some types or under certain conditions, wetlands do not typically evolve into uplands, unless the hydrology is modified by man (or the area is filled), by a climatic change, or by other catastrophic events (e.g., mudslide, volcanic eruption, or earthquake). Several scientists have supported this position (including Nichols, 1915; Heinselman, 1970; Niering, 1989; Tiner, 1993b; 1998). Wetlands are relatively stable spatially in the absence of these significant hydrologic changes. For example, Damman (1979, 1996) believes that many North American peatlands will remain in their present state since they are in equilibrium with the local climate and that their vegetation pattern will change little, unless the hydrology is altered. Thus, most wetlands are generally long-lived features on the landscape, although their vegetation may change due to fire, droughts, heavy grazing, and other factors (Figure 2.6).

Today, more uplands are probably becoming wetlands due to natural processes than the reverse (Tiner, 1993b). This is primarily occurring along the coasts and in northern climes. Rising sea level continues to inundate low-lying areas adjacent to salt marshes and permitting the salt marsh to migrate landward converting upland to wetland or previously nontidal forested wetland to salt marsh. Rising sea levels following the last glaciation drowned river valleys forming estuaries like the Chesapeake and Delaware Bays. Approximately 3000 to 6000 years ago, the rate of sea level rise slowed sufficiently to form barrier beaches and islands, creating conditions favoring the formation of extensive coastal wetlands. As the sea level continues to rise (now accelerating due to global temperature rise attributed to the "greenhouse effect"), lowlands contiguous with salt marshes are being swamped, converted to salt marshes. Direct evidence of this process can be seen by dead trees or stumps in various salt marshes (Figure 2.7). In addition, such processes alter local hydrology by raising water tables thereby promoting the development of freshwater wetlands in

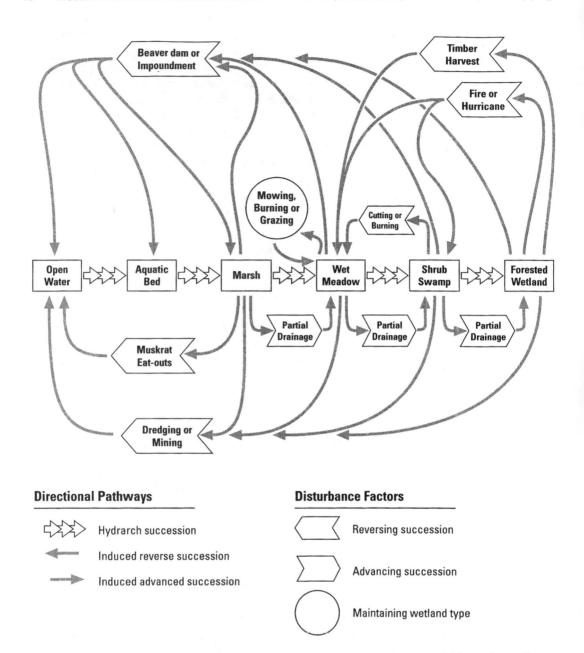

FIGURE 2.6 Wetland hydrarch succession and disturbance factors (i.e., human activities and natural processes) causing changes in wetland types over time.

adjoining low-lying lands. Edmonds et al. (1986) studied Virginia salt marsh soils along the Atlantic Ocean and concluded that the salt-meadow marshes (high marshes) were former uplands because their soils were similar to those of the adjacent uplands. A submerged upland type of salt marsh was identified by Darmody and Foss (1978) in the Chesapeake Bay region. A histic epipedon (shallow layer of peat) has formed over the former silt loam soil.

In boreal regions, paludification processes facilitated by the growth habit of peat mosses (*Sphagnum* spp.) result in the "bogging" of uplands adjacent to existing bogs (Skoropanov, 1961; Crum, 1988). The rolling landscape underlying peatlands in the former Lake Agassiz basin in Minnesota and beneath Poland's Pripet marshes or Poles'ye provide direct evidence of this process

FIGURE 2.7 Dead trees in a salt marsh, evidence of the landward migration of salt marshes, brought about, in part, by rising sea level.

(Heinselman, 1970; Kulczyński, 1949, as reported in Tallis, 1983). In Britain and Ireland, high rainfall creates conditions favoring podzolization and peat accumulation allowing bogs to form on flats or gentle slopes on impermeable soils and once established, the bogs extend downslope changing forests to bogs (Pearsall, 1950). This also can be brought about by clearcutting forests in areas with shallow soils and cool or cold wet climates.

Man and beaver also play an important role in wetland creation. In the arid west of the U.S., irrigation projects (pumping up groundwater) have increased local water tables causing an increase in wetlands (marshes and wet meadows) at the expense of uplands. Drainageways impounded by earthen dams to create ponds often have seepage leaks that may help establish wetlands just below the dams. The creation of ponds on uplands also promotes wetland development along the pond shore. Reservoirs may have similar effects, yet most experience drastic fluctuations in water levels that expose vast acreages of shoreline to drying during the summer. The lack of stable water levels precludes the establishment of significant wetlands under these conditions. Beaver have had a major impact on wetland formation throughout North America. Their dams restrict flow and flood low-lying areas, transforming many uplands to wetlands throughout North America. A series of beaver dams can often be found in local drainage systems, leading to a series of "terraced" wetlands going upslope in the watershed.

Besides major climatic shifts and tectonic activity, natural processes do not convert significant acreages of wetlands to uplands in the short term. Glacial rebound may change wetlands to uplands by changing elevations and local hydrology, as has been reported for Arctic Canada (Tarnocai and Zoltai, 1988) and along the New England coast (Tiner, 1998). Desertification with its migrating dune fields could and probably does convert wetlands to uplands in arid regions of the world, but in North America, it is not a significant process.

The most significant process changing wetlands to uplands in North America is human action. Filling, drainage, channelization, freshwater diversion, and regulating river flows are responsible for converting vast acreages of wetlands to drylands. In the U.S., over half of the nation's original wetlands have been destroyed, mostly due to human activities (Tiner, 1984; Dahl, 1990). Many riparian cottonwood forests in the Southwest represent former wetlands that are no longer flooded sufficiently to be considered wetlands due to river flow alteration.

The above examples amply demonstrate that the ultimate fate of most wetlands is not to become uplands through natural processes. Wetlands should not be viewed simply as transitional habitats, since most wetlands are relatively stable spatially, although the nature of the plant community may change markedly over time due to many factors as do many upland communities.

ARE SOME WETLANDS EPHEMERAL CYCLICAL FEATURES ON THE LANDSCAPE?

While most wetlands probably experience fluctuating water levels and many may appear to be "dryland" at least at the surface during a significant part of the growing season in North America, there are certain wetland types that are dry for years, an event that may drastically alter the vegetation. These wetlands may be called *cyclical wetlands*. A cyclical wetland is one where the plant community changes back and forth between one dominated by wetland species to one dominated by terrestrial species brought about by drastic hydrologic changes due to the natural variation in the hydrologic cycle, to fire in permafrost regions, or to other events which alter the hydrology of the wetland on a periodic (relatively short-term), but not permanent basis. Cyclical wetlands are probably most common and characteristic of arid and semiarid regions where severe droughts occur.

Virtually all prairie pothole wetlands in the upper Midwest and probably playas in the Southwest have dynamic plant communities due to their low precipitation, semiarid climates that are characterized by frequent major droughts. The annual change in spring runoff, summer precipitation, and evapotranspiration causes cyclical fluctuations in the hydrology of these wetlands (Kantrud et al., 1989). The prairies experience a 10- to 20-year cycle of wet to dry conditions. Many potholes lack a permanent open waterbody. During prolonged droughts, annual and perennial terrestrial species may occupy and even dominate the temporarily to seasonally flooded wetlands, while during wet years, obligate wetland species characterize these isolated basins. Consequently, the appearance of wetland plant communities is cyclical, so these wetlands may be considered cyclical from both the vegetative and hydrologic standpoints. Numerous studies have documented vegetative changes during the wet–dry cycle in the Prairie Pothole Region (e.g., Walker, 1959, 1965; Millar, 1973; van der Valk and Davis, 1976; 1978). Despite the shift in vegetation and lack of wetness, the soil retains hydric morphological properties (indicative of wetlands) and therefore can be used to distinguish these wetlands during droughts.

In arid regions such as the southwestern U.S. and much of Australia, the hot, dry climate generally does not favor extensive wetland development. While typical wetlands can be found in these regions in seepage areas, manmade ponds, lakes, reservoirs, and perennial rivers and streams, cyclical wetlands appear on the landscape in other places during wet periods in the hydrologic cycle or simply after very heavy rainfall in local areas or over a broader geographic region. Regional groundwater levels rise during wetter years. Sites for development of these cyclical wetlands may be former lakebeds, in valleys blocked by shifting sands, between dunes, or landward of more typical wetlands. Such sites may be important areas for various water birds (shorebirds, wading birds, and waterfowl). Invertebrates may lie dormant in these areas until favorable conditions appear. Studies have shown that the hyporheic zone (saturated groundwater beneath streams) of streams serves as a refuge for macroinvertebrates during droughts (Hynes, 1958; Griffith and Perry, 1993).

Australian scientists (Paijmans et al., 1985) remarked that low precipitation paradoxically can favor wetland development in a dry, flat continent like Australia due to low gradient streams and low runoff making it easy for streams to be blocked, thereby forming closed depressions where wetlands may develop. They have recognized two general categories of these wetlands, *intermittent* types that are flooded during a series of wet years and *episodic* types that are rarely flooded. The former type occurs in both arid and semiarid places and most have dunes on their eastern side. Clay soils occupy the center of the lake floor, while sandy soils occur along the margins. The basins may be closed or connected to current or former (ancient) drainage systems. Aquatic and emergent plants may occur, but their presence is unstable due to drastic fluctuations in water levels and increasing salinity during dry periods. A thin film of salt often covers these basins when dry. The episodic lakes are dry lakebeds, but also include playas, salinas, pans, and salt lakes that are flooded only during extremely wet conditions. The salts in these areas are derived mostly from groundwater evaporation which often form crusts that prevent infiltration and promote inundation during episodic wet periods. Intermittent and episodic swamps are rare habitats, with the former more frequent than the latter. Intermittent swamps may occur along floodplains at higher elevations than seasonal swamps, or in "terminal drainage depressions" associated with inland dune systems.

Another perplexing type of cyclical wetland occurs in the arctic and subarctic where severe fire both destroys the protective insulation provided by natural vegetation (e.g., spruce forest) and melts the near-surface part of the underlying permafrost. This causes an increase in the depth of the permafrost layer, and thereby improves drainage for some time. Within a period ranging from 30 years to a few hundred years, the vegetative cover insulates the soil and the permafrost layer is gradually restored to its former level (near the surface), thereby impeding drainage and bringing about the return of the black spruce wetland (Ping et al., 1992). The hydrologic change is so drastic that it may be debateable whether the area is or should be considered wetland during the time of recovery. However, it is clear that the ultimate pathway of change leads to a return to wetland conditions within a few decades, at a minimum, provided natural vegetation is allowed to re-establish. Yet, if the area is cultivated, the soils will remain free of permafrost and wetland conditions will not return. This type of cyclical wetland can pose significant questions to both the regulators and the regulated community. In fact, the Swampbuster provision of the Food Security Act has exempted these areas, since they do have high potential for agriculture.

REFERENCES

Ahlgren, C.E. 1957. Phenological observations of nineteen native tree species in northeastern Minnesota. *Ecology,* 38: 622-628.

Bachelard, E.P. and F. Wightman. 1974. Biochemical and physiological studies on dormancy release in trees buds. III. Changes in endogenous growth substances and a possible mechanism for dormancy release in overwintering vegetative buds of *Populus balsamifera.* Can. J. Bot. 52: 1483-1489.

Beaven, G.F. and J.J. Oosting. 1939. Pocomoke Swamp: a study of a cypress swamp on the Eastern Shore of Maryland. *Bull. Torrey Bot. Club,* 66: 367 389.

Bedford, B.L., M. Brinson, R. Sharitz, A. van der Valk, and J. Zedler. 1992. Evaluation of proposed revisions to the 1989 "Federal Manual for Identifying and Delineating Jurisdictional Wetlands." *Bull. Ecol. Soc. Am.,* 73: 14-23.

Bell, J.C, C.A. Butler, and J.A. Thompson. 1996. Soil hydrology and morphology of three mollisol hydrosequences in Minnesota. In the Preliminary Investigations of Hydric Soil Hydrology and Morphology in the United States. Tech. Rep. WRP-DE-13. J.S. Wakely, S.W. Sprecher, and W.C. Lynn (Eds.). U.S. Army Engineer Waterways Experiment Station, Vicksburg, MS. 69-93.

Bernard, J.M. and E. Gorham. 1978. Life history aspects of primary production in sedge wetlands. In *Freshwater Wetlands: Ecological Processes and Management Potential.* R.E. Good, D.F. Whigham, and R.L. Simpson (Eds.). Academic Press, New York. 39-51.

Boggie, R. 1972. Effect of water-table height on root development of *Pinus contorta* on peat in Scotland. *Oikos,* 23: 304-312.

Brinson, M.M. 1993. A Hydrogeomorphic Classification for Wetlands. Wetlands Research Program Tech. Rep. WRP-DE-4. U.S. Army Engineers Waterways Expt. Station, Vicksburg, MS.

Carter, V. 1986. An overview of the hydrologic concerns related to wetlands in the United States. *Can. J. Bot.,* 64: 364-374.

Carter, V., M.S. Bedinger, R.P. Novitzki, and W.O. Wilen. 1979. Water resources and wetlands. In *Wetland Functions and Values — The State of Our Understanding.* P.E. Greeson, J.R. Clark, and J.E. Clark (Eds.). Water Resources Association, Minneapolis, MN. 344-376.

Chabreck, R.A. 1988. *Coastal Marshes Ecology and Management.* University of Minnesota Press, Minneapolis, MN.

Chapin, F.S., III and G.R. Shaver. 1985. Arctic. In *Physiological Ecology of North American Plant Communities*. B.F. Chabot and H.A. Mooned (Eds.). Chapman and Hall, New York. 16-40.

Clark, G.L. 1954. *Elements of Ecology*. John Wiley & Sons, Inc., New York.

Clark, M.H. and C-L. Ping. 1997. Hydrology, morphology, and redox potentials in four soils of south central Alaska. In *Aquic Conditions and Hydric Soils: The Problem Soils*. M.J. Vepraskas and S.W. Sprecher (Eds.). Soil Science Society of America, Inc., Madison, WI. SSSA Spec. Pub. 50: 113-131.

Cooper, D. 1991. Colorado School of Mines. Personal communication.

Costello, D.F. 1936. Tussock meadows in southeastern Wisconsin. *Bot. Gaz.*, 97: 610-648.

Coventry, R.J. and J. Williams. 1984. Quantitative relationships between morphology and current soil hydrology in some Alfisols in semiarid tropical Australia. *Geoderma*, 33: 191-218.

Cowardin, L.M. and H.A. Kantrud. 1991. *An Evaluation Comparing the 1989 and 1991 Federal Manuals for Identifying and Delineating Jurisdictional Wetlands*. Northern Prairie Wildlife Research Center, Jamestown, ND.

Cowardin, L.M., V. Carter, F.C. Golet, and E.T. LaRoe. 1979. *Classification of Wetlands and Deepwater Habitats of the United States*. FWS/OBS-79/31. U.S. Fish and Wildlife Service, Washington, D.C.

Crum, H. 1988. *A Focus on Peatlands and Peat Mosses*. The University of Michigan Press, Ann Arbor, MI.

Dahl, T.W. 1990. *Wetland Losses in the United States 1780's to 1980's*. U.S. Fish and Wildlife Service, Washington, D.C.

Damman, A.W.H. 1979. Geographic patterns in peatland development in eastern North America. In *Classification of Peat and Peatlands*. E. Kivinen, L. Heikurainen, and P. Pakarinen (Eds.). Proceedings Symposium of the International Peat Society, Hyytiälä, Finland. 42-57.

Damman, A.W.H. 1996. Peat accumulation in fens and bogs: effect of hydrology and fertility. In *Northern Peatlands in Global Climatic Change*. R. Laiho, J. Laine, and H. Vasander (Eds.). Academy of Finland, Helsinki, Finland. 213-222.

Daniels, R.B., E.E. Gamble, and S.W. Buol. 1973. Oxygen content in the ground water of some North Carolina Aquults and Udults. In *Field Soil Water Regime*. R. Bruce (Ed.). SSSA Spec. Pub. 5: 153-166. Soil Science Society of America, Inc., Madison, WI.

Darmody, R.G. and J.E. Foss. 1978. *Tidal Marsh Soils of Maryland*. Maryland Agric. Expt. Station, Publication MP 930. University of Maryland, College Park, MD.

Davis McGregor, W.H. and P.J. Kramer. 1963. Seasonal trends in rates of photosynthesis and respiration of loblolly pine and white pine seedlings. *Amer. J. Bot.*, 50: 760-765.

Day, F.P., Jr. and C.D. Mark. 1977. Net primary production and phenology on a southern Appalachian watershed. *Am. J. Bot.*, 64: 1117-1125.

Day, F.P., Jr. and C.V. Dabel. 1978. Phytomass budget for the Dismal Swamp ecosystem. *Virginia J. Sci.*, 29: 220-224.

DeCandolle, A.P. 1865. De la germination sous les degres divers de temperature constante. Biblioth. Univ. et Revue Suisse, Tome XIV: 243-282, as reported in Abbe, C. 1905. A First Report on the Relations between Climates and Crops. USDA Weather Bureau. Washington, D.C. Bulletin No. 36.

DeWald, L.E. and P.P. Feret. 1987. Changes in loblolly pine root growth potential from September to April. *Can. J. For. Res.*, 17: 635-643.

Diers, R. and J.L. Anderson. 1984. Part I. Development of soil mottling. Soil Survey Horizons, Winter 1984: 9-12.

Dise, N.B. 1992. Winter fluxes of methane from Minnesota peatlands. *Biogeochemistry*, 17: 71-83.

Edmonds, W.J., P.R. Cobb, and C.D. Peacock. 1986. Characterization and classification of seaside- salt-marsh soils on Virginia's Eastern Shore. *Soil Sci. Soc. Am. J.*, 50: 672-678.

Environmental Defense Fund and World Wildlife Fund. 1992. How Wet is a Wetland? The Impacts of the Proposed Revisions to the Federal Wetlands Delineation Manual. Washington, D.C.

Environmental Laboratory. 1987. Corps of Engineers Wetland Delineation Manual. Tech. Rpt. Y-87-1. U.S. Army Engineer Waterways Experiment Station, Vicksburg, MS.

Evans, C.V. 1996. Preliminary investigations of hydric soil hydrology and morphology in New Hampshire. In Preliminary Investigations of Hydric Soil Hydrology and Morphology in the United States. J.S. Wakely, S.W. Sprecher, and W.C. Lynn (Eds.). Tech. Rep. WRP-DE-13. U.S. Army Engineer Waterways Experiment Station, Vicksburg, MS. 114-126.

Evans, D.D. and A.D. Scott. 1955. A polarographic method of measuring dissolved oxygen in saturated soil. *Proc. Soil Sci. Soc. Amer.*, 19: 12-16.

Everett, R.L., P.T. Tueller, J.B. Davis, and A.D. Brunner. 1980. Plant phenology in Galleta-Shadescale and Galleta-Sagebrush associations. *J. Range Manag.,* 33: 446-450.

Federal Interagency Committee for Wetland Delineation. 1989. Federal Manual for Identifying and Delineating Jurisdictional Wetlands. Cooperative Technical Publication. U.S. Army Corps of Engineers, U.S. Environmental Protection Agency, U.S. Fish and Wildlife Service, and USDA Soil Conservation Service, Washington, D.C.

Flanagan, P.W. and F.L. Bunnell. 1980. Microflora activities and decomposition. In *An Arctic Ecosystem: The Coastal Tundra at Barrow, Alaska.* J. Brown, P.C. Miller, L.L. Tieszen, and F.L. Bunnell (Eds.). Dowden, Hutchinson, and Ross, Stroudsburg, PA. US/IBP Synthesis Series 12. 291-334.

Flanders, D. 1998. Personal communication.

Ford, J. and B.L. Bedford. 1987. The hydrology of Alaskan wetlands, USA — a review. *Arc. Alp. Res.,* 19: 209-229.

Forsyth, F.R. and I.V. Hall. 1967. Rates of photosynthesis and respiration in leaves of the cranberry with emphasis on rates at low temperatures. *Can. J. Plant Sci.,* 47: 19-23.

Fretwell, J.D., J.S. Williams, and P.J. Redman. 1996. National Water Summary on Wetland Resources. Water-Supply Paper 2425. U.S. Geological Survey, Reston, VA.

Gambrell, R.P. and W.H. Patrick, Jr. 1978. Chemical and microbiological properties of anaerobic soils and sediments. In *Plant Life in Anaerobic Environments.* D.D. Hook and R.M.M. Crawford (Eds.). Ann Arbor Science, Ann Arbor, MI. 375-423.

Gates, F.C. 1926. Plant successions about Lake Douglas, Cheboygan County, MI. *Bot. Gaz.,* 82: 170-182.

Gill, C.J. 1975. The ecological significance of adventitious rooting as a response to flooding in woody species, with special reference to *Alnus glutinosa* (L.) Gaertn. *Flora, Bd.* 164: 85-97.

Gilman, K. 1994. *Hydrology and Wetland Conservation.* John Wiley & Sons, New York.

Griffith, M.B. and S.A. Perry. 1993. The distribution of macroinvertebrates in the hyporheic zone of two small Appalachian headwater streams. *Arch. Hydrobiol.,* 126: 373-384.

Grishkan, I.B. and D.I. Berman. 1993. Microflora of acid organogenic soils in the mountainous subarctic region of north-eastern Asia. In *Joint Russian-American Seminar on Crypopedology and Global Change.* D.A. Gilichinsky (Ed.). Russian Academy of Science, Pushchino, Russia. 272-276.

Groffman, P.M., A.J. Gold, and R.C. Simmons. 1992. Nitrate dynamics in riparian forests: microbial studies. *J. Environ. Qual.,* 21: 666-671.

Haywood, J.D., A.E. Tiarks, and E. Shoulders. 1990. Loblolly and slash pine height and diameter are related to soil drainage in winter on poorly drained silt loams. *New Forests* 4: 81-86.

Head, G.C. 1967. Effects of seasonal changes in shoot growth on the amount of unsuberized root in apple and plum trees. *J. Hortic. Sci.,* 42: 169-180.

Heinselman, M.L. 1970. Landscape evolution, peatland types, and the environment in the Lake Agassiz Peatlands Natural Area, Minnesota. *Ecol. Monogr.* 40: 235-261.

Hillbricht-Ilkowska, A. 1993. Temperate freshwater ecotones: problem with seasonal instability. In *Wetlands and Ecotones: Studies on Land–Water Interactions.* B. Gopal, A. Hillbricht-Ilkowska, and R.G. Wetzel (Eds.). National Institute of Ecology, New Delhi, India. 17-34.

Holland, M.M. 1988. SCOPE/MAB Technical Consultations on Landscape Boundaries: Report of a SCOPE/MAB Workshop on Ecotones. *Biol. Int.* (Special Issue) 17: 47-106.

Holland, M.M. 1996. Wetlands and environmental gradients, chap 3. In *Wetlands: Environmental Gradients, Boundaries, and Buffers.* G. Mulamoottil, B.G. Warner, and E.A. McBean (Eds.). CRC Press/Lewis Publishers, Boca Raton, FL. 19-43.

Hoover, J.J. and K.J. Killgore. 1998. Fish communities, chap. 10. In *Southern Forested Wetlands: Ecology and Management.* M.G. Messina and W.H. Conner (Eds.). Lewis Publishers, Boca Raton, FL. 237-260.

Hu, Shih-Chang and N.E. Linnartz. 1972. *Variations in Oxygen Content of Forest Soils under Mature Loblolly Pine Stands.* Bulletin No. 668. Louisiana State University, Agricultural Experiment Station, Baton Rouge, LA.

Huddleston, J.H. and W. Austin. 1996. Preliminary investigations of hydric soil hydrology and morphology in Oregon. In Preliminary Investigations of Hydric Soil Hydrology and Morphology in the United States. J.S. Wakely, S.W. Sprecher, and W.C. Lynn (Eds.). Tech. Rep. WRP-DE-13. U.S. Army Engineer Waterways Experiment Station, Vicksburg, MS. 127-141.

Hulbert, L.C. 1963. Gates' phenological records of 132 plants at Manhattan, KS, 1926-1955. *Trans. Kansas Acad. Sci.,* 66: 82-106.

Hynes, H.B.N. 1958. The effect of drought on the fauna of a small mountain stream in Wales. *Verh. Int. Verein. Limnol.,* 13: 826-833.

Ingram, H.A.P. 1983. Hydrology, chap. 3. In *Mires: Swamp, Bog, Fen and Moor. General Studies*. A.J.P. Gore (Ed.). Elsevier Scientific Publishing Co., Amsterdam, The Netherlands. 67-158.

Jackson, M.B. and M.C. Drew. 1984. Effects of flooding on growth and metabolism of herbaceous plants. In *Flooding and Plant Growth*. T.T. Kozlowski (Ed.). Academic Press, Inc., Orlando, FL. 47-128.

Johnston, C.E. and C. Figiel. 1997. Microhabitat parameters and life-history characteristics of *Fallicambarus gordoni* Fitzpatrick, a crayfish associated with pitcher-plant bogs in southern Mississippi. *J. Crust. Biol.*, 17(4): 687-691.

Kangas, P.C. 1990. Long-term development of forested wetlands, chap. 3. In *Forested Wetlands. Ecosystems of the World 15*. A.E. Lugo, M. Brinson, and S. Brown (Eds.). Elsevier Science Publishers, Amsterdam, The Netherlands. 25-51.

Kantrud, H.A., J.B. Millar, and A.G. van der Valk. 1989. Vegetation of wetlands of the Prairie Pothole Region. In *Northern Prairie Wetlands*. A.G. van der Valk (Ed.). Iowa State University Press, Ames, IA. 132-187

Klemens, M.W. 1993. *Amphibians and Reptiles of Connecticut and Adjacent Regions*. Bulletin No. 112. State Geological and Natural History Survey of Connecticut, Hartford, CT.

Kozlowski, T.T. (Ed.). 1984. *Flooding and Plant Growth*. Academic Press. New York.

Kramer, P.J. 1942. Species differences with respect to water absorption at low soil temperatures. *Am. J. Bot.*, 29: 828-832.

Kulczyński, S. 1949. Peat bogs of Polesie. *Mem. Acad. Sci., Cracovie B*. Memorial Academy of Science, Poland, 1-356.

Lawrence, W.T. and W.C. Oechel. 1983. Effects of soil temperature on the carbon exchange of taiga seedlings. I. Root respiration. *Can. J. For. Res.*, 13: 840-849.

Leidy, R.A., P.L. Fiedler, and E.R. Micheli. 1992. Is wetter better? *BioScience*, 42: 58-61, 65.

Leitman, H.M., M.R. Darst, and J.J. Nordus. 1991. Fishes in the Forested Floodplain of the Ochlochonee River, Florida During Flood and Drought. Water Resources Investigation Report 90-4202. U.S. Geological Survey, Tallahassee, FL.

Lieffers, V.J. and R.L. Rothwell. 1987. Effects of drainage on substrate temperature and phenology of some trees and shrubs in an Alberta peatland. *Can. J. For. Res.*, 17: 97-104.

Lopushinsky, W. and M.R. Kaufmann. 1984. Effects of cold soil on water relations and spring growth of Douglas fir seedlings. *For. Sci.* 30(3): 628-634.

Lopushinsky, W., and J.A. Max. 1990. Effect of soil temperature on root and shoot growth and on budburst timing in conifer seedling transplants. *New Forests*, 4:107-124.

Lyr, H. and G. Hoffman. 1967. Growth rates and growth periodicity of tree roots. In *International Review of Forestry Research*, vol. 2. J.A.A. Romberger and P. Mikola (Eds.). Academic Press, New York. 181-236.

McKee, W.H., Jr. 1989. A Loblolly Pine Management Guide. Preparing Atlantic Coastal Plain Sites for Loblolly Pine Plantations. General Tech. Rept. SE-57. USDA Forest Service, Southeastern Forest Experiment Station, Asheville, NC.

McKevlin, M.R., D.D. Hook, W.H. McKee, Jr., S.U. Wallace, and J.R. Woodruff. 1987. Loblolly pine seedling root anatomy and iron accumulation as affected by soil waterlogging. *Can. J. For. Res.*, 17: 1257-1264.

Megonigal, J.P., W.H. Patrick, Jr., and S.P. Faulkner. 1993. Wetland identification in seasonally flooded forest soils: soil morphology and redox dynamics. *Soil Sci. Soc. Am. J.*, 57: 140-149.

Megonigal, J.P., S.P. Faulkner, and W.H. Patrick. 1996. The microbial activity season in southeastern hydric soils. *Soil Sci. Soc. Am. J.*, 60: 1263-1266.

Millar, J.B. 1973. Vegetation changes in shallow marsh wetlands under improving moisture regime. *Can. J. Bot.*, 51: 1443-1457.

Mitsch, W.J. and J.G. Gosselink. 1993. *Wetlands*. Van Nostrand Reinhold Co. New York.

Mohanty, S.K. and S. Patnaik. 1975. Effect of submergence on the physico-chemical and chemical changes in different rice soils. 1. Kinetics of pH, Eh, C and N. *Acta Agron. Acad. Sci. Hungary*, 24: 446-451.

Montague, K.A., and F. P. Day, Jr. 1980. Belowground biomass of four plant communities of the Great Dismal Swamp, Virginia. *Am. Midland Nat.*, 103: 83-87.

Moore, A.W. and H.F. Rhoades. 1966. Soil conditions and root distribution in two wet meadows of the Nebraska Sandhills. *Agron. J.*, 58: 563-566.

Moormann, F.R. and H.J.T. van de Wetering. 1985. Problems in characterizing and classifying wetland soils. In *Wetland Soils: Characterization, Classification, and Utilization*. International Rice Research Institute, Los Baños, Laguna, Philippines. 53-68.

Morita, S. 1993. Root system development in rice. In *Abstracts of Platform Presentations and Posters*. B.E. Haissig and T.D. Davis (Eds.). The First International Symposium on the Biology of Adventitious Root Formation (Dallas, TX, April 18-22, 1993). Gen. Tech. Rep. NC-154. USDA Forest Service, North Central Forest Expt. Station, St. Paul, MN. 18.

Muller, R.N. 1978. The phenology, growth and ecosystem dynamics of *Erythronium americanum* in the northern hardwood forest. *Ecol. Monog., 48:* 1-20.

National Research Council. 1995. *Wetlands: Characteristics and Boundaries*. National Academy Press, Washington, D.C.

Nelson, W.M., A.J. Gold, and P.M. Groffman. 1995. Spatial and temporal variation in groundwater nitrate removal in a riparian forest. *J. Environ. Qual.,* 24: 691-699.

Nichols, G.E. 1915. The vegetation of Connecticut. V. Plant societies in lowlands. *Bull. Torrey Bot. Club,* 43: 169-194.

Niering, W.A. 1985. *Wetlands*. Alfred A. Knopf, New York.

Niering, W.A. 1989. Wetland vegetation development. In *Wetland Ecology and Conservation: Emphasis in Pennsylvania*. S.K. Majumdar, R.P. Brooks, F.J. Brenner, and R.W. Tiner, Jr. (Eds.). The Pennsylvania Academy of Sciences, Lafayette College, Easton, PA. 103-113.

Novitzki, R.P. 1982. *Hydrology of Wisconsin Wetlands*. Information Circular 40. Wisconsin Geological and Natural History Survey, Madison, WI.

Novitzki, R.P. 1989. Wetland hydrology, chap. 5. In *Wetlands Ecology and Conservation — Emphasis in Pennsylvania*. S.K. Majumdar, R.P. Brooks, F.J. Brenner, and R.W. Tiner, Jr. (Eds.). Pennsylvania Academy of Sciences, Lafayette College, Easton, PA. 47-64.

Odum, E.P. 1959. *Fundamentals of Ecology,* 2nd ed. W.B. Saunder Co., Philadelphia.

Paijmans, K., R.W. Galloway, D.P. Faith, P.M. Fleming, H.A. Haantjens, P.C. Heyligers, J.D. Kalma, and E. Löffler. 1985. Aspects of Australian Wetlands. Division of Water and Land Resources Technical Paper No. 44. Commonwealth Scientific and Industrial Research Organization, Melbourne, Australia.

Patric, J.H. and W.H. Lyford. 1980. Soil–Water Relations at the Headwaters of a Forest Stream in Central New England. Harvard Forest Paper No. 22. Harvard University, Petersham, MA.

Patrick, W.H., Jr. and I.C. Mahapatra. 1968. Transformation and availability to rice of nitrogen and phosphorus in waterlogged soils. *Adv. Agron.,* 20: 323-359.

Pearsall, W.H. 1950. *Mountains and Moorlands*. New Naturalist Series. Collins, London.

Pickering, E.W. and P.L.M. Veneman. 1984. Moisture regimes and morphological characteristics in a hydrosequence in central Massachusetts. *Soil Sci. Soc. Am. J.,* 48: 113-118.

Ping, C.L., J.P. Moore, and M.H. Clark. 1992. Wetland properties of permafrost soils in Alaska. Proceedings of the Eighth International Soil Correlation Meeting: Characterization, Classification, and Utilization of Wet Soils. J.M. Kimble (Ed.). USDA Soil Conservation Service, Soil Management Support Services, Washington, D.C. 198-205.

Ping, C.L., M. Clark, and G. Michaelson. 1996. Preliminary investigations of hydric soil hydrology and morphology in Alaska. In Preliminary Investigations of Hydric Soil Hydrology and Morphology in the United States. J.S. Wakely, S.W. Sprecher, and W.C. Lynn (Eds.). Tech. Rep. WRP-DE-13. U.S. Army Engineer Waterways Experiment Station, Vicksburg, MS. 142-152.

Powell, S.W. and F.C. Day, Jr. 1991. Root production in four communities in the Great Dismal Swamp. *Am. J. Bot.,* 78: 288-297.

Prentki, R.T., T.D. Gustafson, and M.S. Adams. 1978. Nutrient movements in lakeshore marshes. In *Freshwater Wetlands: Ecological Processes and Management Potential*. R.E. Good, D.F. Whigham, and R.L. Simpson (Eds.). Academic Press, Inc., New York. 169-194.

Reed, P.B., Jr. 1988. National List of Plant Species that Occur in Wetlands: 1988 National Summary. Biol. Rpt. 88(24). U.S. Fish and Wildlife Service, Washington, D.C.

Reich, P.B., R.O. Teskey, P.S. Johnson, and T.M. Hinckley. 1980. Periodic root and shoot growth in oak. *Forest Sci.,* 26: 590-598.

Roper, T.R., E.J. Stang., S.N. Jeffers, D.M. Boone, and D.L. Mahr. 1991. Response to: Draft of St. Paul District Analysis Regarding Section 404 Review of Commercial Cranberry Operations. Depts. of Horticulture, Plant Pathology, and Entomology. University of Wisconsin, Madison, WI.

Roulet, N.T. and M-K. Woo. 1986. Hydrology of a wetland in the continuous permafrost region. *J. Hydrology,* 89: 73-91.

Sculthorpe, C.D. 1985. *The Biology of Aquatic Vascular Plants*. Koeltz Scientific Books. Konigstein, Germany.

Shetler, S.E. and S.K. Wiser. 1987. First flowering dates for spring-blooming plants of the Washington, D.C. area for the years 1970 to 1983. *Proc. Biol. Soc. Wash.*, 100: 933-1017.

Siegel, D.I. 1983. Ground water and the evolution of patterned mires, glacial Lake Agassiz peatlands, northern Minnesota. *J. Ecology,* 71: 913-921.

Siegel, D.I. 1992. Groundwater hydrology. In: *The Patterned Peatlands of Minnesota.* H.E. Wright, Jr., B.A. Coffin, and N.E. Asseng (Eds.). University of Minnesota Press, Minneapolis, MN. 163-172.

Siegel, D.I. and P. H. Glaser. 1987. Groundwater flow in a bog-fen complex, Lost River peatland, northern Minnesota. *J. Ecology,* 75: 743-754.

Simmons, R.C., A.J. Gold, and P.M. Groffman. 1992. Nitrate dynamics in riparian forests: groundwater studies. *J. Environ. Qual.,* 21: 659-665.

Sipple, W.S. 1985. Wetland Identification and Delineation Manual. Draft Report. U.S. Environmental Protection Agency, Office of Federal Activities, Washington, D.C.

Sipple, W.S. 1988. Wetland Identification and Delineation Manual, vol. I. Rationale, Wetland Parameters, and Overview of Jurisdictional Approach, vol. II. Field Methodology. U.S. Environmental Protection Agency, Office of Wetlands Protection, Washington, D.C.

Sipple, W.S. 1992. U.S. Environmental Protection Agency Memorandum to National Technical Committee for Hydric Soils: Literature review of rooting depths of wetland plants. Washington, D.C. August 6, 1992.

Sjors, H. 1991. Phyto- and necromass above and below ground in a fen. *Holarctic Ecol.,* 14: 208-218.

Skoropanov, S.G. 1961. *Reclamation and Cultivation of Peat-bog Soils.* Israel Program for Scientific Translations, Jerusalem, Israel. (Translated from Russian.)

Soil Survey Division Staff. 1993. Soil Survey Manual. Agriculture Handbook No. 18. U.S. Department of Agriculture, Washington, D.C.

Sommerfeld, R.A., A.R. Mosier, and R.C. Musselman. 1993. CO_2, CH_4, and N_2O flux through a Wyoming snowpack and implications for global budgets. *Nature,* 361: 140-142.

Sprecher, S.W., J.S. Wakely, and W.C. Lynn. 1996. Synthesis. In *Preliminary Investigations of Hydric Soil Hydrology and Morphology in the United States.* J.S. Wakely, S.W. Sprecher, and W.C. Lynn (Eds.). Tech. Rep. WRP-DE-13. U.S. Army Engineer Waterways Experiment Station, Vicksburg, MS. 153-162.

Takai, Y., T. Koyama, and T. Kamura. 1956. Microbial metabolism in reduction of paddy soil. *Soil Plant Food,* 2: 63-66.

Tallis, J.H. 1983. Changes in wetland communities, chap. 9. In *Mires: Swamp, Bog, Fen and Moor.* General Studies. A.J.P. Gore (Ed.). Elsevier Scientific Publishing Co., Amsterdam. 311-347.

Tarnocai, C. and S.C. Zoltai. 1988. Wetlands of Arctic Canada. In *National Wetlands Working Group. Wetlands of Canada.* Ecological Land Classification Series No. 24. Sustainable Development Branch, Environment Canada, Ottawa, Ontario, Canada. 27-53.

Teskey, R.O. 1982. Acclimation of *Abies amabilis* to water and temperature in a natural environment. Ph.D. thesis. University of Washington, Seattle.

Tiner, R.W., Jr. 1984. Wetlands of the United States: Current Status and Recent Trends. U.S. Fish and Wildlife Service, Washington, D.C.

Tiner, R.W., Jr. 1987. *A Field Guide to Coastal Wetland Plants of the Northeastern United States.* University of Massachusetts Press, Amherst, MA.

Tiner, R.W. Jr. 1988. Field Guide to Nontidal Wetland Identification. Maryland Dept. of Natural Resources, Annapolis, MD and U.S. Fish and Wildlife Service, Newton Corner, MA. Cooperative publication.

Tiner, R.W. 1991a. The concept of a hydrophyte for wetland identification. *BioScience,* 41: 236-247.

Tiner, R.W. 1991b. How wet is a wetland? *Great Lakes Wetlands,* 2(3): 1-4.7.

Tiner, R.W. 1993a. The primary indicators method — a practical approach to wetland recognition and delineation in the United States. *Wetlands,* 13: 50-64.

Tiner, R.W. 1993b. Wetlands are ecotones: reality or myth? In *Wetlands and Ecotones: Studies on Land–Water Interactions.* B. Gopal, A. Hillbricht-Ilkowska, and R.G. Wetzel (Eds.). National Institute of Ecology, New Delhi, India. 1-15.

Tiner, R.W. 1993c. *Field Guide to Coastal Wetland Plants of the Southeastern United States.* University of Massachusetts Press, Amherst, MA.

Tiner, R.W. 1997. Keys to Landscape Position and Landform Descriptors for U.S. Wetlands (Operational Draft). U.S. Fish and Wildlife Service, Northeast Region, Hadley, MA.

Tiner, R.W. 1998. *In Search of Swampland: A Wetland Sourcebook and Field Guide.* Rutgers University Press, New Brunswick, NJ.

Tiner, R. W. and D.G. Burke. 1995. Wetlands of Maryland. U.S. Fish and Wildlife Service, Hadley, MA and Maryland Dept. of Natural Resources, Annapolis, MD. Cooperative publication.

Tiner, R.W., and P.V. Veneman. 1987. *Hydric Soils of New England*. University of Massachusetts, Cooperative Extension Service, Amherst, MA. Bulletin C-183.

Topa, M.A. and K.W. McLeod. 1986a. Aerenchyma and lenticel formation in pine seedlings: a possible avoidance mechanism to anaerobic growth conditions. *Physiol. Plantarum,* 68: 540-550.

Topa, M.A. and K.W. McLeod. 1986b. Responses of *Pinus clausa, Pinus serotina* and *Pinus taeda* seedlings to anaerobic solution culture. I. Changes in growth and root morphology. *Physiol. Plantarum,* 68: 523-531.

Turner, F.T. and W.H. Patrick, Jr. 1968. Chemical changes in waterlogged soils as a result of oxygen depletion. IX International Congress Soil Science (Adelaide, Australia), Trans. 4: 53-56.

USDA Natural Resources Conservation Service. 1996. National Food Security Act Manual, 3rd ed. U.S. Department of Agriculture, Washington, D.C.

van der Valk, A.G. and C.B. Davis. 1976. Changes in composition, structure, and production of plant communities along a perturbed wetland coenoline. *Vegetatio,* 32: 87-89.

van der Valk, A.G. and C.B. Davis. 1978. The role of seed banks in vegetation dynamics of prairie pothole glacial marshes. *Ecology,* 59: 322-335.

van der Sman, A.J.M., O.F.R. van Tongeren, and C.W.P.M. Blom. 1988. Growth and reproduction of Rumex maritimus and Chenopodium rubrum under different waterlogging regimes. *Acta Bot. Neerl.,* 37: 439-450.

Vepraskas, M.J. 1996. *Redoximorphic Features for Identifying Aquic Conditions*. Tech. Bull. 301. North Carolina Agricultural Research Service, North Carolina State University, Raleigh.

Vepraskas, M.J. and L.P. Wilding. 1983. Aquic moisture regimes in soils with and without low chroma colors. *Soil Sci. Soc. Am. J.,* 47: 280-285.

Vézina, P.E. and M.M. Grandtner. 1965. Phenological observations of spring geophytes in Quebec. *Ecology,* 46: 869-872.

Volobueu, V.R. 1964. *Ecology of Soils*. Academy of Sciences of the Azerbaidzhan, USSR, Institute of Soil Science and Agrochemistry, Jerusalem. 260.

Walker, J.M. 1959. Vegetation studies on the Delta Marsh, Manitoba. Master's thesis. University of Manitoba, Winnipeg.

Walker, J.M. 1965. Vegetation changes with falling water levels in the Delta Marsh, Manitoba. Ph. D. dissertation. University of Manitoba, Winnipeg.

Wakely, J.S., S.W. Sprecher, and W.C. Lynn (Eds.). 1996. Preliminary Investigations of Hydric Soil Hydrology and Morphology in the United States. Tech. Rep. WRP-DE-13. U.S. Army Engineer Waterways Experiment Station, Vicksburg, MS.

Welsch, D.J., D.L. Smart, J.N. Boyer, P. Minkin, H.C. Smith, and T.L. McCandless. 1995. Forested Wetlands: Functions, Benefits, and the Use of Best Management Practices. NA-PR-01-95. U.S. Department of Agriculture, Forest Service, Northeastern Area, Radnor, PA.

Wetzel, R.G. 1990. Land-water interfaces: metabolic and limnological regulators. *Verh. Int. Verein. Limnol.,* 24: 6-24.

Wetzel, R. 1991. University of Alabama. Personal communication.

Williams, A.E. 1992. Memorandum re: Clarification and interpretation of the 1987 manual. U.S. Army Corps of Engineers, Washington, D.C. March 6, 1992.

Winter, T.C. and M-K. Woo. 1990. Hydrology of lakes and wetlands. In *The Geology of North America,* vol. 0-1, Surface Water Hydrology. M.G. Wolman and H.C. Riggs (Eds.). The Geological Society of America, Boulder, CO. 159-187.

Wisconsin State Cranberry Growers Association. 1991. Response to: Draft of St. Paul District Analysis Regarding Section 404 Review of Commercial Cranberry Operations. Wisconsin Rapids, WI. Unpublished letter.

Whigham, D.F. and R.L. Simpson. 1978. The relationship between aboveground biomass of freshwater tidal wetland macrophytes. *Aqua. Bot.,* 5: 355-364.

Woo, M-K. and T.C. Winter. 1993. The role of permafrost and seasonal frost in the hydrology of northern wetlands in North America. *J. Hydrol.,* 141: 5-31.

Zimov, S.A., I.P. Semiletov, S.P. Daviodov, Y.V. Voropaev, S.F. Prosyannikov, C.S. Wong, and Y. H. Chan. 1993. Wintertime CO_2 emission from soils of northeastern Siberia. *Arctic,* 46: 197-204.

3 Plant Indicators of Wetlands and their Characteristics

INTRODUCTION

Thousands of plants have adapted to life in water or wetlands, with a significant proportion of them occurring only in wetlands and shallow water. Many wetlands are readily identified by vegetation and traditional methods relied on plants for wetland identification and delineation (Chapter 6). Plants also serve as a basis for wetland classification (Chapter 8). Yet, despite the value of using plants for these purposes, all wetlands are not simply recognized by vegetation alone. Identifying the point on the gradient at which wetland begins and upland (dryland) ends can be straightforward in areas of high relief or extremely difficult in relatively flat terrain. In the former situations, plants may be used as boundary delineators, but in the latter, soil properties tend to be more indicative of prolonged saturation at or near the surface (Chapter 5). As wetland hydrologic conditions vary from permanent wetness (inundation and/or soil saturation) to periodic or seasonal waterlogging and soil wetness decreases, plant composition changes from a more typical and predictable wetland community to a transitional (ecotonal) community where wetland species intermix with mesic species, making wetland identification challenging and, by plants alone, most difficult and somewhat arbitrary. The multitude of hydrologic regimes associated with wetlands create a diverse set of environmental conditions that require different degrees of adaptation or tolerance of wetness by colonizing plants. This chapter reviews environmental conditions induced by prolonged wetness, describes plant adaptations for living in wetlands (focusing on properties that are important for wetland recognition) and the concept of a hydrophyte, and discusses how plants can be used to identify wetlands and as indicators of associated environmental conditions. A list of wetland plant identification field guides also is included at the end of the chapter.

ENVIRONMENTAL CHANGES DUE TO FLOODING AND WATERLOGGING

Prolonged flooding or waterlogging restricts oxygen movement from the atmosphere to the soil. Diffusion can occur but it is 10,000 times slower in saturated soils than it is in aerated soils (Greenwood, 1961). Upon flooding, respiration by aerobic bacteria and other organisms (aerobes) consume the oxygen remaining in the soil within hours to days (Pezeshki, 1994). Evans and Scott (1955) found that in just 75 min, the concentration of oxygen in a saturated soil dropped to 1/100th of its initial value. The only free oxygen present thereafter is at the soil–water interface — a thin layer (a few millimeters thick) at the soil surface. With free oxygen virtually absent from the soil, facultative and obligate anaerobes multiply in number and utilize oxygen bonded to various soil elements, thereby promoting denitrification and transforming nontoxic (oxidized) forms of manganese, iron, sulfate, and other elements to toxic forms (e.g., manganous, ferrous iron, and hydrogen sulfide). The end result is a reduced soil with low redox potential. Anaerobic* substrates or soils can develop within a

---

* Anaerobic conditions occur when the redox potential of the soil reaches +300 to +350mV (DeLaune and Pezeshki, 1991; Faulkner and Patrick, 1992) as little oxygen diffusion occurs below this value. Aerobic conditions have values usually from +350 to +700mV and higher. Anaerobic values go from around +350 to –300mV. Note that a flooded soil may be aerated if little organic matter is present or no energy source (e.g., soil microbes) exists to consume the oxygen or when temperatures are cold (below biological zero). Redox potential is a measure of the oxidation-reduction status of the soil and the values correspond with changes in the status of various elements (e.g., when iron changes from a stable, oxidized form to an unstable, reduced one at around +100mV).

few days in most soils and redox potential values as low as –250mV can be attained within 2 weeks in soils low in nitrate and manganic and ferric oxides (Ponnamperuma, 1972; 1984).

Soil oxygen deficiency* poses the main ecological problem for plant growth as it affects plant functions such as stomatal opening, photosynthesis, water and mineral uptake, and hormonal balance (Kozlowski, 1984b). Studies have shown that the effects of submergence can be relieved if air or 100% oxygen is supplied to the water (Larqué-Saavedra et al., 1975; MacDonald and Gordon, 1978). This is further supported by the practice of hydroponics where flood-intolerant plants like tomatoes can be grown in an aerated, enriched aqueous solution. In nature, however, most aquatic plants are living in continuously oxygen-deficient substrates. In contrast, most wetland plants grow in a dynamic environment of alternating wet and dry conditions. Significant stresses accompanying anaerobiosis during wet periods are following by drier periods when aerobic conditions are present, allowing for recovery. For example, southern bottomland hardwood wetlands may be oxidized throughout their root zone for 60 to 80% of the growing season (Faulkner and Patrick, 1992). The variety of wetlands makes for a seemingly endless combination of anaerobic/aerobic conditions, with the drier-end wetlands placing the least demand on plants and the wettest ones requiring the greatest adaptations. In permanently saturated wetlands, oxygenated water associated with snowmelt, thawing soils, and early spring rains may provide similar relief from oxygen deficiency. Jackson and Drew (1984) speculated that early-season flush of fresh water may provide a temporal well-oxygenated environment for root growth of bog species and that such conditions are vital to the survival of these species.

Wetland environments also are affected by the type of water flooding them. Wetlands saturated with moving, partly oxygenated waters are more favorable habitats than ones with stagnant water, for the flowing water allows root development and penetration into the substrate, whereas in the latter, roots form just at the soil surface (Jackson and Drew, 1984). Stagnant water imposes a greater stress on wetland plants as evidenced by reduced height growth and dry weight increments in bald cypress (*Taxodium distichum*) (Shanklin and Kozlowski, 1985) and by reduced height and root growth of shortleaf pine (*Pinus echinata*), loblolly pine (*P. taeda*), and pond pine (*P. serotina*) (Hunt, 1951). Seedlings of swamp black gum (*Nyssa sylvatica* var. *biflora*) and water tupelo (*N. aquatica*) both grew taller in moving water treatments vs. stagnant water regimes (Harms, 1973). Cool water dissolves oxygen better than warm water. Flooding by salt water inflicts great hardships on plants beyond anaerobiosis that freshwater wetland plants endure. These stresses include, for example, increased water stress, chloride toxicity, hydrogen sulfide toxicity, and salt uptake (Wainwright, 1984).

Flooding and soil saturation also cause changes in the soil environment which affect plant growth and survival. Soil temperature is lower in flooded soils than in well-drained soils. Bonneau (1982) found that the average difference was 6°C in the surface layer. Lower temperatures may influence soil chemistry, nutrient release, phytotoxin production, organic matter decomposition, and plant growth (Ponnamperuma, 1976). When saturated, soil structure changes as soil colloids expand, especially in soils with high clay content. Swelling is usually completed within 1 to 3 days (Ponnamperuma, 1984). Salty soils swell more than calcareous or acid soils. This swelling may cause soils with low permeability to become impermeable, resulting in ponding of water on the surface for significant periods after rainfall events. Alternate flooding and drying may cause significant cracking in clayey soils. Such activity may affect root growth and restrict root penetration. Flooding also alters the pH of the soil. Acid soils attain a higher pH status when flooded, while flooded alkaline soils decrease in pH (Ponnamperuma, 1984).

---

* Oxygen deficiency may be partial (hypoxia) or complete (anoxia).

Salt marshes offer some rather unique conditions for plant growth due to variable tidal flooding, duration of exposure, and salinity. Armstrong et al. (1985) observed three general elevation zones with differing soil aeration. The first zone was the lower margin of the salt marsh. Here reducing conditions persisted regardless of the length of exposure, with oxidation occurring only near the surface (within 5 cm) and at neap tides. The second zone was on the creek bank and on other better drained areas where low redox potentials (below 200mV) only happened monthly in the upper soil during high spring tides. Wetter years and winter rainfall may lower soil aeration more frequently than in summer. The third zone was the mid and high marsh where the upper soil was aerated most of the summer (even to a depth of 30 cm in July). Reduction took place in the upper soil upon flooding by spring tides, but was present below 20 cm for most of the summer. While flooding caused an almost instanteous 300 to 400mV fall in redox potentials within the upper 10 cm, the rise in redox potential upon natural drainage took several days to happen. Salinity and duration of flooding also are affected by marsh elevation. Plant zonation in salt marshes is well-documented (e.g., Miller and Egler, 1950; Adams, 1963; Teal and Teal, 1969; Tiner, 1987; 1993).

The quality of water entering wetlands may be affected by climatic conditions. In arid, semiarid, and subhumid regions, ground water brought to the surface through capillary action due to high evaporation brings with it various salts (e.g., sodium chloride, calcium carbonate, and magnesium sulfate) that become concentrated upon drying. These evaporites exert salt stress on plants which dramatically affects plant growth; some areas are so salty that they are devoid of vascular plants. Irrigation projects have accelerated the buildup of salts in many areas of the world (Wainwright, 1984). In some cases, lethal concentrations of arsenic, boron, selenium, and other toxics collect in inflow wetlands (Lico, 1992).

PLANT ADAPTATIONS TO FLOODING AND WATERLOGGING

Life in permanently or periodically anaerobic soils or substrates is more difficult than living in mesic soils due to oxygen deficiency, the nature of a highly reduced environment (low redox potential, Eh) with soluble phytotoxins, and other conditions. Grable (1966) reviewed the effect of soil aeration on plant growth. Most vascular plants cannot cope with anaerobiosis resulting from extended flooding or waterlogging. For example, intolerant herbs may quickly succumb to flooding for more than a few days, after passing through a series of worsening symptoms — from a yellowing of their leaves (chlorosis) and/or wilting to shedding their leaves. Prolonged flooding during the growing season typically kills most woody species, with seedlings being most vulnerable. A study of seedlings of 10 forest species by Loucks and Keen (1973) found that all survived 2 weeks of flooding, but longer flooding had significant negative effects on most of the species. Most mature trees die within 3 years of continuous flooding, as they are apparently unable to regenerate vital absorbing roots that they typically need to do every 2 years (Crawford, 1983). In studying a Florida floodplain forested wetland subjected to increased flooding from impoundment, Brown and Lugo (1982) found that a mean water-depth of 2 ft (60 cm) seemed to be the threshold at which the swamp would survive as deeper water brought about mortality.

Plant tolerance of flooding and soil saturation also is affected by the time of year, with increased tolerance early in the growing season. For example, up to 16 weeks of saturation had no effect on budbreak or initial stem growth of sycamore, sweet gum, and Nuttall oak seedlings when this occurred early in the growing season, but 10 to 12 weeks of saturation thereafter (when temperatures were warm and active growth commenced) caused severe reduction in height, root, and stem-diameter growth (Bonner, 1966). For herbaceous plants, different life-stages may have varied tolerances, making them more susceptible to anaerobiosis and submergence at certain times (Van der Sman et al., 1988). Water temperature of flood waters influences plant response, with higher temperatures producing higher metabolic rates. Relatively high water temperatures caused an increase in shoot length in two docks (*Rumex palustris* and *R. maritimus*), while low temperatures led to better preservation of biomass and greater survival after submergence, likely due to lower metabolism (Van der Sman et al., 1993). Study plants were more tolerant of prolonged inundation (i.e., 28 days) in early summer than mid-summer flooding. In mid-summer, only those plants that were able to extend their shoots above water survived, while rosettes of *R. palustris* survived in early summer (Van der Sman et al., 1993).

Aquatic plants are the best adapted and most specialized of the wetland plants, since they spend their entire lives in water. Such an existence necessitated development of a host of adaptations ranging from morphological structures for support, waterproofing, and buoyancy to reproductive adaptations for pollination and germination to physiological mechanisms. Many aquatic species are stimulated by flooding and send their shoots above the flood level (Jackson and Drew, 1984; Ridge and Amarsinghe, 1981; Blom et al., 1996). The adaptations of aquatic plants for life in water are well documented (see Arber, 1920; Gessner, 1955; 1959; Sculthorpe, 1967; Hutchinson, 1975).

Responses of woody and herbaceous plants to flooding and soil saturation have received considerable attention (see Hosner, 1960; Gill, 1970; Teskey and Hinckley, 1977; 1978a; 1978b; Hook and Scholtens, 1978; Armstrong, 1979; Whitlow and Harris, 1979; Crawford, 1983; Hook, 1984a; 1984b; Jackson and Drew, 1984; Kozlowski, 1984a; 1984b; Hook et al., 1988; Pezeshki, 1994 for reviews). Flood tolerance differs by age of plant, with seedlings usually being more vulnerable than mature individuals as mentioned above. Theriot (1993) developed flood tolerance indices for over 300 species in relatively undisturbed southeastern bottomland forests. Despite the wealth of information in these publications, our knowledge is far from complete on this subject, especially in regard to a plant's ability to live under conditions of prolonged waterlogging.

A plant's response to flooding may be quite different than its response to waterlogging. For example, green ash (*Fraxinus pennsylvanica*) was determined to be more flood-tolerant than eastern cottonwood (*Populus deltoides*) (Hosner, 1958), yet the latter was more tolerant of soil saturation (Hosner, 1959). Caution must, therefore, be exercised in extrapolating results of such studies and concluding that one species is more water-tolerant than another because the two situations (flooding and waterlogging) impose some different stresses on plants.

Another problem in interpreting the results of these types of studies is that distinct populations with genotypic or phenotypic differences in flooding tolerance undoubtedly exist (Crawford and Tyler, 1969; Gill, 1970) and in all likelihood, the researchers did not pay attention to this important detail. For example, several studies identified tulip poplar (*Liriodendron tulipifera*) as a flood-intolerant species (Hosner, 1960), yet this species has been frequently observed in floodplain wetlands (Tiner, 1985a; 1985b; 1988; Shear et al., 1997) and was reported growing in a red maple-yellow birch swamp in northern New Jersey (Niering, 1953). Tulip poplar with buttressed trunks have been observed in Virginia's Dismal Swamp (Bill Sipple, personal communication, 1990). These findings suggest possible ecotypes in this species. Keeley (1979) demonstrated the significance of considering ecotypes in flood tolerance studies by experimenting with three phenotypes of black gum (*Nyssa sylvatica*) — upland, swamp, and floodplain. Keeley found that the upland seedlings were very intolerant of flooding and the swamp seedlings highly flood-tolerant, while the floodplain seedlings had intermediate tolerances. These results provide proof of the potential pitfalls brought about by considering only the species level for determining what constitutes a wetland plant (see the wetland ecotypes discussion later in this chapter). Researchers must be particularly mindful of the habitat from which the seeds/seedlings come, as this may affect their response to flood/saturation-tolerances and study findings.

A wide range of adaptations make it possible for plants to grow in water or wetlands. These adaptations include physiological responses, morphological adaptations, behavioral responses, reproductive strategies, and others (Table 3.1; see Ernst, 1990 for a review of physiological mechanisms). These features and processes also affect the flood or saturation tolerance of species, thereby influencing the distribution of plants within wetlands. The least adapted species possess only minor adaptations and consequently are typically restricted to the margins or the highest elevations.

TABLE 3.1
Plant Adaptations or Responses to Flooding and Waterlogging

Morphological Adaptations/Responses	Physiological Adaptations/Responses
Stem hypertrophy (e.g., buttressed tree trunks)	Transport of oxygen to roots from lenticels and/or leaves (as often
Large air-filled cavities in center (stele) of roots and stems	evidenced by oxidized rhizospheres)
Aerenchyma tissue in roots and other plant parts	Anaerobic respiration
Hollow stems	Increased ethylene production
Shallow root systems	Reduction of nitrate to nitrous oxide and nitrogen gas
Adventitious roots	Malate production and accumulation
Pneumatophores (e.g., cypress knees)	Reoxidation of NADH
Swollen, loosely packed root nodules	Metabolic adaptations
Lignification and suberization (thickening) of root	
Soil water roots	**Other Adaptations/Responses**
Succulent roots	Seed germination under water
Aerial root-tips	Viviparous seeds
Hypertrophied (enlarged) lenticels	Root regeneration (e.g., adventitious roots)
Relatively pervious cambium (in woody species)	Growth dormancy (during flooding)
Heterophylly (e.g., submerged vs. emergent leaves on same plant)	Elongation of stem or petioles
Succulent leaves	Root elongation
	Additional cell wall structures in epidermis or cortex
	Root mycorrhizae near upper soil surface
	Expansion of coleoptiles (in grasses)
	Change in direction of root or stem growth (horizontal or upward)
	Long-lived seeds
	Breaking of dormancy of stem buds (may produce multiple stems or trunks)

Source: Tiner, R.W. 1991. *BioScience,* 41: 236–247.

Species with the most effective adaptations are found in the wettest conditions. Besides coping with at least recurrent prolonged anaerobic soil conditions, the presence of soluble toxins (e.g., reduced forms of iron and aluminum, plus hydrogen sulfide), and, for many species, living in water, wetland plants also must adapt to other environmental conditions, including salinity (ocean-derived or inland salts), pH (acidic to alkaline), nutrient availability, substrate types (sand, clay, or other mineral soils), soil temperature, climatic factors, and biological competition. Despite the emphasis on anaerobiosis, the combination of all of these factors influences plant distribution.

MORPHOLOGICAL ADAPTATIONS

Wetland plants exhibit a wide range of morphological features developed in response to frequent prolonged flooding and waterlogging (Table 3.1). These structures have long been recognized as Warming (1909) makes reference to adventitious roots, internal air-containing spaces, aerenchyma, and pneumatophores.* Morphological adaptations are likely manifestations of physiological responses as a flood-induced increase in ethylene production has been linked to the formation of stem hypertrophy (Drew et al., 1979), hypertrophied lenticels (Angeles et al., 1986; Topa and McLeod, 1988), aerenchyma (Drew et al., 1979; Hook, 1984b; Jackson et al., 1985; Topa and McLeod, 1988), and adventitious roots (Drew et al., 1979; Tsukahara and Kozlowski, 1985; Voesenek et al., 1996). Consequently, these features should represent reliable surrogates that confirm physiological adaptations in the species and the best morphological indicators of prolonged anaerobiosis. Some morphological

---

* Extinct trees of the late Carboniferous swamps (330 million years ago) had platform-like flared bases, rhizophores (a meter in diameter), adventitious roots, and aerating tissue (Ingrouille, 1992).

TABLE 3.2
Some Plant Adaptations to Salt Stress

Morphological

Salt-secretion glands (to eliminate excess salt)
Succulent stems and leaves (increased water retention to maintain internal salt balance)
Waxy leaf coatings (to minimize contact with salt water)
Salt concentration in specialized hairs
Reduced leaves (to minimize exposure to salt and evapotranspiration)

Physiological

Salt exclusion (reduced salt uptake by roots)
High ion uptake (lowers osmotic potential of cell sap)
Dilution of salts
Accumulation of salts in cell vacuoles

Other

Stress avoidance (by occupying higher levels of salt marsh or in seepage areas)
Periodic shedding of salt-saturated organs

Sources: Wainwright, S.J. 1984. *Flooding and Plant Growth.* Academic Press, New York.
Tiner, R.W. 1995. *Field Guide to Coastal Wetlands of the Southeastern U.S.,* University
of Massachusetts Press, Amherst.

adaptations like shallow root systems also may be induced by other environmental conditions (e.g., rocky or clayey soils) and may not by themselves be highly reliable wetland indicators.

For wetland identification, morphological adaptations are emphasized over other adaptations, since they may be readily observed, while physiological mechanisms are not because they require laboratory testing or elaborate field assessments beyond the means of most wetland delineators. The following discussion is mostly an overview of morphological adaptations by wetland plants to prolonged flooding and waterlogging, focusing on emergent plants (trees, shrubs, and herbs) and should give readers a better understanding of the validity of using these features as indicators of hydrophytes or of wetland hydrology. It is not an exhaustive review of these plant adaptations; readers wanting more indepth analysis should consult Kozlowski (1984c), Crawford (1983), and Hook and Crawford (1978). For reviews of adaptations of the true aquatic plants found in waters in and contiguous to wetlands, consult Sculthorpe (1967) and Hutchinson (1975). Many wetland plants are halophytic and possess unique adaptations for life in salt-stressed environments (e.g., coastal salt marshes and inland wet saline or alkaline soils), their features are not discussed but are listed in Table 3.2 (see Waisel, 1972; Albert, 1975; Reimold and Queen, 1974; Wainwright, 1984 for more information).

Hypertrophied Stems

Some wetland species growing under extended flooding conditions exhibit a noticeable swelling of the lower stem. Such swelling increases the surface area and often is coupled with the presence of hypertropied lenticels that collectively improve gas exchange (Hook et al., 1970). The diameter of the stem from the ground surface to some distance above is greatly expanded. In herbaceous plants this condition is called simply hypertrophied stems, while in trees, it is called buttressed trunks or buttressing. The earliest swamp trees on Earth exhibited buttressed trunks, related to fluctuating water levels and to unstable substrates (Darrah, 1960; Kangas, 1990).

FIGURE 3.1   Hypertrophied stems in herbaceous plants: (a) rattlebush (*Sesbania drummondii*), note swollen stem on specimen rooted in the pond mud and compare with an upland individual (on left) having a taproot and lacking stem hypertrophy; and (b) water-willow (*Decodon verticillatus*). Note that in both species the hypertrophied stems are composed of soft, spongy tissue.

In herbs, the enlargement of the cortex and collapse of some cells create air-filled spaces (aerenchyma) that lead to an expansion of the stem (Kawase, 1981). Spongy inflated stems have been observed in water-willow (*Decodon verticillatus*), water-purslanes (*Ludwigia* spp.), and rattlebush (*Sesbania drummondii*) (personal observations) (Figure 3.1). In water-willow, these spongy stems may aid in buoyancy (Sculthorpe, 1967).

In trees and shrubs, the swelling is not due to aerenchyma (air-filled tissue), but to larger cells and lower density wood (Penfound, 1934; Kawase, 1981; Hook, 1984a). Kurz and Demaree (1934) show examples of three types of buttressing in pond cypress (*Taxodium ascendens*): (1) shallow buttress (of shallow nonalluvial swamps), (2) cone buttress (of shallow lakes and ponds), and (3) bottle buttress (of deeper waters of lakes and ponds). In studying the effect of prolonged waterlogging on seedlings of loblolly pine, McKevlin et al. (1987) found that hypertrophied stems developed after 3 weeks of flooding. Ethylene production during anaerobiosis may be responsible for this (Kozlowski, 1982). Good and Patrick (1987) noted basal swelling in green ash seedlings after 1 to 2 weeks of flooding. Other species possessing these features include red maple (*Acer rubrum*), white ash (*F. americana*), pond cypress, water gum (*Nyssa aquatica*), and swamp black gum (*N. sylvatica biflora*) (Figure 3.2; Plate 32; Hook, 1984a). Slight stem hypertrophy was observed in overcup oak (*Quercus lyrata*) and laurel oak (*Q. laurifolia*) under waterlogging conditions (Gardiner and Hodges, 1993). Other studies reporting stem hypertrophy include Yamamoto et al. (1987) and Gardiner (1994).

Fluted Trunks

Certain wetland trees exhibit flared bases or fluted trunks, presumably to provide for support in unstable substrates (Figures 3.2a and 3.3). Such features also may create more surface for production of hypertrophied lenticels during periods of prolonged flooding. These structures have been

FIGURE 3.2   Buttressed trunks: (a) bald cypress (*Taxodium distichum*), note trunks also are prominently fluted, and (b) swamp black gum (*Nyssa sylvatica* var. *biflora*). Note shallow roots of windthrown tree.

observed in American elm (*Ulmus americana*), bald cypress, sugarberry (*Celtis laevigata*), pin oak (*Quercus palustris*), cherrybark oak (*Q. falcata* var. *pagodifolia*), overcup oak, and red maple (personal observations). Ingrouille (1992) reported platform-like flared bases in extinct swamp trees as an adaptation for growing in mud of late Carboniferous swamps (330 million years ago).

Hollow Stems

Many wetland herbs possess hollow or chambered stems that favor growth in wetlands. Such stems are typical of grasses and also are present in many rushes and sedges (Crawford, 1983). Hollow or chambered stems may improve aeration to the roots as well as accumulate carbon dioxide important for photosynthesis (Billings and Godfrey, 1967, as reported by Crawford, 1983). Increased photosynthetic activity will produce a corresponding increase in oxygen available for diffusion to

FIGURE 3.3   Fluted trunk of American elm (*Ulmus americana*).

the roots. The presence of hollow stems in these graminoids might help explain why they are often dominant in wetlands.

Shallow Root Systems

High water tables and accompanying anaerobic conditions significantly influence root growth. With few exceptions, plant roots do not grow in anoxic soils or substrates, although once established, the roots can survive such conditions for days or months depending on the species and whether the root is woody or not (Crawford, 1982). Since oxygen is the prime limiting factor in wetlands, the anaerobic environment essentially forces plants to seek oxygen near the surface and thereby also avoid phytotoxins present in the subsoil. As a result, plant roots for most wetland species grow horizontally (diageotropic) or upwardly (negatively geotropic) to form extensive root systems near the surface.

The development of shallow roots is perhaps the most widespread morphological adaptation among wetland species. For trees, this system makes them more susceptible to windthrows than upland trees that are better anchored in the soil (Figure 3.2b). Some plants may be typically shallow-rooted wherever they grow, while others may have adaptable root systems. Both responses favor establishment in wetlands. Eastern hemlock (*Tsuga canadensis*) is an example of a shallow-rooted species (Fowells, 1965) that has successfully colonized wetlands due in part to this feature (Niering, 1953; Huenneke, 1982; Tiner, 1989; 1998). Consequently, it can grow equally well on organic soils that are saturated year-round and better drained shallow soils and rocky soils. In contrast, red maple has an adaptable root system; in swamps, it develops numerous shallow lateral roots to help avoid anaerobic stress, whereas in dry uplands, a deep tap-root is formed (Kramer, 1949). This ability may be responsible for red maple's even distribution in both wetlands and nonwetlands. Development of shallow root systems may be an individual plant's response to a wet environment or a species trait. If the former, timing of germination and the environmental conditions that follow may be crucial to the development of this adaptation. For example, scrub oak (*Quercus ilicifolia*), a typical dry-site species with deep penetrating roots, grows occasionally in wetlands (with seasonal water tables within 1 to 2 ft) where it develops a shallow root system within 6 to 8 in. of the surface (Laycock, 1967). Shallow roots also can form for other reasons, e.g., impenetrable rocky substrates or dense clayey soils. So when considering this feature as a wetland adaptation, existing environmental conditions need to be assessed.

Tussock sedge (*Carex stricta*) has two sets of roots. Deep roots below its tussock and one root at the top of the tussock. When the plant is flooded, the upper root performs typical functions of water and nutrient uptake (Costello, 1936). This adaptation allows the plant to overcome anaerobiosis during periods of extended inundation or soil saturation.

While most wetland plants develop shallow roots, some species can produce fairly deep roots, sending their roots 8 to 20 in. (20–50 cm) below the water table (e.g., *Typha latifolia*, *Sagittaria latifolia*, *Scirpus validus*, and *Eriophorum*) (Emerson, 1921). For other wetland plants, deeper root penetration may occur during dry periods or when soils are otherwise less reduced, allowing the plant to tap nutrients from the subsoil. Once in this zone, metabolic mechanisms and accompanying anatomical structures (e.g., aerenchyma) permit survival when strongly reducing conditions resume. Fisher and Stone (1990) found that tap and sinker roots of slash pine (*P. elliottii*) could withstand weeks or months of saturated, anoxic soils and reported that such roots may penetrate about 3 ft into the seasonal low water table. Roots of sheep laurel (*Kalmia angustifolia*) may penetrate more than 1 ft into the water table in New Jersey's pitch pine lowlands (Laycock, 1967). This portion of the roots had much aerenchyma as evidenced by their lightweight and corky texture (after drying).

Adventitious Roots

Extended inundation causes some plants to develop "adventitious roots" at or just below the water line (close to the air–water interface) where water and oxygen are available (Figure 3.4; Plate 1). These roots usually form as the original root system dies back due to anoxia and the new roots are needed to aid in survival and recovery (Jackson and Drew, 1984). The presence of aerenchyma in these roots makes them more porous than normal roots (Kozlowski, 1982). These roots replace the function of the lost roots in absorbing water and nutrients, vital for plant functions and are positively correlated with the degree of flood tolerance within certain genera (Kozlowski, 1984a). Herbs, like common reed (*Phragmites australis*), water purslanes (*Ludwigia* spp.), purple loosestrife (*Lythrum salicaria*), docks (*Rumex* spp.), corn (*Zea mays*), sunflower (*Helianthus annuus*), rice (*Oryza sativa*), and tomato (*Lycopersicon esculentum*), typically display this adaptation when subjected to these conditions. Adventitious roots in corn seemed important in allowing it to survive flooding for up to 13 days (Wenkert et al., 1981). Van der Sman et al. (1988) found that within 5 days of waterlogging, adventitious roots began growing in golden dock (*Rumex maritimus*), while Wenkert et al. (1981) found that such roots developed in corn within 4 days (Smolders et al., 1990). After these roots formed, the plants recovered their relative growth and transpiration rates.

Some woody plants also have the ability to produce adventitious roots. After an English fen was flooded for nearly an entire year, Compton (1916) noted and took pictures of massive adventitious roots in white and crack willows (*Salix alba* and *S. fragilis*). Bergman (1920) described the formation of adventitious roots in swamp plants. Gill (1970; 1975) reported the formation of adventitious roots in several species — buttonbush (*Cephalanthus occidentalis*), false indigo-bush (*Amorpha fruticosa*), European alder (*Alnus glutinosa*), red alder (*A. rubra*), lodgepole pine (*Pinus contorta*), Sitka spruce (*Picea sitchensis*), western hemlock (*Tsuga heterophylla*) white ash, green ash, black willow (*Salix nigra*), crack willow, Hooker willow (*S. hookerana*), white willow, tulip poplar (*Liriodendron tulipifera*), eastern cottonwood, black cottonwood (*Populus trichocarpa*), redwood (*Sequoia sempervirens*), bald cypress (*Taxodium distichum*), and American elm. Hook (1984a) also found these roots in water gum, swamp black gum, western red cedar (*Thuja plicata*), black poplar (*P. nigra*), and paper-bark (*Melaleuca quinquenervia*). In swamp black gum, adventitious roots may be stimulated by flowing water, but not by stagnant water (Hook, 1984a). Hook speculated that such roots also may serve to trap debris and soil from floodwaters and thereby

FIGURE 3.4 Adventitious roots: (a) hydrophytic graminoid pulled from shallow water in a prairie pothole wetland; and (b) green ash (*Fraxinus pennsylvanica*) with hypertrophied lenticels on roots. Note the moss-lichen line in (b) just above the uppermost roots.

promote the formation of a hummock around the base of the trunk. Gill (1975) commented on the paradoxical occurrence of tulip poplar on stream banks and swamp margins (subject to periodic flooding), despite a reported flood intolerance in the literature.

The ability to produce adventitious roots may be critical for survival of long-term flooding as reported for bald cypress (Pezeshki, 1991). Topa and McLeod (1986b) observed an abundance of adventitious roots in wet-site loblolly and pond pine (*P. serotina*) seedlings subjected to 30 days of anaerobic conditions. In examining the ecological significance of adventitious rooting in European alder, Gill (1975) found that such roots can be produced within 2 to 5 days after bud-burst (depending on temperature) in flooded plants, but do not form during the dormant season. Also experimentally, he found that even when only the lower stem was flooded (and not the roots) that this alder produced some adventitious roots, yet not as many as when the roots were saturated as well.

Soil Water Roots

Prolonged flooding often causes the primary root system of plants to die back and some plants develop a secondary set of roots below ground that aid in oxidizing the rhizosphere. Based on studies of dock species (*Rumex*), Blom (1990) described three root responses to these conditions: (1) an increase in root branching, (2) the formation of adventitious roots, and (3) an altered vertical distribution of lateral roots, with more roots concentrated in the upper layers of soil. In studies of dock species, primary root growth ceased when anaerobic conditions developed in the flooded soil medium, and new lateral roots began forming after 125 to 200 h of submergence (Laan et al., 1991a). These roots were unbranched and thicker (twice as wide in *Rumex maritimus*) than the typical roots. The latter was due to the formation of aerenchyma (see discussion below). After 2 weeks of flooding, curly dock (*R. crispus*) and golden dock (*R. maritimus*) produced about 16 m and 24 m of new lateral roots, thereby replacing 32% and 23% of their original root length, respectively (Laan et al., 1989). During this same study period, clustered dock (*R. conglomeratus*) generated 43 m of new roots — a 44% replacement of its original root system (Laan et al., 1989). Voesenek et al. (1989) described two types of roots forming on the upper part of the tap root of flooded *Rumex*: (1) strongly branched, thin, superficially growing roots, and (2) thick, white, poorly branched roots that penetrated into the waterlogged soils. The former roots exhibited diageotropism (a tendency for horizontal growth) that allowed them to take advantage of oxygen present in the air–water interface.

So called *soil water roots* have been reported for bald cypress and swamp black gum (Harms et al., 1980). Flood-tolerance in woody species may be attributed in part to this ability (Hook and Brown, 1973). Another type of root called *altered roots* may form in some species like sweet gum (*Liquidambar styraciflua*). These roots differ from typical roots by being more fleshy and almost clear. Armstrong and Boatman (1967) described superficial rooting in bog plants.

Pneumatophores

A few tree species develop pneumatophores on their roots (Figure 3.5). Among the most well known of these structures are the "knees" of bald cypress. Knees form in areas subjected to prolonged inundation; cypress trees cultivated in upland parks do not produce them. The taller the knees, the higher the water level. Knees may grow as tall as 10 ft high and in numbers reaching 100 for a single tree (Wilson, 1889). Cypress knees form along the upper roots where increased aeration stimulates cambium activity and they do not seem to function as aerating organs (Kramer and Kozlowski, 1979). Pneumatophores of other species may possess hypertrophied lenticels to improve aeration. Wilson (1889) also reported finding knees of pond pine in Georgia.

Other woody plants with pneumatophores are black mangrove, white mangrove, and water tupelo. Those of the black mangrove are pencil-like projections that extend from the soil about a foot high. These pneumatophores, covered with hypertrophied lenticels, help improve aeration within the plant, allowing this mangrove to colonize anoxic soils. Scholander et al. (1955) demonstrated this for both the red mangrove (*Rhizophora mangle*) and black mangrove (*Avicennia germinans*). Longman and Jenik (1974) described a variety of these pneumorhizae structures found in tropical trees, including lateral knee-roots, serial knee-roots, root knees, peg roots, stilted peg roots, and pneumathodes (reported in Lugo et al., 1990).

Aerenchyma

The presence of aerenchyma (air-filled) tissue and lacunae in many wetland herbs, especially marsh plants, helps these plants grow in anaerobic or anoxic soils (Figure 3.6). An internal system of large air spaces is needed to transport atmospheric oxygen to the roots, thereby creating an oxidized environment around the roots (oxidized rhizosphere). This reduces resistance to oxygen movement

FIGURE 3.5 Pneumatophores: (a) bald cypress knees, (b) black mangrove (*Avicennia germinans*), and (c) white mangrove (*Laguncularia racemosa*). Note the hypertrophied lenticels on the black mangrove.

for respiring cells, decreases the amount of respiring tissue, facilitates diffusion of oxygen-containing air to organs lacking oxygen, and still provides sufficient structural support (DeLaune and Pezeshki, 1991; Voesenek and van der Veen, 1994). It also aids in releasing carbon dioxide and methane to the atmosphere (Wetzel, 1990). If the link to the atmosphere provided by this internal

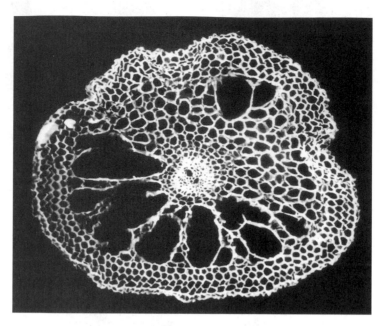

FIGURE 3.6   Aerenchyma tissue in root of smooth cordgrass (*Spartina alterniflora*). (Irv Mendelssohn photo.)

aeration system is eliminated, anoxic conditions quickly result in plant death as demonstrated in cattails (*Typha* spp.) by Sale and Wetzel (1983). Differences in flood tolerances may be caused by differences in internal aeration and the use of aerial and photosynthetic oxygen in root respiration (Blom, 1990). Flood-intolerant *Rumex* plants did not exhibit any internal aeration, whereas flood-tolerant plants did and showed oxygen loss around their roots (Laan et al., 1989). Diffusion of aerial oxygen to roots through aerenchyma tissue accounted for 40 to 50% of root respiration in curly dock and golden dock (Blom, 1990). Smirnoff and Crawford (1983) reported that 10% root porosity was the dividing line between flood-tolerant and intolerant species. Justin and Armstrong (1987) stated that most nonwetland plants have porosities less than 7%, but noted that some wetland species do not produce aerenchyma.

Aerenchyma also promotes root extension into anaerobic soils in some species like rice (Armstrong, 1979). Studies of slash pine (*Pinus elliottii*), perhaps the wettest of the southern pines, found that it extends its roots into the anaerobic zone of waterlogged soils (as much as 3 ft or more below the seasonal low water table) (Schultz, 1972; 1973). This pine can withstand inundation for several months or a few years because its woody root tissue is aerenchymous — with 48 to 69% air content vs. about 15 to 22% in its stemwood (Fisher and Stone, 1990). This tissue likely provides a reservoir of oxygen that permits gas exchange between the atmosphere and the oxygen-depleted root tissues.

Root porosity may differ between flood-tolerant and flood-intolerant species of the same genus. *Rumex martimus* (flood-tolerant) had a root porosity of 20% compared with 8% in *R. thyrsiflora* (flood-intolerant) (Laan et al., 1991b). This has a positive effect on internal aeration.

Many studies have demonstrated an increase in aerenchyma in plants subjected to flooding and to stronger anaerobiosis (Seliskar, 1988; Burdick and Mendelssohn, 1990; Kludze et al., 1994; Kludze and DeLaune, 1994; Kludze and DeLaune, 1996). It is not certain whether the formation of aerenchyma is induced by oxygen deficiency or by accumulation of anaerobiosis-caused phytotoxins and gaseous decomposition products (Kludze and DeLaune, 1996). In some species, increased ethylene production during anaerobiosis enhances aerenchyma development (Drew et al., 1979). Aerenchyma development may be genetically controlled by some species. Rice, wild rice

(*Zizania aquatica*), docks, and bald cypress grown in well-oxygenated environments formed aerenchyma in their roots (Jackson and Drew, 1984; Laan et al., 1989; Kludze et al., 1994).

Aerenchyma formation takes several weeks or months to completely modify root anatomy (Das and Jat, 1977; Keeley and Franz, 1979; Burdick, 1989). This tissue is formed from either cell water separation and collapse (lysigeny) or by cell separation without collapse (schizogeny) (Laan et al., 1989). Even with as much as 50% of the internal tissue being aerenchyma, salt-hay grass (*Spartina patens*) was unable to provide enough oxygen to the roots for complete aerobic respiration to occur under long-term waterlogging (Burdick and Mendelssohn, 1990). If the roots do not receive an adequate supply of oxygen to maintain aerobic respiration, then they will switch to an anaerobic pathway (Pezeshki, 1991). It appears that both anatomical and metabolic responses are important for plant survival. It must be realized, however, that not all wetland plants produce aerenchyma (Smirnoff and Crawford, 1983).

Spatterdock (*Nuphar luteum*) has an interesting internal gas transport system. Dacey (1981) found a pressurized flow-system where air enters young emergent leaves then moves by diffusion through lacunae to the roots, while older leaves serve to release pressure, thereby moving carbon dioxide and methane from the plant to the atmosphere. About 60% of the elongate leaf stalks and up to 40% of the roots and rhizome are filled with air spaces (Dacey, 1979). Coutts and Philipson (1978a,b) found more aerenchyma tissue in the roots of lodgepole pines grown in waterlogged soil than in the roots from ones grown in freely drained soils. Sand-Jensen et al. (1982) found that water lobelia (*Lobelia dortmanna*) had about 50% of its leaves and stems represented by lacunae and a continuous lacunae connection from the leaves to the roots. They believed that this probably led to its high oxygen release from its roots during illumination (more than two times that of the other aquatics which included *Potamogeton* and *Sparganium*). The authors speculated that the function of these anatomical adaptations was to bring in carbon dioxide rather than release oxygen to the anoxic soil.

Hypertrophied Lenticels

Lenticels are the external organs in woody plants that permit gas exchange between the internal parts of the plant and the atmosphere. When flooded, some trees and shrubs develop expanded or enlarged lenticels (with large intercellular spaces) below the water level to a point just above the water line (Plate 1; Figures 3.4b and 3.5b). These hypertrophied lenticels form in a relatively short time from 5 to 10 days after flooding (Kozlowski, 1984a; Good and Patrick, 1987). These organs increase the surface area available for gas exchange, enhance internal aeration, and allow the plant to keep functioning while subjected to anaerobiosis (Gill, 1970; Hook, 1984a; Jackson and Drew, 1984; Tsukahara and Kozlowski, 1985; DeLaune and Pezeshki, 1991). Lenticels are the major entry point for oxygen diffusion to the rhizosphere in some woody plants (Armstrong, 1968; Hook et al., 1971). In some species, hypertrophied lenticels also provide an exit for releasing toxins produced by anaerobiosis (e.g., ethanol, acetaldehyde, and ethylene) (Kozlowski, 1984a). After flood waters recede, these inflated lenticels dessicate, leaving noticeable scars as evidence of their occurrence and abundance (Figure 3.7).

In studying conifer root morphology in relation to soil moisture and aeration, Hahn et al. (1920) found that 17 species developed hypertrophied lenticels when grown in saturated soils. These species included red spruce (*Picea rubens*), jack pine (*Pinus banksiana*), western white pine (*P. monticola*), ponderosa pine (*P. ponderosa*), red pine (*P. resinosa*), pitch pine (*P. rigida*), eastern white pine (*P. strobus*), and eastern hemlock (*Tsuga canadensis*). An abundance of hypertrophied lenticels formed in pine seedlings after 3 weeks of flooding (McKevlin et al., 1987; Topa and McLeod, 1986b). Other species exhibiting these adaptations were green ash, black willow, cottonwood, American elm, bur oak (*Quercus macrocarpa*), and sycamore (*Platanus occidentalis*) (several sources, as reported in Kozlowski, 1984a).

FIGURE 3.7   Hypertrophied lenticel scars on exposed tree root.

OXIDIZED RHIZOSPHERES

While not a morphological property of the plant, an oxidized soil layer around living roots is a feature that can be recognized during wetland delineation and is positive evidence of plant growth under anaerobic conditions. In order to survive such conditions, some plants release oxygen from their roots to the surrounding reduced, anaerobic soils. This radial oxygen loss is believed to be important to detoxify this environment by oxidizing otherwise harmful reduced forms of iron and manganese plus hydrogen sulfide, and to increase nutrient uptake (Armstrong, 1979). It may also promote mycorrhizae establishment important for nutrient uptake, whereas anaerobic conditions typically inhibit them (Hook, 1984a). Mendelssohn (1993) provided a literature review and annotated bibliography of oxidized root channel formation.

Oxidation of the rhizosphere may be the most important mechanism in many plants adapting to an anaerobic environment, since it allows aerobic root respiration that is vital for nutrient uptake and other processes. Leakage of oxygen from roots (root radial oxygen loss) occurs when more oxygen is supplied than required for root respiration, thereby forming oxidized rhizospheres. This zone further aids the plant in oxidizing toxic materials (e.g., fermented organic compounds and hydrogen sulfide) and facilitates the nitrification-denitrification process (oxidizing ammonium to nitrate) which is a vital wetland function. Sand-Jensen et al. (1982) suggest that a shifting zone may occur in the root–soil interface, where aerobic processes dominate by day (due to active oxidation of the rhizosphere) and anaerobic processes by night (no oxidation). This might favor oxygen consumption, mineralization, and nitrification-denitrification rates. Similarly in coastal marshes, the alternating tides also create oscillating soil conditions, with high tide boosting soil anaerobiosis (highly reduced environment) and low tide increasing soil aeration (oxidized condition). This fluctuation promotes plant survivability by allowing time to recover from oxygen deficiency (Pezeshki and DeLaune, 1996). In addition, oxygen diffusion to the roots of some salt marsh plants promotes the oxidation of sulfides in the rhizosphere which is considered a major factor contributing to high sulfide tolerance (Teal and Kanwisher, 1966). This has significant implications on plant distribution and productivity (DeLaune and Pezeshki, 1991).

Oxidized pore linings, root channels, or rhizospheres (iron plaque; Plate 2) develop when oxygen comes in contact with reduced iron (ferrous iron, Fe^{+2}), thereby precipitating iron oxide (ferric iron, Fe^{+3}) in the form of iron oxyhydroxide ($FeOOH$) coatings, also called iron plaque, and increasing the redox potential. These linings may develop in 4 to 8 weeks of ponding or saturation as observed in Texas by Griffin et al. (1996). In rice fields, oxidized rhizospheres were observed after just 7 days of flooding (Chen et al., 1980). Mendelssohn (1993) and Mendelssohn et al. (1995) reviewed factors controlling their formation. These factors include low redox potential, pH, organic matter, and texture (mineral fines), plus the amount of available iron in the soil. The presence of aerenchyma tissue in roots facilitates the formation of oxidized rhizospheres. The buildup of iron plaque is not always related to the amount of radial oxygen loss (ROL) from roots (Mendelssohn, 1993). For example, Laan et al. (1989) found that the species with the least ROL had the most iron plaque among three species of dock studied.

Most, if not all, aquatic species and marsh herbs are capable of moving oxygen to their roots (Bartlett, 1961; Bacha and Hossner, 1977; Joshi and Hollis, 1977; Armstrong, 1964; 1979; Sand-Jensen et al., 1982; Taylor et al., 1984; Kemp and Murray, 1986; Jaynes and Carpenter, 1986; Mendelssohn et al., 1995, among others). Yet, according to Mendelssohn (1993), very few species have been shown to produce oxidized rhizospheres — rice, broad-leaved cattail (*Typha latifolia*), common reed, aster (*Aster tripolium*), smooth cordgrass (*Spartina alterniflora*), cordgrass (*S. maritima*), purple moor-grass (*Molinia caerulea*), buckbean (*Menyanthes trifoliata*), asphodel (*Narthecium* sp.), beaked sedge (*Carex rostrata*), sweet gale (*Myrica gale*), and pondweed (*Potamogeton polygonifolius*). The brevity of this list, however, is probably due to lack of study, as oxidized rhizospheres are frequently observed in wetlands, especially in wet meadows (personal observations). While many wetland plants show evidence of the release of oxygen to the anaerobic substrate around the roots by their oxidized pore linings, this is not a univeral adaptation possessed by all wetland species.

Nonwoody roots of lodgepole, slash, and loblolly pines have large air spaces (lacunae) that oxidize the rhizosphere only under anaerobic conditions (Coutts and Philipson, 1978b; Philipson and Coutts, 1978; Hook and McKevlin, 1988; Fisher and Stone, 1990). Aerenchyma in the primary root tips make this possible. Loblolly pine seedlings grown under flooded, anaerobic conditions developed an internal aeration system composed of aerenchyma, whereas seedlings grown under drained conditions did not (McKevlin et al., 1987). Fisher and Stone (1991) described the oxidized rhizosphere of slash pine and listed references to other species reported to have these manifestations including white oak (*Quercus alba*), water gum, bald cypress, swamp black gum, loblolly pine, lodgepole pine, black spruce (*Picea mariana*), and red spruce (*P. rubens*). Other trees oxidizing their rhizospheres include green ash, crack willow, white willow, sitka spruce, and lodgepole pine (Armstrong, 1968; Chirkova, 1968; Coutts and Philipson, 1978a; Sena-Gomes and Kozlowski, 1980). Dionigi et al. (1985) found that black willow's greater ability to oxidize waterlogged soils allowed it to colonize lower elevations along the Atchafalaya River in Louisiana, while sandbar willow (*S. exigua*) occupied higher sites. Studies of Irish wetland tree species found three species oxidizing their rhizospheres at different levels — willow (*S. cinerea* ssp. *oleifolia* or *S. atrocinerea*), ash (*F. excelsior*), and European alder which positively affected their tolerance of waterlogging (Iremonger and Kelly, 1988). Red and black mangroves also oxidize their rhizospheres. McKee et al. (1988) found that soil around the aerial roots of red mangrove were more oxidized than soil not associated with such roots. Good et al. (1986) concluded that the presence of oxidized rhizospheres in green ash was different enough on wet vs. mesic sites to serve as a useful indicator of site wetness, while Mendelssohn (1993) concluded that they indicated "soil saturation for a sufficient period to produce anaerobic soil conditions" and are a "relatively good indicator of wetland hydrologic conditions."

LIFE-FORM OR HABIT CHANGES

Flood-induced breaking of stem bud dormancy leads to the formation of multiple trunks in some species. Crawford (1983) reported that the European alder developed a "bush" form in frequently flooded wetlands, in contrast to a "pole" form on well-drained sites. The advantage of this adaptation is improved ventilation above the flood level. Similar responses may occur in other species as speckled alder (*A. rugosa*), red maple, silver maple, black willow, swamp privet (*Forestiera acuminata*), and Ogechee gum (*Nyssa ogechee*), and have been observed with multiple trunks in

FIGURE 3.8   Wetland form of saw palmetto (*Serenoa repens*).

periodically flooded wetlands (personal observations; Environmental Laboratory, 1987). McLeod et al. (1988) testing the response of four woody species to flooding with high temperature water found that buttonbush produced several basal stems in lieu of stem hypertrophy. They believed that this mechanism allowed for an increase in the area of stem tissue for potential gas exchange with the atmosphere (the same purpose of stem hypertrophy).

Saw palmetto (*Serenoa repens*) is a southern palm that grows mostly on dry sandy sites such as pine flatwoods, longleaf pine-scrub oak ridges, sand pine-oak scrubs, and coastal dunes. Yet it also common in seasonally wet pine flatwoods (Godfrey and Wooten, 1979). Wells (1942) listed it as one of the community dominants of southeastern shrub bogs. On dry soils, the saw palmetto sends up its palm leaves from a horizontal, usually underground, rhizome. On seasonally flooded sites, however, it develops an upright, often branched stem (Figure 3.8). Godfrey and Wooten (1979) recognized this different growth form and included this species in their book on aquatic and wetland plants of the Southeast.

Germination and Seedling Survival

The seeds of most plants lose viability when inundated for long periods. In contrast, seeds of marsh and aquatic species have prolonged dormancy and remain viable for many years (Leck, 1989). Scientists recently germinated 1200-year-old lotus seed from China (*The Boston Globe*, November 14, 1995, p. 3).

Many wetland plants can germinate underwater. Among these exceptional species, Hook (1984a) listed several wetland plants (including *Peltandra virginica*, *Alisma plantago*, *Typha latifolia*, *Cephalanthus occidentalis*, *Ulmus americana*, *Salix nigra*, and *Populus deltoides*), plus cultivated species (*Phleum pratense*, *Lactuca sativa*, *Oryza sativa*, *Petunia* spp., and *Dianthus* spp.) as good germinators and others as poor submerged germinators (e.g., *Nyssa* spp., *Taxodium distichum*, *Liquidambar styraciflua*, *Fraxinus pennsylvanica*, *F. caroliniana*, *Platanus occidentalis*, *Zea mays, Lolium perenne,* and *Festuca pratensis).* The latter rely on natural drawdown conditions in wetlands for germination, so characteristic southern deepwater swamp species like bald cypress and water gum regenerate at low water accompanying extended droughts. Rapid growth of seedlings allows these species to get their foliage above water, thereby enhancing survival of the species.

Work by Kramer (1949) identified that soil wetness during the first year or two was critical for seedling survival of shallow-rooted species, specifically bald cypress and yellow birch *(Betula alleghenensis).* Experimental studies of germination and seedling survival in lodgepole pine and pondcrosa pine *(P. ponderosa)* by Cochran (1972) in Oregon found significant germination in both species when water tables were below 15 cm and only a few germinants when soil were saturated to the surface, no mortality of seedlings of either species when in saturated soils (duration of 130 days), and most of the seedlings of both species survived 52 weeks of saturation. The conclusion was that the predominance of lodgepole pine (a FAC– species in Oregon) over ponderosa pine (a FACU– species) on wet soils must be related to other factors, possibly the former is a better competitor in sites with grasses and forbs. Tolerance of saturated conditions may change with age, as in loblolly pine; tolerance to oxygen deficiency was found to increase with age (Grablc, 1966, Leyton and Rousseau, 1958, as reported in Hu and Linnartz, 1972).

ACCELERATED STEM GROWTH

Flooding stimulates ethylene production in many aquatic plants which then promotes shoot elongation, allowing some of them to get their shoots or leaf tips above the flood level and thereby continue exchange of gases (mainly through stomata), photosynthesis, and other life processes (Ku et al., 1970; Crawford, 1992; Blom et al., 1996). In some species, adventitious root formation may begin after emergence (Van der Sman et al., 1993). While many aquatic and marsh plants have this adaptive response to flooding, some terrestrial species, including corn *(Zea mays),* also have evolved this mechanism to escape minor submergence (Kozlowski, 1984a; Jackson et al., 1985; Crawford, 1987; Jackson, 1988; Blom, 1990; Laan and Blom, 1990; Voesenek and Van der Veen, 1994). Stem growth can be extremely rapid. This can happen in less than 20 min (Jackson, 1971). Blom (1990) found that certain docks *(Rumex)* increased their petioles 100 to 120% within 4 days of flooding, with one species growing as much as 40 cm within a few days. Variations in plant responsiveness of stem elongation help explain plant distribution at different levels on floodplains (Blom et al., 1996).

McKevlin et al. (1995) suspect that rapid height growth in seedlings of water tupelo *(Nyssa aquatica)* during occasional dry periods facilitates their survival when such sites are reflooded. This species is a dominant species of semipermanently flooded and intermittently exposed wetlands, yet does not germinate underwater. Instead it germinates during drawdowns late in the growing season or during dry years. This late germination also may require rapid stem growth to guarantee that the terminal shoot is above the winter flood level.

WETLAND PLANTS — HYDROPHYTES

Plants growing in wetlands and water are technically called *hydrophytes.* Most wetland plants do not grow strictly in water or very wet soils, but also grow in terrestrial habitats, especially under mesic soil conditions (e.g., the species of Wisconsin's wet-mesic prairies or forests according to Curtis, 1959). Many of these species are more common on the latter sites, but have populations that tolerate varying degrees of soil wetness. Unfortunately, due to the lack of distinctive morphological differences, individuals of these wetland populations can only be recognized as hydrophytes when associated with more typical hydrophytic species or after identification of hydric soils (i.e., periodically anaerobic soils due to excessive wetness) and other reliable signs of wetland hydrology at a given location.

EVOLVING CONCEPT OF A HYDROPHYTE

Today's usage of the term *hydrophyte* is different than its original use. In the 1800s and early 1900s, it was used to define aquatic plants that were plants growing in water (Schouw, 1822, as reported in Warming, 1909) or plants with perennating buds beneath the water (Raunkiaer, 1905; 1934). Warming and Raunkaier were among the earliest of the plant ecologists to use the term *hydrophyte*. Hydrophytes were distinguished from helophytes, which included various wetland plants depending on whose definition was used.

Raunkiaer's life-forms were based on a plant's adaptation to the critical season (e.g., winter), mainly the degree of protection possessed by the dormant buds (Smith, 1913). According to this system, hydrophytes (plants with perennating rhizomes or winter buds) and helophytes (plants with buds at the bottom of the water or in the underlying soil) were the two types of cryptophytes (plants with dormant parts below ground), while other wetland plants were included in other life-forms, such as phanerophytes (trees and shrubs) (Smith, 1913). Raunkiaer's helophytes did not include all typical marsh species.

Warming (1909) was probably the first ecologist to arrange plant communities by the degree of soil wetness. He recognized aquatic plants (water-plants) that spend their entire life submerged or with leaves floating at the surface and terrestrial plants that are mostly exposed to air, including marsh plants. Vegetation was then separated into numerous "oecological classes" based principally on soil properties. The first of the groupings was for soil that was very wet and two classes were listed: class 1 — hydrophytes (formations in water) and class 2 — helophytes (formations in marsh). His concept of helophyte was much broader than Raunkiaer's, including plants with their roots under water or in "soaking" soil with emergent foliage (herbs, shrubs, and trees). The distinction between hydrophytes and helophytes had ecological merit in that the truly aquatic plants (hydrophytes) were separated from those that grew on anaerobic saturated soils (helophytes). In describing marsh and fen vegetation of an English lake, Pearsall (1917) referred to the plants of marshes (near the water) and fens (on peaty or sedimentary soils) as "terrestrial hydrophytic vegetation."

Following the lead of Warming, Clements (1920) also used soil wetness to separate plants into different groupings. He might have been the first ecologist to expand the definition of hydrophyte to include helophytes as a type of hydrophyte. Later in *Plant Ecology*, Weaver and Clements (1929) stated that "typical hydrophytes grow in water, in soil covered by water, or in soil that is usually saturated." They divided hydrophytes into three groups: (1) submerged, (2) floating (including floating-leaved rooted species), and (3) amphibious.

Daubenmire (1947) identified five "morphoecologic" groups of hydrophytes: (1) floating, (2) suspended (e.g., phytoplankton), (3) submerged anchored, (4) floating-leaved anchored, and (5) emergent anchored. Included in the last group are "swamp and bog plants which inhabit soils containing a quantity of water that would prove supraoptimal for the average plant." His hydrophyte definition ("any plant growing in a soil that is at least periodically deficient in oxygen as a result of excessive water content") has been used in various U.S. government wetland delineation manuals and the official federal wetland classification system (Daubenmire, 1968).

While Daubenmire's definition relates to an individual plant, the need to use plant species as indicators of wetlands led to the development of national and regional lists of indicator species for the U.S. (Reed, 1988; see Wetland Plant Lists discussion below). The lists included many typical upland species that had wetland ecotypes or broad ecological amplitudes and were typically adapted for life in wetlands. Recognizing the significance and potential confusion of using wetland ecotypes of common upland species as wetland indicators prompted Tiner (1988) to refine the definition of a hydrophyte in his *Field Guide to Nontidal Wetland Identification* as "an individual plant adapted for life in water or periodically flooded and/or saturated soils (hydric soils) and growing in wetlands and deepwater habitats; may represent the entire population of a species or only a subset of individuals so adapted." This definition embodies the "individualistic" concept of a hydrophyte, recognizing that plant species may exhibit considerable plasticity or ecological amplitude in their adaptations to wet environments. This concept is not bound to the species level in plant taxonomy, but allows, for example, wetland variants of mostly dry-site species to be classified as hydrophytes. Tiner (1991) provides a detailed review of the concept of a hydrophyte as applied to wetland delineation. The EPA wetland delineation manual and interagency federal manual embraced this concept (Sipple, 1988; Federal Interagency Committee for Wetland Delineation, 1989).

PLANT SPECIFICITY

While a couple of thousand species grow exclusively in U.S. wetlands, thousands of others are more wide-ranging. Some of the latter occur mostly but not always in wetlands, whereas others display no particular affinity for wetlands and still others are actually more common in uplands. When the plants are growing under wetland hydrologic conditions, they are hydrophytes regardless of where the majority of individuals of their species occur. The best plant indicators of wetland are clearly those species with the highest affinity for wetlands (see Table 3.3 for common examples).

The affinity of certain species may vary with latitude or, in mountainous areas, with altitude. For example, in the Northeast, labrador tea (*Ledum groenlandicum*) is restricted to wetlands, mainly bogs. Yet in Labrador and Newfoundland, this species also occurs in dry heaths and woods (Ryan, 1989). Mountain holly (*Nemopanthus mucronata*), an exclusive wetland species in most of its U.S. range, occurs in uplands and wetlands in coastal eastern Maine. Presumably the combination of low evapotranspiration, cold climate, moderate frequency of fogs, and cool, moist air (maritime influence) create favorable conditions on drier soils. The colder climates, especially along the coast, may permit some excellent wetland indicator plants to be less reliable indicators locally.

Most plants growing in wetlands have a broad ecological amplitude and are tolerant or adaptable to many environmental conditions (e.g., wet, moist, and sometimes dry). Such species are not particularly useful indicators of any environment, unless they possess morphological adaptations developed in response to prolonged flooding or waterlogging. A flood-tolerance study of red maple (*Acer rubrum*) seedlings from wet sites and from dry sites by Will et al. (1995) demonstrated that this species is highly adaptable regardless of its original habitat. As mentioned earlier, red maple was reported to have a highly adaptable root system (Kramer, 1949). It is a good example of a species demonstrating ecological plasticity. While many species are quite adaptable, some have populations of individuals that are better adapted to one set of environmental conditions than other populations of the same species. These populations have been referred to as *ecotypes*. The existence of ecotypes is well-established in the literature.

WETLAND ECOTYPES

While the definition of *species* is useful for taxonomic reasons and to discuss ecological relationships, it is not without problems. Merrell (1981) commented that up to half of all species of flowering plants are believed to be allopolyploids — the product of hybridization between species plus complications due to asexuality, introgression, geographic variation, and combinations of these factors. Recognizing the varied responses to different habitats, Clements et al. (1950) simply concluded that the typical "species" is the form that occupies the most extensive area. This clearly reveals the possibility for less habitat specificity than the nonscientist might expect from a given plant species.

At the species level, plants do not have exactly the same environmental requirements and individual populations may differ in their tolerance of degrees of waterlogging or flooding. It has long been recognized that a given plant species may include *ecotypes* — a population or group of populations having certain genetically-based morphological and/or physiological characters — but usually prevented from natural interbreeding by ecological barriers (Turesson, 1922a,b; 1925; Barbour et al., 1980). Daubenmire (1968) in his classical treatise on plant communities pointed out that "each habitat type is a distinctive combination of environmental factors, so the different selection pressures in contiguous habitats tend to develop special ecotypes that are homozygous for at least the adaptive characteristics." Ecotypes are more fit for occupying a certain habitat than

TABLE 3.3
Examples of Obligate Hydrophytes that are Widespread or Particularly Common in Certain Wetland Types in the U.S.

Aquatics

Azolla spp. (mosquito-ferns)	*Potamogeton* spp. (pondweeds)
Brasenia schreberi (water-shield)	*Proserpinaca* spp. (mermaid-weeds)
Elodea spp. (waterweeds)	*Ruppia maritima* (widgeon-grass)
Isoetes spp. (quillworts)	*Thalassia testudinum* (turtle-grass)
Lemna spp. (duckweeds)	*Utricularia* spp. (bladderworts)
Myriophyllum spp. (water-milfoils)	*Vallisneria americana* (wild celery)
Najas spp. (naiads)	*Zannichellia palustris* (horned pondweed)
Nuphar spp. (pond lilies)	*Zostera marina* (eel-grass)
Nymphaea spp. (water lilies)	

Emergents (Herbs)

Alisma spp. (water-plantains)	*Osmunda regalis* (royal fern)
Calla palustris (wild calla)	*Peltandra virginica* (arrow arum)
Caltha palustris (marsh marigold)	*Polygonum hydropiperoides* (water pepper)
Carex aquatilis (water sedge)	*Polygonum sagittatum* (arrow-leaved tearthumb)
Carex lenticularis (shore sedge)	*Pontederia cordata* (pickerelweed)
Carex rostrata (beaked sedge)	*Sagittaria* spp. (arrowheads)
Carex stricta (tussock sedge)	*Salicornia virginica* (perennial glasswort)
Carex vesicaria (inflated sedge)	*Scirpus acutus* (hard-stemmed bulrush or tule)
Cicuta maculata (water hemlock)	*Scirpus americanus* (Olney's three-square)
Decodon verticillatus (water-willow)	*Scirpus atrovirens* (green bulrush)
Drosera spp. (sundews)	*Scirpus tabernaemontani* (soft-stemmed bulrush)
Dulichium arundinaceum (three-way sedge)	*Sium suave* (water parsnip)
Eleocharis spp. (spike-rushes)	*Solidago patula* (rough-leaved goldenrod)
Eriophorum spp. (cotton-grasses)	*Solidago uliginosa* (bog goldenrod)
Glyceria spp. (manna grasses)	*Sparganium* spp. (bur-reeds)
Iris versicolor (blue flag)	*Spartina alterniflora* (smooth cordgrass)
Juncus canadensis (Canada rush)	*Symplocarpus foetidus* (skunk cabbage)
Juncus militaris (bayonet rush)	*Triglochin* spp. (arrow-grasses)
Juncus roemerianus (black needlerush)	*Typha* spp. (cattails)
Leersia oryzoides (rice cutgrass)	*Woodwardia virginica* (Virginia chain fern)
Lindernia dubia (water pimpernel)	*Xyris* spp. (yellow-eyed grasses)
Lysichitum americanus (skunk cabbage)	*Zizania aquatica* (wild rice)

Shrubs

Andromeda polifolia (bog laurel)	*Myrica gale* (sweet gale)
Betula pumila (bog birch)	*Rhizophora mangle* (red mangrove)
Cephalanthus occidentalis (buttonbush)	*Rosa palustris* (swamp rose)
Forestiera acuminata (swamp privet)	*Salix sericea* (silky willow)
Lonicera oblongifolia (swamp fly-honeysuckle)	*Vaccinium macrocarpon* (large cranberry)

Trees

Carya aquatica (water hickory)	*Nyssa aquatica* (water gum)
Chamaecyparis thyoides (Atlantic white cedar)	*Planera aquatica* (planer-tree)
Fraxinus profunda (pumpkin ash)	*Quercus lyrata* (overcup oak)
Gleditsia aquatica (water locust)	*Taxodium distichum* (bald cypress)

Note: Genera listed contain all or mostly obligates.

Source: Tiner, R.W. 1991. *BioScience*, 41: 236–247.

individuals from other populations of the species. The significance of ecotypes must not be overlooked for, as noted by Braun-Blanquet (1932), "the most exact indicators are often, indeed, not the 'good Linnaean species' but rather the elementary species or races, the 'ecotypes' of Turesson (1925)" for "these forms require more narrowly circumscribed life conditions and, therefore, are socially more sharply specialized." Recognizing the existence of wetland ecotypes, races, varieties, subspecies, and other variants or simply acknowledging wide wetness tolerances of plant species is vital to understanding how to use plants as indicators of wetlands.

For some plant species, subspecies or varieties that are found in different habitats or with a restricted distribution are morphologically distinguishable (Table 3.4). In some cases, these varieties have been assigned a different indicator status on national and regional wetland plant lists (see

TABLE 3.4
Examples of Species with Recognized Varieties Occurring in Different Habitats

Species (Common Name)	Variety	Nat'l Range of Indicator Status	Habitat
Acer rubrum			
(Red Maple)	*rubrum*	FAC	Swamps, alluvial soils, and moist soil
(Drummond's Red Maple)	*drummondii*	OBL to FACW	Deep swamps
(Trident-leaved Red Maple)	*trilobum*	OBL to FACW+	Forested wetlands
Andropogon virginicus (Broom-sedge)	*virginicus*	FACU to FAC	Dry open soils, thin woods, etc.
	glaucus	Not designated	Dry sandy pine barrens
	tetrastachyus	Not designated	Dry sands, rocks, and pinelands
	glaucopsis	Not designated	Savannas, wet pineland, and swamps
	hirsutior	Not designated	River-swamps, savannas, and marshes
Celtis laevigata (Sugarberry)	*laevigata*	FACW to UPL	Bottomlands and low woods
	smallii	Not designated	Bottomlands and low woods
	texana	Not designated	Bluffs, rocky slopes, dry woods, etc.
Fagus grandifolia (American Beech)	*grandifolia*	FACU	Rich upland soils
	caroliniana[a]	FAC+	Moist or wet lowland soils, especially on or near the coastal plain
Nyssa sylvatica			
(Black Gum)	*sylvatica*	FAC	Low acid woods, swamps, and shores
(Swamp Black Gum)	*biflora*	OBL to FACW+	Inundated swamps and damp sands
	caroliniana	Not designated	Chiefly on uplands of the interior
Panicum virgatum (Switchgrass)	*virgatum*	FACW to FAC	Dry or moist sandy soils, and shores
	spissum	Not designated	Gravelly or sandy fresh to brackish shores and swamps
Quercus falcata			
(Southern Red Oak)	*falcata*	FACU to FACU−	Moist to dry woods
(Cherrybark Oak)	*pagodaefolia*	FACW to FAC+	Chiefly on bottomlands or near streams
Quercus stellata			
(Post Oak)	*stellata*	FACU to UPL	Sandy, gravelly upland soils
	paludosa	FACW	Soils that are alternately waterlogged and hard and dry (southern flatwoods)[b]

[a] Designated as FAC+ only in the Northeast, while this variety also occurs in the Southeast, Midwest, and South Plains (Texas and Oklahoma).

[b] Habitat description is for typical post oak in Fowells (1965), but likely represents the habitat of this variety.

Sources: Range in wetland indicator status in its U.S. distribution based on Reed (1988). Habitat data from Fernald (1950), Gleason and Cronquist (1963), and Fowells (1965).

following section for discussion), especially when their habitats are wetter than the typical species. Their recognizable morphological differences make them particularly useful for identifying wetlands. Yet, it must be understood that ecotypes are typically more distinctive physiologically than morphologically (Daubenmire, 1968). In other words, their differences are not visible, but require scientific study of physiological responses.

Numerous studies have found different responses to flooding within species that occur in wetlands. Keeley's study of black gum (*Nyssa sylvatica*) is perhaps the classic work on this. Keeley (1979) recorded different responses to flooding by seedlings from three distinct populations of black gum (swamp, floodplain, and upland phenotypes). The swamp seedlings easily withstood months of flooding (98% survival) and also did fairly well under drained conditions (65% survival). Upland seedlings were intolerant of flooding (only 27% survival), but had 73% survival in drained situations. Floodplain seedlings did as well as the upland seedlings under drained conditions and produced a flood-tolerant type under flooded conditions. This study demonstrated the existence of wetland ecotypes in black gum.

Wetland ecotypes can evolve due to differences in environmental conditions that reproductively isolate wet-site individuals from dry-site individuals. Timing of flowering may be different thereby limiting cross-fertilization. In their study of ecotypic variation in western white pine (*Pinus monticola*), Squillace and Bingham (1958) found that seed bed moisture served as an ecological barrier to gene flow from contiguous populations. Genetically distinct populations can be created wherever there is a steep gradient in microhabitats, even when gene flow is strong (Liu and Godt, 1983). In the case of western white pine, seedlings from dry-site populations failed to survive wet conditions, while wet-site seedlings could not grow in dry beds (Squillace and Bingham, 1958).

In summarizing literature on plant adaptations to salt water flooding, Wainwright (1984) mentioned the presence of salt-tolerant ecotypes of creeping bentgrass (*Agrostis stolonifera*), red fescue (*Festuca rubra*), and several other typically nonsaline species in salt marshes. Individuals of these two species from salt marsh populations did not grow as vigorously as individuals from inland populations when grown under freshwater conditions (Hannon and Bradshaw, 1968; Tiku and Snaydon, 1971). Intraspecific variation in salt tolerance has also been detected in many common salt marsh plants (*Spartina alterniflora, S. foliosa, S. patens, Sporobolus virginicus,* and *Juncus roemerianus*) (Nester, 1977; Cain and Harvey, 1983; Blits and Gallagher, 1991; Eleutrius, 1989; Pezeshki and DeLaune, 1995; Hester et al., 1996). Salt-hay grass occurs in dunes, interdunal swales, salt marshes, and brackish marshes and these stocks showed different responses to salinity (Silander and Antonovics, 1979; Pezeshki and DeLaune, 1991).

Different populations of species have shown unlike responses to flooding and waterlogging. Lessmann et al. (1997) detected intraspecific variations in three species (*Spartina alterniflora, S. patens,* and *Panicum hemitomom*) and suggested that the more flood-tolerant stocks be used in wetland creation and restoration. Foresters have long recognized different site types to maximize timber production from variable sites. Forestry researchers often use wet-site and dry-site seed sources when studying the responses of seedlings to anaerobic conditions as Topa and McLeod (1986a) did when evaluating such responses in loblolly pine and other pines. In searching for more waterlogging-tolerant genotypes of loblolly pine for forestry improvement, Shear and Hook (1988) subjected seedlings from 11 sources ("families") to three winter waterlogging regimes (waterlogging from 0 to 15 cm, 0 to 30 cm, and 0 to 45 cm) and found significantly different intraspecific responses to the wettest treatments. Gill (1975) noted several examples of intraspecific and intravarietal differences in possessing the ability to produce adventitious roots. Several researchers observed that flood tolerance varied in *Eucalyptus* among ecotypes and provenances (Karschon and Zohar, 1972; Ladiges and Kelso, 1977, as reported in Kozlowski, 1984a). Sahrawat et al. (1996) evaluated numerous rice cultivars and found significantly different responses to iron toxicity tolerance, which could help boost rice production in irrigated and lowland rice paddies. When seaside goldenrod (*Solidago sempervirens*) from a Florida population was grown in New York, it grew poorly, while the New York populations clearly had no trouble growing there (Clausen and Hiesey, 1958). Yellow monkeyflower (*Mimulus guttatus*) has many ecological races related to climatic, altitudinal, and other factors. Even when the distributions of habitats overlap, the races may remain separate due to ecological conditions, with only minor hybridization (Clausen et al., 1941). Darlington (1973) referred to such hybrids as "misfits."

Many species seem to exhibit ecological plasticity — the ability to successfully colonize a wide range of habitats, even within a single geographic region. These are highly adaptive and opportunistic species. An example is pitch pine, the characteristic plant of the New Jersey Pine Barrens, predominating both wetlands and uplands. It grows across a broad continuum of soil moisture from the driest sites (e.g., sand dunes) to the wettest (mucky soils and seasonally ponded sites). Ledig and Little (1979) noted genetic variations in pitch pine, with the dwarf or pygmy form (<4 m tall) occurring on dry sand sites in the New Jersey Pine Barrens and the tallest form (30 m) growing on seasonally wet sites. They admitted that genetic effects are confounded by environmental effects and that genetic variation can occur at several levels in a species — among individuals within stands, among stands within regions, and among physiographic regions.

Persimmon (*Diospyros virginiana*) occurs in permanently flooded sites on the Delmarva Peninsula (Delaware, Maryland, Virginia) and on sand dunes. It coexists with buttonbush (*Cephalanthus occidentalis*) in Delmarva bay or pothole wetlands. Titus (1990) also found this species growing in a Florida hardwood swamp at low elevations, with its mean elevation between that of bald cypress and American elm (*Ulmus americana*). It is likely that a wetland ecotype for persimmon exists.

On the West Coast, lodgepole pine occupies a similarly broad range of habitats (Fowells, 1965), with varietal habitat preferences. The shrubby coastal form (var. *contorta*) grows in peat bogs and muskegs from Puget Sound north, but occupies dry sandy and gravelly sites in the southern part of its range. The inland variety (var. *latifolia*) is typically an upland plant of the interior (e.g., Rocky Mountains).

It is also important to recognize that a portion of the world's vascular (flowering) plants have successfully made the transition back to a fully aquatic existence, despite the origin of land plants from aquatic algae over 400 million years ago (Davy et al., 1990). Cattails and bur-reeds evolved from terrestrial plants (Cronquist, 1968). Curiously, some terrestrial monocots were derived from aquatic ancestors only to return to a watery life as hydrophytes (Crawford, 1983). The first trees on Earth grew in Carboniferous swamps about 250 million years ago, so these swamps are the "ancestors" of terrestrial vegetation (Kangas, 1990). Evolution is still occurring and land plants are continuing to adapt to life in wetlands and water.

Traditionally, plant ecologists have attempted to use certain plants as indicators of specific environmental conditions (e.g., hydrophytes, halophytes, and calciphiles). In seeking to identify wetlands, there are many species that can serve as useful indicators for many wetland types especially the seasonally flooded and wetter types. However, given the transitional nature of many wetlands or portions of wetlands (e.g., ecotonal wetlands between wetter wetlands and mesic uplands), caution must be exercised in deciding what species are used as indicators of either wetlands or uplands. Species in the drier-end wetlands are not likely to be reliable indicators of either wetlands or uplands, so plants may not be useful for identifying wetland boundaries under these conditions. Barbour et al. (1980) summarized the problem aptly, "Plant ecologists would like to use species as deductive tools, as rather precise indicators of certain levels of environmental factors. This may not be a realistic objective for two reasons. First, plants respond to a complex of climatic, edaphic, and biotic factors, and the impact of single factors is difficult to isolate. Second, taxonomic species, whether recognized on morphological, biological, or statistical grounds, are partially artifacts of the human desire to classify." While many botanists and ecologists would like to use the Linnaean species to determine precise limits of wetlands, it must be understood that most wetlands, especially the drier-end ones, cannot be simply identified by plant species alone. The existence of wetland ecotypes of species that are "typically" on drier sites has confounded the situation. Moreover, unless morphological adaptations are present, the only way to recognize these hydrophytic ecotypes is through verifying the presence of undrained hydric soils.

USING PLANTS TO IDENTIFY WETLANDS

Since the beginnings of plant ecology as a modern science, botanists have found certain plant species and communities to be characteristic of wetlands. They have been interested in describing plant communities and explaining the reasons for their establishment. They were not concerned about determining the exact boundaries between the communities as they collected data from representative sites or sites that typified the overall plant community. Typically a plant ecologist would not select a sample plot in the interface between two communities unless he or she was interested in this transitional area. Certain aspects of wetlands have baffled plant ecologists for some time as witnessed by the following quotations.

"...there is no sharp limit between marsh-plants and land-plants...[t]his zone [fresh-water swamp] represents a very gradual transition from terrestrial to lacustrine conditions...[i]t is impossible to establish any sharp distinction between swamp-forests and forests on dry land..." (Warming, 1909, *Oecology of Plants: An Introduction to the Study of Plant-Communities*)

"...amphibious plants have a wide range of adjustment and may grow for a time as mesophytes or partially submerged...[they are the] least specialized of water plants." (Weaver and Clements, 1929, *Plant Ecology*)

"Many sedges and willows, etc. are transitional between this group [emergent anchored hydrophytes] and mesophytes in that they grow in wet soil where the water table is close to the surface." (Daubenmire, 1947, *Plants and Environment: A Textbook of Plant Autecology*)

Today there is great interest in both identification and delineation of wetlands due to passage of laws and promulgation of regulations to protect wetlands or to curtail unnecessary wetland destruction. Consequently, attention is focused on determining the boundaries of wetlands. For the first time in history, it is critical to know the limits of wetlands on individual parcels of land, since many activities (e.g., dredging or filling) require federal or state permits before commencing work. Vegetation plays a major role in wetland identification and delineation, so it is important to know which plants or groups of plants (plant communities) are wetland indicators (see Chapter 9 for examples of U.S. wetland plant communities).

WETLAND PLANT LISTS

To aid in using plants to identify wetlands, a national list of vascular plant species that occur in wetlands has been prepared by the federal government (U.S. Fish and Wildlife Service with cooperation from the Army Corps of Engineers, Environmental Protection Agency, and Natural Resources Conservation Service) (Reed, 1988; Reed, 1997). In reviewing the scientific literature, nearly 7000 species of plants were reported growing in U.S. wetlands. Rather than compiling a simple list of these species, it was realized that the affinity for wetlands varies considerably among plant species and in some cases across regions. Consequently, the roughly 7000 species were assigned to one of four "wetland indicator categories" based on differences in expected frequency of occurrence in wetlands: (1) obligate wetland (OBL), (2) facultative wetland (FACW), (3) facultative (FAC), and (4) facultative upland (FACU) (Table 3.5). Plants not found in wetlands are considered upland plants (UPL). Given that a plant's indicator status may vary across the country, 13 regional lists were developed (Figure 3.9). The national list represents a compilation of regional lists. The regions are still quite broad and further subdivisions also could make the list even more sensitive to species habitat requirements. The proposed 1997 list (Reed, 1997) is the first attempt to address some significant intraregional differences.

TABLE 3.5
Wetland Indicator Categories of Plant Species that Occur
in Wetlands under Natural Conditions

Wetland Indicator Category	Estimated Probability of Occurrence in Wetlands	Estimated Probability of Occurrence in Nonwetlands
Obligate wetland (OBL)	>99% of the time	<1% of the time
Facultative wetland (FACW)	67 to 99% of the time	1 to 33% of the time
Facultative (FAC)	34 to 66% of the time	34 to 66% of the time
Facultative upland (FACU)	1 to 33% of the time	67 to 99% of the time

Note: Plant species that almost always occur in nonwetlands (>99% of the time) are considered upland plants (UPL). Also, in assigning indicator categories to individual plant species, a plus or a minus was added as appropriate; a plus after the category (e.g., FAC+) indicates that the species occurs in the higher portion of the range in wetlands (e.g., 51 to 66% of the time), whereas a minus (e.g., FAC–) indicates the lower portion of the range (e.g., 49 to 34%).

Source: Reed, P.B., Jr. 1988. Biol. Rep. 88(24). Nat'l List of Plant Species that Occur in Wetlands: 1988 National Summary. U.S. Fish and Wildlife Service. Washington, D.C.

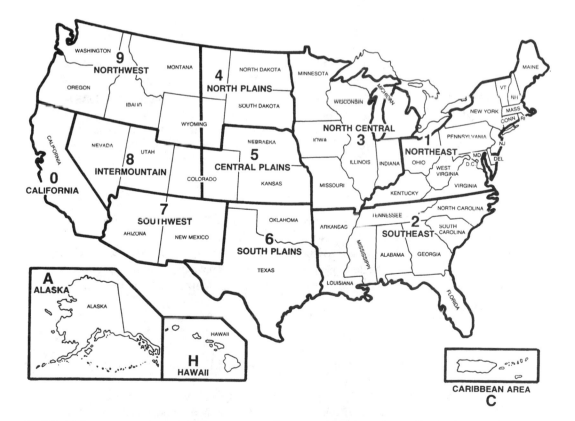

FIGURE 3.9 Map showing regions used to identify wetland indicator status of U.S. plant species. (Reed, P.B., Jr., Nat'l List of Plant Species that Occur in Wetlands: 1988 National Summary, U.S. Fish & Wildlife Service, Washington, D.C., 1988.)

The 1988 national list of wetland plants contains 6728 species out of a total of approximately 22,500 vascular plant species that exist within all habitats in the U.S. and its territories and possessions (Reed, 1988). Although the list is lengthy, it does not contain the majority of U.S. plant species, which are virtually intolerant of flooding or prolonged soil saturation during the growing season. Only 31% of the nation's flora occur often enough in wetlands to be on the list. Roughly one quarter (27%) of the species on the national wetland plant list is represented by OBL species (Tiner, 1991). The majority of listed species, therefore, grow in both wetlands and nonwetlands to at least some extent, which is quite expected for habitats that are, in many ways, transitional between land and water. The 1988 wetland plant list has been technically updated (Reed, 1997) and is scheduled for publication in late 1998 or early 1999.

FIELD INDICATORS OF HYDROPHYTIC VEGETATION

Interpretation of plants as wetland indicators vary according to the approach taken for wetland delineation. To date, the federal government has generally embraced a three-parameter/criteria approach for identifying regulated wetlands (Environmental Laboratory, 1987; Federal Interagency Committee for Wetland Delineation, 1989). This approach tends to define hydrophytic vegetation in broad terms (including FAC-dominated communities) because hydric soils and signs of wetland hydrology are usually required to help make the final wetland determination. Other approaches such as the primary indicators method (PRIMET) focus on plant and other indicators that are unique to wetlands for positive wetland identification (Tiner, 1993). This type of approach has been used to identify wetlands for inventories and mapping projects and is gaining acceptance for regulatory purposes in some states. State agencies may use the predominance of FACW and OBL species as positive indicators of hydrophytic vegetation and wetlands, while other communities may be recognized as hydrophytic upon verification of hydrology or hydric soils (see Chapter 6). Table 3.6 outlines field indicators of hydrophytic vegetation used by several sources. Most methods focus on dominant species (i.e., the most abundant plants in a given community), yet the mere presence of obligate species should be an excellent indicator of wetland.

The best vegetation indicators of wetlands are the genera, species, or subspecies (varieties) that are unique to wetlands — the obligate hydrophytes (see Table 3.3). Whenever these species are common or abundant in an area, the area should be easily recognized as a wetland by vegetation alone. The FACW+ species also are excellent indicators as they are found in wetlands more than 9 times out of 10, on average.

PROBLEMS INTERPRETING FAC– AND FACU SPECIES AS HYDROPHYTES

While both OBL and FACW species are universally recognized as useful indicators of wetlands, FAC and FACU are not reliable wetland indicators and their use in wetland delineation has been contentious (see 56 Federal Register 40446-40480, August 14, 1991). Since they occur in wetlands with some frequency and may even dominate certain types, they have the potential to be hydrophytes (Chapter 6). Hydrophytic members of these species can be recognized in four ways.

1. When associated with OBL and FACW species.
2. When they possess certain morphological adaptations.
3. After verification of undrained hydric soils.
4. By their occurrence in areas with documented wetland hydrology.

FAC species, by definition, have essentially no affinity for wetlands or nonwetlands and, therefore, are not indicative of either. This has led to the development of the so-called "FAC Neutral Rule" for determining the presence of hydrophytic vegetation. This rule does not utilize FAC species (either FAC+, FAC, or FAC–) in assessing the potential for hydrophytic vegetation, but weighs the

TABLE 3.6
Field Indicators of Hydrophytic Vegetation from Various Sources

Indicator	Source
>50% dominants are OBL, FACW, and/or FAC	FI, WI
>50% dominants are OBL, FACW, and/or FAC (excluding FAC–)	CE
Prevalence index less than 3.0	FI, FSA
OBL + FACW species > FACU + UPL species (FAC Neutral Test)	CE, FI, FSA
OBL + FACW species (excluding exotics, vines, and aquatic plants)	FL
>50% of dominants are FACW and/or OBL species and no FACU or UPL species are dominant #	NY
FAC and wetter species (excluding FAC– species)	MA
One or more OBL dominants* #	EPA, PRIMET
OBL perennial species have 10% cover and are evenly distributed throughout the community #	PRIMET, NY
Any plant community dominated by FACU and wetter species occurring on undrained hydric soils (EPA needs hydrology indicator plus no UPL species, except on microsites)	EPA, FI, PRIMET, MA
Dominant species with certain morphological adaptations #	CE, FI, EPA, PRIMET, MA, NY
Plants observed growing in areas flooded or saturated for 10% or more of the growing season	CE
Plants with known physiological adaptations for life in saturated soils	CE, EPA
Plants with known reproductive adaptations for life in wetlands	CE, EPA
Plants reported in the literature growing in wetlands	CE
Patches (expanses) of peat moss (*Sphagnum* spp.)	PRIMET, RI, NY
Plants listed in the wetland protection act	FL, MA, NY, RI
FACW, FAC, and/or FACU species with other indicators of wetland hydrology	RI

Note: An asterisk (*) denotes applicability of indicator in the absence of significant drainage. A pound sign (#) means that the indicator is also used to verify the presence of wetland under PRIMET or NY procedures.

Sources: Federal interagency manual (FI; Federal Interagency Committee for Wetland Delineation, 1989); Corps manual (CE; Environmental Laboratory, 1987); Food Security Act manual (FSA; USDA Natural Resources Conservation Service, 1994); EPA manual (EPA; Sipple, 1988); Primary indicators method (PRIMET; Tiner, 1993); Florida manual (FL; Florida Dept. of Environmental Protection et al., 1995); Massachusetts manual (MA; Jackson, 1995); New York manual (NY; Browne et al., 1996); Rhode Island procedures (RI; Rhode Island Dept. of Environmental Management, 1994); Wisconsin procedures (WI; Wisconsin Coastal Management Program, 1995).

abundance of OBL and FACW species against the abundance of FACU and UPL species. According to the Corps manual, a positive test (OBL + FACW > FACU + UPL) may be used as a secondary indicator of hydrophytic vegetation for plant communities failing the basic rule (i.e., >50% of dominants are FAC and wetter, excluding FAC–). More recently, it also may be used as a secondary indicator of wetland hydrology at the discretion of the local Corps district (Williams, 1992).

While FAC species may not be reliable indicators of wetlands (due to their equal occurrence in wetlands and uplands), they dominate many forested wetlands across the country. Examples of some dominant hydrophytic FAC trees include sweet gum, eastern cottonwood, red maple, balsam fir (*Abies balsamea*), yellow birch, water oak (*Quercus nigra*), black gum, loblolly pine, ironwood (*Carpinus caroliniana*), quaking aspen (*Populus tremuloides*), Engelmann spruce (*Picea engelmannii*), and western red cedar (*Thuja plicata*).* FAC species are well adapted to wetlands. For example, loblolly pine's ability to withstand waterlogging is well known among foresters due to its commercial significance. McKevlin et al. (1987) described seedling adaptations including stem hypertrophy, hypertrophied lenticels (on stem and roots), soil water roots, aerenchyma (large lacunae), and an internal aeration system. Loblolly pine is tolerant of prolonged anaerobic conditions extend-

---

* Some of these species have drier or wetter statuses in some regions.

ing from winter through early spring and achieves some of its best growth under these circumstances. In studying several Louisiana stands, Hu and Linnartz (1972) found that the wettest sites had the highest loblolly pine site index values (most productive sites). The availability of water early in the growing season may be more important for plant growth than an adequate oxygen level in the soil.

FACU species (plants that are typically found in nonwetlands) are more contentious as wetland species, since by definition they occur more in uplands than in wetlands. The national list of wetland plant species includes about 1400 FACU species (21% of the list) (Tiner, 1991). Some species are quite common in wetlands and when growing under such conditions are hydrophytic. In his monograph on the development of the vegetation of New York state, Bray (1930) reported two FACU species, white pine (*Pinus strobus*) and eastern hemlock (*Tsuga canadensis*), as common forested wetland species, commenting that "the fact that very big trees of pine and hemlock do occur in typical swamp forest seems to me noteworthy and to be deserving of special study." Eastern hemlock typifies certain familiar wetland types called hemlock swamps in the Northeast (Niering, 1953; Huenneke, 1982; Tiner, 1989; 1998). In another classic vegetation study, Curtis (1959) reported that white pine occurred in 37% of the tamarack or black spruce bogs in Wisconsin and even dominated some northern stands on more than 10 ft of peat. He also found two other FACU species to be common in wetlands — jack pine (*P. banksiana*) in similar situations and sugar maple (*Acer saccharum*) in some wet-mesic forests dominated by northern white cedar (*Thuja occidentalis*; FACW) and balsam fir (FACW). Crum (1988) observed white pine and jack pine in Michigan bogs, along with beech (*Fagus grandifolia*; FACU) and sugar maple (*Acer saccharum*; FACU) seedlings that failed to persist. He felt that the first two were examples of broad tolerances, while the latter two were freak occurrences. He also noted that white spruce (*Picea glauca*; FACU) was a characteristic peatland species and that some rather large red pines (*P. resinosa*; FACU) were found in Michigan bogs. Pitch pine (*P. rigida*) is a FACU species with a remarkable range in wet and dry tolerances, growing on excessively drained soils to poorly drained sands and gravels, mucks, and peats (Ledig and Little, 1979; Little, 1959; Illick and Aughanbaugh, 1930; personal observations). Subalpine fir (*Abies lasiocarpa*) dominates many western forested wetlands (see Chapter 9, Table 9.8). These plants demonstrate considerable ecological plasticity and create a serious perception problem when attempting to depict these plant species as hydrophytes or wetland plants (Chapter 6). While the general public may have some difficulty understanding that FACU species can be hydrophytes, wetland ecologists should not, recognizing the ecological amplitude of and the possibility of ecotypes for many species. Moreover, the problem with FACU species is an artifact of our attempt to use the species level to identify wetland indicators. The FACU plants growing in wetlands are well-adapted to a wetland existence. Individual plants are adapting to varied environmental conditions in the pursuit of life and continued survival of the species.

HUMAN EFFECTS ON WETLAND PLANT DISTRIBUTION

The occurrence of a plant species on the landscape can be drastically changed by human disturbance or interference. This further complicates the potential use of plants to identify wetlands. Specifically, the distribution and abundance of many plants have been significantly impacted by forestry practices, agricultural activities, urban development, drainage projects, water diversion, pollution, and other human-induced actions. Planted crops, either agricultural or silvicultural, provide little information on the types of plants that would naturally grow in an area. For example, at the time of this country's settlement, in southern New England white pine was probably only abundant in swamps and moist sandy flats and on exposed ridges due to its susceptibility to fire (Bromley, 1935). In Louisiana and probably elsewhere in the Southeast, many dry flatwoods were originally dominated by longleaf pine, whereas loblolly pine and hardwoods characterized the wetter ones (Hu and Linnartz, 1972). Now loblolly pine and hardwoods or loblolly pine plantations probably occupy the bulk of all flatwoods. Forestry practices and the suppression of forest fires have significantly altered the composition of many, if not most, forests in much of the conterminous U.S. Many species now grow on better drained sites where they probably were not particularly abundant before. Consequently, the present distribution of eastern white pine and loblolly pine may be largely attributed to human intervention.

Areas that are annually tilled and planted with row crops offer only limited information on the current wetness of the site. The success of exotic annual weeds associated with agriculture has further complicated the interpretation of vegetation as indicators of wetland. The effects of past drainage on wetlands once used for agriculture but now abandoned creates a sometimes confusing pattern of vegetation to interpret. Recovery of wetland vegetation may take some time in these cases and the present community may be a mixture of drier site species and a few typical wetland species, and may gradually move toward becoming more typically hydrophytic. Wetland delineators must be particularly mindful of these situations or else risk misjudging a plant species' ecological significance. The 20th century landscape can be a most confounding ecological expression to decipher due to the great impact of urban development, agricultural and grazing practices, and natural resource management.

USING PLANTS TO PREDICT ENVIRONMENTAL CONDITIONS IN WETLANDS

Besides their use in wetland identification, plants also may be used as indicators of other environmental conditions associated with certain types of wetlands. Plants with low tolerances or narrow ranges of occurrence are reliable indicators of conditions such as hydrology, water chemistry, and other situations. Knowing the sensitivity of various plants to these conditions may help with attempts to classify wetlands. While certain species of plants may be universally useful predictors of certain environmental conditions, other species may vary in their distribution locally and may have only limited application for indicating a particular environmental condition. An introduction to how plants can be used in this regard follows.

PLANTS AS HYDROLOGY INDICATORS

Many plants occur within a fairly narrow range of hydrologic conditions. These species can be used to distinguish different water regimes in coastal and inland wetlands (Table 3.7). In some cases, their vigor or robustness may be indicative of certain conditions such as the tall form of smooth cordgrass (*Spartina alterniflora*) for daily tidal inundation and the short form for less flooded areas where salt concentration buildup and soils are more reduced (producing the stunted form). Numerous studies have been conducted on the flood-tolerance of plants (e.g., Broadfoot and Williston, 1973; Theriot, 1993). While such studies did not likely take into account the existence of ecotypes, the studies do provide some useful information on the likelihood of occurrence of certain species in flood-prone areas. Other species may be frequently associated with disturbance and their presence may indicate alterations of hydrology (e.g., common reed in tidally restricted wetlands, cattails in lakes with now stabilized water levels such as Lake Ontario, and black cherry (*Prunus serotina*) in partly drained eastern forested wetlands) (Wilcox, 1995; personal observations).

PLANTS AS INDICATORS OF WATER SOURCES

Goslee et al. (1997) have attempted to use plant species to determine whether a wetland is groundwater-fed or surface water-driven. After examining 379 plots in 28 wetlands in central Pennsylvania and comparing vegetation with hydrologic data, they found some species highly correlated with a particular water source, while others showed no preference. Among the indicators of groundwater were yellow birch, black gum, eastern hemlock, rosebay rhododendron (*Rhododendron maximum*), silky dogwood (*Cornus amomum*), marsh St. John's-wort (*Triadenum virginicum*), goldthread

TABLE 3.7
Examples of Potential Indicator Species for Certain Water Regimes in Different Parts of the U.S.

East	Permanently flooded	*Nymphaea odorata, Brasenia schreberi, Nuphar luteum, Potamogeton* spp., *Ceratophyllum, Najas, Myriophyllum*
	Semipermanently flooded	*Scirpus validus, S. acutus, Sparganium eurycarpum, Pontederia cordata, Sagittaria latifolia, Decodon verticillata, Cephalanthus occidentalis*
	Seasonally flooded	*Saururus cernuus, Leersia oryzoides, Acorus calamus, Carex crinita, Thelypteris thelypteroides, Woodwardia areolata, Polygonum sagittatum, Asclepias incarnata, Eupatorium perfoliatum, Chelone glabra, Senecio aureus, Symplocarpus foetidus, Glyceria striata, Rhododendron viscosum, Ilex verticillata, Itea virginica, Magnolia virginiana, Salix nigra, Chamaecyparis thyoides, Nyssa sylvatica biflora, Populus heterophylla, Mikania scandens*
	Temporarily flooded	*Geum canadense, Cryptotaenia canadensis, Polygonum virginicum, Alliaria petiolata, Leersia virginica, Laportea canadensis, Asimina triloba, Platanus occidentalis, Populus deltoides, Carya cordiformis, Acer negundo, Acer saccharinum, Ilex opaca*
	Saturated	*Sarracenia purpurea, Drosera rotundifolia, Rhynchospora alba, Ledum groenlandicum, Vaccinium macrocarpon, Andromeda glaucophylla, Kalmia polifolia, Empetrum nigrum, Carex exilis*
Interior West	Semipermanently flooded	*Scirpus acutus, S. maritimus, S. validus, S. heterochaetus, S. fluviatilis, Typha latifolia, T. glauca, T. angustifolia*
	Seasonally flooded	*Eleocharis palustris, Scirpus americanus, Sium suave, Alisma* spp., *Beckmannia syzigachne, Carex atherodes, Polygonum coccineum, Sparganium eurycarpum, Alopecurus aequalis, Glyceria grandis, Glyceria striata, Scolochloa festucacea, Sagittaria cuneata*
	Temporarily flooded	*Distichlis spicata, Hordeum jubatum, Poa palustris, Juncus balticus, Spartina pectinata, Calamagrostis inexpansa, C. canadensis, Carex sartwellii, C. lanuginosa, C. praegracilis, Agropyron repens, Echinochloa crusgalli, Polygonum lapathifolium, Aster simplex, Bidens frondosa*
	Saturated	*Calamagrostis inexpansa, Pedicularis groenlandica, Caltha leptosephala*
Alaska	Permanently flooded	*Arctophila fulva, Hippuris vulgaris*
	Semipermanently flooded	*Carex aquatilis, Equisetum fluviatile*
	Seasonally flooded	*Eriophorum angustifolium, Carex sitchensis*
	Temporarily flooded	*Populus balsamifera, Equisetum arvense*
	Saturated	*Betula nana, Picea mariana, Eriophorum vaginatum*
East Coast	Regularly flooded*	*Spartina alterniflora* (tall form), *Amaranthus cannabinus, Nuphar luteum, Pontederia cordata, Peltandra virginica, Scirpus validus, Zizania aquatica, Crassula aquatica, Heteranthera reniformis, Alternanthera philoxeroides, Orontium aquaticum, Isoetes riparia, Sparganium americanum, Limosella subulata, Decodon verticillatus, Ascophyllum nodosum, Fucus vesiculosus, Ulva lactuca*
	Irregularly flooded*	*Spartina patens, S. cynosuroides, S. pectinata, S. alterniflora* (short form), *Distichlis spicata, Juncus gerardii, J. balticus, Iva frutescens, Scirpus maritimus, S. robustus, Solidago sempervirens, Aster tenuifolius, Lythrum lineare, Cladium mariscoides, Hibiscus moscheutos, Kosteletzkya virginica, Rosa palustris, Juncus roemerianus* (in salt marsh)
South Florida	Regularly flooded*	*Spartina alterniflora* (tall form)
	Irregularly flooded*	*Batis maritima, Borrichia frutescens, Salicornia virginica, Conocarpus erectus, Monanthochloe littoralis*

Region	Water Regime	Indicator Species
West Coast	Regularly flooded*	*Spartina foliosa*
	Irregularly flooded*	*Monanthochloe littoralis, Frankensia grandifolia, Distichlis spicata, Cressa truxillensis, Cuscuta salina, Salicornia virginica, Jaumea carnosa, Triglochin maritimum, Limonium californicum*
Alaska	Regularly flooded*	*Triglochin maritimum, Salicornia europaea*
	Irregularly flooded*	*Carex subspathacea, C. ramenskii, Potentilla anserina, Elymus arenarius*

Note: Tidal regimes are marked by an asterisk (*); these species include both estuarine and tidal freshwater plants. These species may generally be used to predict water regimes, especially when common or dominant.

Sources: Water regimes are defined by Cowardin et al., 1979. Other sources include Josselyn, 1983; Zedler and Norby, 1986; Kantrud et al., 1989; Tiner and Burke, 1995; C. Elliott and J. Hall, personal communication; author's observations.

(*Coptis groenlandica*), marsh fern (*Thelypteris palustris*), swamp dewberry (*Rubus hispidus*), and sedges (*Carex emoryi, C. folliculata,* and *C. intumescens*). Species associated with seasonal surface water included many drier site species such as timothy (*Phleum pratense*), multiflora rose (*Rosa multiflora*), stinging nettle (*Urtica dioica*), and dandelion (*Taraxacum officinale*), plus poison ivy (*Toxicodendron radicans*). Skunk cabbage (*Symplocarpus foetidus*) also was listed as a surface water indicator, while elsewhere in the Northeast it seems to be widely associated with groundwater seepage (personal observations). The authors did, however, caution readers that the list of indicators needs to be developed on a regional basis. Such "regions" may actually be more local areas having similar geology and soils. Other studies examining the relationship between vegetation and water sources include Grootjans et al. (1988), Vostokova (1965), and Wassen et al. (1989).

In mineral-poor wetlands that receive input of groundwater, there may be a shift in the plant community at and near the source of the minerotrophic water. At the source, more minerotrophic species may be found, whereas further from the groundwater source, oligotrophic species typical of mineral-poor wetlands predominate. Daniel (1981) described this condition for Carolina bay wetlands and the Great Dismal Swamp (Virginia/North Carolina) where minerotrophic forested wetlands dominate the edges and oligotrophic bog-type species abound in the bay interior.

PLANTS AS WATER CHEMISTRY INDICATORS

Water chemistry also can be predicted by the existence of certain species. It may be possible to interpret the average salinity of a given wetland or the general pH level (acid, neutral, alkaline) by noting the occurrence of particular species or groups of species. It is usually best to use a group of species, since most species have some degree of latitude in tolerance. However, those species with rather narrow ranges can by themselves be useful indicators of water chemistry conditions.

Many wetlands are subjected to salt-stress due to flooding by salt-laden tides in coastal regions and by high evaporation in subhumid, semiarid, and arid regions. In coastal areas, tidal waters with ocean-derived salts, mainly sodium chloride, induce this stress. With increasing mixing with freshwater as one moves upstream in tidally influenced rivers, salt stress is reduced and vegetation composition changes. Tidal wetlands have salinities ranging from above sea strength which is roughly 35 parts per thousand (ppt) of salt to fresh (<0.5 ppt). The U.S. federal wetland classification system has six categories for halinity — hyperhaline (>40 ppt), euhaline (30 to 40 ppt), mixohaline (0.5 to 30 ppt), polyhaline (18 to 30 ppt), mesohaline (5 to 18 ppt), oligohaline (0.5 to 5 ppt), and fresh (Cowardin et al., 1979). Soil halinity is quite dynamic due to tidal fluctuations, freshwater inflows, and rainfall. It seems that species occurrence in an area then may be related to its maximum tolerance with the most salt-tolerant species in the pannes (depressions) of salt marshes and the

TABLE 3.8
Examples of Potential Indicator Species for Halinity Regimes for Tidal Wetlands in the U.S.

Halinity Regime	Indicator Species
Hyperhaline to Euhaline	*Salicornia* spp., *Batis maritima*, *Lycium carolinanum*, *Spergularia marina*, *Monanthochloe littoralis*, plus stunted forms of several euhaline/polyhaline species
Euhaline	*Jaumea carnosa**, *Salicornia virginica**, *Lilaeopsis occidentalis**
Euhaline/Polyhaline	*Spartina alterniflora*, *S. patens*, *Distichlis spicata*, *Iva frutescens*, *Juncus gerardii*, *Suaeda maritima*, *Fucus* spp. (algae), *Sporobolus virginicus*, *Spergularia marina*, *Limonium* spp., *Borrichia frutescens*
Mesohaline	*Carex lyngbyei**, *Potentilla anserina**, *Triglochin maritimum**, *Scirpus americanus**, *Cotula coronopifolia**
Mesohaline to Fresh	*Typha latifolia*, *Phragmites australis*, *Hibiscus moscheutos*, *Scirpus americanus*, *Amaranthus cannabinus*, *Eupatorium serotinum*
Oligohaline	*Hippuris tetraphylla**, *Deschampsia cespitosa**, *Grindelia integrifolia**, *Hordeum brachyantherum**, *Potentilla pacifica**
Oligohaline to Fresh	*Nuphar luteum*, *Pontederia cordata*, *Peltandra virginica*, *Scirpus validus*, *Zizania aquatica*, *Zizaniopsis miliacea*, *Panicum hemitomom*, *Erianthus giganteus*, *Juncus canadensis*, *Cladium jamaicense*, *Rhynchospora macrostachya*, *Sagittaria lancifolia*, *Hymenocallis crassifolia*, *Dichromena colorata*, *Aster novi-belgii*, *Polygonum punctatum*, *Boltonia asteroides*, *Sium suave*, *Myrica cerifera*, *Taxodium distichum*
Fresh	*Acorus calamus*, *Sparganium* spp., *Typha latifolia*, *Juncus effusus*, *Impatiens capensis*, *Polygonum pennsylvanicum*, *Rumex verticillatus*, *Saururus cernuus*, *Ludwigia palustris*, *Aster vimineus*, *Bidens laevis*, *Rosa palustris*, *Cyrilla racemiflora*, *Rhododendron viscosum*, *Cephalanthus occidentalis*, and most woody species (except mangroves and limited halophytic shrubs)

Note: Western species (including Alaska) are marked with an asterisk (*), others are eastern species. Many species exhibit a range of tolerance, yet when several of the indicator species predominate, the area's salinity may be reasonably predicted.

Source: Jon Hall and Dennis Peters, personal communication, 1998; author's observations.

least salt-tolerant along the upland edges of salt marshes and in brackish marshes where more dilution occurs. Some examples of common tidal halophytes that may serve as useful predictors of halinity regimes are listed in Table 3.8. Common reed (*Phragmites australis*) may be an indicator of reduced salinity (degraded salt marshes) in the Northeast. It often supplants typical salt marsh species where tidal flow is severely reduced or eliminated.

In dry climates, high evaporation of groundwater in the capillary fringe results in a buildup of four major cations (calcium, magnesium, sodium, and potassium) and three major anions (bicarbonate, sulfate, and chloride) that affect plant distribution. Salinity for inland areas is measured by specific conductance reported in microsiemens per centimeter at 25°C (µS/cm), microhoms (Mhos), or parts per million (ppm). The official U.S. federal wetland classification system identified six categories: hypersaline (>60,000 µS/cm), eusaline (45,000 to 60,000 µS/cm), polysaline (30,000 to 45,000 µS/cm), mesosaline (8000 to 30,000 µS/cm), oligosaline (800 to 8000 µS/cm), and fresh (<800 µS/cm) (Cowardin et al., 1979; LaBaugh, 1989). In the pothole region of North America, Stewart and Kantrud (1972) and Millar (1976) have described vegetation associated with different degrees of salinity. Like their coastal counterparts, inland saline wetlands experience variable levels of salt stress during the year, but they also endure significant annual changes during the course of the wet–dry hydrologic cycle (Lieffers and Shay, 1983). Mesosaline indicator species may include red saltwort (*Salicornia rubra*), Pursh seepweed (*Suaeda depressa*), three-square (*Scirpus americanus*), Nuttall's alkali grass (*Puccinellia nuttaliana*), seaside arrow-grass (*Triglochin maritimum*),

TABLE 3.9
Some Wetland Plant Indicators of Acid and Circumneutral to Alkaline (Calcareous) Conditions in the Glaciated Northern States

Acidic Indicators

Trees: *Picea mariana*
Shrubs: *Vaccinium vitis-idaea, Gaultheria procumbens, G. hispidula, Ledum groenlandicum, Kalmia polifolia, Vaccinium oxycoccus*
Herbs: *Carex trisperma, C. oligosperma, C. pauciflora, Eriophorum spissum, Smilacina trifolia*
Mosses: *Pleurozium schreberi, Dicranum* sp.

Circumneutral to Alkaline Indicators*

Trees: *Thuja occidentalis*
Shrubs: *Potentilla fruticosa, Rhamnus alnifolia, Salix candida, S. pedicellaris, S. serissima, Lonicera villosa, Rubus pubescens, Aronia melanocarpa*
Herbs: *Triglochin maritimum, Carex viridula, C. cephalantha, C. leptalea, C. interior, C. sterilis, C. pseudo-cyperus, Utricularia intermedia, Drosera intermedia, D. linearis, Cladium mariscoides, Potentilla palustris, Geum rivale, Cypripedium calceolus, Parnassia glauca, Muhlenbergia glomerata, Eriophorum viridi-carinatum, Lysimachia thyrsiflora, Campanula aparinoides, Lobelia kalmii*

* Includes mostly calciphiles (calcium-loving species).

Sources: Glaser, P.H. 1987. Biol. Rep. 85(7.14). U.S. Fish and Wildlife Service, Washington, D.C. Tiner, R.W., Jr. 1988. Field Guide to Nontidal Wetland Identification. Maryland Dept. of Natural Resources, Annapolis, and U.S. Fish and Wildlife Service, Newton Corner, MA.

cordgrass (*Spartina gracilis*), and alkali muhly (*Muhlenbergia asperifolia*) (Kantrud et al., 1989). Oligosaline indicators in the prairies are whitetop (*Schlochloa festucacea*), silverweed (*Potentilla anserina*), and biennial wormwood (*Artemisia biennis*), while freshwater indicators include slender bulrush (*Scirpus heterochaetus*), soft-stemmed bulrush (*Scirpus tabernaemontani*), reed mead-owgrass (*Glyceria maxima*), reed canary grass (*Phalaris arundinacea*), and short-awn foxtail (*Alopecurus aequalis*). Some salinity indicator species for western wetlands also are associated with particular water regimes and therefore may be useful predictors of both salinity and water regime (see Kantrud et al., 1989, for a more complete listing of species and their maximum, minimum, and mean specific conductance tolerances).

Another important set of water chemistry variables is pH or hydrogen ion concentration. It is used to separate acidic (pH <5.5) from circumneutral (pH 5.5 to 7.4) from alkaline (pH>7.4) wetlands. Many plants may be sensitive to differences in pH and therefore may be useful for recognizing calcareous conditions of minerotrophic fens or nutrient-poor acidic wetlands (Table 3.9). In studying the vegetation of coastal wetlands along the Great Lakes, Minc and Albert (1998) used various calciphiles to distinguish the northern rich fen type from other wetlands. Key species included yellow sedge (*Carex viridula*) and Kalm's lobelia (*Lobelia kalmii*), plus some others listed in Table 3.9. It is interesting that some calciphiles are typical halophytes, characteristic of northeastern salt and brackish marshes, such as twig-rush (*Cladium mariscoides*), silverweed (*Potentilla anserina*), and seaside arrow-grass (*Triglochin maritimum*). Most ericaceous shrubs are indicators of nutrient-poor conditions. They are well-represented in bogs and pocosins which are low-nutrient, acidic wetlands in the northern and southern regions, respectively. In his study of peatlands in Saskatchewan, Jeglum (1971) found good correlations between vegetation and pH and water levels. He also concluded that groups of species were the most reliable predictors of pH and water levels. Glaser (1992) summarized information on vegetation and water chemistry for Minnesota peatlands. Lagg (moat) vegetation found around many northern bogs is characterized by minerotrophic species like speckled alder and northern white cedar providing evidence of mineral-rich water at the margins in stark contrast to the ericaeous shrubs (e.g., leatherleaf, labrador tea, and sheep laurel) of the nutrient-poor bog interior. The presence of cattails along the margins of some bogs at road and railroad crossings, for example, provide evidence of higher nutrients, perhaps introductions from road runoff.

Changes in water chemistry may be detected by the arrival of new species in watersheds. The best examples of this can be found where naturally acidic, nutrient-poor streams are subjected to nutrient loading from expanding development or agriculture. Significant changes in understory species of Atlantic white cedar swamps in suburbanizing watersheds were reported in the New Jersey Pine Barrens (Ehrenfeld, 1983; Ehrenfeld and Schneider, 1991; 1993). Weedy, cosmopolitan species were replacing native acidic, nutrient-poor indicators such as tussock sedge, round-leaved sundew (*Drosera rotundifolia*), golden-club (*Orontium aquaticum*), bog aster (*Aster nemoralis*), swamp pink (*Helonias bullata*), cotton-grass (*Eriophorum virginicum*), and bladderworts (*Utricularia* spp.). For this area, Zampella and Laidig (1997) identified numerous disturbance indicator species, with water starwort (*Callitriche heterophylla*), lurid sedge (*Carex lurida*), dye bedstraw (*Galium tinctorium*), jewelweed (*Impatiens capensis*), water purslane (*Ludwigia palustris*), climbing hempweed (*Mikania scandens*), arrow-leaved tearthumb (*Polygonum sagittatum*), and Nepal microstegium (*Eulalia viminea*) being among the more common. The latter is an invasive exotic species that has become a dominant herb in many forested wetlands in New Jersey. Disturbance in other naturally acidic areas may be detected by the presence of minerotrophic species.

PLANTS AS SOIL TYPE INDICATORS

Gordon (1940) used certain plant communities as indicators of forest soils in western New York. For wetland soils, he stated, "The presence of yellow birch, ironwood (blue beech), and American elm in a hemlock forest generally indicates poor internal drainage conditions… On lacustrine soils of fine texture and imperfect drainage, American elm, swamp white oak, and beech are favored… Associated dark soils are indicated by the White Pine-American Elm type with an undergrowth of hemlock, yellow birch, black ash, red maple and balsam fir." In local areas there may be strong relationships between certain species and soil types. Wetland plants commonly associated with sandy soils include golden-pert (*Gratiola aurea*) in freshwater areas and salt marsh sand spurrey (*Spergularia marina*), silverhead (*Philoxerus vermicularis*), sea purslane (*Sesuvium portulacastrum*), and common frog-fruit (*Phyla nodiflora*) in estuarine wetlands. Bigler and Richardson (1984) found that the presence of bulrush (*Scirpus* sp.) could be used to predict soil properties in the deep marsh zone of prairie potholes. In the field, many soil scientists use plant communities coupled with landscape position to help prepare soil maps after analyzing soils in selected areas. Russian scientists have developed an entire field of science called "geobotany" based on using plants to indicate a variety of soil and geologic conditions (Viktorov et al., 1965, as reported in Goslee et al., 1997).

As discussed previously, plants have traditionally been used to identify wetlands and many species and certain plant communities are highly reliable indicators of wet soil conditions. Numerous studies have shown an excellent correlation between vegetation and hydric soils (e.g., Scott et al., 1989; Segelquist et al., 1990; Veneman and Tiner, 1990). The OBL and FACW species on the national wetland plant list (Reed, 1988; 1997) remain useful indicators of hydric soils as sites dominated by these species are typically wetlands, whereas a predominance of UPL and FACU species typically indicates nonhydric soils. Moreover, certain species (calciphiles) favoring mineral-rich sites may be used to indicate calcareous or alkaline soils, while acidophylic species can be used to predict likely acid soil conditions noted above.

PLANTS AS INDICATORS OF NATIVENESS

Since European settlement, a tremendous influx of new species have become part of the U.S. flora. Many species were purposefully brought in for agricultural, forestry, or horticultural purposes, while most were simply unintentional migrants. Today, there is increasing concern about conserving and restoring native flora (autochthonous) and plant communities in wetlands and other habitats (e.g., native prairie). In determining priority areas for preservation and for assessing the need for restoration, the presence and predominance of naturalized post-settlement species (allochthonous) has become important.

Wilhelm and Ladd (1988) developed an index called *coefficient of conservatism* based on a given species' habitat specificity. The coefficients are only assigned to native species. Thirteen categories are given: 0 to 10, plus 15 and 20. A species with a 0-rating is ubiquitous over a wide range of disturbances. The 5-value is assigned to species with a pronounced affinity for a certain type of native community. A 10-rating is given to plants that typify stable conditions and that have very specific environmental requirements. A 15-species is like a 10, but is rare, while a 20-species is locally threatened or endangered. The rating system may be useful for wetland evaluations interested maintaining biodiversity of native species. Some examples of species in the Chicago, IL area with different coefficients follow: 0 — *Acer negundo, Ambrosia trifida, Circea quadrisulcata, Echinochloa crusgalli*, and *Equisetum arvense*; 1 — *Bidens frondosa, Calystegia sepium*, and *Cyperus esculentus*; 2 — *Boehmeria cylindrica, Carex vulpinoidea*, and *Claytonia virginica*; 3 — *Aster simplex, Calamagrostis canadensis, Carex tribuloides, Celtis occidentalis*, and *Equisetum hymale intermedium*; 4 — *Alisma trivale, Amphicarpa bracteata, Asclepias incarnata, Aster lateriflorus, Carex lanuginosa*, and *Elymus canadensis*; 5 — *Angelica purpurea, Arisaema atrorubens, Bidens comosa, Caltha palustris, Carex stricta, Cinna arundinacea, Echinochloa walteri, Eleocharis palustris, Elodea canadensis*, and *Elymus virginicus*; 6 — *Alopecurus aequalis, Aster puniceus, Betula nigra*, and *Cornus stolonifera*; 7 — *Acorus calamus, Campanula aparinoides, Carya cordiformis, Cephalanthus occidentalis*, and *Equisetum fluviatile*; 8 — *Alnus rugosa, Bidens connata, Carex lasiocarpa, Carpinus caroliniana, Chelone glabra*, and *Epilobium leptophyllum*; 9 — *Amelanchier humilis*; 10 — *Aster umbellatus, Carex lacustris, Carex rostrata, Carex sartwellii*, and *Dryopteris cristata*; 15 — *Chamaedaphne calyculata, Deschampsia caespitosa, Drosera intermedia*, and *Thelypteris noveboracensis*; 20 — *Cypripedium calceolus* (Wilhelm and Ladd, 1988). Coefficients of conservatism for other areas may be available from state natural heritage programs.

A *natural area index* that reflects the "nativeness" of a particular area can be determined by applying these coefficients to a list of observed species (including the immediate buffer and ecotones). The following equation is used:

$$NI = R/N \times \sqrt{N}$$

where NI = the natural area index, R = the sum of coefficients of conservatism, and N = the number of native species recorded. Following this approach, a single survey during the growing season can usually provide an accurate assessment of site nativeness.

USEFUL GUIDES FOR IDENTIFYING WETLAND PLANTS

Following is a list of some wetland plant manuals and field guides that are useful for identifying these plants across the U.S. The list is arranged by region of the country. Other regional field guides to wildflowers, trees, etc. are not listed but also will be valuable references as will regional flora (taxonomic manuals); they are listed in Appendix A of the interagency wetland delineation manual (Federal Interagency Committee for Wetland Delineation, 1989).

NORTHEAST

Field Guide to Coastal Wetland Plants of the Northeastern United States, by R.W. Tiner (1987), University of Massachusetts Press, Amherst.

Field Guide to Nontidal Wetland Identification, by R.W. Tiner (1988), Maryland Dept. Of Natural Resources, Annapolis, MD and U.S. Fish and Wildlife Service, Newton Corner, MA. (Reprint available from Institute of Wetlands & Environmental Education & Research, P.O. Box 288, Leverett, MA 01054)

Freshwater Wetlands: A Guide to Indicator Plants of the Northeast, by D.W. Magee (1981), University of Massachusetts Press, Amherst.

In Search of Swampland: A Wetland Sourcebook and Field Guide, by R. W. Tiner (1998), Rutgers University Press, New Brunswick, NJ.

SOUTHEAST

Aquatic and Wetland Plants of the Southeastern United States, by R.K. Godfrey and J.W. Wooten (1979/81), Monocotydons and Dicotyledons (2 volumes), University of Georgia Press, Athens.

Identification Manual for Wetland Plant Species of Florida, by R.L. Dressler, D.W. Hall, K.D. Perkins, and N.H. Williams (1987), University of Florida, Institute of Food and Agricultural Sciences, Gainesville.

Field Guide to Coastal Wetland Plants of the Southeastern United States, by R.W. Tiner (1995), University of Massachusetts Press, Amherst.

A Manual of Marsh and Aquatic Vascular Plants of North Carolina With Habitat Data, by E.O. Beal (1977), North Carolina Agricultural Experiment Station, Raleigh, NC.

Wetland Plants of the New Orleans District, by U.S. Army Corps of Engineers (1977), New Orleans District, New Orleans, LA.

A Guide to Selected Florida Wetland Plants and Communities, by U.S. Army Corps of Engineers (1988), Jacksonville District, Jacksonville, FL.

Aquatic and Wetland Plants of Missouri, by D.L. Combs and R.D. Drobney (undated), Missouri Cooperative Fish and Wildlife Research Unit, University of Missouri, Columbia.

Florida Wetland Plants: An Identification Manual, by J.D. Tobe et al. (1998), Florida Department of Environmental Protection, Tallahassee.

MIDWEST

A Manual of Aquatic Plants, by N.C. Fassett (1966), University of Wisconsin Press, Madison.

Wetland Plants and Plant Communities of Minnesota and Wisconsin, by S.D. Eggers and D.M. Reed (1997), U.S. Army Corps of Engineers, St. Paul District, St. Paul, MN.

The Aquatic and Wetland Vascular Plants of North Dakota, by G.E. Larson and W.T. Barker (1980), North Dakota Water Resources Research Institute, North Dakota State University, Fargo.

A Field Guide to the Wetlands of Illinois, by the Illinois Department of Conservation (1988), State of Illinois, Champaign.

Great Lakes Coastal Plants, by W.J. Hoagman (1994), Michigan State University, Tawas City.

INTERIOR WEST

A Handbook of Wetland Plants of the Rocky Mountain Region, by D.J. Cooper (1989), U.S. Environmental Protection Agency, Region VIII, Denver, CO.

Midwestern Wetland Flora-Field Office Guide to Plant Species, by R.H. Mohlenbrock (1991), U.S. Department of Agriculture, Natural Resources Conservation Service, Lincoln, NE.

SOUTHWEST

Aquatic and Wetland Plants of the Southwestern United States, by D.S. Correll and H.B. Correll (1975), Vol. I & II, Stanford University Press, Stanford, CA.

FAR WEST AND PACIFIC NORTHWEST

Common Wetland Plants of Coastal California, by P.M. Faber (1982), Pickleweed Press, Mill Valley, CA.

Lewis Clark's Field Guide to Wildflowers of Marshes and Waterways in the Pacific Northwest, by L.J. Clark (1974), Gray's Publishing, Ltd., Sidney, BC.

Wetland Plants of the Pacific Northwest, by F. Weinmann, M. Boule, K. Brunner, J. Malek, and V. Yoshino (1984), U.S. Army Corps of Engineers, Seattle District, Seattle, WA.

Flora of the Marshes of California, by H.L. Mason (1957), University of California Press, Berkeley, CA.

Wetland Plants of King County and the Puget Sound Lowlands, by V. Crawford (1981), King County, WA.

HAWAII

Wetlands and Wetland Vegetation of Hawaii, by M.E. Elliott and E.M. Hall (1977), U.S. Army Corps of Engineers, Pacific Ocean Division, Ft. Shafter, HI.

ALASKA

The Common Plants of the Muskegs of Southeast Alaska, by O.W. Robuck (1985), USDA Forest Service, Pacific Northwest Forest and Range Experiment Station, Portland, OR.

Alaska Trees and Shrubs, by L.A. Vierek and E.L. Little, Jr. (1972), USDA Forest Service, Washington, D.C. Agricultural Handbook No. 410.

REFERENCES

Adams, D.A. 1963. Factors influencing vascular plant zonation in North Carolina salt marshes. *Ecology,* 44: 445-456.

Albert, R. 1975. Salt regulation in halophytes. *Oecologia,* 21: 57-71.

Anderson, K.L. 1942. A comparison of line transects and permanent quadrats in evaluating composition and density of pasture vegetation on the tall prairie grass type. *J. Amer. Soc. Agron.,* 34: 805-822.

Angeles, G., R.F. Evert, and T.T. Kozlowski. 1986. Development of lenticels and adventitious roots in flooded *Ulmus americana* seedlings. *Can. J. For. Res.,* 16: 585-590.

Arber, A. 1920. *Water Plants: A Study of Aquatic Angiosperms.* University Press, Cambridge, England.

Armstrong, W. 1979. Aeration in higher plants. *Adv. Bot. Res.,* 7: 225-332.

Armstrong, W. 1968. Oxygen diffusion from the roots of woody species. *Physiol. Plantarum,* 21: 539-543.

Armstrong, W. 1964. Oxygen diffusion from the roots of some British bog plants. *Nature,* 204: 801-802.

Armstrong, W. and D.J. Boatman. 1967. Some field observations relating the growth of bog plants to conditions of soil aeration. *J. Ecol.,* 55: 101-110.

Armstrong, W., E.J. Wright, S. Lythe, and T.J. Gaynard. 1985. Plant zonation and the effects of the spring neap tidal cycle on soil aeration in a Humber salt marsh. *J. Ecol.,* 73: 323-339.

Atkinson, R.B., J.E. Perry, E. Smith, and J. Cairns, Jr. 1993. Use of created wetland delineation and weighted averages as a component of assessment. *Wetlands,* 13: 185-193.

Bacha, R.E. and L.R. Hossner. 1977. Characteristics of coatings formed on rice roots as affected by iron and manganese additions. *Soil Sci. Soc. Am. J.,* 41: 931-935.

Barbour, M.G., J.H. Burk, and W.K. Pitts. 1980. *Terrestrial Plant Ecology.* Benjamin/Cummings Publishing Co., Menlo Park, CA.

Bartlett, R.J. 1961. Iron oxidation proximate to plant roots. *Soil Sci.,* 92: 372-379.

Beaven, G.F. and H.J. Oosting. 1939. Pocomoke Swamp: a study of a cypress swamp on the Eastern Shore of Maryland. *Bull. Torrey Bot. Club,* 66: 367-389.

Bergman, H.F. 1920. The relation of aeration to the growth and activity of roots and its influence on the ecesis of plants in swamps. *Ann. Bot.,* 34: 13-33.

Bigler, R.J. and J.L. Richardson. 1984. Classification of soils in prairie pothole wetlands with deep marsh plant species in North Dakota. *Soil Surv. Hor.,* 25: 16-24.

Billings, W.D. and P.J. Godfrey. 1967. Photosynthetic utilization of internal carbon dioxide by hollow-stemmed plants. *Science,* 158: 121-123.

Blits, K.C. and J.L. Gallagher. 1991. Morphological and physiological responses to increased salinity in marsh and dune ecotypes of *Sporobolus virginicus* (L.) Kunth. *Oecologia*, 87: 330-335.

Blom, C.W.P.M. 1990. Responses to flooding in weeds from river areas. In *Biological Approaches and Evolutionary Trends in Plants*. S. Kawano (Ed.). Academic Press, London. 81-94.

Blom, C.W.P.M., H.M. van de Steeg, and L.A.C.J. Voesenek. 1996. Adaptive mechanisms of plants occurring in wetland gradients, chap. 7. In *Wetlands: Environmental Gradients, Boundaries, and Buffers*. G. Mulamoottil, B.G. Warner, and E.A. McBean (Eds.). Lewis Publishers, Boca Raton, FL. 91-112.

Bonneau, M. 1982. Soil temperature. In *Constituents and Properties of Soils*. M. Bonneau and B. Souchier (Eds.). Academic Press, New York. 366-371.

Bonner, F.T. 1966. Survival and First-Year Growth of Hardwoods Planted in Saturated Soils. U.S. Forest Service, Southern Forestry Expt. Station. Research Note SO-32.

Braun-Blanquet, J. 1932. *Plant Sociology. The Study of Plant Communities*. McGraw-Hill, New York.

Bray, W.L. 1930. *The Development of the Vegetation of New York State*. Tech. Pub. No. 29. New York State College of Forestry, Syracuse University, Syracuse, NY.

Broadfoot, W.M. and H.L. Williston. 1973. Flooding effects on southern forests. *J. For.*, 71: 584-587.

Bromley, S.W. 1935. The original forest types of southern New England. *Ecol. Monog.*, 5(1): 61-89.

Brown, S. and A.E. Lugo. 1982. A comparison of structural and functional characteristics of saltwater and freshwater forested wetlands. In *Wetlands Ecology and Management*. B. Gopal, R.E. Turner, R.G. Wetzel and D.F. Whigham (Eds.). National Institute of Ecology and International Science Publications, Jaipur, India. 109-130.

Browne, S., S. Crocoll, D. Goetke, N. Heaslip, T. Kerpez, K. Koget, S. Sanford, and D. Spada. 1996. Freshwater Wetlands Delineation Manual. New York State Dept. Of Environmental Conservation, Division of Fish and Game, Albany, NY.

Burdick, D.M. 1989. Root aerenchyma development in *Spartina patens* in response to flooding. *Am. J. Bot.*, 76: 777-780.

Burdick, D.M. and I.A. Mendelssohn. 1990. Relationship between anatomical and metabolic responses to waterlogging in the coastal grass *Spartina patens*. *J. Exp. Bot.*, 41: 233-238.

Cain, D.J. and H.T. Harvey. 1983. Evidence of salinity-induced ecophenic variation in cordgrass (*Spartina foliosa* Trin.). *Madroño*, 30: 50-62.

Chen, C.C., J.B. Dixon, and F.T. Turner. 1980. Iron coatings on rice roots: morphology and models of development. *J. Soil Sci. Soc. Am.*, 44: 1113-1119.

Chirkova, T.V. 1968. Features of the O_2 supply of roots of certain woody plants in anaerobic conditions. *Fiziol. Rast.* (Moscow), 15: 565-568 (Russian).

Clausen, J. and W.M. Hiesey. 1958. Experimental Studies on the Nature of Species. IV. *Genetic Structure of Ecological Races*. Carnegie Institution of Washington, Washington, D.C.

Clausen, J., D.D. Keck, and W.M. Hiesey. 1941. Regional differentiation in plant species. *Amer. Nat.*, LXXV: 231-250.

Clements, F.E. 1920. *Plant Indicators. The Relation of Plant Communities to Process and Practice*. Carnegie Institution of Washington, Washington, D.C.

Clements, F.E., E.V. Martin, and F.L. Long. 1950. *Adaptation and Origin in the Plant World. The Role of Environment in Evolution*. The Chronica Botanica Co., Waltham, MA.

Cochran, P.H. 1972. Tolerance of lodgepole and ponderosa pine seeds and seedlings to high water tables. *Northwest Sci.*, 46: 322-331.

Compton, R.H. 1916. The botanical result of a fenland flood. *J. Ecol.*, 4: 15-17.

Costello, D.F. 1936. Tussock meadows in southeastern Wisconsin. *Bot. Gaz.*, 97: 610-648.

Coutts, M.P. and J.J. Philipson. 1978a. Tolerance of tree roots to waterlogging. I. Survival of Sitka spruce and lodgepole pine in waterlogged soil. *New Phytol.*, 80: 63-69.

Coutts, M.P. and J.J. Philipson. 1978b. Tolerance of tree roots to waterlogging. II. Adaptation of Sitka spruce and lodgepole pine in waterlogged soil. *New Phytol.*, 80: 71-77.

Cowardin, L.M., V. Carter, F.C. Golet, and E.T. LaRoe. 1979. Classification of Wetlands and Deepwater Habitats of the United States. FWS/OBS-79/31. U.S. Fish and Wildlife Service, Washington, D.C.

Crawford, R.M.M. 1982. The anaerobic retreat as a survival strategy for aerobic plants and animals. *Trans. Bot. Soc., Edinburgh*, 44: 57-63.

Crawford, R.M.M. 1983. Root survival in flooded soils. In *Ecosystems of the World. 4A. Mires: Swamp, Bog, Fen, and Moor.* General Studies. A.J.P. Gore (Ed.). Elsevier Scientific Publishing, Amsterdam. 257-283.

Crawford, R.M.M. 1987. *Plant Life in Aquatic and Ampibious Habitats.* Blackwell Scientific, Oxford, England.

Crawford, R.M.M. 1992. Oxygen availability as an ecological limit to plant distribution. *Adv. Ecolog. Res.,* 23: 93-185.

Crawford, R.M.M., and P.D. Tyler. 1969. Organic acid metabolism in relation to flooding tolerance in roots. *J. Ecol.,* 57: 235-244.

Cronquist, A. 1968. *The Evolution and Classification of Flowering Plants.* Nelson Publishing Co., London.

Crum, H. 1988. *A Focus on Peatlands and Peat Mosses.* University of Michigan Press, Ann Arbor.

Curtis, J. T. 1959. *The Vegetation of Wisconsin.* University of Wisconsin Press, Madison.

Dacey, J.W.H. 1979. Gas circulation through the yellow waterlily. Ph. D. dissertation. Michigan State University, East Lansing.

Dacey, J.W.H. 1981. Pressure ventilation in the yellow waterlily. *Ecology,* 62: 1137-1147.

Daniel, C.C., III. 1981. Hydrology, geology, and soils of pocosins: a comparison of natural and altered systems. In *Pocosin Wetlands.* C.J. Richardson (Ed.). Hutchinson Ross Publishing, Stroudsburg, PA. 69-108.

Darlington, C.D. 1973. *Chromosome Botany and the Origins of Cultivated Plants.* George Allen & Unwin, Ltd., London.

Darrah, W.C. 1960. *Principles of Paleobotany.* Ronald Press, New York.

Das, D.K. and R.L. Jat. 1977. Influence of three soil–water regimes on root porosity and growth of four rice varieties. *Agron. J.,* 69: 197-200.

Daubenmire, R.F. 1947. *Plants and Environment. A Textbook of Plant Autecology.* John Wiley & Sons, Inc., New York.

Daubenmire, R.F. 1968. *Plant Communities: A Textbook of Plant Synecology.* Harper and Row, New York.

Davy, A.J., S.M. Noble, and R.P. Oliver. 1990. Genetic variation and adaptation to flooding in plants. *Aqua. Bot.,* 38: 91-108.

DeLaune, R.D. and S.R. Pezeshki. 1991. Role of soil chemistry in vegetative ecology of wetlands. *Trends in Soil Sci.,* 1: 101-112.

Dionigi, C.P., I.A. Mendelssohn, and V.I. Sullivan. 1985. Effects of soil waterlogging on the energy status and distribution of *Salix nigra* and *S. exigua* (Salicaceae) in the Atchafalaya River Basin of Louisiana. *Amer. J. Bot.,* 72: 109-119.

Drew, M.C., M.B. Jackson, and S. Gifford. 1979. Ethylene-promoted adventitious rooting and development of cortical air spaces (aerenchyma) in roots may adaptive responses to flooding in *Zea mays* L. *Planta,* 147: 83-88.

Ehrenfeld, J.G. 1983. The effects of changes in land-use on swamps of the New Jersey Pine Barrens. *Biol. Conser.,* 25: 353-375.

Ehrenfeld, J.G. and J.P. Schneider. 1991. *Chamaecyparis thyoides* wetlands and suburbanization: effects of nonpoint source water pollution on hydrology and plant community structure. *J. Appl. Ecol.,* 28: 467-490.

Ehrenfeld, J.G. and J.P. Schneider. 1993. Responses of forested wetland vegetation to perturbations of water chemistry and hydrology. *Wetlands,* 13: 122-129.

Eleutrius, L.N. 1989. Natural selection and genetic adaptation to hypersalinity in *Juncus roemerianus* Scheele. *Aqua. Bot.,* 36: 45-53.

Emerson, F.W. 1921. Subterranean organs of bog plants. *Bot. Gaz.,* 72: 359-374.

Environmental Laboratory. 1987. Corps of Engineers Wetlands Delineation Manual. Tech. Rep. Y-87-1. U.S. Army Engineer Waterways Experiment Station, Vicksburg, MS.

Ernst, W.H.O. 1990. Ecophysiology of plants in waterlogged and flooded environments. *Aqua. Bot.,* 38: 73-90.

Evans, D.D. and A.D. Scott. 1955. A polarographic method of measuring dissolved oxygen in saturated soil. *Proc. Soil Sci. Soc. Amer.,* 19: 12-16.

Faulkner, S.P. and W.H. Patrick, Jr. 1992. Redox processes and diagnostic wetland soil indicators in bottomland hardwood forests. *Soil Sci. Soc. Am. J.,* 56: 856-865.

Federal Interagency Committee for Wetland Delineation. 1989. Federal Manual for Identifying and Delineating Jurisdictional Wetlands. Cooperative technical publication. U.S. Army Corps of Engineers, U.S. Environmental Protection Agency, U.S. Fish and Wildlife Service, and USDA Soil Conservation Service, Washington, D.C.

Fernald, M.L. 1950. *Gray's Manual of Botany*. American Book Company, New York.

Fisher, H.M. and E.L. Stone. 1990. Air-conducting porosity in slash pine roots from saturated soils. *For. Sci.,* 36: 18-33.

Fisher, H.M. and E.L. Stone. 1991. Iron oxidation at the surfaces of slash pine roots from saturated soils. *Soil Sci. Soc. Am. J.,* 55: 1123-1129.

Florida Department of Environmental Protection, South Florida Water Management District, St. John's River Water Management District, Suwannee River Water Management District, Southwest Florida Water Management District, and Northwest Florida Water Management District. 1995. The Florida Wetlands Delineation Manual. Florida Dept. Of Environmental Protection, Wetland Evaluation and Delineation Section, Tallahassee.

Fowells, H.A. 1965. Silvics of Forest Trees of the United States. Agric. Handbook No. 271. U.S. Department of Agriculture, Washington, D.C.

Gardiner, E.S. 1994. Physiological responses of four bottomland oak species to root hypoxia. Ph. D. Dissertation. Mississippi State University, Mississippi State.

Gardiner, E.S. and J.D. Hodges. 1993. Physiological, morphological and growth responses to rhizosphere hypoxia by seedlings of North American bottomland oaks. *Ann. Sci. For.,* 53: 303-316.

Gessner, F. 1955. *Hydrobotanik: die Physiologischen Grundlaen der Pflanzenverbreitung im Wasser. I. Energiehaushalt.* VEB Deutscher Werlag der Wissenshaften, Berlin.

Gessner, F. 1959. *Hydrobotanik: die Physiologischen Grundlaen der Pflanzenverbreitung im Wasser. II. Stoffhaushalt.* VEB Deutscher Werlag der Wissenshaften, Berlin.

Gill, C.J. 1970. The flooding tolerance of woody species — a review. *For. Abstr.,* 31: 671-688.

Gill, C.J. 1975. The ecological significance of adventitious rooting as a response to flooding in woody species, with special reference to *Alnus glutinosa* (L.) Gaertn. *Flora. Bd.,* 164: 85-97.

Glaser, P.H. 1987. The Ecology of Patterned Boreal Peatlands in Northern Minnesota: A Community Profile. Biol. Rep. 85(7.14). U.S. Fish and Wildlife Service, Washington, D.C.

Glaser, P.H. 1992. Vegetation and water chemistry. In *The Patterned Peatlands of Minnesota.* H.E. Wright, Jr., B.A. Coffin, and N.E. Aaseng (Eds.). University of Minnesota Press, Minneapolis. 15-26.

Gleason, H.A. and A. Cronquist. 1963. *Manual of Vascular Plants of Northeastern United States and Adjacent Canada.* D. Van Nostrand Co., New York.

Godfrey, R.K. and J.W. Wooten. 1979. *Aquatic and Wetland Plants of the Southeastern United States. Monocotyledons.* The University of Georgia Press, Athens.

Good, B.J. and W.H. Patrick, Jr. 1987. Gas composition and respiration of water oak (*Quercus nigra* L.) and green ash (*Fraxinus pennsylvanica* Marsh.) roots after prolonged flooding. *Plant and Soil,* 97: 419-427.

Good, B.J., S.P. Faulkner, and W.H. Patrick, Jr. 1986. Evaluation of green ash root responses as a soil wetness indicator. *Soil Sci. Soc. Am. J.,* 50: 1570-1575.

Gordon, R.B. 1940. *The Primeval Forest Types of Southwestern New York.* New York State Museum Bull. 321. The University of the State of New York, Albany.

Goslee, S.C., R.P. Brooks, and C.A. Cole. 1997. Plants as indicators of wetland water source. *Plant Ecol.,* 131: 199-206.

Grable, A.R. 1966. Soil aeration and plant growth. *Adv. Agron.,* 18: 57-106.

Greenwood, D.J. 1961. The effect of oxygen concentration on the decomposition of organic materials in soil. *Plant Soil,* 14: 360-376.

Griffin, R.W., L.P. Wilding, W.L. Miller, G.W. Crenwelge, R.J. Tucker, L.R. Drees, and W.C. Lynn. 1996. Preliminary investigations of hydric soil hydrology and morphology on the Texas Gulf Coast Prairie. In *Preliminary Investigations of Hydric Soil Hydrology and Morphology in the United States.* J.S. Wakely, S.W. Sprecher, and W.C. Lynn (Eds.). Tech. Rep. WRP-DE-13. U.S. Army Engineer Waterways Experiment Station, Vicksburg, MS. 9-30.

Grootjans, A.P., R. van Diggelen, M.J. Wassen, and W.A. Wiersirga. 1988. The effects of drainage on groundwater quality and plant species distribution in stream valley meadows. *Vegetatio,* 75: 37-48.

Hahn, G.G., C. Hartley, and A.S. Rhoads. 1920. Hypertrophied lenticels on the roots of conifers and their relation to moisture and aeration. *J. Agric. Research* (Washington, D.C.), XX (4): 253-265.

Hall, T.F., W.T. Penfound, and A.D. Hess. 1946. Water level relationships of plants in the Tennessee Valley with particular reference to malaria control. *J. Tenn. Acad. Sci.,* 21: 18-59.

Hannon, N. and A.D. Bradshaw. 1968. Evolution of salt tolerance in two co-existing species of grass. *Nature,* 220: 1342-1343.

Harms, W.R. 1973. Some effects of soil type and water regime on growth of tupelo seedlings. *Ecology,* 54: 188-193.

Harms, W. R., H.T. Schreuder, D.D. Hook, C.L. Brown, and F.W. Shropshire. 1980. The effects of flooding on the swamp forest in Lake Ocklawaha, FL. *Ecology,* 61: 1412-1421.

Hester, M.W., I.A. Mendelsohn, and K.L. McKee. 1996. Intraspecific variation in salt tolerance and morphology in the coastal grass *Spartina patens* (Poaceae). *Amer. J. Bot.,* 83: 1521-1527.

Hook, D.D. 1984a. Adaptations to flooding with fresh water, chap. 8. In *Flooding and Plant Growth.* T.T. Kozlowski (Ed.). Academic Press, Orlando, FL. 265-294.

Hook, D.D. 1984b. Waterlogging tolerance of lowland tree species of the south. *South J. Appl. For.,* 8: 136-149.

Hook, D.D. and C.L. Brown. 1973. Root adaptations and relative flood tolerance of five hardwood species. *For. Sci.,* 19: 225-229.

Hook, D.D., C.L. Brown, and P.P. Kormanik. 1971. Inductive flood tolerance in swamp tupelo [*Nyssa sylvatica var. biflora* (Walt.) Sarg.]. *J. Exp. Bot.,* 22: 78-89.

Hook, D.D. and R.M.M. Crawford, (Eds.). 1978. *Plant Life in Anaerobic Environments.* Ann Arbor Sci. Publ., Ann Arbor, MI.

Hook, D.D., O.G. Langdon, J. Stubbs, and C.L. Brown. 1970. Effect of water regimes on the survival, growth, and morphology of tupelo seedlings. *For. Sci.,* 16: 304-311.

Hook, D.D. and M.R. McKevlin. 1988. Use of oxygen microelectrodes to measure aeration in the roots of intact tree seedlings. In *The Ecology and Management of Wetlands,* vol. 1. Ecology of Wetlands. D.D. Hook et al. (Eds.). Timber Press, Portland, OR. 467-476.

Hook, D.D. and J.R. Scholtens. 1978. Adaptations and flood tolerance of tree species. In *Plant Life in Anaerobic Environments.* D.D. Hook and R.M.M. Crawford (Eds.). Ann Arbor Sci. Publ., Ann Arbor, MI. 229-331.

Hook, D.D., W.H. McKee, Jr., H.K. Smith, J. Gregory, V.G. Burrell, Jr., M.R. DeVoe, R.E. Sojka, S. Gilbert, R. Banks, L.H. Stolzy, C. Brooks, T.D. Matthews, and T. H. Shear (Eds.). 1988. *The Ecology and Management of Wetlands,* vol. 1. Ecology of Wetlands. Timber Press, Portland, OR.

Hosner, J.F. 1958. The effects of complete inundation upon seedlings of six bottomland tree species. *Ecology,* 39: 371-373.

Hosner, J.F. 1959. Survival, root and shoot growth of six bottomland tree species following flooding. *J. For.,* 59: 927-928.

Hosner, J.F. 1960. Relative tolerance to complete inundation of fourteen bottomland tree species. *For. Sci.,* 6: 246-251.

Hu, Shih-Chang and N.E. Linnartz. 1972. *Variations in Oxygen Content of Forest Soils under Mature Loblolly Pine Stands.* Bulletin No. 668. Louisiana State University, Agricultural Experiment Station, Baton Rouge.

Huenneke, L.F. 1982. Wetland forests of Tompkins County, New York. *Bull. Torrey Bot. Club,* 109: 51-63.

Hunt, F.M. 1951. Effect of flooded soil on growth of pine seedlings. *Plant Physiol.,* 26: 363-368.

Husch, B. and W.H. Lyford. 1956. *White Pine Growth and Soil Relationships in Southeastern New Hampshire.* Tech. Bull. No. 95. Agric. Expt. Station, University of New Hampshire, Durham.

Hutchinson, G.E. 1975. *A Treatise on Limnology,* vol. III. Limnological Botany. John Wiley & Sons, New York.

Illick, J.S. and J.E. Aughanbaugh. 1930. Pitch Pine in Pennsylvania. Res. Bull. No. 2. Pennsylvania Dept. of Forests and Waters, Harrisburg.

Ingrouille, M. 1992. *Diversity and Evolution of Land Plants.* Chapman and Hall, London.

Iremonger, S.F. and D.L. Kelly. 1988. The responses of four Irish wetland tree species to raised soil water levels. *New Phytol.,* 109: 491-497.

Jackson, M.B. 1971. Ethylene and Plant Development with Special Reference to Abscission. Ph. D. thesis. University of Oxford, England.

Jackson, M.B. 1988. Involvement of the hormones ethylene and abscisic acid in some adaptive responses of plants to submergence, chap. 31. In *The Ecology and Management of Wetlands,* vol. 1. Ecology of Wetlands. D.D. Hook et al. (Eds.). Timber Press, Portland, OR. 373-382.

Jackson, M.B., and M.C. Drew. 1984. Effects of flooding on growth and metabolism of herbaceous plants, chap. 3. In Flooding and Plant Growth. T.T. Kozlowski (Ed.). Academic Press, Inc., New York. 47-128.

Jackson, M.B., T.M. Fenning, M.C. Drew, and L.R. Sarker. 1985. Stimulation of ethylene production and gas space formation in adventitious roots of *Zea mays* by small partial pressures of oxygen. *Planta,* 165: 486-492.

Jackson, S. 1995. Delineating Bordering Vegetated Wetlands Under the Massachusetts Wetlands Protection Act. Massachusetts Dept. Of Environmental Protection, Division of Wetlands and Waterways, Boston.

Jaynes, M.L. and S.R. Carpenter. 1986. Effect of vascular and nonvascular macrophytes on sediment redox and solute dynamics. *Ecology,* 67: 875-882.

Jeglum, J.K. 1971. Plant indicators of pH and water level in peatlands at Candle Lake, Saskatchewan. *Can. J. Bot.,* 49: 1661-1676.

Joshi, M.M. and J.P. Hollis. 1977. Interaction of Beggiatoa and rice plant: detoxification of hydrogen sulfide in the rice rhizosphere. *Science,* 195: 179-180.

Josselyn, M. 1983. The Ecology of San Francisco Bay Tidal Marshes: A Community Profile. FWS/OBS-83/23. U.S. Fish and Wildlife Service, Washington, D.C.

Justin, S.H.F. and W. Armstrong. 1987. The anatomical characteristics of roots and plant response to soil flooding. *New Phytol.,* 106: 465-495.

Kangas, P.C. 1990. Long-term development of forested wetlands, chap. 3. In *Forested Wetlands.* A.E. Lugo, M.M. Brinson, and S. Brown (Eds.). Elsevier, Amsterdam. 25-51.

Kantrud, H.A., J.B. Millar, and A.G. van der Valk. 1989. Vegetation of wetlands of the Prairie Pothole Region, chap. 5. In *Northern Prairie Wetlands.* A.G. van der Valk (Ed.). Iowa State University Press, Ames. 132-187.

Karschon, R. and Y. Zohar. 1972. *Effects of flooding on ecotypes of Eucalyptus viminalis.* Isr. Div. For. Leaflet No. 45. Bet Dagan, Israel.

Kawase, M. 1981. Anatomical and morphological adaptation of plants to waterlogging. *Hort.-Sci.,* 16: 30-34.

Keeley, J.E. 1979. Population differentiation along a flood frequency gradient: physiological adaptations to flooding in *Nyssa sylvatica. Ecol. Monogr.,* 49: 89-108.

Keeley, J.E. and E.H. Franz. 1979. Alcoholic fermentation in swamp and upland populations of *Nyssa sylvatica*: temporal changes in adaptive strategy. *Am. Nat.,* 113: 587-591.

Kemp, W.M. and L. Murray. 1986. Oxygen release from roots of the submersed macrophyte *Potamogeton perfoliatus* L.: regulating factors and ecological implications. *Aquat. Bot.,* 26: 271-283.

Kludze, H.K. and R.D. DeLaune. 1994. Methane emissions and growth of *Spartina patens* in response to soil redox intensity. *Soil Sci. Am. J.,* 58: 1838-1845.

Kludze, H.K. and R.D. DeLaune. 1996. Soil redox intensity effects on oxygen exchange and growth of cattail and sawgrass. *Soil Sci. Am. J.,* 60: 616-621.

Kludze, H.K., S. R. Pezeshki, and R.D. DeLaune. 1994. Evaluation of root oxygenation and growth in bald cypress in response to short-term soil hypoxia. *Can. J. For. Res.,* 24: 804-809.

Kozlowski, T.T. 1982. Water supply and tree growth. II. Flooding. *For. Abstr.,* 43: 145-161.

Kozlowski, T.T. 1984a. Responses of woody plants to flooding, chap. 4. In *Flooding and Plant Growth.* T.T. Kozlowski (Ed.). Academic Press, Inc., New York. 126-163.

Kozlowski, T.T. 1984b. Plant responses to flooding of soil. *BioScience,* 34: 162-167.

Kozlowski, T.T. (Ed.). 1984c. *Flooding and Plant Growth.* Academic Press, Inc., New York.

Kramer, P.J. 1949. *Plant and Soil Water Relationships.* McGraw-Hill, New York.

Kramer, P.J. and T.T. Kozlowski. 1979. *Physiology of Woody Plants.* Academic Press, New York.

Ku, H.S., H. Suge, L. Rappaport, and H.K. Pratt. 1970. Stimulation of rice coleoptile growth by ethylene. *Planta,* 90: 333-339.

Kurz, H. and D. Demaree. 1934. Cypress buttresses and knees in relation to water and air. *Ecology,* 15: 36-41.

Laan, P., M.J. Berrevoets, S. Lythe, W. Armstrong, and C.W.P.M. Blom. 1989. Root morphology and aerenchyma formation as indicators of the flood-tolerance of *Rumex* species. *J. Ecol.,* 77: 693-703.

Laan, P. and C.W.P.M. Blom. 1990. Growth and survival responses of *Rumex* species to flooded and submerged conditions: the importance of shoot elongation, underwater photosynthesis and reserve carbohydrates. *J. Expt. Bot.,* 41: 775-783.

Laan, P., J.M.A.M. Clement, and C.W.P.M. Blom. 1991a. Growth and development of *Rumex* roots as affected by hypoxic and anoxic conditions. *Plant and Soil,* 136: 145-151.

Laan, P., M. Tosserams, P. Huys, and H.F. Bienfait. 1991b. Oxygen uptake by roots of *Rumex* species at different temperatures: the relative importance of diffusive resistance and enzyme kinetics. *Plant, Cell Environ.,* 14: 235-240.

LaBaugh, J.W. 1989. Chemical characteristics of water in northern prairie wetlands, chap. 3. In *Northern Prairie Wetlands.* A.G. van der Valk (Ed.). Iowa State University Press, Ames. 56-90.

Ladiges, P.Y. and A. Kelso. 1977. The comparative effects of waterlogging on two populations of *Eucalyptus viminalis Labill.* and one population of *E. ovata Labill. Aust. J. Bot.,* 25: 159-169.

Larqué-Saavedra, A., H. Wilkins, and R.L. Wain. 1975. Promotion of cress root elongation in white light by 3,5-diiodo-4-hydroxybenzoic acid. *Planta,* 126: 269-272.

Laycock, W.A. 1967. Distribution of Roots and Rhizomes in Different Soil Types in the Pine Barrens of New Jersey. Geol. Survey Prof. Paper 563-C. U.S. Geological Survey, Reston, VA.

Leck, M.A. 1989. Wetland seed banks, chap. 13. In *Ecology of Soil Seed Banks*. M.A. Leck, V.T. Parker, and R.L. Simpson (Eds.). Academic Press, San Diego, CA. 283-305.

Ledig, F.T., and S. Little. 1979. Pitch pine (*Pinus rigida Mill.*): ecology, physiology, and genetics. In *Pine Barrens: Ecosystem and Landscape*. R.T.T. Forman (Ed.). Academic Press, New York. 347-371

Lessmann, J.M., I.A. Mendelssohn, M.W. Hester, and K.L. McKee. 1997. Population variation in growth response to flooding of three marsh grasses. *Ecolog. Eng.*, 8: 31-47.

Leyton, L. and L.Z. Rousseau. 1958. Root growth in tree seedlings in relation to aeration. In *The Physiology of Forest Trees*. K.V. Thimann (Ed.). The Ronald Press Co., New York. 467-475.

Lico, M.S. 1992. Detailed Study of Irrigation Drainage In and Near Wildlife Management Areas, West-central Nevada, 1987-90. Part A — Water Quality, Sediment Composition, and Hydrogeochemical Processes in Stillwater and Fernley Wildlife Management Areas. USGS Water-Resources Investigations Report 92-4024A. U.S. Geological Survey, Reston, VA.

Lieffers, V.J. and J.M. Shay. 1983. Ephemeral saline lakes on the Canadian prairies: their classification and management for emergent macrophyte growth. *Hydrobiol.*, 105: 85-94.

Little, S. 1959. Silvical characteristics of pitch pine (*Pinus rigida*). USDA Forest Service, Station Paper No. 119. Northeastern Forest Expt. Stat., Upper Darby, PA.

Liu, E.H. and M.J.W. Godt. 1983. The differentiation of populations over short distances, chap. 5. In *Genetics and Conservation*. C. Schoenwald-Cox, S.M. Chambers, B.MacBryde, and W.L. Thomas (Eds.). Benjamin Cummings, Menlo Park, CA. 78-95.

Longman, K.A. and J. Jenik. 1974. *Tropical Forest and Its Environment*. Longman Group Limited, London.

Loucks, W.L. and R.A. Keen. 1973. Submersion tolerance of selected seedling trees. *J. For.*, 71: 496-497.

Lugo, A.E., S. Brown, and M.M. Brinson. 1990. Concepts in wetland ecology, chap. 4. In *Forested Wetlands*. A.E. Lugo, M.M. Brinson, and S. Brown (Eds.). Elsevier, Amsterdam. 53-85.

MacDonald, I.R. and D.C. Gordon. 1978. An inhibitory effect of excess moisture on the early development of *Sinapis alba* L. seedlings. *Plant Cell Environ.*, 1: 313-316.

McKee, K.L., I.A. Mendelssohn, and M.W. Hester. 1988. Reexamination of pore water sulfide concentrations and redox potentials near the aerial roots of *Rhizophora mangle* and *Avicennia germinans*. *Amer. J. Bot.*, 75: 1352-1359.

McKevlin, M.R., D.D. Hook, and W.H. McKee, Jr. 1995. Growth and nutrient use efficiency of water tupelo seedlings in flooded and well-drained soil. *Tree Physiol.*, 15: 753-758.

McKevlin, M.R, D.D. Hook, W.H. McKee, Jr., S.U. Wallace, and J.R. Woodruff. 1987. Loblolly pine seedling root anatomy and iron accumulation as affected by soil waterlogging. *Can. J. For. Res.*, 17: 1257-1264.

McLeod, K.W., L.A. Donovan, and N.J. Stumpff. 1988. Responses of woody seedlings to elevated flood water temperatures, chap. 36. In *The Ecology and Management of Wetlands*, vol. 1. Ecology of Wetlands. D.D. Hook et al. (Eds.). Timber Press, Portland, OR. 441-451.

Mendelssohn, I.A. 1993. Factors Controlling the Formation of Oxidized Root Channels: A Review and Annotated Bibliography. Wetlands Research Program Tech. Rep. WRP-DE-5. U.S. Army Engineer Waterways Expt. Station, Vicksburg, MS.

Mendelssohn, I.A., B.A. Kleiss, and J.S. Wakeley. 1995. Factors controlling the formation of oxidized root channels: a review. *Wetlands*, 15: 37-46.

Merrell, D.J. 1981. *Ecological Genetics*. The University of Minnesota Press, Minneapolis.

Millar, J.B. 1976. Wetland Classification in Western Canada. Rep. Series No. 37. Canadian Wildlife Service.

Miller, W.B. and F.E. Egler. 1950. Vegetation of the Wequetequock-Pawcatuck tidal marshes, Connecticut, *Ecol. Monogr.*, 20: 143-172.

Minc, L.D. and D.A. Albert. 1998. Great Lakes Coastal Wetlands: Abiotic and Floristic Characterization. Michigan Natural Features Inventory.

Nestler, J. 1977. Interstitial salinity as a cause of ecophenic variation in *Spartina alterniflora*. *Estuar. Coastal Mar. Sci.*, 5: 707-714.

Niering, W.A. 1953. The past and present vegetation of High Point State Park, New Jersey. *Ecol. Monogr.*, 23: 127-147.

Pearsall, W.H. 1917. The aquatic and marsh vegetation of Esthwaite Water. *J. Ecol.*, 5: 180-202; 6: 53-84.

Penfound, W.T. 1934. Comparative structure of the wood in the "knees," swollen bases, and normal trunks of the tupelo gum (*Nyssa aquatica* L.). *Am. J. Bot.*, 21: 623-631.

Pezeshki, S.R. 1991. Root responses of flood-tolerant and flood-sensitive tree species to soil redox conditions. *Trees,* 5: 180-186.

Pezeshki, S.R. 1994. Plant response to flooding, chap. 10. In *Plant-Environment Interactions.* R.E. Wilkinson (Ed.). Marcel Dekker, Inc., New York. 289-321.

Pezeshki, S.R. and R.D. DeLaune. 1991. Ecophenic variations in wiregrass (*Spartina patens*). *J. Aquat. Plant Manage.,* 29: 99-102.

Pezeshki, S.R. and R.D. DeLaune. 1995. Variation in response of two U.S. Gulf Coast populations of *Spartina alterniflora* to hypersalinity. *J. Coastal Res.,* 11: 89-95.

Pezeshki, S.R. and R.D. DeLaune. 1996. Response of *Spartina alterniflora* and *Spartina patens* to rhizosphere oxygen deficiency. *Acta Oecologica,* 17: 365-378.

Philipson, J.J. and M.P. Coutts. 1978. The tolerance of tree roots to waterlogging. III. Oxygen transport in lodgepole pine and Sitka spruce roots of primary structure. *New Phytol.,* 80: 341-349.

Ponnamperuma, F.N. 1972. The chemistry of submerged soils. *Adv. Agron.,* 24: 29-96.

Ponnamperuma, F.N. 1976. Temperature and the chemical kinetics of flooded soils. In *Climate and Rice.* International Rice Research Institute, Los Baños, Philippines. 249-263.

Ponnamperuma, F.N. 1984. Effects of flooding on soils, chap. 2. In *Flooding and Plant Growth.* T.T. Kozlowski (Ed.). Academic Press, New York. 9-45.

Raunkiaer, C. 1905. Types biologiques pour la géographic botanique. *Bull. Acad. Roy. D. Sci. De Danemark,* 1905: 347-437.

Raunkiaer, C. 1934. *The Life Forms of Plants and Statistical Plant Geography.* Oxford University Press, Oxford, England.

Reed, P.B., Jr. 1988. National List of Plant Species that Occur in Wetlands: 1988 National Summary. Biol. Rep. 88 (24). U.S. Fish and Wildlife Service, Washington, D.C.

Reed, P.B., Jr. (compiler). 1997. Revision of the National List of Plant Species that Occur in Wetlands. Department of the Interior, U.S. Fish and Wildlife Service, Washington, D.C.

Reimold, R.J. and W.H. Queen (Eds.). 1974. *Ecology of Halophytes.* Academic Press, New York.

Rhode Island Department of Environmental Management. 1994. Rules and Regulations Governing the Administration and Enforcement of the Freshwater Wetlands Act. Providence, RI.

Ridge, I. and I. Amarsinghe. 1981. Ethylene as a stress hormone: its role in submergence responses. In *Responses of Plants to Environmental Stress and Their Mediation by Plant Growth Substances.* Abstracts. British Plant Growth Regul. Group, Assoc. Of Appl. Biol., Wantage. 13-14.

Ryan, A.G. 1989. Native Trees and Shrubs of Newfoundland and Labrador. Parks Division, Department of Environment and Lands, Province of Newfoundland, St. John's.

Sahrawat, K.L., C.K. Mulbah, S. Diatta, R.D. DeLaune, W.H. Patrick, Jr., B.N. Singh, and M.P. Jones. 1996. The role of tolerant genotypes and plant nutrients in the management of iron toxicity in lowland rice. *J. Agric. Sci.* (Cambridge), 126: 143-149.

Sale, P.J.M. and R.G. Wetzel. 1983. Growth and metabolism of *Typha* species in relation to cutting treatments. *Aquat. Bot.,* 15: 321-334.

Sand-Jensen, K., C. Prahl, and H. Stokholm. 1982. Oxygen release from roots of submerged aquatic macrophytes. *Oikos,* 38: 349-354.

Scholander, P.F., L. Van Dam, and S.I. Scholander. 1955. Gas exchange in the roots of mangroves. *Am. J. Bot.,* 42. 92-98.

Schouw, J.F. 1822. *Grundtrack til en Almindlig Plantegeografie.* German edition, Berlin.

Schultz, R.P. 1972. Root development of intensively cultivated slash pine. *Soil Sci. Soc. Am. Proc.,* 36: 158-162.

Schultz, R.P. 1973. Site treatment and planting method alter root development of slash pine. USDA Forest Service. Res. Paper SE-109.

Scott, M.L., W.L. Slauson, C.A. Segelquist, and G.T. Auble. 1989. Correspondence between vegetation and soils in wetlands and nearby uplands. *Wetlands,* 9(1): 41-60.

Sculthorpe, C.D. 1967. *The Biology of Aquatic Vascular Plants.* Koeltz Scientific Books, Königstein, Germany. (Reprinted in 1985.)

Sena-Gomes, A.R. and T.T. Kozlowski. 1980. Growth responses and adaptations of *Fraxinus pennsylvanica* seedlings to flooding. *Plant Physiol.,* 66: 267-271.

Segelquist, C.A., W.L. Slauson, M.L. Scott, and G.T. Auble. 1990. Synthesis of Soil–Plant Correspondence Data from Twelve Wetland Studies Throughout the United States. Biol. Rep. 90(19). U.S. Fish and Wildlife Service, Washington, D.C.

Seliskar, D.M. 1988. Waterlogging stress and ethylene production in the dune slack plant, *Scirpus americanus*. *J. Exp. Bot.*, 39: 1639-1648.

Shanklin, J. and T.T. Kozlowski. 1985. Effect of flooding of soil on growth and subsequent responses of *Taxodium distichum* seedlings to SO_2. *Environ. Pollut.*, 38: 199-212.

Shear, T.H. and D.D. Hook. 1988. Interspecific genetic variation of loblolly pine tolerance to soil waterlogging. In *The Ecology and Management of Wetlands*, vol. 1. Ecology of Wetlands. D.D. Hook et al. (Eds.). Timber Press, Portland, OR. 489-493.

Shear, T., M. Young, and R. Kellison. 1997. An Old-Growth Definition for Red River Bottom Forests in the Eastern United States. Gen. Tech. Rep. SRS-10. USDA Forest Service, Southern Research Station, Asheville, NC.

Silander, J.A. and J. Antonovics. 1979. The genetic basis of the ecological amplitude of *Spartina patens*. I. Morphometric and physiological traits. *Evolution,* 33: 1114-1127.

Sipple, W.S. 1988. Wetland Identification and Delineation Manual, vol. I. Rationale, Wetland Parameters, and Overview of Jurisdictional Approach. Revised Interim Final. U.S. Environmental Protection Agency, Office of Wetlands Protection, Washington, D.C.

Sipple, B. 1990. Personal communication.

Smirnoff, N. and R.M.M. Crawford. 1983. Variation in the structure and response to flooding of root aerenchyma in some wetland plants. *Ann. Bot.,* 51: 237-249.

Smith, W.G. 1913. Raunkiaer's "life-forms" and statistical methods. *J. Ecol.,* 1: 16-26.

Smolders, A., M. Tosserams, and P. Laan. 1990. Oxidizing activity of the adventitious roots of *Zea mays* L. *J. Genet. Breed.,* 44: 259-262.

Squillace, A.E. and R.T. Bingham. 1958. Localized ecotypic variation in western white pine. *For. Sci.,* 4: 20-33.

Stewart, R.F. and H.A. Kantrud. 1972. Vegetation of Prairie Potholes, North Dakota, in Relation to Quality of Water and Other Environmental Factors. Prof. Paper 585-D. U.S. Geological Survey, Reston, VA.

Taylor, G.J., A.A. Crowder, and R. Rodden. 1984. Formation and morphology of an iron plaque on the roots of *Typha latifolia* L. grown in solution culture. *Am. J. Bot.,* 71: 666-675.

Teal, J. M. and J.W. Kanwisher. 1966. Gas transport in the marsh grass, *Spartina alterniflora. J. Exp. Bot.,* 17: 355-361.

Teal, J. and M. Teal. 1969. *Life and Death of the Salt Marsh*. Ballantine Books, New York.

Teskey, R.O. and T.M. Hinckley. 1977. Impact of Water Level Change on Woody Riparian and Wetland Communities, vol. III. The Central Forest Region. FWS/OBS-77/60. U.S. Fish and Wildlife Service, Washington, D.C.

Teskey, R.O. and T.M. Hinckley. 1978a. Impact of Water Level Change on Woody Riparian and Wetland Communities, vol. IX. Eastern Deciduous Forest Region. FWS/OBS-78/87. U.S. Fish and Wildlife Service, Washington, D.C.

Teskey, R.O. and T.M. Hinckley. 1978b. Impact of Water Level Change on Woody Riparian and Wetland Communities, vol. V. Northern Forest Region. FWS/OBS-78/88. U.S. Fish and Wildlife Service, Washington, D.C.

Theriot, R.F. 1993. Flood Tolerance of Plant Species in Bottomland Forests of the Southeastern United States. Wetlands Research Program Tech. Rep. WRP-DE-6. U.S. Army Engineer Waterways Expt. Station, Vicksburg, MS.

Tiku, B.L. and R.W. Snaydon. 1971. Salinity tolerance within the grass species *Agrostis stolonifera* L. *Plant Soil,* 35: 421-431.

Tiner, R.W., Jr. 1985a. Wetlands of New Jersey. U.S. Fish and Wildlife Service, National Wetlands Inventory, Newton Corner, MA.

Tiner, R.W., Jr. 1985b. Wetlands of Delaware. U.S. Fish and Wildlife Service, Newton, MA and Delaware Dept. of Natural Resources and Environmental Control, Dover. Cooperative publication.

Tiner, R.W., Jr. 1987. *A Field Guide to Coastal Wetland Plants of the Northeastern United States*. University of Massachusetts Press, Amherst.

Tiner, R.W., Jr. 1988. Field Guide to Nontidal Wetland Identification. Maryland Dept. of Natural Resources, Annapolis and U.S. Fish and Wildlife Service, Newton Corner, MA. Cooperative publication. (Reprinted by Institute for Wetland & Environmental Education & Research, Leverett, MA.)

Tiner, R. W. 1989. Wetlands of Rhode Island. U.S. Fish and Wildlife Service, National Wetlands Inventory, Newton Corner, MA.

Tiner, R.W. 1991. The concept of a hydrophyte for wetland identification. *BioScience,* 41: 236-247.

Tiner, R.W. 1993. The primary indicators method — a practical approach to wetland recognition and delineation in the United States. *Wetlands*, 13: 50-64.

Tiner, R.W. 1998. *In Search of Swampland: A Wetland Sourcebook and Field Guide*. Rutgers University Press, New Brunswick, NJ.

Tiner, R.W. and D.G. Burke. 1995. Wetlands of Maryland. U.S. Fish and Wildlife Service, Hadley, MA and Maryland Department of Natural Resources, Annapolis, MD. Cooperative technical publication.

Titus, J.H. 1990. Microtopography and woody plant regeneration in a hardwood floodplain swamp in Florida. *Bull. Torrey Bot. Club*, 117: 429-437.

Topa, M.A. and K.W. McLeod. 1986a. Aerenchyma and lenticel formation in pine seedlings: a possible avoidance mechanism to anaerobic growth conditions. *Physiol. Plantarum*, 68: 5540-550.

Topa, M.A. and K.W. McLeod. 1986b. Responses of *Pinus clausa*, *Pinus serotina* and *Pinus taeda* seedlings to anaerobic solution culture. I. Changes in growth and root morphology. *Physiol. Plantarum*, 68: 523-531.

Topa, M.A. and K.W. McLeod. 1988. Promotion of aerenchyma formation in *Pinus serotina* seedlings by ethylene. *Can. J. For. Res.*, 18: 276-280.

Tsukahara, H. and T.T. Kozlowski. 1985. Importance of adventitious roots to the growth of flooded Platanus occidentalis seedlings. *Plant Soil*, 88: 123-132.

Turesson, G. 1922a. The species and the variety as ecological units. *Hereditas*, 3: 100-113.

Turesson, G. 1922b. The genotypical response of the plant species to the habitat. *Hereditas*, 3: 211-350.

Turesson, G. 1925. The plant species in relation to habitat and climate. *Hereditas*, 6: 147-236.

USDA Natural Resources Conservation Service. 1994. National Food Security Act Manual. U.S. Department of Agriculture, Washington, D.C.

Van der Sman, A.J.M., C.W.P.M. Blom, and G.W.M. Bardendse. 1993. Flooding resistance and shoot elongation in relation to developmental stage and environmental conditions in *Rumex maritimus* L. and *Rumex palustris* Sm. *New Phytol.*, 125: 73-84.

Van der Sman, A.J.M., O.F.R. Van Tongeren, and C.W.P.M. Blom. 1988. Growth and reproduction of *Rumex maritimus* and *Chenopodium rubrum* under different waterlogging regimes. *Acta Bot. Neerl.*, 37: 439-450.

Veneman, P.L.M. and R.W. Tiner. 1990. Soil–Vegetation Correlations in the Connecticut River Floodplain of Western Massachusetts. Biol. Rep. 90(6). U.S. Fish and Wildlife Service, Washington, D.C.

Viktorov, S.V., E.A. Vostokova, and D.D. Vyshkivin. 1965. Some problems in the theory of geobotanical indicator research. In *Plant Indicators of Soils, Rocks and Subsurface Waters*. A.G. Chikichev (Ed.). Conference on Indicator Geobotany, Moscow, 1961. Consultants Bureau, New York. 1-4.

Voesenek, L.A.C.J., M. Banga, J.G.H.M. Rijnders, E.J.W. Visser, and C.W.P.M. Blom. 1996. Hormone sensitivity and plant adaptations to flooding. *Folia Geobot. Phytotax.*, 31: 47-56.

Voesenek, L.A.C.J., C.W.P.M. Blom, and R.H.W. Pouwels. 1989. Root and shoot development of *Rumex* species under waterlogged conditions. *Can. J. Bot.*, 67: 1865-1869.

Voesenek, L.A.C.J. and R. Van der Veen. 1994. The role of phytohormones in plant stress: too much or too little water. *Acta Bot. Neerl.*, 43: 91-127.

Vostokova, E.A. 1965. The present state of hydrologic indicator research. In *Plant Indicators of Soils, Rocks and Subsurface Waters*. A.G. Chikichev (Ed.). Conference on Indicator Geobotany, Moscow, 1961. Consultants Bureau, New York. 5-10.

Wainwright, S.J. 1984. Adaptations of plants to flooding with salt water, chap. 9. In *Flooding and Plant Growth*. T.T. Kozlowski (Ed.). Academic Press, New York. 295-343.

Waisel, Y. 1972. *Biology of Halophytes*. Academic Press, New York.

Warming, E. 1909. *Oecology of Plants. An Introduction to the Study of Plant-communities*. Clarendon Press, Oxford, England. (Updated English translation of 1986 text.)

Wassen, M.J., A. Barendreght, J.L. Bootsma, and P.P. Schot. 1989. Groundwater chemistry and vegetation of gradients from rich fen to poor fen in the Naardermeer (the Netherlands). *Vegetatio*, 79: 117-132.

Weaver, J.E. and F.E. Clements. 1929. *Plant Ecology*. McGraw-Hill, New York.

Wells, B.W. 1942. Ecological problems of the southeastern United States coastal plain. *Bot. Rev.*, 8: 533-561.

Wenkert, W., N.R. Fausey, and H.D. Watters. 1981. Flooding response of *Zea mays* L. *Plant Soil*, 62: 351-366.

Wetzel, R.G. 1990. Land-water interfaces: metabolic and limnological regulators. *Verh. Int. Verein. Limnol.*, 24: 6-24.

Whitlow, T.H., and R.W. Harris. 1979. Flood Tolerance in Plants: A State-of-the-Art Review. Tech. Rep. E-79-2. U.S. Army Engineers Waterways Exp. Sta., Vicksburg, MS.

Wilcox, D.A. 1995. Wetland and aquatic macrophytes as indicators of anthropogenic hydrologic disturbance. *Nat. Areas J.,* 15: 240-248.

Wilhelm, G. and D. Ladd. 1988. *Natural area assessment in the Chicago region.* Trans. 53rd North American Wildlife and Natural Resources Conference. 361-375.

Will, R.E., J.R. Seiler, P.P. Feret, and W.M. Aust. 1995. Effects of rhizosphere inundation on the growth and physiology of wet and dry-site *Acer rubrum* (red maple) populations. *Am. Midl. Nat.,* 134: 127-139.

Williams, Major General A.E. 1992. U.S. Army Corps of Engineers memorandum on clarification and interpretation of the 1987 manual. March 6, 1992.

Wilson, W.P. 1889. The production of aerating organs on the roots of swamp and other plants. *Proc. Acad. Nat. Sci. Phil.,* 1889: 67-69.

Wisconsin Coastal Management Program. 1995. Basic Guide to Wisconsin's Wetlands and Their Boundaries. State of Wisconsin, Dept. of Administration, Madison.

Yamamoto, F., T.T. Kozlowski, and K.E. Wolter. 1987. Effect of flooding on growth, stem anatomy, and ethylene production of Pinus halepensis seedlings. *Can. J. For. Res.,* 17: 69-79.

Zampella, R.A. and K.J. Laidig. 1997. Effect of watershed disturbance on Pinelands stream vegetation. *J. Torrey. Bot. Soc.,* 124(1): 52-66.

Zedler, J.B. and C.S. Nordby. 1986. The Ecology of Tijuana Estuary, California: An Estuarine Profile. Biol. Rep. 85 (7.5). U.S. Fish and Wildlife Service, Washington, D.C.

4 Vegetation Sampling and Analysis for Wetlands

INTRODUCTION

Various techniques have been used to characterize wetland vegetation. Some are simply visual observations for general descriptions (e.g., for state wetland reports by the U.S. Fish and Wildlife Service's National Wetlands Inventory) or for routine wetland determinations for regulatory purposes (see Chapter 6). For ecological studies or more qualitative regulatory assessments, several vegetation sampling techniques are used, including plot-based methods and plotless methods. The following review summarizes the more commonly used techniques, with particular emphasis on their use for wetland identification and delineation. Some tips on vegetation sampling in wetlands are offered. Examples of plant community data are presented to demonstrate how to analyze such data for identification of potential hydrophytic vegetation. Accompanying exercises are given at the end of the chapter to help readers make sure that analysis procedures are fully understood. For more complete reviews of scientific methods for characterizing vegetation, consult Mueller-Dombois and Ellenberg (1974), Barbour et al. (1980), Gauch (1982), and Kent and Coker (1992). This chapter is intended to give readers a good understanding of the techniques that form the basis for vegetation sampling used in various wetland delineation manuals (Chapter 6) and a better idea of how to select representative areas and to use certain measures for determining dominant species and for characterizing plant communities.

When considering vegetation sampling, three approaches are possible: (1) locating samples subjectively in representative stands (used by trained observers), (2) subjective identification of stands and random sampling within (including by regular placement of samples), and (3) random selection of stands and random sampling within (Barbour et al., 1980). The latter is inefficient, very time consuming, and of little practical utility. Derivations of the first approach are typically used for wetland delineation in the U.S., while the second approach is better for research studies where statistical analysis is desired. The first method, often called the relevé method, is attributed to the Swiss ecologist Josias Braun-Blanquet who did much of the early work on the vegetation of Europe. Although widely used in most countries, the relevé method has been sparingly used in the U.S. for ecological research (Barbour et al., 1980). Conard (1935) was one of the few American ecologists to use this method. He described vegetation types on Long Island, including some wetland types.

In its basic form, the technique is applied by a scientist familiar with regional or local plant ecology who first makes a list of plant communities for study and then locates typical stands for analysis. The analysis is done by walking through a stand, compiling a species list of plants observed, and then selecting a representative area for plot sampling. A number of small plots are analyzed in a nested sampling quadrant (Figure 4.1a). As data are collected, a species-to-area curve (the occurrence of new species with increased area sampled) is plotted. As the size of the sampled area increases, new species are seen until a point is reached where nearly all of the species associated with the community (e.g., 90 to 95% of the species) have been observed in at least one sample. The species-area curve will tend to flatten out at this point and the minimum area is represented by the point of inflection in the curve (Figure 4.1b). The area sampled should always be a little larger than the minimum area. Ecological studies involving minimum area determinations have provided the foundation for selecting plot sizes for evaluating different vegetation strata for federal wetland delineation purposes.

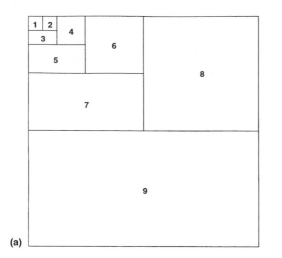

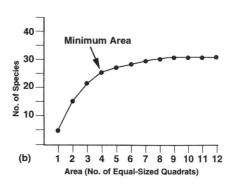

FIGURE 4.1 (a) Nested plot, (b) species-area curve.

STAND IDENTIFICATION

When sampling vegetation, the first decisions are what and where to sample. A stand is an assemblage of plants of relatively the same structure (floristic composition) occurring on one soil type in a particular landscape position. All parts of the stand are subjected to basically the same environmental conditions. There may be minor differences in plant composition due to microrelief, but overall the plant community is relatively uniform. Before going into the field, it is worthwhile to examine aerial photographs of the study area (see Chapter 10) to detect distinctive patterns of vegetation. Stereoscopic analysis will reveal changes in topography that are likely to produce differences in plant communities related to variations in soil type and soil moisture.

Once in the field, a walk through the parcel of land should help validate and refine the photointerpretation results. For plant ecology research studies, most American plant ecologists subjectively identify stands, and then locate several random plots within them for sampling (random quadrat method in plant ecology; Barbour et al., 1980). Most wetland delineations involve similar stand identification, but require locating one or more representative (not random) plots for sampling. Stands are recognized by considering both the overstory and understory species in an attempt to identify homogeneous areas. This admittedly is a somewhat subjective process, but when done without perceived bias, one can readily identify individual plant communities* by quickly assessing dominant species in various strata and fully considering topographic and other environmental variables (e.g., soils). Places of low topographic relief (e.g., coastal and glaciolacustrine plains) may pose some difficulty in locating plant community breaks, but rapid assessment of soils and examination of groundcover are particularly helpful in separating plant communities.

---

* In this book, the term "plant community" is used in a general sense, essentially equivalent to a stand (an assemblage of plants growing together in a similar landscape position, on the same soil, and subject to the same hydrologic and environmental conditions). Such usage is not intended to embrace the "climax community" concept of Clements (1916) nor to reject the "continuum concept" of Gleason (1926) or Whittaker (1956; 1978). Although no two plant communities are exactly alike, many plant species commonly occur together, forming recognizable combinations (*community, vegetation type, habitat type,* or *association*) under specific environmental conditions.

LOCATING SAMPLING AREAS

After identifying individual stands for a wetland project site, the investigator choses representative areas within each stand for sampling. This "preferential sampling" is the most commonly used technique by plant ecologists for descriptive purposes (Gauch, 1982). Some useful points to consider when selecting sample plots are (1) keep away from roads and other disturbances where fill or sedimentation may have raised the original ground levels or produced a localized effect on hydrology; (2) avoid sampling near the perceived wetland boundary (sample first in the recognizable wetland community, next in the obvious upland, and then analyze the transitional community inbetween); (3) given that for wetland delineation the assessment of vegetation is directed more at yielding the dominant species, absolute values for determining dominants are of little interest and more accurate means of assessing vegetation will not likely produce different dominant species; (4) selection of sampling attributes (cover, basal area, etc.) may yield different dominant species in some cases; and (5) different dominants and decisions regarding the presence of hydrophytic vegetation also may result from differences in sampling techniques (plot-based assessments vs. prevalence indices) as observed by Wakely and Lichvar (1997). The latter two situations require that the investigator use sound professional judgment in selecting the appropriate means for analysis.

TIMING OF SAMPLING

For ecological studies, seasonal patterns of vegetation must be taken into account. This is most significant for herbaceous species. For example, spring ephemeral herbs like spring beauty (*Claytonia virginica*) will not persist above ground throughout the summer. Vegetation sampling should be performed two or three times during the growing season to account for this variability. Such sampling should attempt to evaluate vegetation in unusually wet and dry years to assess plant response to these extreme conditions.

When making wetland determinations, vegetation sampling is typically accomplished during a single visit lasting one or more days (weather permitting). Sampling period can make a difference when assessing hydrophytic vegetation, especially in meadows. In many forested wetlands, the herbs best reflect site wetness, and seasonal changes in dominant species can occur. For example, skunk cabbage (*Symplocarpus foetidus*), a dominant spring and early summer herb in forested wetlands in the Northeast, will not be seen in late summer, except perhaps in small holes where new growth (spathe) is beginning to form. In some wetlands, species with wetter indicator statuses may predominate earlier in the growing season, with drier site species dominating in late summer and fall due to seasonal differences in near-surface wetness. Consequently, sampling period may affect the presence of a positive indicator of hydrophytic vegetation. In such cases, spring sampling may produce more accurate results for determining site wetness.

SAMPLING STRATA

For wetland delineation work, the plant community is usually analyzed by strata within sample plots, with the exception being where point sampling is performed. Rather than emphasize the dominant overstory species, all layers are evaluated. In many cases, the herbaceous species may be the most indicative of current site wetness, while woody species once established may persist under a variety of wetness regimes. The herbaceous stratum also is more likely to quickly change in response to altered hydrology.

Strata are typically defined by life form (tree or woody overstory, sapling and shrub or woody understory, herb or groundcover, woody vine, and bryophyte). Table 4.1 defines strata recommended by different federal manuals. The 5.0-in. diameter at breast height (dbh) break for trees vs. saplings

TABLE 4.1
Sampling Strata for Wetland Delineation Based on the Corps (CE) and Federal Interagency (FI) Wetland Delineation Manuals

Manual	Stratum	Definition
CE	Woody overstory	Woody plants with dbh ≥ 3.0 in.
	Woody understory	Woody plants less than 3.0 in. dbh
	Woody vines	Climbing woody plants
	Groundcover	Herbaceous species, including seedlings of woody plants less than 1 m tall, plus woody trailing species
FI	Tree	Woody plants ≥5.0 dbh and ≥20 ft tall
	Sapling	Woody plants <5.0 dbh and ≥20 ft tall
	Shrub	Woody plants <20 ft tall (and generally ≥3.0 ft)
	Woody vines	Climbing woody plants
	Herb	Nonwoody plants, including seedlings of woody plants less than 3.0 ft tall, plus woody trailing species

Note: There are 4 possible strata under the CE and 5 potential strata in the FI. "dbh" = diameter at breast height.

Source: Environmental Laboratory, 1987; Federal Interagency Committee for Wetland Delineation, 1989, respectively.

represents the dbh split between poletimber vs. sapling as used by the U.S. Forest Service in their forest vegetation sampling (e.g., Beltz et al., 1992).

SAMPLING VARIABLES

With strata defined, different variables are then measured or estimated to determine the dominant species. For wetland delineation, the more frequently used variables are areal cover, density, and basal area, and frequency (summarized below). Other variables used in ecological studies include vitality (plant vigor), periodicity (seasonal importance), and sociability (individual dispersal within a plot or stand) (see Mueller-Dombois and Ellenberg, 1974; Barbour et al., 1980; Gauch, 1982; Kent and Coker, 1992 for discussions of these variables).

AREAL COVER

Areal cover or simply cover is the sum total of areas within a sample plot that is covered by the canopy of a particular species (areal coverage of the species) or the total coverage by all plants of a given stratum in the plot (areal coverage of trees, shrubs, or herbs, for example). It is estimated as a percentage of the sample plot so covered. In ecology, estimates are often preferred over direct measurements due to time considerations, seasonal variations in cover, and that exact measurement may yield a sense of "false precision." Seasonal and annual fluctuations in grasslands, for example, resulted in cover values differences of 20% or more, especially in droughts and very wet years (Brown, 1954). The noted biostatistician Slobodkin once said that the ecological world is a sloppy place (Slobodkin, 1974, as reported in Barbour et al., 1980). Also given that measurement requires more effort, it is far better to make more estimates in a stand or in different stands than to spend an exhaustive amount of time taking measurements at fewer sites. Examples of cover percentages is given in Figure 4.2. Cover estimates have been used for all strata — herb and other groundcover, shrubs and saplings (woody understory), trees (woody overstory), and woody vines, although estimating cover for the latter is made difficult by the foliage of its host. For ecological studies, cover should be estimated for all strata for comparative analysis.

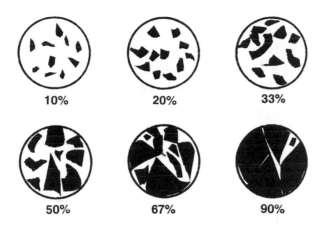

FIGURE 4.2 Examples of cover estimates in circular plots.

TABLE 4.2
Estimating Cover Based on Canopy Diameter Within a 15- and 30-ft Radius Plot

Shrub Cover Estimates Based on Canopy Diameter

Diameter (Ft)	Approximate Cover for 30-Ft Radius Plot	Approximate Cover for 15-Ft Radius Plot
3–5	0.5%	2%
6–10	2%	7%
11–15	5%	19%
16–20	10%	36%
21–25	15%	59%
26–30	20%	87%

Note: For plants on the edge of the plot, only the portion of the canopy lying within the plot should be included in the estimate.

There are three options for estimating cover — actual cover, relative cover, and cover classes. *Actual cover* is the percent of a plot that is covered by individuals of a given species and can include overlap. When overlap is included, the total cover for all species in a particular stratum may exceed 100% as commonly occurs in forests. Table 4.2 provides a correlation table where the areal cover of individual plants or clumps of plants is estimated by the canopy diameter, looking up the applicable canopy diameter range which is then matched to the area of a 15-ft radius plot and a 30-ft radius plot for a percent cover estimate. An alternative estimator is *relative cover* where the cover for the stratum is set at 100% and one estimates the proportion or percentage of the whole that is represented by each species (the sum of the cover of all species is always 100%). Recognizing that people differ in their ability to estimate and that estimates are approximate anyway,* plant ecologists realized that it may be more appropriate to have people estimate a range of cover rather than a single number estimate for a given species. Consequently, ecologists developed cover classes

* Individuals tend to overestimate the cover of species in flower or that are otherwise conspicuous, while underestimating others (Kent and Coker, 1992).

TABLE 4.3
Suggested Cover Classes for Use
in Wetland Delineation

Cover Class	Range of Cover	Midpoint Value
Trace	<1%	0.0%
1	1–4%	2.5%
2	5–15%	10.0%
3	16–25%	20.5%
4	26–39%	32.5%
5	40–60%	50.0%
6	61–74%	67.5%
7	75–84%	79.5%
8	85–95%	90.0%
9	96–99%	97.5%
10	100%	100.0%

Note: Midpoint values tend to correspond with familiar percentages or fractions (e.g., 1/10, 1/5, 1/3, 1/2, 2/3, 4/5, and 9/10).

which allow estimating whether a particular species falls within a certain range of values. Cover classes for wetland delineation (FICWD 1989) are adapted from Daubenmire (1968): trace = <1% (0), 1 = 1 to 5% (3.0), 2 = 6 to 15% (10.5), 3 = 16 to 25% (20.5), 4 = 26 to 50% (38.0), 5 = 51 to 75% (63.0), 6 = 76 to 95% (85.5), and 7 = 96 to 100% (98.0). The values in parentheses represent midpoints of the cover classes that are used as the estimated cover value for determining dominant species.

Despite the frequent use of Daubenmire's cover classes, experience in testing and training over 1000 students in wetland delineation has shown me that most students have difficulty making the 50% break between cover classes 4 and 5. This is largely due to the scatter of vegetation within a plot, since plant coverage is not neatly arranged like slices of pie on a pie pan (which would make it easy to separate >50% from <50%), but is instead often scattered about within the plot (see Figure 4.2). In my opinion, most students can better estimate that a cover is about 50%, so this should be a midpoint value for a class defined in the range from 40 to 60%, rather than the separation between cover classes. Table 4.3 contains suggested cover classes based on these findings.

When estimating cover, the following tips may be helpful. First list all the species in the plot. Then begin with one that has fairly significant cover and estimate what percent of the total plot or cover it represents. Where patches of the species are scattered throughout the plot, one has to mentally group them together to get a sense of what fraction of the whole they represent. A way of improving one's estimating ability when using a sampling frame (e.g., for herb stratum) is to stand over the frame (e.g., a 1 m × 1 m square or a 3 ft × 3 ft square, the latter made by using a collapsible 6-ft carpenter's ruler), and determine how many fists fill the frame (e.g., 25 fists). In this example, one fist then equals 1/25 of the plot or 4% cover. So, if there are four patches of a single species that are covered by 6 fists, the cover is 24%. Afterwards, one may decide to group the species into cover classes and use the midpoints for determining dominant species, if desirable.

BASAL AREA

Basal area is the cross-sectional area of a plant and is usually expressed in units such as square in. or square ft per acre or square m per ha. Basal area can be calculated through direct measurements or estimates. To calculate "actual basal area", the diameter at breast height (dbh) is measured with a diameter tape. Breast height is standardized at 4.5 ft (1.37 m) above the ground. If a diameter tape is not available, the circumference of the trunk can be measured at this point with a conventional measuring tape and the dbh and basal area approximated from a correlation table (Table 4.4). Basal area (BA) is determined for each specimen in a sample plot (BA = πd2/4, where d is the diameter, dbh). The basal areas of all the individuals of a given species are then totaled for the plot. The sum total of the basal areas for all species represents the total basal area of the plot. This metric may be used in wetland delineations requiring use of the comprehensive method (Chapter 6). It also is useful for forested wetlands with high diversity and much-overlapping canopies. Although measurement of basal area of herbs has been done with calipers and other devices for ecological studies, it is too time-consuming for wetland delineations and is not recommended.

Basal area also may be estimated. The most accurate "estimated basal area" or "relative basal area" would be derived by using a sampling device such as a basal area factor (BAF) prism, an angle-gauge, panama sighting tube, or similar device that can be purchased from a forestry supply store like Forestry Suppliers of Jackson, MS or Ben Meadows of Atlanta, GA. When using these devices, the area sampled is not confined to a predefined plot, such as a 30-ft radius plot, but rather is variable being related to the sampling device, the dbh of the tree, and its distance from the observation point (see Bell and Dilworth, 1978 for discussion of the theory behind this approach). It is sometimes called the *variable plot* method. Since it is not a fixed plot method, it is often called a plotless approach, since technically every tree in view is evaluated.

BAF prisms are cut to a specific dimension for selectively sampling trees of a particular dbh relative to the distance from the observer. In the eastern U.S., the BAF 10 prism is commonly used. The 10 factor means that trees counted as "hits" will represent a basal area of 10 square ft per acre. This device will usually lead to about 6 to 12 trees being counted for analysis of the tree stratum for wetland delineations. In the Pacific Northwest, for example, where much larger trees abound, a 20 or 40 factor would be more appropriate. The intention is to get an adequate representation of the resident trees for determining dominant species which usually means about 6 to 12 individuals. The observer holds the prism over the sampling point (center of sample "plot"), points the prism at the dbh point on the first tree seen (in forests with dense understory, observations may be made at 16-ft up on the trunks rather than at dbh), and looks through the lens to see if the tree should be tallied (see Figure 4.3 for what is tallied vs. what is not). The observer then moves the prism in a sweeping fashion covering an entire 360° arc, looking at each tree in view to see if it is tallied. If a tree is hidden from view by another tree or other object, the observer may move slightly to get it into view, being careful to maintain the same distance from the sample point to the tree. At the end of a sweep, a list of species and number of tallies will be available for calculating dominant species.

The angle-gauge is a metal object with rectangular cutouts, attached to a chain. The gauge consists of four "angles" for sampling: a 5-angle (one tally being equivalent to a basal area of 5 square ft/acre), a 10-angle (10 square ft/acre/tally), a 20-angle (20 square ft/acre/tally), and a 40-angle. The observer holds the gauge in one hand while the end of the chain is either grasped by the teeth or held to the chin with the other hand. The observer stands at the center of the observation area, looking through the appropriate angle at the dbh point on each tree, tallying selected individuals (Figure 4.3), and making a 360° sweep of the plant community. The critical distance is from the eye of the observer to the specimen seen.

For wetland delineations,* another way to determine relative basal area is simply to estimate the dbhs of trees in a 30-ft radius circular plot. A value of "x" can be assigned to a particular dbh, and all trees in the plot evaluated relative to this one. For example, a tree with twice the diameter would be assigned "2x", while one that is 3.5 times as wide would be given a value of "3.5x", and so forth. The "x" could then be measured and the corresponding basal area of each estimated dbh

---

* Since vegetation assessments for wetland delineation are more generalized than for quantitative ecological studies, more simplified estimators of basal area are acceptable. Dominant species can be accurately and efficiently identified by various estimating procedures.

TABLE 4.4
Basal Area Conversion Chart

Circumference (ft)	Diameter (in.)	Basal Area (square in.)	Circumference (ft)	Diameter (in.)	Basal Area (square in.)
1.3	5.0	19	6.2	23.7	440
1.4	5.3	22	6.3	24.1	455
1.5	5.7	26	6.4	24.4	469
1.6	6.1	29	6.5	24.8	484
1.7	6.5	33	6.6	25.2	499
1.8	6.9	37	6.7	25.6	514
1.9	7.3	41	6.8	26.0	530
2.0	7.6	46	6.9	26.4	546
2.1	8.0	51	7.0	26.7	561
2.2	8.4	55	7.1	27.1	578
2.3	8.8	61	7.2	27.5	594
2.4	9.2	66	7.3	27.9	611
2.5	9.5	72	7.4	28.3	628
2.6	9.9	77	7.5	28.6	645
2.7	10.3	84	7.6	29.0	662
2.8	10.7	90	7.7	29.4	679
2.9	11.1	96	7.8	29.8	697
3.0	11.5	103	7.9	30.2	715
3.1	11.8	110	8.0	30.6	733
3.2	12.2	117	8.1	30.9	752
3.3	12.6	125	8.2	31.3	771
3.4	13.0	132	8.3	31.7	789
3.5	13.4	140	8.4	32.1	809
3.6	13.8	149	8.5	32.5	828
3.7	14.1	157	8.6	32.8	848
3.8	14.5	165	8.7	33.2	867
3.9	14.9	174	8.8	33.6	887
4.0	15.3	183	8.9	34.0	908
4.1	15.7	193	9.0	34.4	928
4.2	16.0	202	9.1	34.8	949
4.3	16.4	212	9.2	35.1	970
4.4	16.8	222	9.3	35.5	991
4.5	17.2	232	9.4	35.9	1013
4.6	17.6	242	9.5	36.3	1034
4.7	18.0	253	9.6	36.7	1056
4.8	18.3	264	9.7	37.1	1078
4.9	18.7	275	9.8	37.4	1101
5.0	19.1	286	9.9	37.8	1123
5.1	19.5	298	10.0	38.2	1146
5.2	19.9	310	10.1	38.6	1169
5.3	20.2	322	10.2	39.0	1192
5.4	20.6	334	10.3	39.3	1216
5.5	21.0	347	10.4	39.7	1239
5.6	21.4	359	10.5	40.1	1263
5.7	21.8	372	10.6	40.5	1288
5.8	22.2	385	10.7	40.9	1312
5.9	22.5	399	10.8	41.3	1337
6.0	22.9	413	10.9	41.6	1361
6.1	23.3	426	11.0	42.0	1387

Circumference (ft)	Diameter (in.)	Basal Area (square in.)	Circumference (ft)	Diameter (in.)	Basal Area (square in.)
11.1	42.4	1412	12.8	48.9	1877
11.2	42.8	1437	12.9	49.3	1907
11.3	43.2	1463	13.0	49.7	1937
11.4	43.5	1489	13.1	50.0	1967
11.5	43.9	1515	13.2	50.4	1997
11.6	44.3	1542	13.3	50.8	2027
11.7	44.7	1569	13.4	51.2	2058
11.8	45.1	1596	13.5	51.6	2088
11.9	45.5	1623	13.6	51.9	2119
12.0	45.8	1650	13.7	52.3	2151
12.1	46.2	1678	13.8	52.7	2182
12.2	46.6	1706	13.9	53.1	2214
12.3	47.0	1734	14.0	53.5	2246
12.4	47.4	1762	14.1	53.9	2278
12.5	47.7	1790	14.2	54.2	2311
12.6	48.1	1819	14.3	54.4	2343
12.7	48.5	1848	14.4	55.0	2376

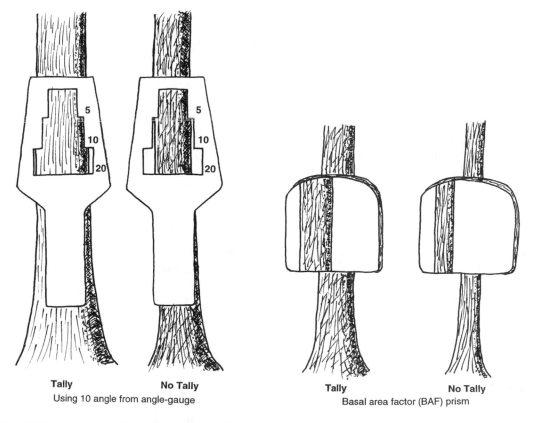

Tally   No Tally
Using 10 angle from angle-gauge

Tally   No Tally
Basal area factor (BAF) prism

FIGURE 4.3 Use of the angle-gauge and BAF prism. Note that in using the angle-gauge, the tree on the left is tallied when utilizing either the 5 or 10 angle, but not the 20; whereas the tree on the right is only tallied if using the 5 angle for sampling.

located in a diameter/basal area correlation table (Table 4.4). Basal areas would then be summed for each species to determine dominants. Alternatively, one might simply total the number of "x"s per species to determine dominant species; this approach should yield the same dominants as the other option requiring determination of the basal area for all species, with rare exceptions.

HEIGHT CLASSES

The height of vegetation is often used to separate the community into different strata (see Table 4.1). The 20-ft level separates the tree and sapling strata from the shrub stratum in the 1989 interagency wetland delineation manual (FICWD 1989). The Corps manual suggests using six height classes for determining sapling/shrub (woody understory) dominants: (1) 1 to 3 ft (midpoint = 2), (2) 3 to 5 (m = 4), (3) 5 to 7 (m = 6), (4) 7 to 9 (m = 8), (5) 9 to 11 (m = 10), and (6) >11 (m = 12) (Environmental Laboratory, 1987). Unfortunately, the latter class may include plants to 30 to 40 ft tall, so the number of height classes are not sufficient and seem to favor lower growing species. This feature has not been widely used in wetland delineation work.

For wetland plant community descriptions, it may be worthwhile identifying height classes for each life form and for individual species. The following classes are recommended: (1) herbs and shrubs — very low (<1 ft), low (1 to 3 ft), medium-height (3 to 6 ft), tall (6 to 12 ft), and very tall (12 to 20 ft); and (2) trees and saplings — low (20 to 40 ft), medium-height (40 to 100 ft), and tall (>100 ft). This provides some useful data on the structure of the community.

DENSITY

Density is the number of individuals per unit area. Armstrong (1907; reported by Brown, 1954) may have been the first to use this measure. Stem counts are particularly useful when the individuals are the same size or for generating a weighted statistic (e.g., importance value) when used in combination with two variables (cover and frequency). For evaluating the woody vine and sapling (woody plant 20 ft or taller and less than 5.0 in. diameter at breast height) strata, counting the number of stems is a convenient, rapid assessment measure. If all the trees are approximately the same size, density also is useful since it should yield the same dominant species as would be derived from measuring the dbh and calculating basal area in much less time. *Relative density* is the density of one species expressed as a percent of the total plant density, with the sum of the individual species' densities equaling 100. When the plants to be counted are small and numerous, however, determining density can be time-consuming as in the case of counting stems of graminoids in wet meadows or open fields. For ecological studies of grasslands, it is a worthwhile statistic, but for wetland delineation in open lands, it is not recommended as other measures provide more cost-effective means of determining dominant species. In various ecological studies, several classes of abundance have been defined — very abundant, abundant, frequent, occasional, and rare (Tansley and Chipp, 1926) or very numerous, numerous, not numerous, sparse, and very sparse (Braun-Blanquet, 1932). Usually these terms are poorly defined and therefore are subjective. Better definitions, such as giving them density values, would make them more meaningful (Brown, 1954).

FREQUENCY

Frequency is another useful attribute in plant ecological studies. Reportedly first introduced by the Danish botanist Raunkiaer (Brown, 1954), it is defined as the number of occurrences of a particular species in a series of samples. It has been combined with density and cover to yield an index called *importance value* (Curtis and McIntosh, 1951; Mueller-Dombois and Ellenberg, 1974). In wetland delineation, frequency is used only in the 1989 manual's point intercept sampling technique, which consists of a series of point samples along transects (see below). Frequency then is the number of

Ch. 3, Plate 1. Adventitious roots and hyper-trophied lenticels in a flooded willow.

Ch. 5, Plate 5. Hydric Spodosol showing dark brownish spodic horizon and bleeding of spodic into E horizon above.

Ch. 3, Plate 2. Oxidized rhizospheres along living roots in a depleted soil matrix.

Ch. 5, Plate 6. Well-drained soil with no redox depletions or concentrations near the surface (Paxton series, Oxyaquic Dystrudepts).

Ch. 5, Plate 3. Peat (Waskish series, Typic Sphagnofibrists).

Ch. 5, Plate 7. Redox depletions (gray colors) and redox concentrations (orangish colors).

Ch. 5, Plate 4. Muck (Carlisle series, Typic Medisaprists).

Ch. 5, Plate 8. Oxidized pore linings with diffuse boundary.

Ch. 5, Plate 11. Thin muck over sand (A8).

Ch. 5, Plate 9. Organic soil from salt marsh (Ipswich series, Typic Sulfihemists).

Ch. 5, Plate 12. Blotchy, sandy soil (S6) with brownish ortstein (cemented layer) below.

Ch. 5, Plate 10. Histic epipedon (indicator A2) with depleted matrix (F3) left.

Ch. 5, Plate 13. Umbric epipedon (F13) with depleted matrix below (Pocomoke series, Typic Umbraquults).

Ch. 5, Plate 14. Mollic epipedon (F4) with depleted matrix below (Chobee series, Typic Argiaquolls).

Ch. 5, Plate 17. Loamy gleyed matrix (F2), bluish gray in color.

Ch. 5, Plate 20. Depleted matrix (F3) below dark surface (F4).

Ch. 5, Plate 15. Redox dark surface (F6) with remains of invertebrate shells evident.

Ch. 5, Plate 18. Gleyed matrix (F2) below plowed A-horizon.

Ch. 5, Plate 21. Mottled depleted matrix (F3).

Ch. 5, Plate 16. Loamy gleyed matrix (F2).

Ch. 5, Plate 19. Depleted matrix (F3).

Ch. 5, Plate 22. Iron and manganese masses (F12).

Ch. 5, Plate 23. Marl layer (F10) begins at the surface in this South Florida soil.

Ch. 5, Plate 24. Hydric nonsandy soil from Appalachian region (Nolo series, Typic Fragiaquults).

Ch. 5, Plate 25. Hydric nonsandy soil from the coastal plain (Perquimans series, Typic Ochraquults).

Ch. 5, Plate 26. Hydric nonsandy soil from the coastal plain (Hyde series, Typic Umbraquults).

Ch. 5, Plate 27. Hydric soil and adjacent nonhydric soil (left) from a northeastern meadow.

Ch. 5, Plate 28. Waterborne sediment deposits on rocks in floodplain wetland.

Ch. 5, Plate 29. Aquatic snails, duckweeds, and matted, water-stained leaves.

Ch. 5, Plate 30. Blackened, water-stained leaves from a depressional wetland.

Ch. 5, Plate 31. Iron precipitate film on water.

Ch. 7, Plate 34. Hemlock swamp in New York.

Ch. 7, Plate 37. Hydric sandy soil from coastal Texas (Mustang series, Typic Psammaquents).

Ch. 5, Plate 32. Moss-lichen line on buttressed trunk of swamp black gum.

Ch. 7, Plate 35. Hydric sandy soil with muck surface (A8), dark surface (S7), and blotchy subsoil.

Ch. 7, Plate 38. Hydric Spodosol from a northern evergreen forest.

Ch. 5, Plate 33. Peat moss, a reliable indicator of wetland hydrology in most of the eastern U.S.

Ch. 7, Plate 36. Hydric sandy soil with stripped matrix (S6) at 3-5 in. and sandy redox (S5) at 5-6 in.

Ch. 7, Plate 39. Hydric Spodosol from coastal Massachusetts.

Ch. 7, Plate 40. Hydric Spodosol from North Carolina (Lynn Haven series, Typic Alaquod).

Ch. 7, Plate 41. Hydric Mollisol from Illinois (Typic Argiaquolls).

Ch. 7, Plate 42. Hydric Mollisol from North Carolina (Picture series, Vertic Argiaquolls).

Ch. 7, Plate 43. Hydric soil (on right) and adjacent nonhydric soil from coastal plain flatwood.

Ch. 9, Plate 44. Northern fen (sedge meadow).

Ch. 9, Plate 45. Western wet meadow.

Ch. 9, Plate 46. Northern shrub bog (floating mat).

Ch. 9, Plate 47. New Jersey Pine Barrens' wetland (cedar swamp in background).

Ch. 9, Plate 48. Southern swamp.

Ch. 9, Plate 49. Northeastern salt marsh along a tidal river.

Ch. 10, Plate 50. Spring color-infrared photo (scale 1:12,000) showing red maple swamp (dark bluish area) and common reed marsh (light grayish blue area above landfill). Reddish clusters within the swamp are white pine.

Ch. 10, Plate 51. Spring color-infrared photo (scale 1: 58,000) showing deciduous forested wetlands (dark gray areas), cranberry bogs (orangish areas), and marshes (light gray areas along streams).

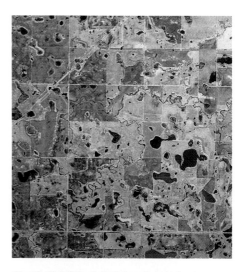

Ch. 10, Plate 52. Color-infrared photo (scale 1:65,000) showing North Dakota pothole marshes at high water in April.

Ch. 10, Plate 53. Color-infrared photo (scale 1:58,000) showing coastal plain flatwoods. The dark blue areas are seasonally flooded forested wetlands, whereas the lighter blue areas are temporarily flooded flatwoods. Red areas are cropland.

Ch. 10, Plate 54. Color-infrared photo (scale 1:58,000) showing mixed forested/shrub wetlands and riparian habitat along Nebraska's Platte River.

Ch. 10, Plate 55. True color photo (scale 1:14,400; December) showing Atlantic white cedar swamp (smooth green signature; top right), pitch pine lowlands (left), and brackish marsh (light brown to grayish areas along river).

times that individuals of a given species are counted or the percentage of total samples that a species occurs in the plots. For example, if a species is observed 10 times in 100 samples, its frequency of occurrence is 10%.

DOMINANCE

Dominant species may be defined differently for research studies than for wetland delineation assessments. In general terms, dominance may be referred to as the species representing the uppermost stratum having the greatest cover, for example. The dominant species in a forest may simply be the most abundant trees or the trees with the highest cover with little regard to understory or groundcover species. For example, the Society of American Forests has named forest types by their dominant trees (Eyre, 1980). In a savanna where trees are not particularly abundant, the dominant species may be the understory grasses or shrubs (Barbour et al., 1980). In other studies, dominants may include both the predominant overstory and understory species (e.g., red maple–tussock sedge association).

For wetland delineations, dominant species are identified for each stratum present in the community. Although a minimum cover was not specified in the federal manual, the Federal Interagency Committee for Wetland Delineation had planned on specifying 10% as the threshold coverage that would be significant enough to determine a dominant species in a stratum. In other words, if the stratum had 10% or more cover in the plot, dominants would be determined, and if not, the stratum was not evaluated since it is poorly represented. Dominant species were defined differently in the three manuals. The definition now used is from the 1989 interagency federal manual (FICWD 1989): "…dominant species are the most abundant plant species (when ranked in descending order of abundance and cumulatively totaled) that immediately exceed 50% of the total dominance measure (e.g., basal area or areal coverage) for the stratum, plus any additional species comprising 20% or more of the total dominance measure for the stratum." This definition employs the so-called "50/20 rule" to determine dominant species as shown in the example in Table 4.5. These species are the most "abundant" members of the community and are used, in the three-parameter/criteria methods to help interpret site wetness. Once dominant species are determined, they are treated equally to determine the presence of hydrophytic vegetation. Note that tree species may appear as a dominant in more than one stratum depending on their height and dbh. The U.S. Forest Service also uses this approach for defining dominant species in wetlands for its forest inventory and analysis work (Mark Brown, personal communication, 1998).

WEIGHTED AVERAGES

A vegetation index can be developed for evaluating the response of a plant community to an environmental gradient. To identify possible wetlands, species abundance and their wetland indicator status (their frequency of occurrence in wetlands) can be used to generate a statistic of site "wetlandness," a measure of the likelihood that an area is a wetland. Michener (1983) was the first to propose a wetland site index using the abundance of plant species and their wetland indicator status as defined by the U.S. Fish and Wildlife Service's regional wetland plant panels. Wentworth and Johnson (1986) further examined the use of weighted averages for making wetland determinations. In this context, a weighted average for a stand or plant community (or plot) is established by taking the importance value (e.g., cover) of each species, multiplying it by a numerical value representing its ecological indicator value (i.e., its corresponding wetland indicator status as defined by Reed, 1988; 1997; OBL = 1.0, FACW = 2.0, FAC = 3.0, FACU = 4.0, and UPL = 5.0), summing these values and then dividing by the sum of the importance values (cover) of all species in the stand. The following equation can be used for calculating weighted averages.

TABLE 4.5
Example of Determining Dominant Species Following the 50/20 Rule in the Federal Interagency Manual

Species	Wetland Indicator	% Cover
Trees		
Acer rubrum	FAC	70
Liquidambar styraciflua	FAC	20
Quercus alba	FACU	10
Fraxinus pennsylvanica	FACW	30
Total cover =		130
(Dominance threshold) 50% of Total cover =		65
20% of Total cover =		26
Shrubs		
Ilex opaca	FACU+	20
Vaccinium corymbosum	FACW–	5
Acer rubrum	FAC	10
Quercus alba	FACU–	10
Total cover =		45
(Dominance threshold) 50% of Total cover =		22.5
20% of Total cover =		9
Herbs		
Thelypteris noveboracensis	FAC	5
Mitchella repens	FACU	20
Maianthemum canadense	FAC–	40
Total cover =		65
(Dominance threshold) 50% of Total cover =		32.5
20% of Total cover =		13

Applying the 50/20 rule: Dominant species are determined by first seeing if any single species represents more than half of the total cover (= dominance threshold) for a stratum. So, for the tree stratum, *Acer rubrum* is a dominant (70% > 65%); then check to see if any other species represent 20% or more of the total cover for the tree stratum; then *Fraxinus pennsylvanica* also becomes a dominant. For the shrub stratum, there is no one species that represents more than half of the total shrub cover, so more than one species is needed to exceed the dominance threshold. Starting with top-ranked shrub *Ilex opaca*, then next ranked, *Acer rubrum* and *Quercus alba* have the same cover (tied), so both are taken to exceed the dominance threshold number. Three shrubs are dominant, *Ilex opaca*, *Acer rubrum*, and *Quercus alba*. The remaining shrub does not have more than 20% of the total shrub cover (9%). For herbs, *Maianthemum canadense* alone represents more than half of the herb cover; then check for species with more than 20% of the total herb cover (=13%); so *Mitchella repens* also is a dominant herb.

Total number of dominants	= 7
Number of dominants with FAC, FACW, or OBL status including FAC–	= 4 (57.1% of dominants)
Number of dominants with FAC, FACW, or OBL status (excluding FAC–)	= 3 (42.9% of dominants)
Is this community an indicator of hydrophytic?	Yes, following federal interagency manual, No, following Corps manual.

$$W_j = \frac{\sum_{i=1}^{p} I_{ij} E_i}{\sum_{i=1}^{p} I_{ij}}$$

where: W_j = weighted average for stand j

   I_{ij} = importance value for species i in stand j (e.g., cover or frequency)

   E_i = ecological index for species i

   p = the number of species occurring in the stand

This analysis will yield a weighted average value for the community somewhere between 1.0 and 5.0 (see example in Table 4.6).

   This approach can be used for plot data (cover) and point data (frequency analysis). The weighted average index is also referred to as the prevalence index especially when point data are used (see Point Intercept Sampling discussion).

   A plant community with a weighted average index (prevalence index) of 3.0 (±0.5) therefore is equivalent to a FAC species that occurs equally in wetlands and nonwetlands. Such communities (2.5 through 3.5) are inconclusive regarding their wetland status as assessed by vegetation analysis alone; in other words, other features must be examined to determine whether they are wetland or not. As the index value moves toward 1.0, the likelihood of the community being a wetland increases, and as the index goes toward 5.0, its tendency for being wetland diminishes, while its probability for being an upland increases. Plant communities with index values of less than 2.0 should be wetlands (as they are dominated by OBL species) and communities with values greater than 4.0 should be uplands (dominated by UPL species), with rare exceptions. Communities with values between 2.0 and 2.5 have a good probability of being wetlands but their soils and hydrology should

---

TABLE 4.6

Sample Calculation of Weighted Average for a Wetland Plant Community Derived from Areal Cover Estimates of Species Within a Representative Plot

Species	% Cover in Plot	Ecological Index	Product (Cover × Index)
Acer rubrum	90	3	270
Fraxinus pennsylvanica	20	2	40
Vaccinium corymbosum	30	2	60
Viburnum recognitum	10	2	20
Carex stricta	25	1	25
Leersia oryzoides	25	1	25
Symplocarpus foetidus	5	1	5
Peltandra virginica	1	1	1
Total	206		446

Note: Weighted Average = Sum of the products/sum of the covers = 446/206 = 2.165. Ecological index is assigned based on the wetland indicator status of the species (OBL = 1; FACW = 2; FAC = 3; FACU = 4; UPL = 5).

be confirmed to verify the wetland designation. Likewise, most communities with values between 3.5 and 4.0 are probably uplands, with few exceptions, yet soils and hydrology should be evaluated especially in landscape positions favoring wetlands (Wentworth and Johnson, 1986; Wentworth et al., 1988). Atkinson et al. (1993) used the weighted averaging approach to evaluate the effectiveness of forested wetland creation projects in Virginia and suggested that pre-impact vegetation analysis (or analysis of reference wetland) may be useful in monitoring success of such projects.

BASIC VEGETATION SAMPLING TECHNIQUES

Sampling may be done within subjectively established communities for either research or wetland delineation purposes. Plant communities are identified with no perceived bias and their vegetation and other characteristics analyzed. When interested in detecting vegetation changes along some type of environmental gradient, the transect approach is usually followed. Transects are lines or a series of subplots within an elongate rectangular plot running down the gradient of interest. Transects may represent the sampling unit itself as in line intercept method or belt transect (strip transect) method (Lindsey, 1955), but more frequently represent the lines along which quadrat (plot) samples will be taken either at fixed intervals or in different plant communities encountered along the line. For wetland delineation, transects are oriented downslope, since sampling along a topographic gradient will likely reveal difference in plant composition related to changing soil types and hydrology.

The vegetation sampling methods fall into three general categories — plot-based methods, point-based methods, and plotless methods. The former requires analyzing vegetation in a specified sampling area called a plot, whereas point sampling is a type of plot sampling since the smallest plot size is a point. Plotless methods involve making observations along transects or at random points in the entire plant community. For wetland delineation, plot-based methods are most commonly used. These methods are generally summarized below, with specific wetland delineation procedures described in Chapter 6.

PLOT OR QUADRAT SAMPLING

The quadrat sampling method is the most popular method for ecological research and wetland delineation purposes. Quadrat sampling involves making observations within an area that is representative of the community as a whole. It requires some knowledge of the overall plant community, sufficient enough to locate one or more areas within a stand that are typical of the whole. Two English botanists, F.W. Oliver and A.G. Tansley (1904) were among the first plant ecologists to use this sampling technique: for analyzing salt marsh vegetation in the Bouche d'Erquay estuary. In the early 1900s, American ecologist J.W. Harshberger used this technique to describe vegetation in a number of habitats, most notably for the New Jersey Pine Barrens (Harshberger, 1970). His plot size was a 10 m × 10 m square. He actually mapped the individual plants by species found in smaller 1 m × 1 m squares within the plot (100 subplots for the entire nested plot) which was quite time consuming as he could do only three 100 m² plots a day.

Sample and Plot Sizes

Sample size relates to the type of plant community being evaluated, specifically its diversity. Plot size varies according to the complexity of the vegetation pattern. In general, the more diverse and interspersed the vegetation, the smaller the plot size for efficient and accurate estimating. The number of plots required should be sufficient to at least sample the so-called "minimum area" based on minimum area determinations for the plant community in question (see discussion of minimum area in Number of Plots next). Table 4.7 compares recommended sample sizes from federal wetland delineation manuals with recommended minimum sizes based on ecological studies of different

TABLE 4.7
**Recommended Minimum Sampling Areas for Different Plant Communities Based
on Minimum Area Determinations**

Plant Community or Stratum	Minimum Sampling Area	Source
Bryophytes/lichens	0.25 m^2	Kent and Coker, 1992
	0.1 to 4 m^2	Barbour et al., 1980
	1 to 4 m^2	Gauch, 1982
Dune grasslands	0.13 m^2	Barbour et al., 1980
	1 to 10 m^2	Gauch, 1982
Salt marshes	2 to 10 m^2	Gauch, 1982
Pastures	25 to 50 m^2	Cain and de O. Castro, 1959
Hay meadows	25 to 100 m^2	Cain and de O. Castro, 1959
	10 to 25 m^2	Gauch, 1982
Alpine meadows	10 to 50 m^2	Gauch, 1982
Dry temperate grasslands	50 to 100 m^2	Barbour et al., 1980
Grasslands, dwarf heaths	1 to 4 m^2	Kent and Coker, 1992
Wet meadows	5 to 10 m^2	Barbour et al., 1980
Herb stratum[a]	78.5 ft^2 (7.3 m^2)	Environmental Laboratory, 1987; FICWD, 1989
Shrub heaths, tall herbs and grasses	4 to 16 m^2	Kent and Coker, 1992
Shrub heaths	10 to 25 m^2	Barbour et al., 1980
Dwarf shrubland	10 to 50 m^2	Gauch, 1982
Scrublands	50 to 250 m^2	Cain and de O. Castro, 1959
Scrublands, shrubby woodlands	100 m^2	Kent and Coker, 1992
Shrub stratum[a]	78.5 ft^2 (7.3 m^2)	Environmental Laboratory, 1987
	314.2 ft^2 (29.2 m^2)	Environmental Laboratory, 1987
	2,826 ft^2 (263 m^2)	FICWD, 1989
English woodlands	100 m^2	Barbour et al., 1980
Boreal coniferous and temperate deciduous forests	100 to 500 m^2	Cain and de O. Castro, 1959; Gauch, 1982
Mixed deciduous forests (North America)	200 to 800 m^2	Gauch, 1982
Forests	400 to 2500 m^2	Kent and Coker, 1992
Forests	375 to 500 m^2	Various U.S. Forest Service Reports (e.g., Cooper et al., 1991)
Subtropical mixed forests	200 to 1000 m^2	Cain and de O. Castro, 1959
Tree overstory	200 to 500 m^2	Barbour et al., 1980
Tree understory	50 to 200 m^2	Barbour et al., 1980
Tree stratum[a]	2826 ft^2 (263 m^2)	Environmental Laboratory, 1987; FICWD, 1989
Sapling stratum[a]	2826 ft^2 (263 m^2)	FICWD, 1989
Tropical high rainforests	800 to 40,000 m^2	Cain and de O. Castro, 1959
Rainforests valley bottoms	20,000+ to 50,000 m^2	Barbour et al., 1980
Ridges	1000 m^2	Barbour et al., 1980
Tropical swamp forest	2000 to 4000 m^2	Gauch, 1982

[a] Federal wetland delineation manual plot size. For herbs the federal interagency manual also permits sampling of an area slightly larger than the "minimum area" determined for the specific community being evaluated (FICWD, 1989).

plant communities. The plot method should work for temperate, arid, boreal, alpine, and arctic communities, but may not be appropriate for subtropical and tropical communities, especially rainforests, due to the enormous minimum area required to adequately sample these environments.

Federal wetland delineation manuals recommend plot sizes that correspond to different strata. The federal interagency manual notes that plot sizes may be changed to suit site conditions. For example, larger plot sizes should be evaluated for more diverse communities. Using the Corps manual, the recommended plot size for the woody understory is a 10-ft radius plot in the comprehensive method or a 5-ft radius plot for the routine method. The rationale for this difference in not known, yet the 5-ft radius plot is likely to be too small, especially for sampling woody understory in forests, although it may be appropriate for some shrub communities such as bogs (see Table 4.7). In contrast, the 1989 federal interagency manual recommends a 30-ft radius plot for the shrub stratum. This plot size would be too large for a shrub bog and result in inefficient sampling, so plot size may be reduced accordingly. Recommended plots for evaluating the herb stratum are (1) a single 5-ft radius plot (the routine methods of both manuals and the intermediate method of interagency manual); (2) a series of 1-square m quadrats at fixed intervals along a transect (number variable depending on the length of the plant community; comprehensive methods of both the Corps and interagency manuals); (3) a series of 1.64-ft radius plots at fixed intervals along a transect (Corps comprehensive determination); and (4) multiple, randomly located small-plots within the 30-ft radius plot (the number of small plots based on minimum area determination for the community; interagency manual comprehensive method). The federal interagency manual lists three recommended sizes for the small plots: 8 × 20 in. quadrat, 20 × 20 in., or 40 × 40 in. This sampling technique also is useful for sampling shrub bogs with diverse flora.

Number of Plots

The number of plots in ecological studies is either determined by an assessment of minimum area (or using results from similar studies for the habitat in question) or by ensuring that 1 to 20% of the stand is sampled (Barbour et al., 1980). The former approach is more accurate and probably more efficient. One unbiased representative plot (greater than the minimum area) is satisfactory for general classification purposes, but multiple small plots (in total, greater than the minimum area) are best for ecological studies requiring statistical analysis.

For wetland delineation, the number of plots is usually left to the observer, except for multiple small-plot sampling of the herb stratum where the minimum area is determined to establish the appropriate number of sample plots (comprehensive method of the 1989 manual). In most cases, one 30-ft radius plot is used to characterize the community for wetland delineation purposes in forests. One or more additional plots may be necessary for evaluating plant communities with essentially the same overstory species but significantly different understory occurring on the same soil type and apparent hydrologic conditions. To compute dominant species, the results from the plots in the community are totaled and averaged. Although this would not be done for ecological studies, it is acceptable for wetland delineation, since the goal is to separate wetlands from uplands and not one wetland plant community from another.

Using federal wetland delineation methods (comprehensive approach), the number of small plots for the herb stratum required for sampling varies from one community to the next. It is determined by collecting multiple samples (of the same size) until approximately 80 to 90% of the species in the stratum are sampled. This requires plotting the number of new species observed in each sample on a graph and constructing a species-area curve. The number of species observed will increase as the area sampled increases (with analysis of additional small-plot samples) until nearly all of species associated with this community are tabulated. At this point, the number of new species becomes zero or significantly drops as only rare or uncommon species are encountered. Since the species-area curve charts the number of new species observed as the area of sampling is increased, the curve will rise (with additional sampling) until flattening out (Figure 4.1b). The point

FIGURE 4.4   Estimating areal cover in a 3 × 3 ft plot (using carpenter's rule).

of inflection on the curve defines the minimum area. The minimum area is the smallest area necessary to adequately represent the structure and composition of the plant community. One always samples a little more than the minimum area to characterize the community (e.g., more than 80% of the species in temperate wetland communities).

Small-plot samples are taken within the 30-ft radius plot as follows. First, the 30-ft radius plot is divided into four equal-sized quadrants, then a sample frame (such as a 6-ft collapsible carpenter's ruler) is randomly tossed in the first quadrant (Figure 4.4). Data recorded include a list of species observed in each plot and the estimated cover of each (Table 4.8). The number of species observed in the first sample is plotted on the species-area graph. After completing the first sample, other random samples are taken in the other quadrants with new species added to the previous total observed and that point plotted on the graph. This procedure is continued until the species-area curve flattens out. Although this analysis is recommended for use in the comprehensive method, it is also useful for sampling diverse herbaceous communities and in forests with diverse or scattered herb cover where a 5-ft radius plot would not adequately reflect the herb community.

For ecological studies, a supplemental list of species observed in the plant community but not found in the plots should be compiled. Some of these species may be useful indicators for classifying vegetation. Such list also may be advisable for wetland delineation work as it provides a more complete description of the plant community.

Shape of Plots

Shape of plots may have an effect on the results and are a major concern for ecological studies. Accuracy of sampling may decrease with increased plot length due to the edge effect (e.g., more decisions to make regarding a plant is "in" or "out" of the plot, hence more opportunities for mistake) (Barbour et al., 1980). Given this observation, circular plots are recommended since a circle has the smallest perimeter per given area. Some researchers also claim that such plots are easier to lay out in the field with a tape and center stake. There may be increased uncertainty when determining whether a plant is "in" or "out" of the plot with a circular plot vs. a square or rectangular plot, since the arc will have to be interpreted. Typically, four flags are used to mark the boundary of a circular plot which may cause some uncertainty in defining the arc of the circle between flagged portions of the circumference. Where this is a problem, a total of eight flags should be used. Federal manuals recommend the use of mostly circular plots, except for small-plot samples where square plots are suggested. The 1989 manual notes that plot size and shape may be varied to meet site conditions. Rectangular plots are appropriate for evaluating linear communities such

TABLE 4.8
Data from Multiple-Small Plot Sampling in a Coastal Bog in Eastern Maine Used to Construct the Species-Area Curve in Figure 4.5

Species	Percent Cover in Quadrats										
	Q1	Q2	Q3	Q4	Q5	Q6	Q7	Q8	Q9	Q10	$\overline{X}$
Ledum groelandicum	1.0	—	2.0	20.0	5.0	15.0	5.0	4.0	2.0	—	5.4
Chamaedaphne calyculata	5.0	10.0	30.0	10.0	25.0	20.0	20.0	70.0	13.0	30.0	23.3
Nemopanthus mucronata	1.0	—	—	—	—	—	—	—	—	7.0	0.8
Kalmia angustifolia	2.0	40.0	5.0	1.0	25.0	45.0	5.0	10.0	—	35.0	16.8
Aronia melanocarpa	12.5	—	0.5	—	1.0	5.0	1.0	3.0	—	3.0	2.6
Larix laricina	0.5	—	—	—	—	—	—	5.0	—	—	0.6
Kalmia polifolia	1.0	—	—	—	10.0	20.0	2.0	—	—	—	3.3
Picea mariana	0.5	—	0.5	25.0	1.0	—	—	1.0	75.0	3.0	10.6
Gaylussacia dumosa			20.0	40.0	1.0	—	80.0	—	—	—	14.1
Myrica gale			2.0	—	10.0	—	—	—	8.0	—	2.0
Vaccinium angustifolium			0.5	—	—	—	—	—	—	—	—
Empetrum nigrum					50.0	50.0	40.0	30.0	—	10.0	18.0
Drosera rotundifolia					1.0	5.0	1.0	5.0	t	—	1.2
Sarracenia purpurea					5.0	7.0	—	2.0	t	5.0	1.9
Vaccinium oxycoccus					1.0	1.0	—	1.0	—	t	0.3
Scirpus cespitosus					20.0	—	—	—	—	30.0	5.0
Solidago uliginosa							1.0	—	—	5.0	0.6
Trientalis borealis									t	3.0	0.3
Aster nemoralis									t		—
Myrica pennsylvanica										3.0	0.3

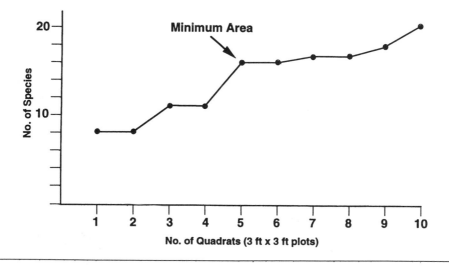

Note: In this sample, 8 quadrats should be sufficient (85% of species tallied, 72 ft² sampled). (t = trace.)

as ridge and swale communities or plant communities in narrow valleys. Hansen et al. (1991) used 2.5 m × 20 m plots to sample wetland–riparian communities fringing streams in Montana.

POINT INTERCEPT SAMPLING

The origins of the point intercept sampling are not well defined. Brown (1954) notes that in its earliest form, the points were identified by the "toe-cap of a boot" in studies of tussock vegetation in New Zealand. Later, Levy and Madden (1933) refined the technique by using knots on a string to identify sample points. This technique is very effective for evaluating grassland and rangeland vegetation and also has been used to assess tree canopy cover (Hays et al., 1981; Higgins et al., 1994).

The point intercept method is one of the methods recommended by the 1989 manual. It is a plot-based method with a point representing the smallest possible plot — a series of points are sampled at fixed intervals along a line (transect). The technique originated from grassland ecology studies where a portable sampling frame with numerous pins (or rods) was placed over the vegetation. The pins would then be lowered and any plant touched would be tallied as a "hit." Some researchers counted only those plants whose bases were touched. The frame would be randomly located in the stand. Percent cover by species can be determined by dividing the number of tallies by the number of pins moved and multiplying this figure by 100. The height of the frame was adjusted to the vegetation; for low grasslands, the frame was 8 in. high; for taller grasses, it was 12 in. or more in height (Brown, 1954). Levy and Madden (1933) considered 100 points to be sufficient where information on dominants was the study focus, and where information on less common species was sought, 400 to 500 points were recommended. Other researchers concluded that 200 points were suitable for dominant determinations (Crocker and Tiver, 1948). The point intercept technique has been claimed to be the "most trustworthy and objective method available" for measuring low vegetation (Goodall, 1957, as reported in Barbour et al., 1980). Although the device seems cumbersome, the pin system does reduce the subjectivity of simply looking down at the vegetation at sample points. Besides grasslands (including wet meadows), the technique may be suitable for conducting research studies of ericaceous shrub bogs and other habitats of low shrubs.

For wetland delineation, the use of the sampling frame is not recommended. Reasonable results can be gained by simple observations without the pinpoint accuracy of the frame. Moreover, the method is not restricted to open grasslands and low shrubs, but also may be used in forests. As used for wetland delineation, the method generates a frequency analysis of species along multiple transects. The frequency of occurrence of species is determined by taking point samples at 2-ft intervals along three or more randomly selected 200-ft transects. A total of 100 observations are made per transect and at least 300 per stand. Counts are made by the number of individuals by species that intersect a sample point extending from the tip of one's boot or shoe or the marked sample point on a 200-ft tape. If two trees of the same species intersect the point, two tallies for the species are recorded. At the end of sampling a transect, one has compiled a list of species with the number of "hits" or tallies (see example in Table 4.9). These data are then used to calculate a prevalence index for the transect and then a mean prevalence index for the plant community (by averaging the indices for the transects). The number of transects analyzed is usually three, but may be more where the standard error of the mean prevalence index is greater than 0.20 or 20% of the mean value. A mean prevalence index of less than 3.0 is typically used as a positive indicator of hydrophytic vegetation, with analysis of soils and hydrology still required for wetland determinations.

OTHER SAMPLING METHODS

Line intercept sampling was reportedly developed for grassland studies (Brown, 1954). Prominent American ecologists advancing this method were Canfield (1941), Anderson (1942), Bauer (1943), and Parker and Savage (1944). Most of these ecologists used the technique for grasslands, while Bauer used it in his studies of California's chaparral. This technique also can be applied to forests, but is most efficient for open areas with scattered trees and shrubs rather than for dense forest stands.

The line intercept method is commonly used for analyzing vegetation along a gradient to record changes in the plant community with topographic changes. All species covering the line are recorded and the percent of the line covered by a given species is its cover, so total cover can exceed 100% for the total line segment. When used to assess vegetation along an environmental gradient, one can easily see where species enter and depart the community. These shifts are correlated to various

TABLE 4.9
Sample Calculation of Prevalence Index (PI) Based on Weighted Averages Derived from the Frequency of Occurrence of Species Along Three Transects for a Coastal Bog in Maine

| | | Frequency of Occurrence of Identified Plants with Known Indicator Status | | | | |
| | | | | Frequency of Occurrence | | |
Species	Total for Species along Transect	Freq. OBL (F_O)	Freq. FACW (F_{fw})	Freq. FAC (F_f)	Freq. FACU (F_{fu})	Freq. UPL (F_u)
Empetrum nigrum	38		38			
Chamaedaphne calyculata	42	42				
Sarracenia purpurea	4	4				
Gaylussacia dumosa	17			17		
Myrica gale	4	4				
Osmunda cinnamomea	1		1			
Nemopanthus mucronata	5	5				
Kalmia angustifolia	26			26		
Picea mariana	7		7			
Kalmia polifolia	3	3				
Scirpus cespitosus	15	15				
Ledum groelandicum	8	8				
Aronia melanocarpa	4			4		
Larix laricina	4		4			
Thuja occidentalis	5		5			
Trientalis borealis	1			1		
Vaccinium oxycoccus	1	1				
Subtotals	185	82	55	48	0	0

$$PI_i = \frac{\left(1 \times F_O\right) + \left(2 \times F_{fw}\right) + \left(3 \times F_f\right) + \left(4 \times F_{fu}\right) + \left(5 \times F_u\right)}{\left(F_O + F_{fw} + F_F + F_{fu} + F_u\right)}$$

$$PI_1 = \frac{\left(1 \times 82\right) + \left(2 \times 55\right) + \left(3 \times 48\right) + \left(4 \times 0\right) = \left(5 \times 0\right)}{185} = \frac{82 + 110 + 144}{185} = 1.82$$

PI for Transect 1 = 1.82
PI for Transect 2 = 1.88
PI for Transect 3 = 2.10
5.73   Mean PI_M = 1.93

Compute standard deviation (s) and standard error (sx̄) for the mean prevalence index:

$$s = \sqrt{\frac{\left(PI_1 - PI_M\right)^2 + \left(PI_2 - PI_M\right)^2 + \left(PI_3 - PI_M\right)^2}{N-1}}$$

TABLE 4.9 (continued)
Sample Calculation of Prevalence Index (PI) Based on Weighted Averages Derived from the Frequency of Occurrence of Species Along Three Transects for a Coastal Bog in Maine

Transect	PI_i	PI_M	$(PI_i - PI_M)$	$(PI_i - PI_m)^2$
1	1.82	1.93	-0.10	0.0100
2	1.88	1.93	-0.05	0.0025
3	2.10	1.93	+0.17	0.0289
				0.0414

Total

$$s = \sqrt{\frac{0.0414}{3-1}} = \sqrt{\frac{0.0414}{2}} = \sqrt{0.0207} = 0.144$$

$$s\bar{x} = \frac{s}{\sqrt{N}} = \frac{0.144}{\sqrt{3}} = \frac{0.144}{1.73} = 0.083$$

Since 0.083 does not exceed 0.20, no additional sampling is required, so the mean prevalence index (PI_M) of the community is 1.93.

Note: The standard deviation and standard error are calculated to determine variance; if the standard error is above 0.20, additional sampling is required. (OBL = Obligate Wetland; FACW = Facultative Wetland; FAC = Facultative; FACU = Facultative Upland; UPL = Upland.)

environmental variables such as slope, soils, and water tables. Sometimes this approach is combined with plot sampling to periodically collect information on density and cover in addition to the frequency data gathered by line intercept sampling. When the plots run continously along the line, the method is called the *belt transect method*, which has been frequently used in forest studies (Lindsey, 1955, as reported in Barbour et al., 1980). Cooper et al. (1997) used a line perpendicular to the slope to locate and analyze vegetation in Montana's Beaverhead National Forest. Out of the 23 community types described, four were designated as wetlands following 1989 manual procedures.

The *plotless methods* do not utilize specific plots or points for sampling. They are used for analyzing forest vegetation, namely the tree stratum. Most involve randomly locating an "observation point" and recording the distance to and the species of either the nearest tree (nearest individual method), the plant nearest to the closest plant (nearest neighbor method), or the nearest tree in each of four quadrants (point-centered quarter method), and determining the basal area (see plant ecology texts for details).

Only one plotless method is referred to in the federal wetland delineation manuals — the Bitterlich variable plot method that was listed in the EPA manual (Sipple, 1988) and the 1989 manual (FICWD, 1989; plotless comprehensive method). This method is used for evaluating the tree stratum. The technique was developed by a German researcher named Bitterlich who used a 1 m-long sighting stick with a 1.4 cm crosspiece to selectively tally individual trees based on their distance from the point of observation and their diameter at breast height (dbh) (Barbour et al., 1980). Today, the stick has been replaced by angle-gauges ("cruz-all") or basal area factor (BAF) prisms (see Figure 4.3). The method is plotless because the observer looks at all trees in the stand (regardless of their distance from the observation point), making a full sweep (360°). He or she tallies only those trees whose dbh exceeds the width of the crosspiece or width of the angle, or whose trunk image is not completely offset from the trunk above and below the prism. Borderline trees (e.g., look to be equal to the width of the angle, but can't tell for certain) should be counted as 0.5 (half a tree). Using these devices, a number of individual trees are tallied. The number of tallied individuals are separated by species and total basal area for the species can be estimated. For example, each tree tallied with a 10 BAF prism or the 10-angle on the angle-gauge has an estimated basal area of 10-square ft/acre.

EXERCISES IN DETERMINING THE PRESENCE OF HYDROPHYTIC VEGETATION

A few sample problems are given below for determining dominant species and assessing whether the plant community represents a positive indicator of hydrophytic vegetation according to either the Corps manual (Environmental Laboratory, 1987) or the federal interagency manual (Federal Interagency Committee for Wetland Delineation, 1989).

Exercise 1. Using the plot sampling data below, answer the following questions:

1. What are the dominant species in each stratum?
2. Do more than 50% of the dominants have an indicator status of FAC or wetter? (Federal interagency manual = hydrophytic vegetation)
3. Do more than 50% of the dominants have an indicator status of FAC or wetter, excluding FAC–? (Corps manual = hydrophytic vegetation indicator)

Stratum	Species (Wetland Indicator Status)	Sampling Variable (% cover)
Tree	Acer rubrum (FAC)	80
	Liquidambar styraciflua (FAC)	50
	Pinus taeda (FAC–)	30
Shrub	Clethra alnifolia (FAC)	30
	Acer rubrum (FAC)	30
	Vaccinium corymbosum (FACW–)	25
	Viburnum dentatum (FAC)	20
	Ilex verticillata (FACW+)	5
Herb	Chasmanthium laxum (FAC)	35
	Carex lupuliformis (FACW+)	10
	Mitchella repens (FACU)	5

Exercise 2. Using the plot sampling data below, answer the following questions:

1. What are the dominant species in each stratum?
2. Do more than 50% of the dominants have an indicator status of FAC or wetter? (Federal interagency manual = hydrophytic vegetation)
3. Do more than 50% of the dominants have an indicator status of FAC or wetter, excluding FAC–? (Corps manual = hydrophytic vegetation indicator)

Stratum	Species (Wetland Indicator Status)	Sampling Variable
Tree	Acer rubrum (FAC)	650 square in. (basal area)
	Liriodendron tulipifera (FACU)	425 square in.
	Liquidambar styraciflua (FAC)	210 square in.

Stratum	Species (Wetland Indicator Status)	Sampling Variable
	Ilex opaca (FACU)	135 square in.
	Quercus alba (FACU–)	50 square in.
Sapling	*Acer rubrum* (FAC)	5 stems
	Liquidambar styraciflua (FAC)	10 stems
	Ilex opaca (FACU)	2 stems
	Carpinus caroliniana (FAC)	1 stem
Shrub	*Viburnum prunifolium* (FACU)	25% cover
	Acer rubrum (FAC)	5% cover
	Ilex verticillata (FACW+)	2% cover
	Gaylussacia baccata (FACU)	10% cover
Herb	*Gaultheria procumbens* (FACU)	10% cover
	Mitchella repens (FACU)	1% cover
Woody Vine	*Toxiocodendron radicans* (FAC)	5 stems
	Parthenocissus quinquefolia (FACU)	10 stems
	Smilax rotundifolia (FAC)	25 stems

Exercise 3.   Using the point intercept data given, answer the following questions:

1. What is the prevalence index for each transect?
2. What is the mean prevalence index for the community?
3. What is the standard error of the mean prevalence index?
4. Is additional sampling required to get standard error below 0.20?
5. Does the data suggest that this community is a positive indicator for hydrophytic vegetation?

Transect 1: 10 tallies of OBL plants, 50 tallies of FACW, 100 tallies of FAC, 20 tallies of FACU, and 0 tallies of UPL.

Transect 2: 5 tallies of OBL, 50 tallies of FACW, 120 tallies of FAC, 10 tallies of FACU, and 2 tallies of UPL.

Transect 3: 0 tallies of OBL, 60 tallies of FACW, 90 tallies of FAC, 5 tallies of FACU, and 0 tallies of UPL.

Exercise 4.   Using the point intercept data given, answer the following questions:

1. What is the prevalence index for each transect?
2. What is the mean prevalence index for the community?
3. What is the standard error of the mean prevalence index?
4. Is additional sampling required to get standard error below 0.20?
5. Does the data suggest that this community is a positive indicator for hydrophytic vegetation?

Transect 1: 0 tallies of OBL plants, 50 tallies of FACW, 60 tallies of FAC, 37 tallies of FACU, and 0 tallies of UPL.

Transect 2: 0 tallies of OBL, 30 tallies of FACW, 45 tallies of FAC, 30 tallies of FACU, and 2 tallies of UPL.

Transect 3: 0 tallies of OBL, 15 tallies of FACW, 70 tallies of FAC, 20 tallies of FACU, and 0 tallies of UPL.

ANSWERS

Exercise 1.   Total tree cover is 160%, so 50% of total cover = 80%; *Acer rubrum* has the highest cover (80%) and is a dominant, but it does not represent more than half of the tree cover, so the next ranked species is needed to exceed the "dominance threshold" (>50% of the total cover); *Liquidambar styraciflua* is another dominant tree. Any remaining species with a cover of ≥20% of total cover (32% in this case) is also dominant. *Pinus taeda* is not dominant because its cover is less than 20% of the total cover. Total shrub cover is 110%, so 50% of this is 55%; no shrub species has >55%, so one must take the top-ranked shrubs. *Clethra alnifolia* and *Acer rubrum* when combined (60%) exceed the dominance threshold (55%), so both are considered dominant shrubs. Now, check the 20% threshold (20% of total shrub cover = 22%); only one of the remaining shrubs has more than this, so *Vaccinium corymbosum* (25%) is also listed as a dominant shrub. Total cover for the herb stratum is 50%; 50% of total cover then is 25%. *Chasmanthium laxum* exceeds this so it is a dominant herb; check for 20% threshold (20% of total herb cover = 10%), so *Carex lupiliformis* is also dominant. The dominants for the community are *Acer rubrum* (tree), *Liquidambar styraciflua* (tree), *Clethra alnifolia* (shrub), *Acer rubrum* (shrub), *Vaccinium corymbosum* (shrub), *Chasmanthium laxum* (herb), and *Carex lupuliformis* (herb). All seven dominants have a wetland indicator status of FAC or wetter, so there is a positive indicator of hydrophytic vegetation using either the Corps or federal interagency manuals.

Exercise 2.   Dominants are *Acer rubrum* (tree), *Liriodendron tulipifera* (tree), *Liquidambar styraciflua* (sapling), *Acer rubrum* (sapling), *Viburnum prunifolium* (shrub), *Gaylussacia baccata* (shrub), *Gaultheria procumbens* (herb), *Smilax rotundifolia* (vine), and *Parthenocissus quinquefolia* (vine). Only 4 (or 44%) of the 9 dominants are FAC or wetter, so the plant community is not a hydrophytic vegetation indicator by either Corps or federal interagency manuals.

Exercise 3.   PI for transect 1 = 2.72; PI for transect 2 = 2.75; PI for transect 3 = 2.65; Mean PI for the community = 2.71; standard error = 0.051; no additional sampling is required as the standard error is less than 0.20; this community is a positive indicator for hydrophytic vegetation.

Exercise 4.   PI for transect 1 = 2.91; PI for transect 2 = 3.04; PI for transect 3 = 3.05; Mean PI for the community = 3.00; standard error = 0.078; no additional sampling is required as the standard error is less than 0.20; this community is not a positive indicator for hydrophytic vegetation following the basic rule of the federal interagency manual (PI must be <3.0), but it may be a problem wetland, so soils should be examined.

REFERENCES

Anderson, K.L. 1942. A comparison of line transects and permanent quadrats in evaluating composition and density of pasture vegetation on the tall prairie grass type. *J. Amer. Soc. Agron.,* 34: 805-822.

Armstrong, S.F. 1907. The botanical and chemical composition of the herbage of pastures and meadows. *J. Agric. Sci.,* 2: 283-304.

Atkinson, R.B., J.E. Perry, E. Smith, and J. Cairns, Jr. 1993. Use of created wetland delineation and weighted averages as a component of assessment. *Wetlands,* 13: 185-193.

Barbour, M.G., J.H. Burk, and W.K. Pitts. 1980. *Terrestrial Plant Ecology.* Benjamin/Cummings Publishing Co., Menlo Park, CA.

Bauer, H.L. 1943. The statistical analysis of chaparral and other plant communities by means of transect samples. *Ecology,* 24: 45-60.

Bell, J.F. and J.R. Dilworth. 1978. *Variable Probability Sampling. Variable Plot and Three P.* Oregon State University Book Stores, Inc., Corvallis.

Beltz, R.C., D.F. Bertelson, J.L. Faulkner, D.M. May. 1992. Forest Resources of Arkansas. Res. Bull. SO-169. USDA Forest Service, Southern Forest Expt. Station, New Orleans, LA.

Braun-Blanquet, J. 1932. *Plant Sociology. The Study of Plant Communities.* McGraw-Hill, New York.

Brown, D. 1954. Methods of Surveying and Measuring Vegetation. Bulletin 42. Commonwealth Bureau of Pastures and Field Crops, Hurley, Berks. Commonwealth Agricultural Bureaux, Farnham Royal, Bucks, England.

Brown, M. 1998. Personal communication.

Cain, S.A. and G.M. de Oliveiro Castro. 1959. *Manual of Vegetation Analysis*. Harper & Brothers, New York.

Canfield, R.H. 1941. Application of the line interception method in sampling range vegetation. *J. For.*, 39: 388-394.

Clements, F.E. 1916. *Plant Succession: An Analysis of the Development of Vegetation*. Carnegie Institute of Washington, Washington, D.C.

Conard, H.S. 1935. The plant associations on central Long Island: a study in descriptive sociology. *Am. Midl. Nat.*, 16: 433-516.

Cooper, S.V., P. Lesica, and D. Page-Dumroese. 1997. Plant Community Classification for Alpine Vegetation on the Beaverhead National Forest, Montana. Gen. Tech. Rep. INT-GTR-362. USDA Forest Service, Intermountain Research Station, Ogden, UT.

Cooper, S.V., K.E. Neiman, and D.W. Roberts. 1991. Forest Habitat Types of Northern Idaho: A Second Approximation. Gen. Tech. Rep. INT-236. U.S. Forest Service, Intermountain Research Station, Ogden, UT.

Crocker, R.L. and N.S. Tiver. 1948. Survey methods in grassland ecology. *J. Brit. Grassland Soc.*, 3: 1-26.

Curtis, J.T. and R.P. McIntosh. 1951. An upland forest continuum in the prairie–forest border region of Wisconsin. *Ecology*, 32: 476-498.

Daubenmire, R.F. 1968. *Plant Communities: A Textbook of Plant Synecology*. Harper and Row, New York.

Environmental Laboratory. 1987. Corps of Engineers Wetlands Delineation Manual. Tech. Rep. Y-87-1. U.S. Army Engineer Waterways Experiment Station, Vicksburg, MS.

Eyre, F.H. (Ed.). 1980. *Forest Cover Types of the United States and Canada*. Society of American Foresters, Washington, D.C.

Federal Interagency Committee for Wetland Delineation. 1989. Federal Manual for Identifying and Delineating Jurisdictional Wetlands. Cooperative technical publication. U.S. Army Corps of Engineers, U.S. Environmental Protection Agency, U.S. Fish and Wildlife Service, and USDA Soil Conservation Service, Washington, D.C.

Gauch, H.G., Jr. 1982. *Multivariate Analysis in Community Ecology*. Cambridge University Press, Cambridge, England.

Gleason, H.A. 1926. The individualistic concept of the plant association. *Bull. Torrey Bot. Club*, 53: 7-26.

Goodall, D.W. 1957. Some considerations in the use of the point quadrat methods for the analysis of vegetation. *Aust. J. Biol. Sci.*, 5: 1-41.

Hansen, P., K. Boggs, R. Pfister, and J. Joy. 1991. Classification and Management of Riparian and Wetland Sites in Montana. Draft Version 1. Montana Riparian Association, Montana Forest and Conservation Expt. Station, University of Montana, Missoula, MT. (Current version is 1995; Misc. Publication No. 54)

Harshberger, J.W. 1970. *The Vegetation of the New Jersey Pine-Barrens. An Ecologic Investigation*. Dover Publications, Inc., New York. (Reprint of a 1916 publication by Christopher Sower Co., Philadelphia, PA.)

Hays, R.L., C. Summers, and W. Seitz. 1981. Estimating Wildlife Habitat Variables. FWS/OBS-81/47. U.S. Fish and Wildlife Service, Washington, D.C.

Higgins, F.G., J.L. Oldemeyer, K.J. Jenkins, G.K. Clambey, and R.F. Harlow. 1994. Vegetation sampling and measurement. In *Research and Managment Techniques for Wildlife and Habitats*. T.A. Bookhout (Ed.). The Wildlife Society, Bethesda, MD. 567-591.

Kent, M. and P. Coker. 1992. *Vegetation Description and Analysis. A Practical Approach*. CRC Press, Boca Raton, FL.

Levy, E.B. and E.A. Madden. 1933. The point method of pasture analysis. *N.Z. J. Agric.*, 46: 267-279.

Lindsey, A.A. 1955. Testing and line-strip method against full tallies in diverse forest types. *Ecology*, 36: 485-495.

Magee, D.W. 1981. *Freshwater Wetlands. A Guide to Common Indicator Plants of the Northeast*. University of Massachusetts Press, Amherst.

Michener, M.C. 1983. Wetland site index for summarizing botanical studies. *Wetlands*, 3: 180-191.

Mueller-Dombois, D. and H. Ellenberg. 1974. *Aims and Methods of Vegetation Ecology*. John Wiley & Sons, New York.

Oliver, F.W. and A.G. Tansley. 1904. Methods of surveying vegetation on a large scale. *New Phytolog.*, III: 228-237.

Parker, K.W. and D.A. Savage. 1944. Reliability of the line interception method in measuring vegetation on the Southern Great Plains. *J. Amer. Soc. Agron.*, 36: 97-110.

Reed, P.B., Jr. 1988. National List of Plant Species that Occur in Wetlands: 1988 National Summary. Biol. Rep. 88 (24). U.S. Fish and Wildlife Service, Washington, D.C.

Reed, P.B., Jr. (compiler). 1997. Revision of the National List of Plant Species that Occur in Wetlands. Department of the Interior, U.S. Fish and Wildlife Service, Washington, D.C.

Sipple, W.S. 1988. Wetland Identification and Delineation Manual. Volume I. Rationale, Wetland Parameters and Overview of Jurisdictional Approach. Revised Interim Final. U.S. Environmental Protection Agency, Office of Wetlands Protection, Washington, D.C.

Slobodkin, L.B. 1974. Comments from a biologist to a mathematician. S.A. Leven (Ed.). *Proceedings of SIAM-SIMS Conference,* Alta, UT. 318-329.

Tansley, A.G. and T.F. Chipp (Eds.). 1926. Aims and Methods in the Study of Vegetation. Brit. Emp. Veg. Committee and Crown Agents for the Colonies, London.

Wakely, J.S. and R.W. Lichvar. 1997. Disagreements between plot-based prevalence indices and dominance ratios in evaluations of wetland vegetation. *Wetlands,* 17: 301-309.

Wentworth, T.R. and G.P. Johnson. 1986. Use of Vegetation in the Designation of Wetlands. North Carolina State University, School of Agriculture and Life Sciences, Raleigh. Report for the U.S. Fish and Wildlife Service's National Wetlands Inventory, St. Petersburg, FL.

Wentworth, T.R., G.P. Johnson, and R.L. Kologiski. 1988. Designation of wetlands by weighted averages of vegetation data: a preliminary evaluation. *Water Res. Bull.,* 24: 389-396.

Whittaker, R.H. 1956. Vegetation of the Great Smoky Mountains. *Ecol. Monogr.,* 26: 1-80.

Whittaker, R.H. 1978. Approaches to classifying vegetation. In *Classification of Plant Communities.* R.H. Whittaker (Ed.). Dr. W. Junk bv Publishers, The Hague, The Netherlands. 1-31.

5 Soil Indicators of Wetlands

INTRODUCTION

Since the beginnings of scientific study of wetlands, soils have been recognized as an important feature of wetlands. Plant ecologists and geologists alike found that the nature of the soils had a profound effect on plant growth and the formation of peat deposits (a mineral resource of considerable economic value). In the 1950s, the U.S. Fish and Wildlife Service's first wetland classification system made reference to hydromorphic, halomorphic, and alluvial soils being associated with the nation's wetlands (Shaw and Fredine, 1956). Today, the predominance of undrained hydric soil is a key attribute for identifying wetlands for wetland inventories and federal regulation (Cowardin et al., 1979; Environmental Laboratory, 1987; Federal Interagency Committee for Wetland Delineation, 1989). Following the lead of the federal government, many states are using soils in combination with vegetation and hydrology indicators to identify and delineate potentially regulated wetlands (e.g., Florida, Maine, Maryland, Massachusetts, Michigan, New Hampshire, New Jersey, New York, Pennsylvania, Vermont, Washington, and Wisconsin), although there is variation in the use of these indicators. Connecticut has relied strictly on soil types for identifying inland wetlands for regulatory purposes since 1972. Local governments in New Hampshire also use soil types to define wetlands for local ordinances due to technical support from NRCS soil and water conservationists.

Hydric soils are used to verify the presence of wetland hydrology for sites that are not significantly hydrologically altered, especially where the vegetation has been removed (see Chapter 6 for details). In general, there is excellent agreement between hydric soils and hydrophytic vegetation, with a few exceptions (Scott et al., 1989; Segelquist et al., 1990).

Several factors exert a major influence on soil development — climate, time, topography, parent material, biota, and human actions. Many of these factors affect the amount of water in the soil and water plays a major role in the formation of different soil horizons (Simonson, 1967). Humid climates generate more water than arid climates. The longer the presence of water in a given area and the more frequently such event occurs, the more pronounced its effect on the soils. Topography greatly affects the flow of water and collection of water. Development of wet soils is favored by certain landscape positions and landforms, such as depressions, floodplains, toes of slopes, drainageways, seepage slopes, and along the margins of coastal and inland waterbodies. Depressional (concave) landforms retain more water than sloping (convex) landforms. Broad flat terrain in humid areas tends to support wetland formation as drainage outlets are lacking. Such terrain is characteristic of coastal plains and glaciolacustrine plains. In such landscapes, the width of the flat often affects drainage, as wider plains tend to lack outlets for drainage with wet soils tending to form on the wider flats rather than on more dissected landscapes (with better drainage). Daniels et al. (1971a) found a relationship between the width of the interfluve (interstream divide) on North Carolina's coastal plain and seasonal high water tables and resultant soil morphology. If the interfluve was 2 miles wide or more, the high water table reached the surface and the water table would be within 20 in. about half of the year in the center of the divide and within 5 in. about 25% of the time. In contrast, if the interfluve was only 0.2 miles wide, the seasonal high water table rose only to about 18 in. from the surface. For interfluves from 0.4 to 2 miles, there was little change in water table relations. The largest change came when the interfluve was less than 0.4 mile wide, when drainage markedly improved due to the proximity of streams and increased slope of the land.

Parent materials also influence soil wetness. Clayey and silty soils retain much moisture in their small pores, while the larger pores of sandy soils provide better internal drainage. In humid areas, clayey soils may produce wet soils on sloping landscapes. Beaver can affect the presence of water in soil by damming streams and flooding former dryland soils. Human actions also influence soil wetness, e.g., diversion of fresh water and drainage projects make soils drier while dam/impoundment construction increases soil wetness. Land clearing and construction of impervious surfaces increases the amount of surface water runofff, often leading to more flooding of low-lying soils.

Differences in the frequency, duration, and seasonality of water typically cause a variety of recognizable features to develop in soils. Properties associated with recurrent prolonged wetness serve as reliable predictors of the long-term hydrology. This makes interpretation of soil morphology (e.g., soil color) vital for wetland identification and delineation, especially in the more problematic situations.

This chapter has several primary purposes: (1) to provide an overview of factors causing the formation of diagnostic properties in flooded and waterlogged soils, (2) to discuss the evolving concept of hydric soils and its origins, (3) to describe recent development on soil indicators of wetlands, and (4) to discuss how soils can be used to aid in wetland identification and delineation. Some background on soils (e.g., definition, composition, and soil color) will be presented at the outset since these factors have bearing on hydric soil properties later described.

BASIC SOIL PROPERTIES

Soil Definition

For the past 20 years, "soil" has been defined in the U.S. as the collection of natural bodies (mineral and organic materials) that support or are capable of supporting the growth of land plants "out of doors" (Soil Survey Staff, 1975). The upper limit of soil was air or shallow water, while the lower limit of soil was the depth at which soil-forming processes (pedogenesis) cease to take place. Land plants (i.e., self-supporting free standing plants, such as trees, shrubs, and robust nonwoody species like grasses, sedges, and flowering herbs) had to be present or capable of growing in these areas; areas that lacked these plants and were not able to support such growth were not considered to have soils. Consequently, permanently flooded areas that did not support the growth of emergent (free-standing) plants did not possess soils, rather their bottoms were considered hydric substrates. Even shallow water areas covered by floating-leaved aquatic beds of water lilies* did not have soils, since they did not support emergent growth.

In 1998, the soil definition was revised to stress pedogenesis and horizonation that characterizes most soils. Soil is now defined as "a natural body comprised of solids (minerals and organic matter), liquid, and gases that occurs on the land surface, occupies space, and is characterized by one or more of the following: horizons, or layers, that are distinguishable from the initial material as a result of additions, losses, transfers, and transformations of energy *or* the ability to support rooted plants in a natural environment" (Soil Survey Staff, 1998). This soil definition expands the concept of soil beyond where "land plants" grow to wherever there is rooted plant growth. Although the accompanying text mentions 2.5 m water depth as "too deep ... for the growth of plants," the discussion of soil suggests that permanently flooded substrates deeper than this may be considered soil where they support or are capable of supporting aquatic species. This change is apparently the result of recent work by the Maryland Soil Survey Program which completed the first subaqueous (underwater) soil survey in Sineputxent Bay (Demas 1998). Characteristics of these soils are controlling factors affecting the distribution of submerged aquatic vegetation (SAV) that has serious implications for success of SAV restoration projects. Other countries have, for some time, included the bottoms of lakes, rivers, and estuaries in their concept of soil. For example, in Europe these submerged substrates are called "subhydric" soils (Moormann and van de Wetering, 1985).

---

* Water lilies are not self-supporting plants; when water is absent, they simply lie flat on the muddy substrates.

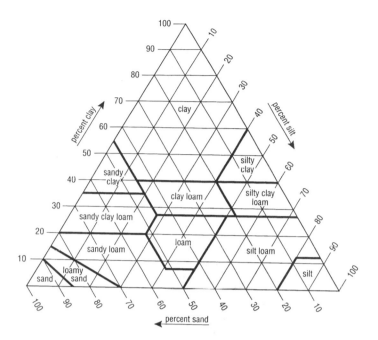

FIGURE 5.1 Textural triangle for determining soil texture based on the percent of sand, silt, and clay. (Soil Survey Division Staff, USDA Handbook No. 18, Washington, D.C., 1993.)

Unconsolidated materials lacking any vegetation cover are nonsoil. Nonsoil areas include glaciers, rock outcrops, salt flats, sandy beaches and bars, muddy shores, badlands (rocky plateaux), and barren lands. Tidal mud flats, sandy intertidal beaches, and rocky shores are examples of nonvegetated wetlands occurring on nonsoils.

SOIL COMPOSITION

Soil is usually made up of a combination of sand, silt, clay, and organic material, with some soils having various amounts of gravel, stones, and rocks present. Soils are separated into two general types based on the amount of organic matter in the upper layer — organic soils and mineral soils. Organic soils are dominated by the remains of plants (e.g., leaves, stems, twigs, and roots) that accumulate in significant amounts at the soil surface. These soils are commonly called mucks and peats. Mineral soils are mainly composed of mixtures of sand, silt, and clay, often with some enrichment of the surface layer with organic matter. They are further classified by texture based on the relative proportions of sand, silt, and clay (Figure 5.1). Texturing can be done in the field by feel (Figure 5.2).

Soil texture has a significant effect on the soil's ability to retain water. The size and amount of pore space in the soil play vital roles in holding water under tension. This means that water can be held in pores above the free water table. The zone above the water table that is saturated (or nearly saturated) due to water held under tension (capillary action) is called the *capillary fringe*. The thickness of the capillary fringe is strongly correlated with soil texture. For example, fine-textured soils have small pores and can hold water under tension at higher levels than coarse-textured soils (Table 5.1). Clayey soils have the smallest pores and can form wetlands on considerable slopes in high rainfall regions. Coarse-textured, sandy soils possess large pore spaces and good internal drainage, and are not usually associated with wetlands. Wet sandy soils typically result from poor external drainage, high regional water tables, or their location along a waterbody where they are subjected to frequent inundation or high groundwater levels.

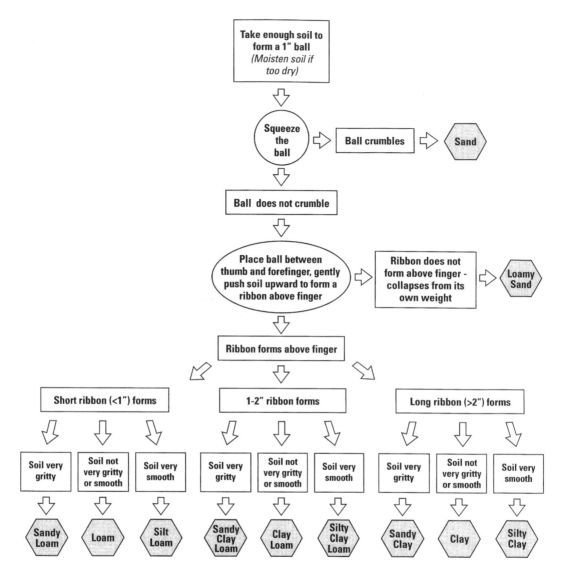

FIGURE 5.2 Soil texturing by feel. (Adapted from Thien, S.J., *J. Agro. Educ.*, 8: 54–55, 1979.)

Soil Color

Soil color is an important soil property that is used in soil classification. Given the importance of soil color in distinguishing wetland (hydric) from nonwetland soils, a brief introduction to soil color provides a needed foundation for the later discussion of hydric soil properties. Simonson (1993) presents an overview of the history of the use and standardization of colors for describing soils.

In the U.S., early descriptions (in the late 1800s) of soils emphasized texture, while only minimal reference was made to soil color. Yet, shortly after the initiation of the federal soil survey program (around 1900), the value of soil color in separating and describing different soils was evident, as within 7 years, color was used as a distinguishing feature in the definition of soil series. While color was included in these early soil descriptions, there was no standardization in the definition of colors across the country.

As early as the 1920s, Russian scientists began to describe soil color by three qualities — tonality (the degree of the main shade of color), brightness (the degree of darkness), and saturation

TABLE 5.1
Thickness of the Capillary Fringe is, in Large Part, a Function of Soil Texture

Soil Texture	Estimated Range in Thickness of the Fringe	
	(in.)	(cm)
Coarse sand	0.4–2.8	1–7
Sand	0.4–3.6	1–9
Fine sand	1.2–4.0	3–10
Very fine sand	1.6–4.8	4–12
Loamy coarse sand	2.0–5.6	5–14
Loamy very fine sand	4.0–8.0	10–20
Coarse sandy loam	3.2–7.2	8–16
Sandy loam	4.0–8.0	10–20
Very fine sandy loam	6.2–10.2	16–26
Loam	8.0–12.0	20–30
Silt loam	10.0–16.0	25–40
Silt	14.0–20.0	35–50
Sandy clay loam	8.0–12.0	20–30
Clay loam	10.0–14.0	25–35
Silty clay loam	14.0 22.0	35–55
Sandy clay	8.0–12.0	20–30
Silty clay	16.0–24.0	40–60
Clay	10.0–16.0	25–40

Source: Mausbach, M.J. 1992b. Proc. 8th Intl. Soil Correlation Meeting. USDA Soil Conservation Service, National Soil Survey Center, Lincoln, NE. 172–178.

(the degree of dilution of the main shade by white) (Zakharov, 1927). The earliest efforts to standardize soil colors were laboratory analyses, yet some scientists recognized that soil colors could change over time and that soil descriptions needed to be done in the field (Conrey, 1924, as reported in Simonson, 1993). In the 1930s, color technology rapidly improved and the U.S. Department of Agriculture worked with the Munsell Company to prepare a set of soil color charts much like those in use today. In the 1951 soil survey manual (Soil Survey Staff, 1951), the Munsell soil color charts were adopted for use in soil mapping (see Simonson, 1993 for details of the administrative process established to standardize soil colors in the U.S.). Of interest to wetland delineators, the gley page was added in 1958. Also due to expanded use of the charts for wetland determinations and the significance of separating chroma 2 from chroma 3 soils, chroma 3 chips were added to the 1990 version of the Munsell soil color charts.* The use of standardized procedures for measuring and describing soil color has greatly improved both the precision and accuracy of soil color descriptions in soil mapping (Schulze et al., 1993).

The Munsell color system uses three features to describe color: (1) hue (related to the spectral wavelength of the color), (2) value (lightness or darkness), and (3) chroma (the purity or strength of the color). These elements are similar to those developed by Russian soil scientists in the 1920s.

---

* The initial recommendation for this addition was made by Ralph Tiner and Peter Veneman, professors in the Plant and Soil Sciences Department at the University of Massachusetts, who in teaching wetland delineation recognized that students were having some difficulty separating chroma 2 from chroma 3 soils using the existing charts, which went from a chroma 2 column to chroma 4. The USDA Soil Conservation Service concurred with this recommendation and the Munsell Color Company added the chips and republished its Munsell Soil Color Charts in 1990.

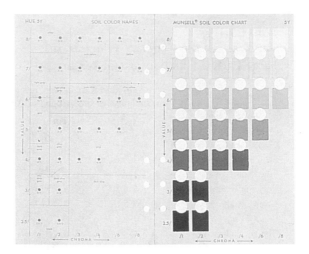

FIGURE 5.3 Black and white photograph of the Munsell soil color chart for hue 5Y (soil color page on right and soil name page on left). (Peter Veneman photo, courtesy of Munsell Color.)

Hues are red (R), yellow red (YR), yellow (Y), green yellow (GY), green (G), blue green (BG), blue (B), purple blue (PB), and purple (P). R, YR, and Y are common soil colors for both hydric and nonhydric soils, while B, BG, PB, and G are gleyed colors typically asssociated with hydric soils. Hues are further divided based on the percent of the intensity of the hue color, in increments of 25%. For example, 10Y equals 100% yellow; 5Y is 50% yellow, 2.5Y is 25% yellow, and 10YR has no yellow, but 100% yellow red. Common hue notations are 10R, 2.5YR, 5 YR, 7.5 YR, 10YR, 2.5Y, and 5Y. Separate charts for these hues are presented in the typical Munsell soil color book. Pages for other hues (e.g., 10Y) can be purchased separately. Values range from black to white on a scale from 0 to 10, whereas the soil color charts tend to show values from 2 to 8. Chroma also varies from 0 to 10, with the charts for the red and yellowish hues showing chroma 1 to 8 (chips for 1, 2, 3, 4, 6, and 8), and separate gley charts for mostly chroma 1 colors in the green, green yellow, blue, blue green, and purple blue hues. Chroma 0 represents "neutral colors" (N) that essentially lack hue; they are shown in one column on chart 1 for gleyed colors.

The Munsell system now serves as the international standard for describing soil colors. Today, soil scientists use standardized color notations and names to describe soils. The Munsell soil color charts produced by Kollmorgen Corporation, Middletown, NY have been the standard tool for the nation's soil scientists for decades (Macbeth Division, 1994). Each chart contains color chips associated with a particular hue (Figure 5.3). There are as many as 42 10 × 15 mm color chips per page, with about 35 per page being more typical. A new soil color book, *EarthColors Soil Color Book: A Guide to Soil and Earthtone Colors*, was published in 1997 by Color Communications of Poughkeepsie, NY. Also based on the Munsell color system, this book includes pages for specific hues and chromas with usually 7 larger (40 × 15 mm) color chips per page. Standard pages are for 10Y/1, 5GY/1, 5G/1, 5G/2, 5BG/1, 5B/1, 5PB/1, N/, 5R, 7.5R, 10R, 2.5YR, 5YR, 7.5YR, 10YR, 2.5Y, and 5Y. Chromas 1, 2, 3, 4, 6, and 8 are included for each hue.

Despite the standardization of soil color and nomenclature, the act of describing soil colors is not an exact science because many factors come into play. First, individuals using color charts must not be color blind. Three major factors affect a person's ability to perceive color — light source,* surface characteristics of the object, and the spectral response of the human eye (Melville and Atkinson, 1985). The condition of the Munsell charts also may be a factor (Cooper, 1990) as color chips may fade when exposed to sunlight or due to wetting/drying after use in heavy rains. Soil

---

* When evaluating soil colors, do not wear sunglasses. Also, periods of low sun angle (e.g., late afternoon in the fall) should be avoided as the light source has a significant effect on one's ability to properly describe soil color.

color also varies with the moisture content of the soil, the degree of crushing and its effect on soil particle coatings, and differences in individual's perception of color (Simonson, 1993). The difference in dry vs. moist condition of the soil is particularly important. Hue and chroma change very little due to moisture differences, yet value may change from 0.5 to 2.0 units (Post et al., 1993). In some cases, another important consideration is that anaerobic soil color may change to varied degrees when exposed to air (see Plate 7B in Schwertmann, 1993; Figure 7 in Vepraskas, 1996). Despite these issues, soil color has been used successfully by soil scientists for decades and is a major diagnostic feature used to separate hydric soils from nonhydric soils.

WATERLOGGING EFFECTS ON SOIL PROPERTIES

Extended flooding and prolonged waterlogging significantly influence soil-forming (pedogenic) processes, resulting in a set of unique and recognizable soil properties. Such conditions fill the soil pore spaces replacing air with water. This may cause clayey soil to expand reducing a soil's permeability. Chemical processes also are altered as the supply of oxygen is depleted, with a noticeable increase in carbon dioxide levels and the pH of the flooded soils moving toward neutral (Henderson and Patrick, 1982).

Soils that are saturated with water and chemically reduced to the point that dissolved oxygen is virtually absent, have an "aquic moisture regime" (Soil Survey Staff, 1975). Many of these soils are associated with wetlands. Due to some problems applying this concept consistently in the field, a more explicit term, *aquic conditions*, was developed (Vepraskas, 1996). It is has been applied to soils experiencing saturation and reduction and displaying morphological evidence (i.e., redoximorphic features) of these conditions. Three types of saturation have been defined to identify processes causing saturation — endosaturation, episaturation, and anthric saturation (Vepraskas, 1996). Endosaturation occurs where periodic saturation occurs from 2 m below the soil surface to some layer above; in other words, it is wet from the bottom up. This saturation is brought about by groundwater. Episaturation may be considered wet from the top down, with saturated horizons occurring above the 2 m depth and with one or more unsaturated layers below the saturated zone. The saturated layers are often called *perched layers* and form where water tables are perched above an impermeable layer (e.g., hard pan, clay pan, or dense basal till). Where flooding is controlled by human activities, such as in cranberry and rice farming, saturation is considered anthric. This is a type of episaturation where reduction occurs in a saturated, ponded surface layer, but oxidation processes take place in the underlying unsaturated zone (Soil Survey Staff, 1996).

Whether caused by natural forces or human actions, inundation of the soil effectively eliminates gas exchange between the atmosphere and the soil. Existing oxygen in soil pores is usually consumed quickly by soil microbes, creating low oxygen (anaerobic) conditions. Some gas exchange may occur at the soil–water surface, but gas exchange between water in the pores and the soil is extremely slow — 10,000 times slower than gas exchange from air-filled pores to the soil (Greenwood, 1961). Anaerobic conditions can develop within 1 day in some soils and within a few days in many, if not most, soils (Ponnamperuma, 1972). Temperature affects the rate at which such conditions become established — the colder the temperature, the slower the rate. Soil scientists have long considered 41°F (5°C) to be "biological zero," the temperature at which biological activity in the soil ceases. Studies in Alaska showed that permafrost soils never rise above this temperature at a depth of 20 in. (due to year-round frozen soil at this level), yet the soils above permafrost still developed typical hydric soils when saturated for prolonged periods (Ping et al., 1992). Soil microbial activity also has been detected in these soils and at lower temperatures even below 2°C, leading to the conclusion that reduction was more related to soil water content than to soil temperature (Ping et al., 1996). Bell et al. (1996) recorded low redox potentials in Minnesota soils when the soil temperature was 2.6°C and inferred that microbial activity takes place below 5°C in these soils. Pickering and Veneman (1984) reported similar results for southern New England. The National Research Council (1995) concluded that the concept of biological zero is not appropriate for use in permafrost and some temperate wetlands. Furthermore, the concept needs revision since soil is biologically active to some degree whenever it is not frozen (see earlier discussion of this subject under growing season section in Chapter 2).

Anaerobic bacteria are important agents in the formation of soil properties associated with prolonged and repeated saturation. These microbes are well adapted to the low oxygen conditions resulting from sustained wetness. Microbes derive their energy from the oxidation of organic matter. Organic matter may occur in the soil proper or along the rhizospheres of plants where excretory soluble organic materials and sloughed-off root tissue are found (Reddy et al., 1986). Soil microbes play a major role in a series of biochemical events called *oxidation-reduction (redox) reactions* that take place in flooded or saturated soils. In a flooded soil, the oxygen that was present prior to flooding is quickly consumed by obligate aerobic microbes in a couple of days provided there is organic matter present for them to digest. Once oxygen is consumed and obligate aerobic micro-organisms are decimated, facultative anaerobes and obligate anaerobic microorganisms become active, with a significant reduction in the rate of organic matter mineralization (Gambrell et al., 1991). The soil becomes chemically reduced. Anaerobic metabolism increases ferrous iron, manganous manganese, methane, ammonia, hydrogen sulfide, and other materials. Chemical reduction affects availability of nutrients and plant toxins in the soil. Various elements are reduced and become more soluble in a sequential path, beginning with nitrate. After converting nitrate to free nitrogen in a process called *denitrification* (recognized as an important process for water quality renovation provided by wetlands), selected microbes reduce manganese from manganic (MnIV, the oxidized form) to manganous (MnIII, the reduced form). Iron is the next element to be reduced from ferric iron (FeIII, the oxidized form) to ferrous iron (FeII, the reduced form). The reduction process continues in saturated soils with microbes reducing the sulfates and carbon and producing hydrogen sulfide (giving the smell of rotten eggs) and methane (odorless), respectively as by-products. Methane and hydrogen sulfide production generally is initiated after long-term saturation, i.e., 1 to 2 months (Peter Veneman, personal communication, 1998). Reduced compounds are often soluble in water and available for plant uptake. These and other mobilized elements (e.g., aluminum) are toxic to most plants in large quantities, so they limit the types of plants that can live in wetlands. The reduction process, therefore, has a profound effect on plant composition as well as on soil chemical and morphological processes.

There may be three major periods of reduction during the year when the soil is moist to wet — late winter through early spring, in the fall, and after heavy rains in the summer (Griffin et al., 1996). These periods seem to coincide with increased organic matter availability. Hudnall and Szögi (1996) detected extreme reduction in many Louisiana soils after heavy rains following an extended dry period.

Flooding and soil saturation for a couple of months or more during the growing season create long-term anaerobic soil conditions sufficient to prevent aerobic decomposition or oxidation of leaves, stems, roots, and other dead plant parts. The aerobic bacteria responsible for the oxidation of these materials cannot survive under these wet conditions. They are replaced by facultative anaerobes and obligate anaerobes that are less efficient in assimilating organic matter during decomposition, leading to the accumulation of organic materials at the surface (Gambrell et al., 1991). This buildup causes the formation of peats and mucks (Plates 3 and 4) as well as generally higher organic matter content in the surface layer of wetland soils than in adjacent upland soils.* Annual rates of peat accumulation range from about $1/12$ to $1/6$ inch (2 to 4 mm) in some freshwater wetlands (Craft and Richardson, 1993). In the Mississippi Delta, marsh peat accretion was estimated at 0.4 to 0.6 mm per year in a subsoil peat layer that formed 2250 to 1050 years ago (Kosters et al., 1987). Peat accumulation in Canadian subarctic wetlands is an order of magnitude less (0.2 to 0.5 mm/yr) (Ovenden, 1990). Leisman (1953; 1957) recorded 14 to 15 mm/year buildup rate during

---

* Acid conditions do not create favorable environments for many microbes; these sites tend to accumulate organic matter due to a low microbial activity.

an 8-year study of a Minnesota bog. A number of wetland types have peat accumulation rates around 1 mm/year (Lugo et al., 1990). Botch and Masing (1983) reported rates of 0.6 to 0.8 mm/year in boreal raised bogs in the former U.S.S.R. and a phenomenal rate of up to 20 mm/year for the Rioni lowland swamps. They also indicated that for boreal bogs the highest rate of accumulation occurred during the Subatlantic period with up to 3.0 mm/year buildup. In 10,000 years, peat layers 50 ft (15.2 m) or more thick have developed in some waterlogged depressions in the northeastern U.S., changing lakes to forested wetlands (see discussion of hydrarch succession in Chapter 2).

In some wet sandy soils, organic matter is transported by percolating acidic water (fulvic acids) from the surface to a subsoil horizon called a *spodic horizon* that becomes rich in dark reddish brown humate (designated as a Bh-horizon) (Plate 5). This also takes place on upland soils in cold climates (e.g., high elevations in northern New England). Factors responsible for the precipitation of carbon are poorly understood, but chemical and mineral bonding mechanisms are likely agents (Daniels et al., 1976; Buurman, 1984; Buurman and Van Reeuwijk, 1984). On the Atlantic–Gulf coastal plain, Bh-horizons about 1 m or less thick occur above gray sands (Daniels et al., 1976). In North Carolina, 3.5 to 47.0 g of carbon may accumulate per square m annually (Holzhey et al., 1975) in these horizons. At this rate, spodic horizons as thick as 7 m could form in less than 30,000 years, other conditions notwithstanding. Much thinner spodic horizons have developed in northern glaciated regions where the soils are young (less than 12,000 years old) and rates of accumulation are different.

Shorter periods of wetness are not sufficient to stop oxidation of organic materials. The alternate wetting and drying, however, lead to the development of other unique soil properties. The degree of flooding and/or soil saturation affects the amount of leaching, reduction-oxidation, and the accumulation of precipitates of iron and calcium (Richardson and Daniels, 1993). Gambrell and Patrick (1978) provide an overview of chemical and microbial processes and their effects on anaerobic soils. Where soils are not wet long enough to promote the buildup of thick organic deposits, mineral soils typically develop. They exhibit a wide range of properties related to differences in parent material, climate, topography, age, and other factors. Changes in the frequency, duration, and depth of soil saturation produce soil characteristics that help separate wetlands from uplands. Wetter mineral soils subject to frequent extended inundation often develop thick organic surface layers (up to 16 in. thick), yet not thick enough to be considered organic soils. Most mineral soils, however, are not wet long enough for these layers to develop. Instead organic matter enriches the surface layer, but does not accumulate as a separate peat or muck layer. This often produces a characteristic very dark brown to blackish surface layer in wetland soils. Also, below the topsoil, the soil (subsoil) is typically grayish colored.

Soil morphological properties now called *redoximorphic features* (formerly *mottles*) indicate varying amounts of soil wetness in mineral soils. The biochemical processes causing iron and manganese reduction have a great effect on soil color and morphology. Iron is typically the most abundant chemical element in the soil. In its oxidized (FeIII) state, iron gives well-drained soils their characteristic yellowish, reddish, orangish, or brownish colors (Plate 6). Prolonged saturation of mineral soil converts iron from its stable oxidized (ferric) form to its mobile reduced (FeII) state. When iron is reduced, it is soluble and usually moves within or out of the soil, leaving the parts of the soil grayish in sandy soils or bluish, grayish, or greenish (gleyed) in finer grained soils. These splotches of color are considered "redox depletions" since they are evidence of iron depletion (Plate 7; Vepraskas, 1996). When the dominant color of a horizon is the result of iron depletion, it is called a *depleted matrix*. The process causing these depletions is frequently called *gleization*. When the reduced compounds are completely removed from the soil, the soil usually takes on a grayish color and is classified as a *gleyed matrix* when the color is one shown on the gley charts (Veneman et al., 1976; see later discussion of hydric soil field indicators). The gray color reflects the natural color of soil particles (sand, silt, and clay), while the bluish colors usually indicate the presence of ferrous sulfide compounds. If the soil conditions are such that free oxygen is present, organic matter is absent, or temperatures are too low (below freezing) to sustain microbial activity, gleization will not begin and redox depletions (gleyed colors) will not form, even though the soil may be saturated for long periods (Diers and Anderson, 1984). Gleization and redox feature formation processes are strongly influenced by the activity of certain soil microorganisms. They reduce iron when the soil environment is anaerobic, that is, when virtually no free oxygen is present, and when the soil contains organic matter (their energy or food source).

Redox depletions also may result from depletion of clay by similar processes (Vepraskas, 1996). Clay depletions appear as gray-colored coatings either lining channels or on the outer surface of natural soil macroparticles (peds). The adjacent soil matrix and the underlying soil layers will have higher clay content. In the latter, clay coatings on outer layers of peds are often present. In studying Belgium Alfisols, Eswaran (1968) found that the amount of clay coatings increased with wider groundwater fluctuations.

Iron mobilized during reduction moves laterally or downward in the soil where it may be reprecipitated. The rate depends on physiochemical conditions in the soil. In calcareous soils, reprecipitation may occur within 3 days, whereas in clayey soils it may take more than 1 month (Stoops and Eswaran, 1985). The rate of reprecipitation is reflected in the redoximorphic feature pattern: slow reprecipitation yields a coarse pattern — while fast reprecipitation produces a fine diffuse pattern.

The pattern of the redoximorphic features may reveal information on variable saturation within the soil profile. For example, reddish colors on ped exteriors and gray colors in the ped interior suggest extended wetness in the interior and better drained ped surfaces (Richardson and Daniels, 1993). This pattern also was observed in a Wisconsin soil with finer textured materials overlying coarser substrates and the researchers concluded that the larger exterior pores likely drained freely to the coarser materials below (Vepraskas et al., 1974; Veneman et al., 1976). A saturated soil layer above a restrictive layer may have redox depletions around large pores (macropores) and reddish to orangish colors (redox concentrations) in the ped interiors may indicate water movement along the pores leading to reduction along the pores (Vepraskas, 1996). Care must be exercised in older, nonglaciated landscapes as such patterns may be indicative of past and not current conditions (Richardson and Daniels, 1993). Studies identifying relict soil color patterns include Ruhe et al. (1955), Daniels et al. (1973), Vepraskas and Wilding (1983), and Childs and Clayden (1986).

Sandy soils may exhibit no consistent pattern of redoximorphic features in relation to macropores, since water may move through both macropores and the soil matrix. It also must be recognized that although general relationships between redoximorphic features and degrees of soil saturation have been established and tend to work well in many place, there are invariably exceptions to these correlations.

In studying upper Midwest soils, Anderson (1984) determined soil matric potential using tensiometers and described five classes of soil wetness and associated soil morphology. Type 1 soil was saturated less than a day with short periods near saturation, resulting in high chroma (>2) color* throughout profile. Type 2 soil was also saturated less than a day, but experienced longer periods of near saturation leading to high chroma ped interiors and some black manganese coatings on ped surfaces. Type 3 soil was saturated for a few days and near saturation for several months producing some chroma 2 inside the peds, few black manganese coatings, and common red iron oxide coatings on ped surfaces. Type 4 soil was saturated continuously for several months creating chroma 1 ped interiors, few black manganese coatings (nodules or concretions) and gleyed zones around the pores (gleyed zone around a pore may be banded with ring of iron oxide). Type 5 soil was nearly continuously saturated throughout the year producing chroma 1 throughout and no evidence of manganese or iron.

---

* Chroma is a property of soil color: high chroma (chroma >2) colors are brighter colors (e.g., oranges, reds, or yellows), while low chroma colors (chroma ≤2) are duller colors (e.g., grays, blues, greens, black, white, or dark browns). The latter colors are typically characteristic of the subsoil of wet soils or the topsoil layer of soils with significant amounts of organic matter. See Soil Color discussion for more information.

Richardson and Daniels (1993) proposed a similar sequence of soil coloration for a soil with 18 to 35% clay ranging from unmottled high chroma, well-drained soils through variously mottled soils to gleyed hydric soils with oxidized rhizospheres. Mausbach and Richardson (1994) also offered a five-step sequence of soil color changes: (1) high chroma B-horizon (well drained soils); (2) iron depletions along pores and crushed colors of chroma 3 or higher (well aerated soils with occasional or temporary saturation); (3) redox depletions along roots and ped surface, with matrix chroma from 2 to 3 (somewhat poorly drained soils); (4) thick redox depletions along ped surfaces with redox concentrations in ped interior, with wetter phases of this having some gley colors in the ped interior (poorly drained soils); and (5) matrix dominated by neutral and gley hues with oxidized rhizospheres (iron plaque or pipestems) and/or redox concentrations along ped surfaces (very poorly drained soils).

Mineral soils that are wet for most of the year usually have reduced (low chroma) colors dominating the subsoil layer immediately below the surface layer and often within 1 ft of the surface. Soils exposed to shorter periods of wetness are variously colored. The subsoils of the wetter soils are grayish with spots or blotches of yellow, orange, or reddish brown (high chroma mottles). These brighter colors represent concentration of iron oxides — *redox concentrations* indicative of a fluctuating water table (Plate 7). Many times, these concentrations develop along root channels in wet soils because plant roots are leaking oxygen and the soluble (ferrous) iron combines with the oxygen to form ferric iron. These orangish-coated channels, called *oxidized rhizospheres*, are evidence of a plant living under anaerobic conditions (see Chapter 3; Plate 2). In other cases, oxidation may occur along open pores in the soil matrix, producing an orangish halo extending into the soil matrix. These features are called *oxidized pore linings*. They may be evidence of a reduced matrix with oxygen entering pores and oxidizing the ferrous iron around the pores (Plate 8). The drier soils subjected to brief periods of wetness are brighter colored overall (red, yellow, brown, or orange) with grayish mottles (low chroma mottles). Soils with short-term wetness may only have high chroma mottles, while soils lacking significant wetness typically are not mottled.

In climates where evapotranspiration (ET) exceeds precipitation (P) (i.e., arid to subhumid regions), an accumulation of salts (evaporites) may be associated with many wetlands. Miller et al. (1985) reported that soil salinity in Canadian prairie wetlands increased with decreased landscape elevation. Depressional wetlands higher in the landscape were typically recharge wetlands and water movement was downward with no salt accumulation. Saline and carbonated soils were found in wetlands at lower elevations that received discharge from surrounding uplands. Arndt and Richardson (1986) also noted this condition in similar wetlands in North Dakota attributing the increased salts due to mineral dissolution and transport from the higher elevation wetlands ("stepped wetlands") to the lower ones. Knuteson et al. (1989) detected carbonates in a recharge wetland in eastern North Dakota. They reported three conditions favoring this: (1) the capillary fringe from a water table approaching the root zone, (2) ET exceeding P during the summer, and (3) bicarbonate and sulfate anions dominating the shallow groundwater. These anions promoted the formation of calcite and gypsum. Steinwand and Richardson (1989) found concentrations of gypsum along the edges of semipermanently flooded pothole wetlands in this same region and described four landscapes where such conditions were observed. Gypsum formation strongly correlated with salinity of the wetland, with very saline wetlands having a continuous band of evaporites. Soils with fluctuating water tables tended to be nonsaline since salts were flushed out. They also noted that freshwater discharge into a "brackish" inland wetland can prevent local accumulation of gypsum. Other studies reporting similar findings in the prairies were Bigler and Richardson (1984) and Richardson and Bigler (1984). Al-Janabi and Lewis (1982) found saline soils in depressions on the floodplain of the Platte River in central Nebraska. They attributed the increased salinity to the precipitation of salts from fluctuating groundwater levels.

Changes in soil color with respect to changing topography and hydrology have been noted in a number of studies. Some of the more significance ones include Daniels and Gamble (1967), Simonson and Boersma (1972), Fanning et al. (1973), Moore (1974), Guthrie and Hajek (1979),

Richardson and Hole (1979), Veneman and Bodine (1982), Franzmeier et al. (1983), Richardson and Lietzke (1983), Michalyna and Rust (1984), Pickering and Veneman (1984), Zobeck and Ritchie (1984), Evans and Franzmeier (1986), and Wakely et al. (1996). Other studies correlating soil morphology with seasonal high water tables include Spaeth and Diebold (1938), Gile (1958), Holstener-Jørgensen (1959; reported in Lyford, 1964), McKeague (1965a,b), Siuta (1967), Latshaw and Thompson (1968), Crown and Hoffman (1970), Daniels et al. (1971b), Boersma et al. (1972), Moore (1974), Vepraskas et al. (1974), Daniels et al. (1975), Schwertmann and Fanning (1976), MacIntosh and Van der Hulst (1978), Bouma (1983), Vepraskas and Wilding (1983), Coventry and Williams (1984), Coombs et al. (1985), Couto et al. (1985), Roman et al. (1985), Hyde and Ford (1989), Watts and Hurt (1991), Cogger and Kennedy (1992), Daniels and Buol (1992), Faulkner and Patrick (1992), Vepraskas and Guertel (1992), Zampella (1994), and Vepraskas and Sprecher (1997). Wakely et al. (1996) present the results of some recent studies examining the relationship between hydrology and hydric soil morphology.

OTHER EFFECTS OF FLOODING ON SOIL DEVELOPMENT

While prolonged inundation causes waterlogging and its accompanying influences on soil genesis, flooding also has other consequences. Soils prone to frequent flooding develop other properties that can be readily observed or measured. Floodwaters often carry sediments and debris with them and upon spreading out over floodplains, they deposit these materials on the soil surface. Sediment accretion rates from a single event can be as high as 6 ft (180 cm) as recorded on North Dakota floodplains in 1952 (Johnson et al., 1976). The annual deposition rates, however, tend to be much less than this catastrophic flood, perhaps 1 to 4 cm/year or less depending on the watershed characteristics (Lugo et al., 1990). Louisiana coastal forests had sedimentation rates of 6.3 mm (±3.5) per year (DeLaune et al., 1987), while Mississippi Delta marshes may be accreting at an annual rate of 10.0 mm (Kosters et al., 1987). Salt marsh and mangrove accretion rates may be a 1 to 3 mm/year (Steers 1977). Vertical accretion rates averaged 7.8 mm/year since 1954 for southwest Louisiana's coastal marshes, with higher rates (10.2 mm) for the period 1954 to 1963, possibly due to heavy sedimentation by Hurricane Audrey in 1957 (DeLaune et al., 1983). In another Louisiana study, sediments were accumulating at 13.5 mm per year near a tidal stream, whereas the interior marsh accretion rate was 7.5 mm (DeLaune et al., 1981). An annual accretion rate of 2.5 mm/year has been reported for Chesapeake Bay marshes (Darmody and Foss, 1978). Similar marsh accretion rates (2.0 to 2.5 mm/year) were found in Connecticut since 1938, which is equivalent to local sea level rise (Warren and Niering, 1993). In the Pacific Northwest, Thom (1992) reported a mean accretion rate of 3.6 mm per year which exceeds the estimated annual rise in sea level of 1.3 mm. The 1938 hurricane deposited a sand layer about 15.0 mm thick. Submergence of the Mississippi Delta is viewed as the primary factor causing higher accretion than in many other coastal regions (DeLaune et al., 1983). A study in south San Francisco Bay found salt marsh accretion rates ranging from 4.0 to 42.0 mm per year, with the higher rate in an area of high subsidence. In The Netherlands, annual rates of 4.0 to 15.0 mm have been reported (Oenema and DeLaune, 1988).

Alluvial soils (e.g., Fluvaquents and Udifluvents) are typified by an irregular distribution of carbon in the soil profile. Frequent deposition of waterborne sediments buries the A-horizon, so numerous such events over the course of decades produce several buried A-horizons below the current soil surface. Since the A-horizon is the layer with high organic matter content, the distribution of the various A-horizons now buried in the subsoil yields the characteristic irregular carbon content with depth. This carbon content can be measured in the laboratory, although it is often possible to see a number of these buried A-horizons due to their darker color than the deposited material above and below them. Even if the A-horizons are not visible, many of these soils tend to have recognizable stratified layers due to these frequent depositions. Sometimes these layers have different textures.

FACTORS AFFECTING THE COLOR OF SOIL

As mentioned above, soils subjected to prolonged saturation and/or inundation usually develop colors that are different from dryland soils. While reduction processes exert strong influences on soils, the actual color of soil is influenced by many factors including iron state (oxidized or reduced) and its form (e.g., type of oxide), manganese oxides, organic matter, moisture status of the soil, and parent material (Daniels et al., 1961; Schwertmann, 1993). Taylor (1982) provides a review of the factors affecting soil color. Iron oxides are important coloring agents in aerobic soils. They also produce the conspicuous oxidized rhizospheres found along the roots of many hydrophytes. The lack of iron oxides typically gives anaerobic soils their dull colors. Soil color may be masked by black manganese oxides (Schwertmann, 1993) or dark colors associated with organic matter.

Hematite, goethite, lepidocrocite, maghemite, and ferrihydrite are immobile iron oxides that produce colors from yellow to red (Schwertmann, 1993). The amount of hematite is usually responsible for the redness of a soil. It is typical of aerobic soils of the tropics and subtropics, but also may develop in temperate regions in soils formed from well-drained calcareous material. Hematitic soils have hues ranging from 7.5YR or 5YR to 5R (Schwertmann and Lentze, 1966). Goethite is produced in all soils wherever weathering occurs and is, therefore, the most common soil iron oxide. It typically yields hues from 10YR to 7.5YR, so many U.S. soils have colors in these hues. When formed as oxidized rhizospheres, goethite may be brownish yellow. Lepidocrocite is found in fluctuating aerobic and anaerobic environments typical of most wetlands, and is especially common in humid temperate regions and temperate areas within subtropical regions of Australia and South Africa. It is usually 7.5YR in color, but may range from 5YR (if the mineral is poorly crystalline) to 10YR (if low in concentration <2%) (Schwertmann, 1993). Lepidocrocite is the second most abundant form of iron oxide. It requires reduced iron (FeII) and is often found in clayey, noncalcareous, periodically anaerobic soils. It is the iron oxide form that typically produces orange-colored mottles in hydric soils and the orange-colored oxidized rhizospheres common along the roots of certain hydrophytes in noncalcareous wetlands. Maghemite is associated with fire-prone areas, especially in the tropics and subtropics. Its hue ranges between 2.5YR and 5YR — intermediate between goethite and hematite. Ferrihydrite forms in wet environments where rapid oxidation takes place. It is typical of gleyed soils and in the B-horizon of Spodosols (Schwertmann, 1993). Its hue ranges between 7.5YR and 2.5YR and its value is less than 6 (lepidocrocite has a value of 6 or more). More red-colored oxidized rhizospheres may be due to this form of iron oxide.

Anaerobic soils usually lack the yellow to red colors since iron oxides are partly or completely removed during reduction, resulting in grayish to greenish colors. The latter usually form in ferriferous clay minerals (Schwertmann, 1993). The so-called "green rust", an iron compound, may be responsible for greenish colors (Van Breemen, 1988). Anaerobic soils also may be bluish and sometimes the bluish color changes to a more grayish color upon exposure to air. Daniels et al. (1961) in studying ferrous iron content and soil color noticed a color change in Iowa stream sediments from a greenish gray (5G 5/1) color of freshly excavated sediments to olive brown and/or dark grayish brown colors (2.5Y 4/4 to 10YR 4/2) after exposure to air. Such observation provides direct evidence of the presence of reduced iron (ferrous iron) in the soil. Neutral hues and greenish gray colors may develop in soils with more than .002% ferrous iron content (Daniels et al., 1961). The bluish color seen in some wetland soils may be due to the presence of an iron-phosphate mineral called vivianite (Postma, 1981).

Organic matter affects soil color, producing brown to black colors. The so-called "topsoil" (A-horizon) is typically darker than the underlying subsoil due to enrichment by organic matter. Organic matter can build up in extremely wet areas where leaf-litter slowly decomposes or in dry areas where extensive root systems eventually die and cause a darkening of the soil. The former is typical of wetland environments and the latter typical of prairie or steppe regions (Mollisols). Organic-coated sand grains are black in color, in marked contrast to the whiter, uncoated sand grains. The abundance of the former may help separate hydric sandy soils from mesic sandy soils. Organic-enriched layers also may form in the subsoil of Spodosols (formerly podzols). Some spodic horizons (Bh- or Bhs-horizons) are dark colored due to the accumulation of organic matter (humus), while those lacking such matter tend to be more reddish brown due to the presence of sequioxides. Buried A-horizons can be easily recognized by their darker color (or higher organic matter content) in frequently flooded soils typical of floodplains (alluvial soils). The number of such buried horizons gives a strong indication of the frequency of major flooding events. Generally, the darker the color, the more organic matter present. Franzmeier (1993) and Blume and Helsper (1987) concluded that Munsell value and chroma were needed to accurately predict organic matter content, whereas Schulze et al. (1993) demonstrated that value alone was sufficient to predict this for soils of similar texture in a given landscape.

Drainage activities may affect soil color, especially in tidal marsh soils. The latter typically have sulfidic mineral materials with chroma ≤ 1 and values ≤ 4 and sulfidic organic materials of chroma 2 or more (Fanning et al., 1993). Estuarine bottom soils with monosulfides (e.g., FeS) are black-colored and when exposed to air change from black (N2) to a grayish color (e.g., 5Y 6/2) (Fanning et al., 1993). Drainage of these soils produces sulfuric acid that affects soil development (e.g., the formation of acid-sulfate soils in drained tidal marshes), as well as plant and animal life. Oxidized layers of such soils may have chroma of 4 or more. The yellow mottles typically observed in these drained soils are jarosite. Prasittikhet and Gambrell (1989) provided a review of acid-sulfate soils.

Richardson and Daniels (1993) described the influence of parent materials, landscape position (geomorphology), and hydrology on soil colors. They noted that soil colors may reflect current conditions related to present-day climate or may reflect past climatic conditions, especially in eroded sites on dissected landscapes. Given the degree of hydrologic modification in the U.S., the existence of relict hydric soils is not insignificant (see Chapter 7 for discusssion).

Prolonged soil saturation has an enormous effect on soil color as noted in the preceding sections. In particular, depleted matrices and redox depletions (mottles of chroma 2 or less) are usually well correlated to such conditions, with the extent of low chroma colors in the subsoil typically related to increased duration of frequent saturation (e.g., low chroma matrix is wetter longer than a high chroma matrix with low chroma mottles). Northeast hydric mineral soils have low chroma matrix subsoils that better drained soils do not (Tiner and Veneman, 1987; 1995). Daniels et al. (1971b) found that soils with high water tables on North Carolina's Coastal Plain had low chroma matrices. Low chroma mottles started appearing where the soil profile was saturated 50% of the time. In studying Pacific Northwest soils, Simonson and Boersma (1972) found similar patterns, noting that hues of 2.5Y and 5Y coupled with low chroma matrices and mottles were useful predictors of soil waterlogging. They also concluded that chroma appeared to be a manifestation of the amount of time that each of their study soils was saturated. In studying semiarid tropical soils in Australia, Coventry and Williams (1984) reported that low chroma colors were directly related to depths where the soils were saturated for 5 weeks, with dominant gray colors occurring in strongly mottled subsoils that were saturated for at least 21 weeks. These colors were associated with wetness conditions that had a frequency of once every 3 years.

WETLAND SOILS

Over time, several terms have been used to describe wetland soils. In the early 1900s, American soil scientists referred to wet soils by terms such as swamp, glei, alluvial, peat, and tundra soils (Marbut, 1921). Hydromorphic, halomorphic, and alluvial soils were mentioned in the U.S. Fish and Wildlife Service's 1950s wetlands inventory report (Shaw and Fredine, 1956). Hydromorphic soils included waterlogged soils associated with marshes, wet meadows, swamps, and bogs — peats, mucks, and gleyed soils. Alluvial soils were frequently inundated floodplain soils, whereas the halomorphic soils were the wet soils of coastal salt marshes and inland alkaline wetlands (arid

West). European soil classifications also used the term *hydromorphic* as the generic word for wet soils and have terms for specific types including peat soils, ground-water gley (gley-soils), and surface-water gley (pseudo-gley) soils (Moormann and van de Wetering, 1985). The German soil classification system identifies various types of surface water soils (terrestrial soils with redoximorphic surface water features), semiterrestrial soils (mineral soils influenced by groundwater within the upper 40 cm, including numerous types of gley soils), subhydric soils (lake or sea bottom soils), and peat soils (Blume and Schlichting, 1985). The soil classification system used in the U.S. in the 1930s recognized many soils with poor drainage (Baldwin et al., 1938). Most of these were listed under "intrazonal" soils, including solonchak (inland saline soils), wiesenboden (wet meadow soils), half bog soils, bog soils, alpine meadow soils, planosols, ground-water podzols, and ground-water laterites (wet tropical soils). Other wet soils were classified under zonal soils (poorly drained tundra soils) and azonal soils (poorly drained alluvial soils).

In characterizing soils for agricultural uses, American soil scientists have divided soils into seven drainage classes of soil wetness (Soil Survey Division Staff, 1993)

1. Excessively drained.
2. Somewhat excessively drained.
3. Well drained.
4. Moderately well drained.
5. Somewhat poorly drained.
6. Poorly drained.
7. Very poorly drained.

The latter two types are characteristic of wetlands. Poorly drained soils remain waterlogged at or near the surface long enough during the growing season that most crops cannot be grown unless the land is artificially drained. Very poorly drained soils are saturated for extended periods, are frequently inundated, and have the same limitation for crop production. Under natural conditions, both of these soil types typically support the growth of hydrophytes (wetland vegetation). The other drainage classes are too dry to produce wetlands, with two exceptions — better drained soils that are frequently flooded or ponded for long periods and somewhat poorly drained, fine-textured soils in lower landscape positions that occasionally are wet enough for wetland establishment. The latter soils are typically wet only for short periods usually early in the growing season. It is also important to recognize that these drainage classes were not developed for wetland determinations; rather they were created to separate different soils into groups with certain wetness limitations for agriculture. As such, the terms have been interpreted locally, so a soil that one state considers to be somewhat poorly drained may be poorly drained from another state's view (varied interpretations also may be found within a state). This has caused some confusion in their use for wetland identification. Huddleston and Austin (1996) and Tiner and Veneman (1987; 1995) have recognized this problem. Today, wetland soils are called *hydric soils* in the U.S.

THE EVOLVING HYDRIC SOIL DEFINITION AND TECHNICAL CRITERIA

The term *hydric soil* was coined by the U.S. Fish and Wildlife Service (FWS) in its national wetland classification system (Cowardin et al., 1979). The "predominance of undrained hydric soils" is one of the indicators for identifying wetlands, according to the FWS wetland definition. Hydric soil was originally defined as "soil that is wet long enough to periodically produce anaerobic conditions, thereby influencing the growth of plants" (Cowardin et al., 1979).

After creating the term, the FWS received assistance from the USDA Soil Conservation Service — now the Natural Resources Conservation Service (NRCS) — in better defining the term and for developing a national list of these soils to aid in wetland identification following the FWS classification system. Blake Parker, an NRCS soil scientist, was assigned to the FWS's National Wetlands

Inventory Project in St. Petersburg, FL to formalize the concept of hydric soils and coordinate the preparation of a national list of hydric soils. From 1977 to 1980, he worked informally with NRCS soil scientists on this task.

Parker (1992) summarized the development of the hydric soil definition. One of the earliest definitions was "hydric soils are either (a) saturated at or near the soil surface with water that is virtually lacking free oxygen (O_2) for significant periods during the growing season, or (b) flooded frequently for long periods during the growing season." In this early version, there were no technical criteria, but instead procedures for identifying hydric soils were given. The indicators were gray colors, iron and manganese concretions near the surface, and the wetness state of the soil. The position on the landscape also was cited by Parker as an indicator for hydric soil, but no specifics were given. Gray colors were essentially low chroma colors associated with the aquic moisture regime: chroma 2 or less with mottles or chroma 1 with/without mottles within 25 cm of the surface. While not mentioned by Parker, organic soils except Folists were also recognized as hydric soils.

In 1979, the definition was revised to read, "A hydric soil is a soil that in its undrained condition is saturated, flooded, or ponded long enough during the growing season to develop anaerobic conditions that favor the growth and regeneration of hydrophytic vegetation." A list of hydric soils entitled *Wet Soils of the United States* was printed in 1979 (Parker 1992). It included all Histosols (except Folists), all soils with Aquic suborders, all poorly or very poorly drained soils, all soils with capability class and subclass of IVw through VIIIw, and all soils with a water table less than 2 ft from the surface. After doing field work evaluating the correspondence between these soil characteristics and wetland vegetation, the water table depth requirement was changed to 1.5 ft. In 1980, the term *undrained condition* was replaced with *natural condition*.

In 1982, the hydric soil was defined as soil that was either saturated at or near the soil surface with water that is virtually lacking free oxygen for significant periods during the growing season or flooded frequently for long periods (i.e., more than 7 consecutive days) during the growing season (USDA Soil Conservation Service, 1982). A year later, a list of soils with actual or high potential for hydric conditions was published by the NRCS for field testing to aid in the agency's decision of whether or not to adopt the FWS's wetland classification system (Flach, 1983). A hydric soil condition was defined as "when the soil in its natural undrained state is saturated at or near the surface during much of the growing season." Three features commonly associated with hydric conditions were listed — aquic moisture regime, a deficiency of oxygen at or near the surface during much of the growing season, and flooding or ponding of long duration during the growing season.

Controversy over the proposed national list of hydric soils (e.g., some state NRCS offices proposed deleting all soils except Histosols and those flooded daily by the tides) led to the establishment of the National Technical Committee for Hydric Soils (NTCHS) (Parker, 1992). This committee would finalize the definition and national list of hydric soils (see Mausbach, 1994 for the history of this group and details on the evolution of the hydric soils list).*

In 1985, two definitions and sets of technical criteria were published as the concept of hydric soil was being refined by the NTCHS. In March 1985, of hydric soil was defined as follows: "an undrained hydric soil is saturated, flooded, or ponded long enough to produce anaerobic conditions (lacking oxygen) that affect the growth of plants" (USDA Soil Conservation Service, 1985a). Such soils were further characterized by soil saturated at or near the soil surface with water that is virtually lacking free oxygen for a significant period during the growing season, or frequently flooded or ponded for a long period during the growing season. The first set of technical criteria were established (Table 5.2). The land capability referred to in the criteria relates to suitability to grow crops. Classes 4W through 8W have very severe limitations due to wetness, while classes 2W and 3W have moderate and severe limitations, respectively. Technical criteria were established for the primary purpose of extracting a list of soil series meeting these criteria from NRCS's national

---

* The NTCHS was formalized as a committee in April 1985 and was charged with completing the national hydric soils list by July 1, 1985.

TABLE 5.2
Technical Criteria for Hydric Soils (March 20, 1985)

1. All Histosols except Folists.
2. Aquic suborders, Albolls, or Salorthids that have water tables less than 1.5 ft during the growing season and which are either (a) poorly drained or very poorly drained and have a land capability classification of 2W-8W, or (b) are somewhat poorly drained and have a land capability classification of 4W-8W.
3. Soils with frequent flooding or ponding of long duration or very long duration that occurs during the growing season and have a land capability classification of 4W-8W.

Note: Growing seasons were defined by soil temperature regimes: hyperthermic (February-December), thermic (March-October), mesic (April-October), frigid (June-September), cryic (June-August), pergelic (July-August), isohyperthermic (January-December), isothermic (January-December), and isomesic (January-December).

Source: USDA Soil Conservation Service, 1985a.

soils database for compilation of the national list of hydric soils, yet they served in combination with the hydric soils list as an aid to field interpretations of hydric soils until recently (when replaced by the field indicators list). Later in 1985, the NTCHS published the first edition of *Hydric Soils of the United States* containing the following hydric soil definition: "a soil that in its undrained condition is saturated, flooded, or ponded long enough during the growing season to develop anaerobic conditions that favor the growth and regeneration of hydrophytic vegetation" (USDA Soil Conservation Service, 1985b). The intent of this definition is quite similar to the original definition in the FWS wetland classification system, except for the referenced to "undrained condition". The 1985 definition does not, therefore, recognize drained hydric soils as being separate from undrained hydric soils (i.e., a hydric soil remains "hydric" even when drained). This change in the definition was intended to prevent people from concluding that a drained hydric soil would not meet the definition of hydric soils (Mausbach, 1994) which in itself caused some confusion since according to the FWS wetland definition, "undrained hydric soils" are wetland indicators, whereas drained hydric soils are not.

The 1985 technical criteria are given in Table 5.3. Note that the water table depths vary based on soil permeability and that the duration of the high water table is "at some time" which was interpreted as 1 week or more. The flooding criterion (number 4) specified frequency (more than 50 years out of 100) and duration (long or very long duration — 1 week or more) of flooding for

TABLE 5.3
October 1985 Hydric Soil Criteria

1. All Histosols except Folists.
2. Soils in Aquic suborders, Aquic subgroups, Albolls suborder, Salorthids great group, or Pell great groups of Vertisols that are
   a. Somewhat poorly drained and have water table less than 0.5 ft from the surface for a significant period (usually a week or more) during the growing season, or
   b. Poorly drained or very poorly drained and have either
      (1) Water table at less than 1.0 ft from the surface for a significant period (usually a week or more) during the growing season if permeability is equal to or greater than 6.0 in./h in all layers within 20 in., or
      (2) Water table at less than 1.5 ft from the surface for a significant period (usually a week or more) during the growing season if permeability is less than 6.0 in./h in all layers within 20 in.
3. Soils that are ponded for long duration or very long duration during the growing season.
4. Soils that are frequently flooded for long duration or very long duration during the growing season.

Source: USDA Soil Conservation Service, 1985b.

TABLE 5.4
1991 Hydric Soil Criteria

1. All Histosols except Folists.
2. Soils in Aquic suborder, Aquic subgroups, Albolls suborder, Salorthids great group, Pell great groups of Vertisols, Pachic subgroups, or Cumulic subgroups that are
    a. Somewhat poorly drained and have a frequently occurring water table at less than 0.5 ft from the surface for a significant period (usually more than 2 weeks) during the growing season, or
    b. Poorly drained or very poorly drained and have either:
        (1) A frequently occurring water table at less than 0.5 ft from the surface for a significant period (usually more than 2 weeks) during the growing season if textures are coarse sand, sand or fine sand in all layers within 20 in. for other soils
        (2) A frequently occurring water table at less than 1.0 ft from the surface for a significant period (usually more than 2 weeks) during the growing season if permeability is equal to or greater than 6.0 in./h in all layers within 20 in., or
        (3) A frequently occurring water table at less than 1.5 ft from the surface for a significant period (usually more than 2 weeks) during the growing season if permeability is less than 6.0 in./h in any layer within 20 in.
3. Soils that are frequently ponded for a long duration or very long duration during the growing season.
4. Soils that are frequently flooded for a long duration or very long duration during the growing season.

Source: USDA Soil Conservation Service, 1991.

hydric soils, while the ponding criterion (number 3) made no reference to duration or frequency. Despite the oversight by the latter omission, it was interpreted to be consistent with that of the flooding criterion. Parker (1992) recalls that technical criterion 2a (for somewhat poorly drained soils) was added at the request of Dick Johnson of the NRCS's Midwestern National Technical Center, who informed Parker that many formerly poorly drained prairie soils were effectively drained to the point that NRCS changed their drainage class to somewhat poorly drained. Yet any undrained soils in these series remaining would have a water table listed at 0.5 ft and a drainage class of somewhat poorly drained. Since these soils would morphologically and taxonomically meet the hydric soil criteria, the suggestion was made to include somewhat poorly drained soils with water tables less than 0.5 ft as hydric. The first national list of hydric soils was published in October 1985 (USDA Soil Conservation Service, 1985b).

In 1987, the list was revised and republished (USDA Soil Conservation Service, 1987) and again in 1991 (USDA Soil Conservation Service, 1991). The 1987 list contained a revised definition of hydric soils. "A hydric soil is a soil that is saturated, flooded, or ponded long enough during the growing season to develop anaerobic conditions in the upper part." The 1991 list also contained this definition. Comparing it to the 1985 definition, one notes that the references to "undrained condition" and to "hydrophytic vegetation" were removed and the latter was replaced by "in the upper part." Reference to hydrophytic vegetation was eliminated to make the definition independent of such vegetation and, therefore, more based on certain hydrologic conditions that caused anaerobiosis or a reducing soil environment near the surface.

The 1991 list also had a revised set of technical criteria (Table 5.4). They included a significant change in the duration of the seasonal high water table and clarified the frequency and duration of the ponding criterion (number 3). The water table duration was increased from 1 week to "usually more than 2 weeks" based on a unpublished review of recent literature and research on wet soils (Mausbach, 1994). This change, however, did not cause any changes in the list since the soils recorded have high water tables for several weeks during the growing season. It did, however, change field interpretations of hydric soils in areas with hydrologic observations.

TABLE 5.5
Current Technical Criteria for Hydric Soils

1. All Histosols except Folists.
2. Soils in Aquic suborders, great groups, or subgroups, Albolls suborder, Aquisalids, Pachic subgroups, or Cumulic subgroups that are
   a. Somewhat poorly drained with a water table equal to 0.0 ft from the surface during the growing season, or
   b. Poorly drained or very poorly drained and have either:
      (1) Water table equal to 0.0 ft during the growing season if textures are coarse sand, sand, or fine sand in all layers within 20 in.

      Por for other soils

      (2) Water table at less than or equal to 0.5 ft from the surface during the growing season if permeability is equal to or greater than 6.0 in./h in all layers within 20 in., or
      (3) Water table at less than or equal to 1.0 ft from the surface during the growing season if permeability is less than 6.0 in./h in any layer within 20 in.
3. Soils that are frequently ponded for a long duration or very long duration during the growing season.
4. Soils that are frequently flooded for a long duration or very long duration during the growing season.

Source: National Resources Conservation Service, 1995.

In 1991, the NTCHS contemplated adopting a list of fidelity categories for hydric soils that would complement what was done for plants (wetland plant indicator status regarding the affinity for wetlands). Four categories were proposed (Mausbach, 1992a as reported in NRC, 1995):

Class 1. Obligate wet hydric soils (very poorly drained soils including Histosols, Histic subgroups, Hydraquents, Sulfa great groups, and Sulfi subgroups of Aquic suborders, plus very poorly drained soils in Aquic suborders, Pell great groups, and Albolls).

Class 2. Facultative wet hydric soils (Salorthids and poorly drained soils in Aquic suborders, Pell great groups, and Albolls)

Class 3. Facultative hydric soils (somewhat poorly drained soils in Aquic suborders, Pell great groups, and Albolls).

Class 4. Facultative upland hydric soils (Aquic subgroups and soils from the previous groups that were not listed in the other three classes)

Although not adopted, the NRC concluded that such listing would facilitate hydrologic inference from the soil type.

In 1994, the definition of hydric soil was revised to "a soil that formed under conditions of saturation, flooding, or ponding long enough during the growing season to develop anaerobic conditions in the upper part" (NRCS, 1994). Later, in 1995, the technical criteria also were revised (NRCS, 1995; Table 5.5). References to the duration and frequency of the high water table were removed from criterion number 2 since the water table frequency and duration data were not in the map unit interpretation records database and were not selection criteria. While the criteria were admittedly not intended for field use, they did represent useful guidelines. Some interpretation problems resulted from the lack of field-oriented criteria. Huddleston and Austin (1996) noted somewhat poorly drained soils in their study of Oregon soils had "morphology and hydrology more like the hydric soils" but were not called hydric simply because of a drainage class designation. They suggested deleting the drainage class reference in the hydric soil criterion. Hudnall and Szögi (1996) experienced a similar problem with a somewhat poorly drained soil (Commerce) in Louisiana. The removal of the frequency and duration requirements may pose problems as many seasonally wet nonhydric soils in a Louisiana study exhibited some reduction in the upper part of the soil, especially after heavy rains following a long dry period (Hudnall and Szögi). These scientists recommended a duration requirement for defining aquic conditions in *Soil Taxonomy* and the same suggestion could apply to the current hydric soil criteria.

Mausbach (1994) indicated that the criteria were not intended for field use because they were derived to compile a list of the nation's hydric soils from the national soils database at Ames, IA. While the latter statement is true, in the absence of a list of hydric soil field indicators, the criteria also served as guidance in terms of where to look in the soil profile for hydric soil indicators. Before the criteria were established characteristics associated with an aquic moisture regime were used to identify hydric soils (Tiner, 1986). Since 1994, the NTCHS has been developing indicators that can be used to identify wetlands in the field (see section on hydric soil field indicators). Presently, field indicators have replaced the criteria for field identification of hydric soils.

During the past 20 years, NRCS has worked on the concept of hydric soil — refining the term and mostly, revising the list and technical criteria. Despite changes in technical criteria and changes in the list, the overall concept of hydric soil has changed little from its origins in the late 1970s, except perhaps for sandy soils. The fundamental problem regarding changes in the technical criteria reflect the paucity of long-term hydrologic data for the variety of wetlands (and hydric soils) that exist across the country. The lack of these data makes it difficult to develop a time-tested threshold for wetland hydrology and the upper limits of seasonal water tables for hydric soils. The National Research Council (1995) has provided interim guidelines that state that wetlands should have a water table within 1 ft of the land surface for 2 weeks or more during the growing season in most years. This is the scientific standard that should be used until such time as regionally focused long-term data demonstrate otherwise (see Chapter 2).

GENERAL CONCEPT OF HYDRIC SOILS

Hydric soils have standing water for significant periods and/or are saturated at or near the surface for extended periods during the growing season. The duration and depth of soil saturation (i.e., seasonal high water table) are essential criteria for identifying hydric soils and wetlands. Hydric soils are either inundated for a week or more, or saturated within 1 ft of the surface for more than 2 weeks during the growing season in most years. These soils also are usually wet for long periods during the nongrowing season. If the full year is taken into account, these soils are likely to be saturated for 6 to 8 weeks or more. They may be inundated by river overflow, tidal action, direct precipitation, or surface water runoff. Soil saturation results from low-lying topographic position, ground-water seepage, or presence of a slowly permeable layer (e.g., clay, confining bed, fragipan, or hardpan), or direct connection to the underlying water table.

Some hydric soils are submerged for the greater part of the year, while those hydric soils intergrading toward the better drained upland soils are wet for shorter periods and dry at the surface for much of the growing season. Hydric soils typically consist of very poorly drained and poorly drained soils having a water table within a foot of the surface for 2 weeks or more during the growing season. Very poorly drained soils are flooded for much of the year or permanently saturated. Poorly drained soils are usually saturated to the surface at some time during the year, especially in winter in much of the coterminous U.S. Hydric soils also include soils ponded or frequently flooded for 7 or more consecutive days during the growing season. When flooded for a week, a soil is likely to be saturated within a foot for at least another week and usually much longer, particularly for nonsandy soils. Biogeochemical processes involved in hydric soil formation are discussed in detail by Mausbach and Richardson (1994).

Many hydric soils are drained by open ditches, tile drains, or combinations of dikes, ditches, and pumps. Water also is diverted from wetlands through other means, such as by government-financed channelization projects, upstream dams, or ground-water extraction (e.g., municipal or industrial well fields). Former hydric soils that are now effectively drained (e.g., no longer having wetland hydrology and, therefore, not functioning like wetlands) are considered drained hydric soils and are not wetlands. Where drainage system failure occurs, such soils usually revert to hydric conditions and return to wetland, although land leveling operations have significantly altered the microtopography of many sites (an important characteristic of many wetlands, especially forested wetlands). This simple fact makes wetland restoration of these sites feasible, cost effective, and highly successful, especially when contrasted with wetland creation projects.

Soils that were not naturally wet, but are now subject to periodic flooding or soil saturation for specific management purposes (e.g., waterfowl impoundments and rice paddies) or flooded by accident (e.g., highway-created impoundments or leakage from irrigation projects) meet the hydric soils criteria (see criteria 3 and 4 in Table 5.5). Formerly better drained soils now flooded by impoundment from beaver dams also are considered hydric, provided the dams will remain in place. Newly built beaver dams or beaver-blockage of culverts may create flooding of farmlands and private property that may be viewed as a hazard or nuisance and the dam or blockage is likely to be corrected with normal flowage restored. In these cases, the flooded condition is temporary and the soils are not to be considered hydric. The changed hydrology must be more or less long-lasting in order to create a hydric soil from a better drained one. Recently flooded, formerly nonhydric soils do not show strong indicators of gleying or mottling, since it typically takes a long time to develop the characteristic low chroma colors. Obligate hydrophytes are usually present in these created wetlands, so they are easily recognized. Better-drained soils that are frequently flooded for short intervals (usually less than 1 week) during the growing season, or are saturated for less than 2 weeks during the growing season are not hydric soils or wetlands.

Not all wetland soils are anaerobic and have reduced conditions. Exceptions include sands saturated with moving (aerated) water along mountain streams and perhaps for seepage areas on slopes (see Chapter 7). Daniels et al. (1973) found that certain saturated soils had variable amounts of oxygen in North Carolina Coastal Plain soils. However, their study detected significant decreases in soil oxygen content when soluble carbon was transported from the forest litter to the subsoil by heavy rains, further documenting the importance of organic matter in soil oxidation-reduction processes. Where organic matter is not uniformly present in the soil, pockets of organic matter in the soil may create anaerobic microzones in some basically aerobic soils when saturated (Greenland, 1962). Griffin et al. (1996) noted considerable microsite reduction within the capillary fringe of some Texas Gulf Coast Prairie soils.

Some hydric soils do not exhibit strong evidence of gleying and mottling. This is particularly true for red parent material soils and some sandy soils (see Chapter 7). Vepraskas and Wilding (1983) found that for some Texas soils periods of saturation may not overlap with periods of reduction due to water table relationships. Soils saturated by rainfall and seepage interflow above the argillic horizon were saturated longer than they were reduced. Some ponded soils were not saturated throughout the upper part of the soil, yet experienced reduction for extended periods (>6 months), probably due to reduction in capillary pores within the unsaturated zones.

Five key points to remember about hydric soils are

1. Soils classified as hydric must be saturated or flooded long enough during the growing season to produce anaerobic conditions that damage most plants but favor hydrophytes (unless drained).
2. Soils that are well drained but frequently flooded or saturated for short periods of time are not hydric, while those soils inundated for a week or more during the growing season every other year, on average, are hydric.
3. Soils that were formerly wet but are now effectively drained are considered drained hydric soils and the areas are not considered wetlands.
4. The absence of hydric soil should not be construed to mean that the area is always nonwetland, since some wetlands exist on hydric substrates or develop under prolonged saturated conditions where typical hydric soils do not form, according to the National Research Council (1995).
5. While most hydric soils do possess typical properties, some do not.

LISTS OF HYDRIC SOILS AND HYDRIC SOIL MAP UNITS

The need for a national list of hydric soils evolved from the U.S. Fish and Wildlife Service's (FWS) wetland definition. The list would be used to help people understand what soils were considered hydric, just like the list of hydrophytes was in recognizing wetland plants. The Natural Resources Conservation Service (NRCS) worked with the FWS to develop a national list of hydric soils. From the late 1970s, Blake Parker (NRCS soil scientist assigned to the FWS's National Wetlands Inventory Program) worked with FWS and NRCS personnel across the country to compile the list. Preliminary lists were sent in 1981 to state NRCS offices for review and comment. A 1983 list contained 1206 hydric soil series, representing about 10.5% of the nation's soil series (Parker, 1992). This list included soils that consistently displayed hydric properties and soils with a high potential for possessing these properties. Field investigations were always required to verify that the latter series had hydric properties and supported hydrophytic vegetation (Flach, 1983). The list was generated for review and comment purposes to help the NRCS decide whether or not to adopt the FWS wetland classification for its programs.

At the outset, the national hydric soil lists were quite controversial. Some state NRCS offices proposed deleting all soils except Histosols and those flooded daily by the tides (Parker, 1992). This resulted in the establishment of the National Technical Committee for Hydric Soils (NTCHS) to provide technical oversight of the process and to develop a scientifically sound list.

Several lists were published in 1985, leading to the publication of the first national list approved by the newly formed NTCHS (USDA Soil Conservation Service, 1985b). Since then, the list has undergone several revisions and continues to be refined as new series are described and old series names are dropped from the list. The last hard-copy publication of the list was in 1991 (USDA Soil Conservation Service, 1991). Since then the list has been reviewed and updated annually, with updates posted on the Internet along with state lists of these soils (see *http://www. statlab.iastate.edu/soils/hydric*). For individuals lacking access to the Internet, copies of the state and county lists can be obtained from the local NRCS office.

The national list summarizes (in tabular form) certain characteristics of each designated hydric soil. Soils are listed by series name and only those soils that meet the hydric soils criteria are given. Other soils observed in the field with hydric properties may not fit within the allowable range of the characteristics for a given hydric soil series as noted by Segal et al. (1995). These taxadjuncts are not on the list. In addition, newly described series may not be on the list, hence the list should be used with caution. Information provided on the list includes series name; taxonomic subgroup; temperature regime; drainage class; seasonal high water table (depth/time of year); flooding characteristics (frequency, duration, and time of year); and hydric soil criteria met (Table 5.6).

In 1984, the FWS and NRCS cooperated to produce the first official state list of hydric soils (Tiner and Kirkham, 1984). This list was prepared for New Jersey to aid in drafting a chapter on the state's hydric soils for inclusion in the FWS's first state wetland report summarizing the results of its National Wetlands Inventory mapping (Tiner, 1985). The list contained soil series and land types that were associated with wetlands. The list was basically a list of soil map units that tend to support wetlands. It was recognized that while some map units and soil series are almost always associated with wetlands, others are not and include wetlands either most of the time or only some of the time. The list of soil series/land types was therefore divided into three groups: Group 1 — soils that nearly always display consistent hydric conditions, Group 2 — soils displaying consistent hydric conditions in most places, but additional verification is needed, and Group 3 — soils displaying hydric conditions in few places and additional verification is needed. This list was somewhat similar in construction to the wetland plant list, but in this case, relating to a soil's likelihood to exhibit hydric properties. The Group 1 soils are essentially obligate hydric soils, Group 2 soils tend to be facultative wetland hydric soils, and Group 3 tend to be facultative upland hydric soils. Interestingly, this approach is similar to that recommended by the National Research Council (1995) for improving the current national list. "NTCHS should consider developing a system for assigning hydric soils to fidelity categories" (recommendation 13, page 1146). The NTCHS considered this in 1991, but did not adopt such system as mentioned earlier (Mausbach, 1992b).

In response to the Swampbuster provision of the 1985 Food Security Act (Farm Bill), the NRCS published lists of hydric soils for each state. Sometime thereafter, the NRCS compiled county lists of hydric soil map units to aid in using data in soil survey reports for making preliminary wetland determinations. These lists enumerate both hydric soil map units and nonhydric soil map units that may have hydric soil inclusions. These lists are most helpful for locating potential wetlands based on soil types and are the primary lists that should be used to aid in wetland delineation (see Chapter 10 regarding use of county soil survey reports for preliminary wetland identification). In the absence of such lists, it is possible to make interpretations of the likelihood of a soil map unit being hydric by interpreting its taxonomy from the soil classification table given in all soil survey reports (see discussion on Interpreting Soil Taxonomy to Identify Potential Hydric Soils at the end of this chapter).

MAJOR CATEGORIES OF HYDRIC SOILS

Soils fall into two major categories on the basis of soil composition — organic soils (*Histosols*) and mineral soils. In general, sandy soils having 20% or more organic material by weight in the upper 16 in. are considered organic soils as are clayey soils with 30% or more of this material. Soils with less organic content are mineral soils. Organic soils also include soils with thinner organic deposits extending from the surface to underlying bedrock, gravel, cobbles, or other rocky material. For a technical definition, the reader is referred to *Keys to Soil Taxonomy* (Soil Survey Staff, 1996; 1998). Mineral soils are dominated by various combinations of sand, silt, and clay with lesser amounts of organic matter.

Soil morphology features are widely used to interpret long-term soil moisture (Bouma, 1983). Fanning et al. (1973) found a good correlation between certain soil morphology and water tables in Worcester County, MD. Soil mottling has long been recognized as one of the best indicators of poor drainage (Haswell, 1938). For distinguishing hydric soils from nonhydric ones, the three most widely recognized morphological features are the accumulation of organic matter (peat or muck), gleying, and mottling (redoximorphic features).

ORGANIC SOILS

When aerobic environments change to anaerobic environments due to prolonged inundation and soil saturation, organic matter mineralization slows due to less efficient microbial action. These saturated conditions impede aerobic decomposition (or oxidation) of the bulk organic materials, such as leaves, stems and roots, with the net result over time being an accumulation of organic matter. This organic matter is the peat or muck that forms in the wettest wetlands. Organic soils are dominated by the remains of plants. Their organic carbon content exceeds 12 to 18% depending on the clay fraction of the soil (Soil Survey Staff, 1975). Organic soils have more than 12% organic carbon when clay is absent, more than 18% organic carbon if there is more than 60% clay, and between 12 to 18% organic matter with less clay (0 to 60%) (Figure 5.4). Most organic soils are characterized as "very poorly drained" soils. They typically form in waterlogged depressions, in low-lying areas along streams and coastal waters where flooding is frequent, and in cold, humid climates with low evapotranspiration and high groundwater levels.

TABLE 5.6
Sample Page from the Hydric Soil List

Series and Subgroup	Temperature	Drainage Class	High Water Table Depth	High Water Table Months	Perm. within 20 in.	Flooding Frequency	Flooding Duration	Flooding Months	Hydric Criteria Number	Capability Critical Phase Criteria	Capability Class and Sub-Class
Chincoteague (VA0223) Typic sulfaquents	Thermic	VP	+3.–0	Jan–Dec	<6.0	Frequent	V Brief	Jan–Dec	2B3,3	All	8W
Chippeny (MI0133) Lithic borosaprists	Frigid	VP	+1–1.0	Sep–May	<6.0	None			1	All	7W
Chippewa (NY0068) Typic fragiaquepts	Mesic	P	+.5–0.5	Nov–May	<6.0	None			2B3,3	All	4W
Chippwa, stony (NY0069) Typic fragiaquepts	Mesic	P	+.5–0.5	Nov–May	<6.0	None			2B3,3	All	7S
2/Chivato (NM0222) Cumulic haplaquolls	Mesic	P	3.0–5.0	Jul–Oct	<6.0	Frequent	V Long	Jul–Oct	4	Freq	6W
Chobee (FL0062) Typic argiaquolls	Hyperthermic	VP	0–0.5	Jun–Oct	<6.0	None			2B3	All	3W
Chobee, depressional (FL0412) Typic argiaquolls	Hyperthermic	VP	+2–0	Jun–Mar	<6.0	None			2B3,3	All	7W
Chobee, flooded (FL0040) Typic argiaquolls	Hyperthermic	VP	0–0.5	Jun–Oct	<6.0	Rare–Common	Brief–V Long	Jun–Feb	2B3,4	Rare, Occas Freq	3W 5W
Chobee, limestone substratum (FL0450) Typic argiaquolls	Hyperthermic	VP	+2–0	Jun–Mar	<6.0	None–Common	V Long	Jan–Dec	2B3,3,4	All	7W
Chock (OR0156) Andaqueptic cryaquents	Cryic	P	1.0–2.5	Mar–Jun	<6.0	Frequent	Brief	Mar–May	2B3	All	
Chocorua (NH0022) Terric borohemists	Frigid	VP	+1–0.5	Jan–Dec	<6.0	None			1,3	Undrained Drained	8W 4W
Chowan (NC0146) Thapto-histic fluvaquents	Thermic	VP	0–0.5	Nov–May	<6.0	Frequent	V Long	Nov–Apr	2B3,4	All	7W
Chummy (CA1647) Typic humaquepts	Frigid	P	0.5–1.5	Jan–Jun	<6.0	None			2B3	0–2% 2–3%	5W 6W
Cieno (TX0903) Typic ochraqualfs	Hyperthermic	P	+1–1.5	Sep–Jun	<6.0	None			2B3,3	All	4W
Cisne (IL0126) Mollic albaqualfs	Mesic	P	0–2.0	Feb–Jun	<6.0	None			2B3	All	3W

Source: USDA Soil Conservation Service, 1991.

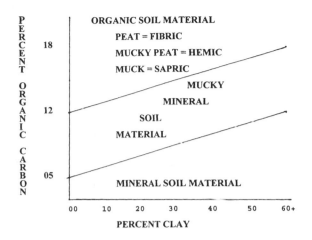

FIGURE 5.4 Separating organic soils from mucky mineral and mineral soil material based on the percent of organic carbon and percent of clay. (Hurt, G.W., Whited, P.M., and Pringle, R.F., Eds., *Field Indicators of Hydric Soils in the U.S.*, USDA Natural Resources Conservation Service, Fort Worth, TX, 1988.)

Organic soil materials can be further subdivided into three groups based on the fraction of identifiable plant material in the soil after gentle rubbing (Soil Survey Staff, 1980).

1. Muck (*Saprists*) where more than five sixths (83.3%) of the fibers are decomposed and less than one sixth is identifiable.
2. Peat (*Fibrists*) with less than one quarter decomposed and three quarters or more identifiable.
3. Mucky peat or peaty muck (*Hemists*) where between one sixth and three quarters is identifiable as fibers (greater than 0.15 mm and less than 2 cm).

A fourth group of organic soils, *Folists*, occur in boreal and tropical mountainous areas, but they do not develop under hydric conditions. All organic soils, with the exception of the Folists, are hydric soils. For more information on organic soils, the reader is referred to Aandahl et al. (1974) and Soil Survey Staff (1998).

MINERAL HYDRIC SOILS

According to the most recent version of *Keys to Soil Taxonomy* (Soil Survey Staff, 1998), there are 11 orders containing mineral soils.

- Alfisols (deciduous forest soils)
- Andisols (volcanic soils)
- Aridisols (desert soils)
- Entisols (floodplain/sandy soils)
- Gelisols (permafrost or extremely cold soils)
- Inceptisols (young soils)
- Mollisols (grassland soils)
- Oxisols (tropical soils)
- Spodosols (evergreen forest soils)
- Ultisols (older temperate forest soils)
- Vertisols (high shrink-swell clayey soils)

The wetter of these soils typically have "aquic conditions" (formerly, aquic moisture regime; replaced in 1992). These soils experience continuous or periodic saturation and reduction (which can be measured) as indicated by the presence of redoximorphic features (Soil Survey Staff, 1996). The duration of saturation usually is at least a few days as oxygen must be virtually removed from the soil, but duration is not specified since it is variable among soils. At the suborder level in soil taxonomy, the wetter mineral soils are designated by "aqu" as in Aquents (wet Entisols), Aquepts (wet Incepti-sols), Aquods (wet Spodosols), Aqualfs (wet Alfisols), Aquults (wet Ultisols), Aquerts (wet Vertisols), and Aquolls (wet Mollisols) (see discussion on Soil Taxonomy later in this chapter). At this level, soils are reduced and saturated within 50 cm of the surface (Soil Survey Staff, 1996). Diagnostic properties of the aquic suborders are given in Table 5.7. Traditionally, soil colors have been used to determine whether soils have an aquic moisture regime (Vepraskas, 1996), due to the effort required to measure hydrology and reduction and the good correlation between soil color and wetness. In general, low chroma (≤2) (grayish) colors were used as indicators of soil saturation and reduction, but in some soils, a combination of 2.5Y or 5Y hues and iron mottles were used (Vepraskas, 1996).

Under saturated and reduced conditions, mineral soils form where organic surface layers do not exceed 16 in. in thickness and where varying proportions of sand, silt, and clay predominate. Mineral soils exhibit a wide range of properties related to differences in parent material, climate, topography, age, and other factors. Mineral soils that are predominantly gray directly below the A-horizon with brown, red, or yellow mottles are usually anaerobic and reduced for long periods during the growing season and are classified as hydric. Soils that are predominantly brown or yellow beneath the A-horizon with a few gray (low chroma) mottles are reduced for shorter periods and are usually not hydric. If numerous gray mottles extend to within 6 in. of the surface *and* a low chroma matrix occurs within 18 in. of the soil surface, the soil may be hydric in mesic temperate regions. Mineral soils that are never reduced are usually brightly-colored and are not mottled. Figure 5.5 illustrates common properties of hydric and adjacent nonhydric soils for the Northeast. In some hydric soils, mottles may not be visible due to masking by organic matter (Parker et al., 1984). Vepraskas (1996) provides a detailed technical review of the formation and characteristics of soils with "aquic conditions." A forthcoming book edited by Richardson and Vepraskas (1999) will offer a comprehensive review of wetland soils.

LISTS OF U.S. HYDRIC SOIL INDICATORS

While the lists of hydric soils and the technical criteria were valuable contributions to wetland delineation, a comprehensive list of hydric soil field indicators was the principal need for site-specific investigations. Since the concept of hydric soils was new to soil scientists, there were considerable differences in the interpretation of the list and criteria. Some scientists said soils had to be classified to series and then if on the national list, they were hydric. Other scientists used the technical criteria as guidelines by looking for soil morphological properties to verify a seasonal water table within the specified depth. The use of drainage classes in the technical criteria posed particular problems since the interpretation of drainage classes is not standardized but open to local interpretation. The interpretation can and often does differ from state to state and even within a state as these classes were intended to help group soils with similar problems or limitations for various uses, especially crop production. Consequently, the hydric soil criteria were not true standards due to this flexible interpretation, but were simply developed to extract a list of potential hydric soils out of a national soils database (Mausbach, 1994). This problem became evident when the Corps and EPA adopted the accepted hydric soil criteria in the 1989 interagency manual for use in identifying federally regulated wetlands throughout the country (Federal Interagency Committee for Wetland Delineation, 1989).* The varied interpretations of the hydric soils criteria soon became evident during 1989 and 1990 when used for regulatory determinations.

---

* This was the first time that standardized scientifically based methods were put into practice for identifying and delineating regulated wetlands in the U.S. Before this time, Corps districts used a host of methods including locally derived techniques and the Corps wetland manual (Environmental Laboratory, 1987) to establish the presence of a regulated wetland. Consequently, identification and delineation of regulated wetlands was inconsistent and not technically sound in most areas of the country prior to 1989. Although the federal interagency manual is not used for federal determinations today due to largely political considerations, it did establish a basis and impetus for improving interpretation of the Corps manual which is currently used for such determinations.

TABLE 5.7
Diagnostic Properties of Wet Mineral Soils (with Aquic Moisture Regimes) for Some Major Orders at the Suborder Level

Order	Properties of Aquic Suborders
Alfisols	Have, in one or more horizons within 50 cm of the mineral soil surface, aquic conditions (other than anthraquic conditions) for some time in most years (or artificial drainage), and have one or both of the following: 1. Redoximorphic features in all layers between either the lower boundary of an Ap horizon or a depth of 25 cm below the mineral soils surface, whichever is deeper, and a depth of 40 cm; and one of the following within the upper 12.5 cm of the argillic, natric, glossic, or kandic horizon: – Fifty percent or more redox depletions with a chroma of 2 or less on faces peds, and redox concentrations within peds – Redox concentrations and 50% or more redox depletions with a chroma of 2 or less in the matrix – Fifty percent or more redox depletions with a chroma of 1 or less on faces of peds or in the matrix, or both 2. In the horizons that have aquic conditions, enough active ferrous iron to give a positive reaction to α, α′-dipyridyl at a time when the soil is not being irrigated.

Entisols	Have one or more of the following:
	1. Aquic conditions and sulfidic materials within 50 cm of the mineral soil surface
	2. Permanent saturation with water, and a reduced matrix in all horizons below a depth of 25 cm from the mineral soil surface
	3. In a layer above a densic, lithic, or paralithic contact or in a layer between 40 and 50 cm from the mineral soil surface, whichever is shallower, aquic conditions for some time in most years (or artificial drainage) and one or more of the following:
	– A texture finer than loamy fine sand and, in 50% or more of the matrix, one or more of the following:
	  a. A chroma of 0
	  b. A chroma of 1 or less and a color value, moist, of 4 or more
	  c. A chroma of 2 or less, and redox concentrations
	– A texture of loamy fine sand or coarser and, in 50% or more of the matrix, one or more of the following:
	  a. A chroma of 0
	  b. A hue of 10YR or redder, a color value, moist, of 4 or more, and a chroma of 1
	  c. A hue of 10YR or redder, a chroma of 2 or less, and redox concentrations
	  d. A hue of 2.5Y or yellower, a chroma of 3 or less, and distinct or prominent redox concentrations
	  e. A hue of 2.5Y or yellower and a chroma of 1
	  f. A hue of 5GY, 5G, 5BG, or 5B
	  g. Any color if it results from uncoated sand grains
	– Enough active ferrous iron to give a positive reaction to α, α′-dipyridyl at a time when the soil is not being irrigated.

Inceptisols	Have one or more of the following: 1. In a layer above a densic, lithic, or paralithic contact or in a layer between 40 and 50 cm from the mineral soil surface, whichever is shallower, aquic conditions for some time in most years (or artificial drainage) and one or more of the following: – A histic epipedon – A sulfuric horizon that has its upper boundary within 50 cm of the mineral soil surface – A mollic, an ochric, or an umbric epipedon that is underlain directly, or within 50 cm of the mineral soil surface, by a horizon that has, on faces of peds or in the matrix if peds are absent, 50% or more chroma of either 2 or less if there are redox concentrations or 1 or less. 2. An exchangeable sodium percentage (ESP) of 15 or more (or a sodium adsorption ratio, SAR, of 13 or more) in half or more of the soil volume within 50 cm of the mineral soil surface, and a decrease in EPS (or SAR) values with increasing depth below 50 cm, and groundwater within 100 cm of the mineral soil surface for some time during the year. 3. Within 50 cm of the mineral soil surface, enough active ferrous iron to give a positive reaction to α, α′-dipyridyl at a time when the soil is not being irrigated.

TABLE 5.7 (continued)
Diagnostic Properties of Wet Mineral Soils (with Aquic Moisture Regimes) for Some Major Orders at the Suborder Level

Order	Properties of Aquic Suborders

Mollisols Have in a layer above a densic, lithic, or paralithic contact or in a layer between 40 and 50 cm from the mineral soil surface whichever is shallower, aquic conditions for some time in most years (or artificial drainage) and one or more of the following:

1. A histic epipedon overlying the mollic epipedon.
2. An exchangeable sodium percentage (ESP) of 15 or more (or a sodium adsorption ratio, SAR, of 13 or more) in the upper part of the mollic epipedon, and a decrease in ESP (or SAR) values with increasing depth below 50 cm from the mineral soil surface.
3. A calcic or petrocalcic horizon that has its upper boundary within 40 cm of the mineral soil surface.
4. One of the following colors:
   - A chroma of 1 or less in the lower part of the mollic epipedon, and either
     a. Distinct or prominent redox concentrations in the lower part of the mollic epipedon
     b. Either directly below the mollic epipedon, or within 75 cm of the mineral surface if a calcic horizon intervenes, a color value, moist, of 4 or more and one of the following:
       - Fifty percent or more chroma of 1 on faces of peds or in the matrix, a hue of 10YR or redder, and redox concentrations
       - Fifty percent or more chroma of 2 or less on faces of peds or in the matrix, a hue of 2.5Y or yellower, and redox concentrations
       - Fifty percent or more chroma of 1 on faces of peds or in the matrix, and a hue of 2.5Y or yellower
       - Fifty percent or more chroma of 3on or less on faces of peds or n the matrix, a hue of 5Y, and redox concentrations
       - Fifty percent or more chroma of 0 on faces of peds or in the matrix
       - A hue of 5GY, 5G, 5BG, or 5B
       - Any color if it results from uncoated sand grains
   - A chroma of 2 in the lower part of the mollic epipedon, and either:
     a. Distinct or prominent redox concentrations in the lower part of the mollic epipedon
     b. Directly below the mollic epipedon, *one of* the following matrix colors:
       - A color value, moist, of 4, a chroma of 2, and some redox depletions with a color value, moist, of 4 or more and a chroma of 1 or less
       - A color value, moist, of 5 or more, a chroma of 2 or less, and redox concentrations
       - A color value, moist, of 4 and a chroma of 1 or less
5. Between 40 and 50 cm from the mineral soil surface, enough active ferrous iron to give a positive reaction to α, α′-dipyridyl at a time when the soil is not being irrigated.

Spodosols Have aquic conditions for some time in most years (or artificial drainage) in one or more horizons within 50 cm of the mineral soils surface, and one or both of the following:
1. A histic epipedon
2. Within 50 cm of the mineral soil surface, redoximorphic features in an albic or spodic horizon.

Ultisols Have aquic conditions for some time in most years (or artificial drainage) in one or more horizons within 50 cm of the mineral soil surface, and one or both of the following:
1. Redoximorphic features in all layers between either the lower boundary of an Ap horizon or a depth of 25 cm from the mineral soil surface, whichever is deeper, and a depth of 40 cm; and one of the following within the upper 12.5 cm of the argillic or kandic horizon:
   - Redox concentrations, and 50% or more redox depletions with a chroma of 2 or less either on faces of peds or in the matrix
   - Fifty percent or more redox depletions with a chroma of 1 or less either on faces of peds or in the matrix
   - Distinct or prominent redox concentrations and 50% or more hue of 2.5Y or 5Y in the matrix, and also a thermic, isothermic, or warmer soil temperature regime
2. Within 50 cm of the mineral soil surface, enough active ferrous iron to give a positive reaction to α, α′-dipyridyl at a time when the soil is not being irrigated.

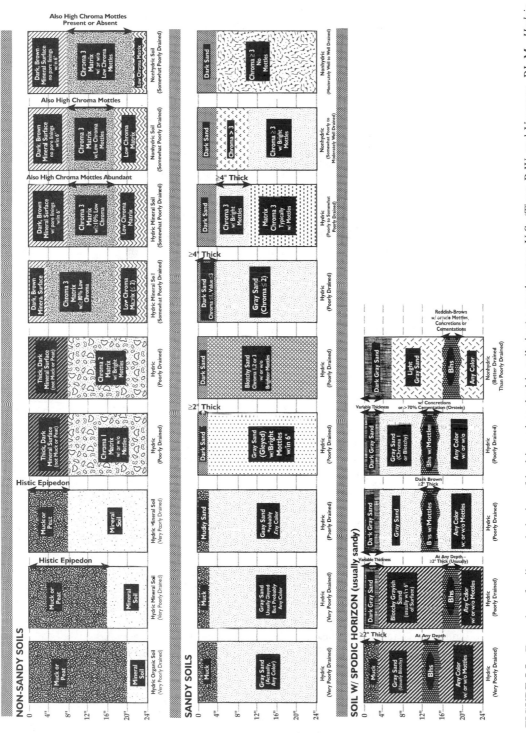

FIGURE 5.5 Examples of soil morphologies of hydric and nonhydric soils in the northeastern U.S. (Tiner, R.W. and Veneman, P.L.M., *Hydric Soils of New England*, Bulletin C-183R, University of Massachusetts Extension, Amherst, MA, 1995.)

Before the technical committee (Federal Interagency Committee for Wetland Delineation) reconvened to revise the interagency manual in 1990, NRCS soil scientists (including Maury Mausbach) and FWS wetland ecologist Ralph Tiner met at the National Soil Survey Center in Lincoln, NE to prepare the first draft of a comprehensive national list of positive indicators of hydric soils. This list was compiled by evaluating aquic suborder requirements from *Soil Taxonomy* (Soil Survey Staff, 1975). Flow charts were developed to aid in the decision process (Figure 5.6). These charts were used to help train soil scientists and nonsoil scientists alike in hydric soil recognition. Later, Regional NRCS technical centers refined and expanded these lists as the national list was evolving.

After field testing a couple of versions of a national list of hydric soil indicators (including a 1993 published version), NRCS, in cooperation with the NTCHS, published *Field Indicators of Hydric Soils in the United States: A Guide for Identifying and Delineating Hydric Soils* in 1996 (Hurt et al., 1996). This publication represents the collective efforts of numerous soil scientists (especially from the NRCS) across the country and reflects the indicators that are consistently associated with hydric soils. On a regional or local level, there may be other useful indicators, but they may not have universal application. Field indicators are listed for four soil categories: all soils, sandy soils, loamy and clayey soils, and test indicators. The indicators are briefly described and some are illustrated with color plates. The descriptions specify whether they can be used throughout the country or only in certain geographic areas. A CD-ROM is available on hydric soil indicators from the NRCS (Hurt and Richardson, 1998). The NRCS uses this list to identify wetlands subject to the Food Security Act, yet when making wetland determinations for the Section 404 program of the Clean Water Act, NRCS scientists are restricted to using the list of hydric soil indicators published in the Corps manual (Environmental Laboratory, 1987) which is more conservative. (See Table 5.8 for comparison.) An updated version of the national list containing minor changes was published in 1998 (Hurt et al., 1998).

The national guide was not the first publication to address hydric soil indicators. All three federal wetland delineation manuals contained lists of rather basic hydric soil indicators (Environmental Laboratory, 1987; Sipple, 1988; Federal Interagency Committee for Wetland Delineation, 1989). Hurt et al. (1996, 1998), however, provide the most comprehensive listing of hydric soil indicators for the nation to date. Additional technical hydric soil indicator publications include *Field Indicators for Identifying Hydric Soils in New England*, printed by the New England Interstate Water Pollution Control Commission (NEIWPCC Wetlands Work Group, 1995; 1998), *Soil and Water Relationships of Florida's Ecological Communities*, produced by the USDA Natural Resources Service (Florida Soil Survey Staff, 1992), and *Field Indicators of Hydric Soils in the Mid-Atlantic United States* (Mid-Atlantic Hydric Soil Committee, 1998). Hurt and Brown (1995) explained the development of hydric soil indicators for Florida; a list of these indicators was needed by the Florida Department of Environmental Regulation to use as a tool for identifying wetlands regulated by the state.

The first publication specifically addressing hydric soil identification was a regional guidebook entitled *Hydric Soils of New England*, published jointly by the University of Massachusetts Cooperative Extension Service and the U.S. Fish and Wildlife Service (Tiner and Veneman, 1987, revised in 1995). Designed for the nonsoil scientist, this guidebook described the concept and recognition of hydric soil in layman's terms and illustrated diagnostic properties through the use of color plates. The 1995 revision contained a set of diagrams to further aid readers in interpreting hydric soils (Figure 5.5). Other wetland publications containing information on hydric soil indicators (usually illustrations or photographs of hydric soils) include *Field Guide to Nontidal Wetland Identification*, published by the FWS and the State of Maryland (Tiner, 1988), *Maine Wetlands and Their Boundaries*, issued by the State of Maine (Tiner, 1991; 1994), *The Florida Wetlands Delineation Manual*, issued by the state of Florida (Florida Dept. of Environmental Protection et al., 1995), and *In Search of Swampland*, published by Rutgers University Press (Tiner, 1998).

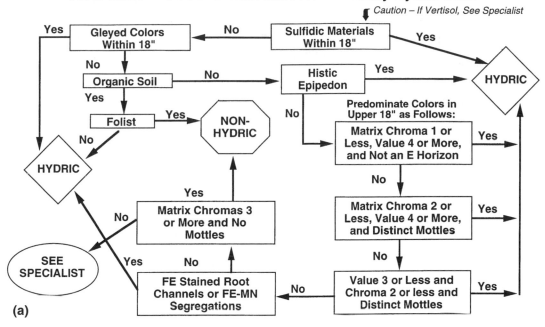

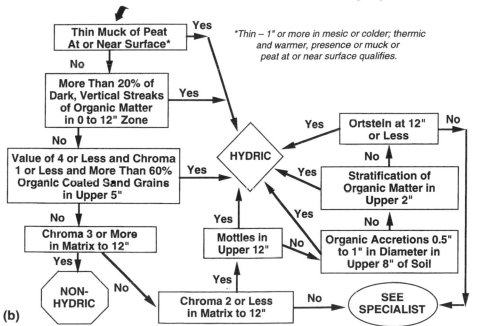

FIGURE 5.6 Flow charts developed in 1990 for identifying hydric soils: (a) nonsandy soils and (b) sandy soils. (USDA Natural Resources Conservation Service.)

TABLE 5.8

Correspondence Between National Hydric Soil Indicators and Corps Manual Hydric Soil Indicators

Corps Manual	1998 Indicators
Nonsandy Soils	
a. Organic soils (Histosols)	A1 (Histosols)
b. Histic epipedon	A2 (Histic epipedon)
	A3 (Black histic)
c. Sulfidic material	A4 (Hydrogen sulfide)
d. Aquic or peraquic moisture regime	
e. Reducing soil conditions	
f(1). Gleyed soils (gray color)	F2 (Loamy gleyed matrix)
	F14 (Alaska redox gleyed)
	F15 (Alaska gleyed pores)
f(2). Soils with bright mottles and/or low matrix chroma	F3 (Depleted matrix)
	F8 (Redox depressions)
	F9 (Vernal pools)
	F11 (Depleted ochric)
	F16 (High plains depressions)
	TF8 (Redox spring seeps)
	TF9 (Delta ochric)
	TF10 (Alluvial depleted matrix)
g. Soil appearing on the hydric soils list	
h. Iron and manganese concretions	F12 (Iron/manganese masses)
	TF3 (Alaska concretions)
Not listed in the 1987 Manual	A5 (Stratified layers)
	A6 (Organic bodies)
	A7 (5 cm mucky mineral)
	A8 (Muck presence)
	A9 (1 cm muck)
	A10 (2 cm muck)
	F1 (Loamy mucky mineral)
	F4 (Depleted below dark surface)
	F5 (Thick dark surface)
	F6 (Redox dark surface)
	F7 (Depleted dark surface)
	F10 (Marl)
	F13 (Umbric surface)
	TA1 (Playa rim stratified layers)
	TF1 (? cm mucky peat or peat)
	TF2 (Red parent material)
	TF4 (Y below dark surface)
	TF5 (Y below thick dark surface)
	TF6 (Calcic dark surface)
	TF7 (Thick dark surface 2/1)
Sandy Soils	
a. Organic soils (Histosols)	A1 (Histosols)
b. Histic epipedon	A2 (Histic epipedon)
	A3 (Black histic)
c. Sulfidic material	A4 (Hydrogen sulfide)
d. Aquic or peraquic moisture regime	

Corps Manual	1998 Indicators
e. Reducing soil conditions	
f. Iron and manganese concretions	S5 (Sandy redox)
	TS5 (Chroma 3 sandy redox)
g. High organic matter content in the the surface horizon	A6 (Organic bodies)
	A7 (5 cm mucky mineral)
	A8 (Muck presence)
	A9 (1 cm muck)
	A10 (2 cm muck)
	S1 (Sandy mucky mineral)
	S2 (3 cm mucky peat or peat)
	S3 (5 cm mucky peat or peat)
	S7 (Dark surface)
	TA2 (Structureless muck)
	TS2 (Thick sandy dark surface)
h. Streaking of subsurface horizons by organic matter	S6 (Stripped matrix)
	S8 (Polyvalue below surface)
i. Organic pan	S9 (Thin dark surface)
j. Soils appearing on the hydric soils list	
Not listed in the 1987 Manual	A5 (Stratified layers)
	S4 (Sandy gleyed matrix)
	S10 (Alaska gleyed)
	TA1 (Playa rim stratified layers)
	TS1 (Iron staining)
	TS4 (Sandy neutral surface)
Problem Soils	
Recently deposited sandy materials	A5 (Stratified layers)
	S6 (Stripped matrix)
	S8 (Polyvalue below surface)
	TA1 (Playa rim stratified layers)
Soils with thick dark A horizons	F4 (Depleted below dark surface)
	F5 (Thick dark surface)
	F6 (Redox dark surface)
	F7 (Depleted dark surface)
	F13 (Umbric surface)
	F16 (High plains depressions)
	TF4 (Y below dark surface)
	TF5 (Y below thick dark surface)
	TF6 (Calcic dark surface)
	TS2 (Thick sandy dark surface)
	TF7 (Thick dark surface 2/1)
Soils with red parent material	F8 (Redox depressions)
	F9 (Vernal pools)
	TS1 (Iron staining)
	TF2 (Red parent material)
	TF8 (Redox spring seeps)
Soils with low chroma parent material	S4 (Sandy gleyed matrix)
	S10 (Alaska gleyed)
	F10 (Marl)
	TS4 (Sandy neutral surface)

Source: Environmental Laboratory. 1987. U.S. Army Engineer Waterways Expt. Station, Vicksburg, MS. Hurt, G.W., et al., 1998. Field Indicators of Hydric Soils in the United States. A Guide for Identifying and Delineating Hydric Soils (ver. 4). USDA Natural Resources Conservation Service, Fort Worth, TX.

HYDRIC SOIL FIELD INDICATORS IN THE U.S.

The various federal wetland delineation manuals listed a few simple readily identifiable hydric soil properties.

1. A peat or muck surface layer 8 in. or thicker.
2. Dominant colors in the mineral soil matrix of chroma of 2 or less, if there are mottles (usually orangish, yellowish, or reddish brown) present.
3. Dominant colors in the mineral soil matrix of chroma of 1 or less, if there are no mottles present.
4. Organic streaking or blotchiness of sandy soils.
5. The smell of hydrogen sulfide in the upper foot of the soil.
6. Observed reduction (ferrous iron test, see section near end of this chapter) (Table 5.8).

Other hydric soil indicators listed in *Field Indicators of Hydric Soils in the United States: A Guide for Identifying and Delineating Hydric Soils* (Hurt et al., 1996; 1998) are outlined in Table 5.9 and most are included in the following discussion. This list is fairly comprehensive and includes the most typical and reliable ones, yet some wetlands will undoubtedly possess soils lacking these features (see Chapter 7 for discussion). Also some of the indicators on the national list that are restricted to some regions may actually be equally applicable in neighboring regions (e.g., indicators for resource region T, the Coastal Plain, also should be reliable in the northern part of the Coastal Plain extending into New Jersey and New York; see map in Figure 5.7 for regional boundaries). Although the national list contains many region-specific indicators, some regions (e.g., New England) are developing supplemental lists to include other regional properties typical of the region's hydric soils. Hurt and Carlisle (1998) have further separated the national indicators into two basic groups — those useful for wetland identification vs. those useful for boundary determinations (see Table 5.9). It is important to realize that most hydric mineral soils have more than one of these indicators. Although the indicators focus on the upper 6 to 12 in., other signs of hydric soils occur below this level. Two of the most common subsoil features are a gleyed matrix and a depleted matrix with redox concentrations. Plates 3 through 27 show examples of hydric soil properties. Plate 27 shows an example of a hydric soil and an adjacent nonhydric soil. See Chapter 7 for discussion of problem hydric soils and Plates 35 through 43 for examples.

HISTIC MATERIALS

As mentioned previously, under extremely wet conditions, organic matter (histic material) builds up. The presence of histic materials in appreciable amounts may, therefore, be a useful indicator of hydric soils. Such materials are usually associated with the wetter wetlands. Where predominate, organic soils form and where lesser amounts of histic materials are present, hydric mineral soils are often present. *Caution:* Peat and muck may form under nonhydric conditions in cryic regions.

Organic Soils

Thick deposits of histic materials (i.e., organic soils) typically form in waterlogged depressions (e.g., glacially formed depressions called kettleholes, river ox-bows, or lake margins) where peat or muck deposits range today from about 1.5 ft to 50 ft or more deep; in cold, wet climates like northern New England; and in low-lying areas along coastal waters where tidal flooding is frequent and saturation is nearly continuous. In the humid subarctic, the cold climate lowers evaporation and plant transpiration allowing for development of organic soils on broad lowlands such as the Hudson Bay Lowlands in Canada and former Glacial Lake Agassiz from Minnesota north. In these regions, organic soils may cover many miles of the landscape, even rolling terrain with blanket bogs.

TABLE 5.9
Brief Descriptions of Hydric Soil Field Indicators for the United States

Soil Type (Code)	Indicator	Brief Description (including applicable regions)
A1	Histosol	40 cm (16 in.) or more of the upper 80 cm (32 in.) as organic soil material. These materials include muck (sapric soil material), mucky peat (hemic soil material), or peat (fibric soil material). For use in all LRRs.
A2	Histic epipedon	Most histic epipedons are surface horizons 20 cm (8 in.) or more thick of organic soil material. For use in all LRRs except W, X and Y; for testing in LRRS W, X, and Y.
A3	Black histic	A layer of peat, mucky peat, or muck 20 cm (8 in.) or more thick starting within the upper 15 cm (6 in.) of the soil surface having hue 10YR or yellower, value 3 or less, and chroma 1 or less. For use in all LRR except W, X, and Y; for testing in LRRS W, X, and Y.
A4	Hydrogen sulfide	A hydrogen sulfide odor within 30 cm (12 in.) of the soil surface. For use in all LRRs.
A5*	Stratified layers	Several stratified layers starting within the upper 15 cm (6 in.) of the soil surface. One or more of the layers has value 3 or less with chroma 1 or less and/or it is muck, mucky peat, peat, or mucky modified mineral texture. The remaining layers have value 4 or more and chroma 2 or less. For use in LRRs F, K, L, M, N, O, P, R, S, T, and U; for testing in LRRs V and Z.
A6*	Organic bodies	Presence of 2% or more organic bodies of muck or a mucky modified mineral texture, approximately 1 to 3 cm (0.5 to 1 in.) in diameter, starting within 15 cm (6 in.) of the soil surface. For use in LRRs P, T, U, and Z.
A7*	5 cm mucky mineral	A mucky modified mineral surface layer 5 cm (2 in.) or more thick starting within 15 cm (6 in.) of the soil surface. For use in LRRs P, T, U, and Z.
A8*	Muck presence	A layer of muck with value 3 or less and chroma 1 or less within 15 cm (6 in.) of the soil surface. For use in LRRs U, V, and Z.
A9*	1 cm muck	A layer of muck 1 cm (0.5 in.) or more thick with value 3 or less and chroma 1 or less starting within 15 cm (6 in.) of the soil surface. For use in LRRs D, F, G, H, P, and T; for testing in LRRs I, J, and O.
A10*	2 cm muck	A layer of muck 2 cm (0.75 in.) or more thick with value 3 or less and chroma 1 or less starting with 15 cm (6 in.) of the soil surface. For use in LRRs M and N; for testing in LRRs A, B, C, E, K, L, R, S, U, W, X, Y, and Z.
S1*	Sandy mucky mineral	A mucky modified mineral layer 5 cm (2 in.) or more thick starting within 15 cm (6 in.) of the soil surface. For use in all LRRs except W, X, and Y.
S2*	3 cm mucky peat or peat	A layer of mucky peat or peat 2.5 cm (1 in.) or more thick with value 4 or less and chroma 3 or less starting within 15 cm (6 in.) of the soil surface. For use in all LRRs G and H.
S3*	5 cm mucky peat or peat	A layer of mucky peat or peat 5 cm (2 in.) or more thick with value 3 or less and chroma 2 or less starting within 15 cm (6 in.) of the soil surface. For use in LRRs F and M; for testing in LRR R.
S4	Sandy gleyed matrix	A gleyed matrix which occupies 60% or more of a layer starting within 15 cm (6 in.) of the soil surface. For use in all LRRs except W, X, and Y.
S5*	Sandy redox	A layer starting within 15 cm. (6 in.) of the soil surface that is at least 10 cm (4 in.) thick, and has a matrix chroma 2 or less with 2% or more distinct or prominent redox concentrations as soft masses and/or pore linings. For use in all LRRs except V, W, X, and Y.
S6*	Stripped matrix	A layer starting within 15 cm (6 in.) of the soil surface in which iron/manganese oxides and/or organic matter have been stripped from the matrix exposing the primary base color of soil materials. The stripped areas and translocated oxides and/or organic matter form a diffuse splotchy pattern of two or more colors. The stripped zones are 10% or more of the volume; they are rounded and approximately 1 to 3 cm (0.5 to 1 in.) in diameter. For use in all LRRs except V, W, X, and Y.

TABLE 5.9 (continued)
Brief Descriptions of Hydric Soil Field Indicators for the United States

Soil Type (Code)	Indicator	Brief Description (including applicable regions)
S7*	Dark surface	A layer 10 cm (4 in.) or more thick starting within the upper 15 cm (6 in.) of the soil surface with a matrix value 3 or less and chroma 1 or less. At least 70% of the visible soil particles must be covered, coated, or similarly masked with organic material. The matrix color of the layer immediately below the dark layer must have value 4 or more and chroma 2 or less. For use in LRRs N, P, R, S, T, U, V, and Z.
S8	Polyvalue below surface	A layer with value 3 or less and chroma 1 or less starting within 15 cm (6 in.) of the soil surface underlain by a layer(s) where translocated organic matter unevenly covers the soil material forming a diffuse splotchy pattern. At least 70% of the visible soil particles in the upper layer must be covered, coated, or masked with organic material. Immediately below this layer, the organic coating occupies 5% or more of the soil volume and has value 3 or less and chroma 1 or less. The remainder of the soil volume has value 4 or more and chroma 1 or less. For use in LRRs R, S, and T; for testing in LRRs K and L.
S9	Thin dark surface	A layer 5 cm (2 in.) or more thick within the upper 15 cm (6 in.) of the surface, with value 3 or less and chroma 1 or less. At least 70% of the visible soil particles in this layer must be covered, coated, or masked with organic material. This layer is underlain by a layer(s) with value 4 or less and chroma 1 or less to a depth of 30 cm (12 in.) or to the spodic horizon, whichever is less. For use in LRRs R, S, and T; for testing in LRRs K and L.
S10*	Alaska gleyed	Dominant hue N, 10Y, 5GY, 10GY, 5G, 10G, 5BG, 10BG, 5B, 10B, or 5PB, with value 4 or more in the matrix, within 30 cm (12 in.) of the mineral surface, and underlain by hue 5Y or redder in the same type of parent material. For use in LRRs W, X, and Y.
F1*	Loamy mucky mineral	A mucky modified mineral layer 10 cm (4 in.) or more thick starting within 15 cm (6 in.) of the soil surface. For use in all LRRs except V, W, X, and Y.
F2	Loamy gleyed matrix	A gleyed matrix that occupies 60% or more of a layer starting within 30 cm (12 in.) of the soil surface. For use in all LRRs except W, X, and Y.
F3*	Depleted matrix	A layer at least 15 cm (6 in.) thick with a depleted matrix that has 60% or more chroma 2 or less starting within 25 cm (10 in.) of the surface For use in all LRRs except W, X, and Y.
F4*	Depleted below dark surface	A layer at least 15 cm (6 in.) thick with a depleted matrix that has 60% or more chroma 2 or less starting within 30 cm (12 in.) of the surface. The layer(s) above the depleted matrix have value 3 or less and chroma 2 or less. For use in all LRRs except LRRs W, X, and Y; for testing in LRRS W, X, and Y.
F5	Thick dark surface	A layer at least 15 cm (6 in.) thick with a depleted matrix that has 60% or more chroma 2 or less (or a gleyed matrix) starting below 30 cm (12 in.) of the surface. The layer(s) above the depleted or gleyed matrix have hue N and value 3 or less to a depth of 30 cm (12 in.) and value 3 or less and chroma 1 or less in the remainder of the epipedon. For use in all LRRs except LRRs W, X, and Y; for testing in LRRs W, X, and Y.
F6*	Redox dark surface	A layer at least 10 cm (4 in.) thick entirely within the upper 30 cm (12 in.) of the mineral soil that has: – matrix value 3 or less and chroma 1 or less and 2% or more distinct or prominent redox concentrations as soft masses or pore linings – matrix value 3 or less and chroma 2 or less and 5% or more distinct or prominent redox concentrations as soft masses or pore linings For use in all LRRs except LRRs W, X, and Y; for testing in LRRs W, X, and Y.
F7*	Depleted dark surface	Redox depletions, with value 5 or more and chroma 2 or less, in a layer at least 10 cm (4 in.) thick entirely within the upper 30 cm (12 in.) of the mineral soil that has: – matrix value 3 or less and chroma 1 or less and 10% or more redox depletions – matrix value 3 or less and chroma 2 or less and 20% or more redox depletions For use in all LRRs except LRRs W, X, and Y; for testing in LRRs W, X, and Y.

Soil Type (Code)	Indicator	Brief Description (including applicable regions)
F8*	Redox depressions	In closed depressions subject to ponding, 5% or more distinct or prominent redox concentrations as soft masses or pore linings in a layer 5 cm (2 in.) or more thick entirely within the upper 15 cm (6 in.) of the soil surface. For use in all LRRs except LRRs W, X, and Y; for testing in LRRs W, X, and Y.
F9*	Vernal pools	In closed depressions subject to ponding, presence of a depleted matrix in a layer 5 cm (2 in.) thick entirely within the upper 15 cm (6 in.) of the soil surface. For use in LRRs C and D.
F10*	Marl	A layer of marl with a vlaue 5 or more starting with 10 cm (4 in.) of the soil surface. For use in LRR U.
F11*	Depleted ochric	A layer(s) 10 cm (4 in.) or more thick that has 60% or more of the matrix with value 4 or more and chroma 1 or less. The layer is entirely within the upper 25 cm (10 in.) of the soil surface. For use in LRR O.
F12*	Iron/manganese masses	On floodplains, a layer 10 cm (4 in.) or more thick with 40% or more chroma 2 or less, and 2% or more distinct or prominent redox concentrations as soft iron/manganese masses with diffuse boundaries. The layer occurs entirely within 30 cm (12 in.) of the soil surface. Iron/manganese masses have value 3 or less and chroma 3 or less; most commonly black. The thickness requirement is waived if the layer is the mineral surface layer. For use in LRRs N, O, P, and T; for testing in LRR M.
F13*	Umbric surface	On concave positions of interstream divides and in depressions, a layer 15 cm (6 in.) or more thick starting within the upper 15 cm (6 in.) of the soil surface with value 3 or less and chroma 1 or less immediately underlain by a layer 10 cm (4 in.) or more thick with chroma 2 or less. For use in LRRs P and T.
F14*	Alaska redox gleyed	A layer that has dominant matrix hue 5Y with chroma 3 or less, or hue N, 10Y, 5GY, 10 GY, 5G, 10G, 5BG, 10 BG, 5B, 10B, or 5PB, with 10% or more redox concentrations as pore linings with value and chroma 4 or more. The layer occurs within 30 cm (12 in.) of the soil surface. For use in LRRs W, X, and Y.
F15*	Alaska gleyed pores	Presence of 10% hue N, 10Y 5GY, 10GY, 5G, 10G, 5BG, 10BG, 5B, 10B, or 5PB with value 4 or more in the matrix or along channels containing dead roots or no roots within 30 cm (12 in.) of the soil surface. The matrix has dominant chroma 2 or less. For use in LRRs W, X, and Y.
F16*	High plains depressions	In closed depressions subject to ponding, the presence of a layer at least 10 cm (4 in.) thick within the upper 35 cm (13.5 in.) of the mineral soil that has chroma 1 or less and: − 1% or more redox concentrations as nodules or concretions − redox concentrations as nodules or concretions with distinct or prominent corona For use in MLRAs 72 and 73 of LRR H; for testing in other MLRAs of LRR H.
TA1	Playa rim stratified layers	Stratified layers starting within the upper 15 cm (6 in.) of the soil surface. At least one layer has value 3 or less and chroma 1 or it has value 2 or more and chroma 2 or less with 2% or more distinct or prominent redox concentrations as soft masses or pore linings. The upper 15 cm (6 in.) has dominant chroma 2 or less. For testing in LRR D.
TA2	Structureless muck	Starting within 15 cm (6 in.) of the soil surface on concave positions or in depressions, a layer of muck 2 cm (0.75 in.) or more thick that has no soil structure. For testing in MLRAs 141, 143, 144b, 145, and 146 of LRR R.
TS1	Iron staining	A continuous zone, 3 cm (1 in.) or more thick, of iron staining with value 4 or more and chroma 6 or more within 15 cm (6 in.) of the soil surface. The zone is immediately below a horizon in which iron/manganese oxides have been removed from the matrix and exposed the primary base color of the silt and sand grains. For testing in LRRs W, X, and Y.
TS2	Thick sandy dark surface	A layer at least 15 cm (6 in.) thick with a depleted matrix that has 60% or more chroma 2 or less or a gleyed matrix starting below 30 cm (12 in.) of the soil surface. The layer(s) above the depleted or gleyed matrix have hue N and value 3 or less; or hue 10YR or yellower with value 2 or less and chroma 1 to a depth of 30 cm (12 in.) and chroma 1 or less in the remainder of the epipedon. For testing in LRR F.

TABLE 5.9 (continued)
Brief Descriptions of Hydric Soil Field Indicators for the United States

Soil Type (Code)	Indicator	Brief Description (including applicable regions)
TS3	Dark surface 2	A layer 10 cm (4 in.) or more thick starting within 15 cm (6 in.) of the soil surface with matrix value 2 or less and chroma 1 or less. At least 70% of the soil materials are covered, coated, or masked with organic material. The matrix color of the layer immediately below the dark surface must have value 4 or more and chroma 2 or less. For testing in LRR G.
TS4	Sandy neutral surface	A layer at least 10 cm (4 in.) thick with a depleted matrix that has 60% or more chroma 2 or less or a gleyed matrix starting within 30 cm (12 in.) of the soil surface. The layer(s) above the depleted or gleyed matrix have hue N and value 3 or less. For testing in LRR M.
TS5	Chroma 3 sandy redox	A layer starting within 15 cm (6 in.) of the soil surface that is at least 10 cm (4 in.) thick and has a matrix chroma 3 or less with 2% or more distinct or prominent redox concentrations as soft masses and/or pore linings. For testing in LRRs F, G, H, K, L, and M.
TF1	? cm mucky peat or peat	A layer of mucky peat or peat ? cm thick with value 4 or less and chroma 3 or less starting within 15 cm (6 in.) of the soil surface. For testing in LRRs F, G, H, and M.
TF2	Red parent material	In parent material with a hue of 7.5YR or redder, a layer at least 10 cm (4 in.) thick with a matrix value 4 or less and chroma 4 or less and 2% or more redox depletions and/or redox concentrations as soft masses and/or pore linings. The layer is entirely within 30 cm (12 in.) of the soil surface. For testing in LRRs with red parent material.
TF3	Alaska concretions	Within 30 cm (12 in.) of the soil surface redox concentrations as nodules or concretions greater than 2 mm in dismeter that occupy more than approximately 2% of the soil volume in a layer 10 cm (4 in.) or more thick with a matrix chroma 2 or less. For testing in W, X, and Y.
TF4	2.5Y/5Y below dark surface	A layer at least 15 cm (6 in.) thick with 60% or more hue 2.5 Y or yellower, value 4 or more, and chroma 1; or hue 5Y or yellower, value 4 or more, and chroma 2 or less starting within 30 cm (1 in.) of the soil surface. The layer(s) above the 2.5Y/5Y layer have value 3 or less and chroma 2 or less. For testing in LRRs F, M, and S.
TF5	2.5Y/5Y below thick dark surface	A layer at least 15 cm (6 in.) thick with 60% or more hue 2.5Y or yellower, value 4 or more, and chroma 1; or hue 5Y or yellower, value 4 or more, and chroma 2 or less starting below 30 cm (12 in.) of the soil surface. The layer(s) above the 2.5Y/5Y layer have hue N and value 3 or less; or have hue 10YR or yellower with value 2 or less and chroma 1 or less to a depth of 30 cm (12 in.) and value 3 or less and chroma 1 or less in the remaining of the epipedon. For testing in LRRs D, F, and M.
TF6	Calcic dark surface	A layer with an accumulation of calcium carbonate or calcium carbonate equivalent, occurs within 40 cm (16 in.) of the soil surface. It is overlain by a layer(s) with value 3 or less and chroma 1 or less. The layer of calcium carbonate accumulation is underlain by a layer within 75 cm (30 in.) of the surface 15 cm (6 in.) or more thick having 60% or more by volume one or more of the following: depleted matrix, gleyed matrix, or hue 2.5Y or yellower value 4 and chroma 1. For testing in LRRs F, G, and M.
TF7	Thick dark surface 2/1	A layer at least 15 cm (6 in.) thick with a depleted matrix that has 60% or more chroma 2 or less (or a gleyed matrix) starting below 30 cm (12 in.) of the soil surface. The layer(s) above the depleted or gleyed matrix have hue 10YR or yellower, value 2.5 or less to a depth of 30 cm (12 in.) and value 3 or less and chroma 1 or less in the remainder of the epipedon. For testing in all LRRs except LRRs O, P, T, U, and Z.
TF8	Redox spring seeps	A layer with value 5 or more and chroma 3 or less with 2% or more distinct or prominent redox concentrations as soft masses or pore linings. The layer is at least 5 cm (2 in.) thick and is within the upper 15 cm (6 in.) of the soils surface. For testing in LRR D.
TF9	Delta ochric	A layer 10 cm (4 in.) or more thick that has 60% or more of the matrix with value 4 or more and chroma 2 or less with no redox concentrations. This layer occurs entirely within the upper 30 cm (12 in.) of the soil surface. For testing in LRR O.

Soil Type (Code)	Indicator	Brief Description (including applicable regions)
TF10	Alluvial depleted matrix	On frequently flooded flood plains, a layer with a matrix that has 60% or more chroma 3 or less with 2% redox concentrations as soft iron masses, starting within 15 cm (6 in.) of the soil surface and extending to a depth of more than 30 cm (12 in.). For testing in LRRs M, N, and S.

Note: Those marked by an asterisk (*) are particularly useful for wetland boundary determinations according to Hurt and Carlisle (1998). A = all soils; S = sandy soils; F = loamy and clayey soils; TA = test all soils; TS = test sandy soils; TF = test loamy clayey soils; LLR = Land Resource Region and upper cases letters following LRR are for specific regions shown on the map in Figure 5.10. Reference to layer(s) means a combination of continuous layers can satisfy the requirement.

Source: Hurt, G.W., et al., 1998. Field Indicators of Hydric Soils in the United States. A Guide for Identifying and Delineating Hydric Soils (ver. 4). USDA Natural Resources Conservation Service, Fort Worth, TX.

Peats and mucks are not the same, although nonscientists commonly use the terms interchangeably (Tiner, 1998). Muck (Saprists) consists of organic matter that breaks down into a greasy mass upon rubbing, less than one sixth of the material can be identified. After rubbing, three fourths or more of the organic material is identifiable (leaves, stems, roots) in peats (Fibrists) (Hurt et al., 1998). There are intergrades between the two, mucky peats and peaty mucks (Hemists), depending on the amount of identifiable material. A fourth group of organic soils (Folists) is nonhydric. They form in high mountains in the tropics and in boreal and arctic regions. In northern New England, Folists are generally thin organic soils on bedrock in landscape positions that are obviously not associated with wetlands.

Organic soils can easily be recognized by their characteristic black muck or black to orange brown peat which is usually thicker than 16 in. (Plates 3, 4, and 9; hydric soil indicator A1). Shallow organic soils of variable thickness exist over bedrock. If the organic layer is less than 16 in. thick and overlies mineral material, the soil is classified as mineral. Perhaps the easiest way to identify the presence of organic soil, besides sinking in the muck, is to take a shovel or auger and try to push it into the soil. If the shovel is easily pushed 16 in. or deeper, the soil is generally an organic soil. If the depth of penetration is less, the soil may still be an organic soil if over bedrock, but it is more likely that the soil is mineral with a shallow organic layer on top.

Peats can be separated from mucks by a simple finger rubbing test. When rubbed between the fingers, mucks have almost all of the plant remains decomposed beyond recognition and feel somewhat greasy. Peats are slightly decomposed and when rubbed between the fingers, most of the plant materials are not destroyed, but can be recognized as parts of grasses, sedges, and mosses, or types of wood. Also, when squeezing a ball of saturated peat, a somewhat clear, brownish-colored liquid results, whereas pressing muck produces a dark-colored solution (Peter Veneman, personal communication, 1998).

An alternative rubbing technique is used to evaluate the proportion of rubbed fibers (Sid Pilgrim, personal communication, 1998). Place a small soil sample in the palm of one hand and rub 10 times with the thumb from the opposite hand. Make a small ball from the rubbed sample, then break it in half. Now estimate the proportion of fibers (diameter > 0.15 mm), excluding live roots, on the exposed face with a 5x or 10x hand lens. If less than one sixth of the volume is composed of fibers, the material is sapric (muck). If fibers represent three quarters or more of the volume, it is fibric (peat). If not, the material is hemic (mucky peat or peaty muck).

Histic Materials in Mineral Soils

The presence of significant amounts of organic matter in mineral soils is often diagnostic of hydric soils. For example, some mineral soils have thick organic surface layers of muck or peat on their

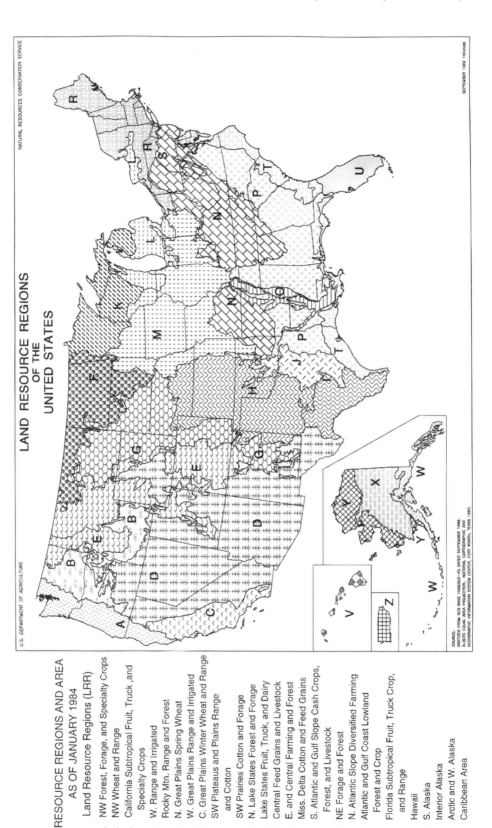

FIGURE 5.7 Map showing major land resource regions used to determine applicability of certain hydric soil indicators. (Courtesy of USDA Natural Resources Conservation Service.)

FIGURE 5.8   Organic bodies comprised of muck or mucky modified mineral texture is a hydric soil indicator in the southern U.S.

surfaces. When these layers are 8 to 16 in. thick, they are called *histic epipedons* (Plate 10; indicators A2 and A3). Such soils form in areas with an abundance of soil moisture due to heavy seasonal rainfall and/or high water tables (Ponnamperuma, 1972).

Histic materials also may accumulate on the surface of other wet soils. A 0.75-in. (2 cm) layer of muck should be a reliable indicator of hydric soils in many areas, especially when overlying a gleyed subsoil. An exception might be in cold, humid regions where organic matter may build up due to high rainfall and low evapotranspiration (e.g., northern New England). Yet for temperate forest regions and arid and semiarid grasslands, a thin layer of muck should identify a hydric soil, even on brighter subsoils, especially chroma 3 sandy loams (Plate 11; indicators A8, A9, A10, S2, S3, and TF1). Obligate wetland plants may be present on such soils. Even a thinner layer of muck (0.5 in. or 1 cm) on top of a sandy soil should indicate that the site is very wet, at least seasonally. Wet, sandy soils also may have streaks of organic matter in the subsoil; also indicative of hydric soils if within 6 in. of the soil surface (Plate 12; indicator S6). Other wet sandy soils may have organic bodies forming along the roots (Figure 5.8; indicator A6). These mucky or mucky-modified mineral bodies are the product of frequent (e.g., annual) die-back of some roots when exposed to prolonged saturation. They are indicators of hydric soils when they are at least 0.5 in. wide and account for 2% of more of the soil within the upper 6 in. (15 cm) of soil (Hurt et al., 1996).

Mucky mineral surface layers are also indicative of hydric soils (indicators A7, S1, and F1). These soils feel somewhat greasy and much of the organic material can be broken down by gentle rubbing. The layer should be a surface layer at least 2 in. (5 cm) or more in thickness (Table 5.9). A mucky mineral layer of this thickness should be a reliable indicator of hydric soils. A mucky sand may be detected by putting a piece of soil between one's thumb and forefinger and gently rubbing it twice; if light-colored sand grains are not exposed, there are sufficient organic coatings to consider, the soil to be a mucky sand (Peter Veneman, personal communication, 1998). While the current national list of indicators generally limits the use of A7 (generally to the Southeast coastal plain), it also may be useful farther north into New England, perhaps excluding northern New England and higher elevations in the Northeast where colder temperatures may facilitate buildup of organic matter under nonhydric conditions (e.g., Borofolists).

Thin layers of organic material or predominantly organic-coated soil particles also may be indicators of hydric soils. Many alluvial or floodplain soils have stratified layers due to frequent depositions of water-borne materials. In order to be considered hydric, however, the stratified layers must occur within 6 in. of the soil surface and at least one layer must be organic, mucky modified mineral texture, or mineral of chroma 1 or less and a value of 3 or less (indicator A5). The emphasis is on finding at least one layer of any thickness (even less than 1 in. or 2.5 cm) with 70% or more soil grains having organic coatings. One must exercise caution as not to include alluvial soils with stratified layers that are not anaerobic during the growing season. The dominance of histic materials in at least one layer provides ample evidence of anaerobiosis. (See Chapter 7 for additional considerations regarding the use of histic materials with high chroma soils.)

THICK, DARK SURFACE LAYERS

When organic matter content is not high enough to form organic soils or mucky mineral soils, wet mineral soils may develop thick, dark mineral surface layers. Organic matter makes these layers blackish to dark brown-colored (Plates 13-14). These layers may be classified as *umbric* or *mollic* epipedons. Umbric epipedons are thick, dark surface horizons with base saturation of less than 50% (Soil Survey Staff, 1975) or a pH ≤ 5.5. In the Northeast, the presence of an umbric epipedon is typically interpreted as a hydric soil indicator, whereas in the Southeast, it may occur in some uplands so it is not as useful (Steven Sprecher, personal communication, 1998). The "thick, dark surface layer indicator" can be found in sandy and nonsandy soils (see descriptions in Table 5.9 for indicators S7, F5, F13, TS2, TS3, TF6, and TF7).

Mollic epipedons are thick, dark surface layers, rich in humus, and with high base saturation (Soil Survey Staff, 1996). They are characteristic horizons for Mollisols, soils of subhumid grasslands. In these regions, even nonwetlands have thick, relatively dark topsoils because the dominant grasses developed extensive deep root systems to reach the underlying water table. The decomposition of these roots darkens the soil profile for a considerable depth. In these grassland regions, thick, dark surface layers alone are often not sufficient to identify hydric soils. Typically one needs to look for features within or below the A-horizon that might indicate prolonged wetness, e.g., oxidized rhizospheres or redox depletions (see Chapter 7).

In other dry regions, such as in Texas, soils with high clay content called Vertisols have formed. These soils frequently shrink and swell, creating cracks in the soil. These cracks permit organic matter and other soil to fall into the cracks. When wetted, the soils swell. The churning of these soils and mixing of organic materials also cause these soils to be dark for some depth. Again, the thick, dark surface horizon is not particularly useful in these regions (see Chapter 7). The thick, dark surface indicator, while useful in many regions, also does not automatically indicate the presence of hydric soil in Alaska, although it is being tested there.

REDOX DARK SURFACE LAYERS

Where the dark surface is not quite thick enough (i.e., 4 to 6 in.) to qualify as a thick, dark surface, the presence of redox concentrations is needed to verify hydric soils (Plate 15; indicator F6). When the dark color is chroma 1, a minimum of 2% redox concentrations are needed to be a hydric soil indicator. If the dark color is chroma 2, then 5% redox concentrations are required. In both cases, the features must be distinct or prominent.* This indicator is used throughout the conterminous U.S., but is being used on a test basis in Alaska.

GLEYED MATRIX

Some of the wettest hydric mineral soils are typically gleyed in the subsoil. Gleyed soils are predominantly neutral gray in color and occasionally greenish or bluish gray (Plates 16–18). These colors are shown on two gley charts in the Munsell soil color book. They are universally associated with many mineral soil wetlands. In Europe and Russia, gley (groundwater-driven) and pseudogley

---

* Distinct means the feature has moderate contrast from the matrix color, while prominent represents the most obvious difference between a mottle and the matrix (see Hurt et al., 1996 for technical criteria; they differ slightly from those in *Soil Survey Manual*, Soil Survey Division Staff, 1993).

(surfacewater-driven) soils are classic wetland soils (Moormann and van de Wetering, 1985). The former have dominant low chroma colors throughout, with rust-colored mottles along ped surfaces and root channels. The pseudogleys develop mainly in areas with perched water tables and typically have browner interiors due to wetting from the surface downward, with reduction occurring in cracks and wide pores and iron diffusing to the ped interior. They are gleyed in the upper part, but their underlying horizons lack gley colors. In Maryland, Fanning and Reybold (1968) noted that poorly drained Coastal Plain soils could be recognized in summer by their gray-colored subsoils. Gleyed soils in Maryland have soil profiles or horizons that are predominantly chroma 2 or less and may contain up to 40% high chroma colors. The early concept of gleyed soils included both the depleted matrix and gleyed matrix of today (Sid Pilgrim, personal communication, 1998).

Mineral soils that are always saturated are generally uniformly gleyed throughout the saturated area, with sandy stratified soils as possible exceptions (Peter Veneman, personal communication, 1998). Soils gleyed to the surface layer (A-horizon) are hydric soils (indicators S4, S10, F2, and F14). Soils with a gleyed matrix have one of the following characteristics and are not glauconitic (greensand — hydrous silicate of potassium, iron, aluminum, or magnesium) (Hurt et al., 1996).

- Neutral (N) hue and value of 4 or more
- 10Y, 5GY, 10GY, 10G, 5BG, 10BG, 5B, 10B, or 5PB hue with value 4 or more and chroma of 1
- 5G with value of 4 or more and chroma of 2 or less
- Possibly, 5Y with value of 4 and chroma of 1 (for testing)

These soils often show evidence of oxidation only along root channels (i.e., oxidized rhizospheres; Plate 2). The color of some gleyed soils may fade when exposed to air (e.g., going from a bluish color to a more grayish color). This usually means that reduced (ferrous) iron is present. The minimum hydric soil requirements for a gleyed matrix are that the matrix make up at least 60% of the layer within 12 in. (30 cm) of the surface.

DEPLETED MATRIX

Most hydric mineral soils are characterized by low chroma subsoils — a depleted matrix (Plates 19-21; indicators F3, F4, F7, F11, and TF 10). This matrix usually occurs immediately below and within 12 in. (30 cm) of the surface and must be at least 6 in. (15 cm) thick. These matrices have chromas of 2 or less and values of 4 or more and are often variously mottled with brighter, higher chroma colors (>2 chroma, redox concentrations). A soil with a depleted matrix has one of the following characteristics (Hurt et al., 1998).

- Matrix value 5 or more and chroma 1 or less
- Matrix value 6 or more and chroma 2 or less
- Matrix value 4 or 5 and chroma 2 with ≥2% distinct or prominent redox concentrations as soft masses and/or pore linings
- Matrix value 4 and chroma 1 with ≥2% distinct or prominent redox concentrations as soft masses and/or pore linings

In New England, a soil with a matrix value of 4 or more and chroma 1 or less also is considered to have a depleted matrix (NEIWPCC Wetlands Working Group, 1998). The abundance, size, and color of the mottles usually reflect the duration of the saturation period and indicate whether or not the soil is hydric. In general, the more gray present and the closer the depleted matrix is to the surface, the wetter the soil and the more likely the soil is hydric (with Spodosols being a major exception, see Chapter 7). Soils with only a few gray mottles near the surface are usually not wet enough to be considered hydric.

Mottled Depleted Matrix

Many hydric soils are characterized by a thick, dark surface layer (black or dark brown), a depleted matrix (predominantly grayish, low chroma subsoil) marked with yellow, orange, brown, or reddish mottles and sometimes, iron oxide concretions near the surface (Plate 21). Some hydric mineral soils have reddish brown to orange mottles lining the root channels (oxidized rhizospheres or pore linings). Black concretions of manganese oxide also may be present near the surface. The abundance of the brighter mottles reflects increasing oxidizing conditions, so the higher percentage of high chroma mottles, the drier the soil. The presence of iron and manganese masses coupled with at least 40% low chroma colors indicates the presence of hydric soil on some floodplains (Plate 22, indicator F12).

The pattern of mottling may reflect differences in water table fluctuations as observed by Crown and Hoffman (1970) in Ontario gleysols. For example, horizontal banding of mottles was associated with soils with the highest frequency of water table fluctuation through a relatively narrow range in depth, whereas vertical streaking of mottles were found where the water table fluctuated only once during 4 months of observation. They also noted that almost permanently saturated horizons seemed to have large, somewhat indistinct mottles, while rarely saturated horizons had smaller mottles with distinct boundaries.

Sulfidic Materials

Some hydric soils have high sulfur content. This is especially true of salt marshes and some marshes and swamps on organic soils. Salt and brackish waters contain sulfates which are reduced by sulfate-reducing bacteria (e.g., *Desulfovibrio desulfuricans*) in these soils to produce hydrogen sulfide. Consequently, these hydric soils exude a noticeable hydrogen sulfide odor (the smell of rotten eggs) immediately upon excavating the soil. When this odor is detected within 12 in. of the soil surface, it represents a reliable hydric soil indicator. Richardson and Daniels (1993) noted that hydrogen sulfide is produced in open water in spring, but is not detected in deep marshes until late summer. Drainage of high sulfur soils such as tidal marshes creates acid-sulfate conditions which are extremely toxic to plantlife. Such conditions cause jarosite (iron sulfate) to form from the oxidation of sulfide-rich mineral or organic soils, thereby creating a sulfuric horizon (Soil Survey Staff, 1996). Jarosite is represented by yellowish to yellow-orange mottles in these soils. The formation of a sulfuric horizon usually takes a few years following drainage but may occur as quickly as within a few weeks.

Depressional Indicators

There are three types of depressional situations recognized by Hurt et al. (1998) that produce hydric soil properties other than those cited above — frequently ponded closed depressions, vernal pools, and high plains depressions. The first type occurs throughout the U.S. The second type is found in California and most of the arid Southwest, whereas the third type is associated with playas in western Kansas, southwestern Nebraska, eastern Colorado, and southeastern Wyoming. Frequently ponded closed depressions with loamy very fine sand or finer textures having 5% or more redox concentrations in a layer at least 2 in. (5 cm) thick within the upper 6 in. (15 cm) of the surface are hydric soil indicators (indicator F8). The concentrations must be distinct or prominent. Examples of wetlands having these indicators are vernal pools, playas, rainwater basins, "Grady" ponds, and potholes. The vernal pool indicator (F9) is a ponded closed depression that has a depleted matrix at least 2 in. (5 cm) thick within the upper 6 in. (15 cm) of the soil surface. The high plains depression indicator (F16) requires the presence of a thick chroma 1 layer (4 in. or greater) within the upper 13.5 in. (35 cm) of the mineral soil and either at least 1% redox concentrations (i.e., nodules or concretions), or nodules or concretions with distinct or prominent corona (usually reddish brown, brown, or yellowish brown halos). The nodules or concretions are usually black or reddish black. A 10x lens will be helpful to identify the halos. These features reportedly occur on the Ness and Pleasant series. See discussion below for hydrology indicators of ponded conditions.

OTHER HYDRIC SOIL INDICATORS

Iron and manganese masses may be hydric soil indicators on floodplains in the Southeast (except Florida) and are being tested in the lower Midwest (Plate 22; indicator F12). To be a hydric indicator, a layer must be 4 in. (10 cm) or thicker having a depleted matrix (at least 40% chroma 2 or less) and 2% distinct or prominent soft iron and manganese masses (redox concentrations) with diffuse boundaries. The soft masses are typically black, but must have values and chromas of 3 or less. If the surface layer is mineral, there is no thickness requirement. This indicator has been reported for southern river floodplain wetlands along the Apalachicola (Florida), Congaree (South Carolina), Mobile (Alabama), Savannah (Georgia/South Carolina), and Tennessee rivers.

Marl forms in water as calcium carbonate precipitates from algae. It has a moist color value of 5 or more and reacts with hydrogen chloride to produce carbon dioxide. If marl is present within 4 in. (10 cm), it is a hydric soil indicator (Plate 23; indicator F10). This indicator is used in most of Florida. Marl also may form in other soils, but is usually deeper in the profile. The author has observed it below organic soil (Saprists) in New York state.

Table 5.9 lists and briefly describes some other indicators including ones that are being tested in various regions (refer to Hurt et al., 1998 for more information). Plates 24 through 27 show some examples of hydric soils in the U.S. Chapter 7 contains other hydric soil indicators for more problematic soils including sandy soils, red parent material soils, dark-colored parent materials, and others. It is important to recognize that some hydric soils do not have any of the morphologies described above and in Table 5.9. The technical criteria for hydric soils also identifies soils that are frequently ponded or flooded soils for long or very long duration during the growing season as hydric soils. These soils can be recognized as hydric by known hydrology or by assessing their landscape position, vegetation, and other signs of wetland hydrology (see following subsection).

INDICATORS OF FLOODED OR PONDED SOILS

The hydric soil criteria include two criteria that establish hydric soils by hydrology alone — frequent flooding or ponding for long or very long duration (criteria 3 and 4) (Figure 5.9). While these criteria may be best evaluated with long-term hydrologic data, it is unlikely that NRCS has gathered such data to support designations of frequently flooded or ponded conditions. Short-term studies were likely conducted in many instances, while in other cases, interpretations may have been based on more casual observations. There are several indicators that may be used to identify flooded or ponded soils. Besides observed surface water, the federal wetland delineation manuals recognized the following indicators (Environmental Laboratory, 1987; Sipple, 1988; Federal Interagency Committee for Wetland Delineation, 1989; Plates 1 and 28-33; Figure 5.9):

- Water-carried debris (drift lines)
- Sediment deposits (including silt lines)
- Scoured or bare areas
- Wetland drainage patterns (reflecting movement of surface water)
- Moss lines on tree trunks (upland mosses plus lichens occur above flood level or aquatic mosses and liverworts below this level)
- Plant morphological adaptations (see Chapter 3)
- Water marks
- Water-stained leaves

Segal et al. (1987) used lichen lines and buttressing as indicators of inundation. While the above indicators provide evidence of flooding or ponding, most simply signify that an event has occurred

FIGURE 5.9 Hydrologic indicators of flooding and ponding: (a) water-carried debris, (b) scoured floodplain with sediment deposits and bare areas, (c) silt lines, (d) wetland drainage patterns on floodplain with evidence of recent deposition, (e) crayfish chimney, and (f) water marks on rocks along a tidal river.

FIGURE 5.9 (continued)

with little indication of the frequency and duration of the event. The latter four indicators require some time to form (e.g., at least 1 week), so significant duration may be interpreted. Frequency cannot be established with any certainty, although the presence of certain morphological adaptations (e.g., buttressing, fluting, and abundant hypertrophied lenticel scars) may suggest sufficient recurrent prolonged wetness associated with wetlands. Other possible indicators of inundation (ponding or flooding) of some duration include algal encrustations or mats (see Figure 7.2), surface films of iron precipitates on the ground, crayfish chimneys (Figure 5.9), aquatic invertebrates and their remains (on wetland plants, e.g., casts of dragonflies, on the ground surface or in the upper part of the soil; Plates 15 and 29), aquatic and other wetland plants indicative of prolonged inundation (Plate 29; see Table 3.7 for examples), and peat moss in depressions in forests (in temperate regions from southern New England south, especially on the Atlantic–Gulf coastal plain; Plate 33).

There is apparently no published literature on the time required to create water marks (blackened tree trunks or dark lines on bridge abutments and rocks) or water-stained leaves (Godsalk, 1993) as they have not been topics for scientific research. Despite this lack of research, water marks and water-stained leaves are common features in most, if not all, wetlands subject to prolonged inundation. These features represent one of the more obvious indirect signs of inundation across the country — in the early spring, water is observed and when the water disappears, water marks and/or water-stained leaves are present, depending on the duration of inundation. The presence of water-stained leaves in depressional wetlands is evidence that water was on the surface for some time earlier in the year. As far as the validity of the indicator as a sign of wetland hydrology, it is no less an indicator than the presence of water itself. In fact, it is a better sign than water itself as it reflects some duration of surface water, while the presence of water on the ground yields little information on how long it has been there or will persist. Since uplands may be flooded or ponded on occasion, it is typically the plants and soils that provide information on the approximate duration and frequency of inundation.

Preliminary results from studies of water-staining in submerged leaves conducted by Peter Veneman and Ralph Tiner at the University of Massachusetts showed that the inundated leaves must be in contact with soil to develop dark colors. At 38°F, blackening of the leaf surface was detected in red maple leaves submerged and in contact with the soil. Considerable discoloration was noted after 4 weeks. More work on this indicator would help validate its use as a wetland indicator of ponded conditions, although this indicator has been seen by the author in countless wetlands across the country in the past 25 years.

INTERNATIONALLY RECOGNIZED HYDRIC SOIL PROPERTIES

In Europe, soil scientists ascribe four categories of features to wetland (hydromorphic) soils — halomorphic, gypsimorphic, calcimorphic, and redoximorphic, depending on the nature and size of the watershed (Blume and Schlichting, 1985). The first three features form in arid and semiarid regions where salts form, while the latter is found universally in soils with a reducing environment. Blume and Schlichting comment that hydromorphic features "often coincide with but are not identical to redoximorphic features," e.g., when soil water is oxygen-rich or organic matter-poor. They also indicated that some upland soils can have redoximorphic features when parent material is rich in reductates (e.g., shales and mudstones) or the soil air contains much carbon dioxide or methane that excludes or consumes oxygen (e.g., landfills). Blume and Schlichting identified four types of hydromorphic soils — submerged soils (permanent and total saturation), floodwater soils (temporary and total saturation), ground-water soils (permanent saturation in the subsoil), and surface-water soils (permanent to temporary saturation in the topsoil). Some soils are combinations of gleys and pseudogleys. They described typical morphology of wetland soils in central Europe.

Submerged soils are called *subhydric* soils. These "soils" are actually substrates (nonsoils) by the American definition of soil. Subhydric soils are characterized by organic deposits, black-brown muds, liver-colored mud (gyttja), and black hydrogen sulfide-rich mud (sapropel; e.g., tidal flats).

Alluvial or floodwater soils may lack redoximorphic features except for some rusty mottles. They do, however, possess stratified layers that help separate them from adjacent upland soils.

Groundwater (gley) soils may include blue-green loams and clays, blackish sulfur-rich soils, whitish carbonate-rich soils (due to calcite or siderite), and white to light gray sands. Rusty colors may represent 5% of the upper soil and are especially common in animal and root channels. Some soils may be organic (e.g., in fens or Niedermoor). The wetter gley soils have thick organic layers (histic epipedons). Fluvioglacial sands may exhibit cementation of iron oxides (bog iron gley).

Surfacewater (pseudogley) soils tend to have bleached ped surfaces of clayey subsoil and ped interiors with brownish iron oxides and black manganese oxides or orange lepidocrocite in very clayey lime-free slightly acid soils. The longer the duration of wetness, the more bleaching (gleying) and more rusty mottling. Iron concretions may be present.

MEASURING REDUCTION OR SOIL REDOX POTENTIAL

Two methods have been frequently employed to assess the reduced nature of the soil — a colormetric test using α, α'-dipyridyl solution for detecting ferrous iron, and measuring redox potential with platinum probes. Both tests yield a one-time measure of soil reduction and require repeated measurements to determine the frequency and duration of reducing conditions. Both tests also have significant problems that may affect the readings.

The colormetric test is administered when the soil is moist since this is the time that reduced iron would be present. The presence of ferrous iron can be observed by noticing a pink color shortly after wetting the freshly broken soil with 0.2% α, α'-dipyridyl in 10% acetic acid or with a 0.2% α, α'-dipyridyl solution buffered with 1 N ammonium acetate (Childs, 1981; Hudnall and Szögi, 1996). A number of situations may give false positive or false negative readings. In order for the test to be valid, the soil must contain iron and a sufficient amount of organic matter must be present for microbial reduction. Without organic carbon, there will be no iron reduction, hence no positive test, even though the soil may be anaerobic and at low redox potential. Sandy soils have low iron content and do not usually give positive results even when anaerobic (Segal et al., 1995). Organic soils low in iron also are not recommended soils for use of this test (Federal Interagency Committee on Wetland Delineation, 1989). The species of iron can make a difference as some iron minerals have stronger bonds than others. For example, the iron mineral hematite requires a great deal of energy to break the Fe-O bonds. The presence of hematite in a soil with neutral pH and free calcium carbonate and gypsum led Hudnall and Szögi to conclude that the anaerobic soil was reduced but could not be validated by the colormetric test. Vepraskas (1996) states that a false positive reading also may occur if the soil has been in contact with steel augers, probes, or knives, or if the soil had been previously treated with 10% hydrochloric acid to test for carbonates.

The test itself is prone to some problems in the reaction with acetic acid and in the timing of the observed color. Griffin et al. (1996) in applying this test in Texas Coastal Plain soils found ferric iron reduction occurred in the acetic acid (at pH of 4) and not necessarily as a result of reduction in the moistened soil. They adjusted the pH to 7 to eliminate this problem (Note: Hudnall and Szögi, 1996 also had to do this). They also looked for the pink color within 30 to 60 seconds to avoid photochemical reduction of ferric-organic complexes resulting from longer exposures. Since the dye is sensitive to light, special care should be taken to keep it out of sunlight. Vepraskas (1996) gives an example of a Spodosol with ferric iron and oxidizable organic compounds giving a false positive test due to iron reduction by sunlight after application of the α, α'-dipyridyl solution. The chemicals have a limited shelf life (1 to 2 years) and probably less under field conditions, thereby needing replacement annually or semiannually (Mike Whited, personal communication, 1998). Given the above concerns, the observed reduction test is not highly recommended.

Rather than employing the colormetric test, the intensity of soil reduction may be measured by redox potentials which reflect electron availability. Seasonally flooded soils have redox potentials that vary during the year from +700mV to +400mV when oxided to −250mV to −300mV when saturated and reduced (Patrick et al., 1996). Redox potential is measured by installing platinum electrodes (probes) at the depth of interest (e.g., within 12 in. in nonsandy soils and within 6 in. in sandy soils). Guidance in construction, installation, and use of these probes is given in Patrick et al. and Faulkner et al. (1989). Certain soil conditions pose particular problems for accurate measurements using these platinum microelectrodes (probes). Spatial variability within soils may affect readings as instruments yield a point sample of soil conditions. The presence of animal burrows (krotovina) could affect readings as well as actual water levels. Natural cracks in the soil creates similar consequences, especially with the platinum probes. Caution must be exercised when using these instruments and making interpretations of their results (see Patrick et al., 1996).

INTERPRETING SOIL TAXONOMY TO IDENTIFY POTENTIAL HYDRIC SOILS

Soil classification like plant taxonomy involves examining numerous properties, many readily observable and others requiring laboratory analysis, and giving a name to the soil. The U.S. system for soil classification is a dynamic system that continues to be modified. The current system is based on *Soil Taxonomy* (Soil Survey Staff, 1975; as amended). The system uses six main levels of classification — order, suborder, great group, subgroup, family, and series. Each of these levels are divided into separate categories. Soil names are typically composed of one "adjective" and one "noun", as in the example Typic Epiaqualf. In interpreting this name, one reads from right to left: "alf" is the order (Alfisols), "aqu" indicates the suborder "aqualfs," "epi" is the great group, and "Typic" is the subgroup. A Typic Epiaqualf is a typical wet Alfisol, whereas an "Aeric Epiaqualf" is a somewhat drier one.

For wetland studies it is not crucial, although helpful, to be fluent in soil taxonomy, but familiarity with key elements in the codes is vital to identifying potentially hydric soils when reviewing a list of series with their taxonomic designation (Table 5.10). A close look at the hydric soil criteria will reveal such terms as Aquic suborders and Aquic subgroups. This designation relates to soils with the aquic moisture regime, an important distinguishing feature of soils. At the suborder level, the profile must be saturated to within 50 cm to be designated as "aqu", whereas at the subgroup level only the lower horizons need to be saturated. In soil taxonomy, "aqu" appears as a suborder as in *aqu*ents (wet *Ent*isols), *aqu*alfs (wet *Alf*isols), *aqu*olls (wet *Moll*isols), *aqu*ods (wet Sp*od*osols), *aqu*ults (wet *Ult*isols), and *aqu*epts (wet Inc*ept*isols). These soils are at least seasonally wet for some time. When these soils have a Typic, Humic, or Histic subgroup designation, they are definitely hydric soils (they are reduced due to wetness for prolonged periods) in their undrained condition. However, when an aquic suborder has an "Aeric" subgroup (e.g., Aeric Haplaquept), it may or may not be hydric because it is drier in the upper horizons and has a reducing environment in only the lower part of the soil. Series designated as Aeric soils usually are not hydric, but the range in properties ascribed to these series may include some soils exhibiting hydric soil properties. These Aeric hydric soils are typically found in lower landscape positions (e.g., toe of slopes), while the nonhydric ones occur upslope. Sulfaquents are wet Entisols associated with salt marshes, definitely hydric. When "aquic" appears at the subgroup level, such as in Aquic Dystrochrept, it means that the soil has some wetness typically between 50 and 75 cm deep, but is usually not hydric. Soils with a "Fluv" at the great group or "Fluvaquentic" at the subgroup are floodplain soils that may be hydric; Typic Fluvaquents are usually hydric. Some soils lacking an "aqu" may be hydric. The exceptions are soils that are frequently flooded or ponded for long or very long duration, and organic soils. Histosols are organic soils and all Histosols except Folists are hydric; they do not have "aqu" in their descriptor. Table 5.11 shows the number and percent of hydric soils within each soil order in the U.S.

TABLE 5.10
Examples of Interpreting Potential Hydric Soils from their Taxonomic Designation

Taxonomic Classification	Aquic Suborder?	Aeric Subgroup?	Aquic Subgroup?	Probability of Being Hydric
Mollic Albaqualfs	yes	no	no	very high
Typic Hapludalfs	no	no	no	essentially zero
Typic Fragiaqualfs	yes	no	no	very high
Aeric Epiaqualfs	yes	yes	no	possibly
Humic Haplustands	no	no	no	essentially zero
Calcic Aquisalids	yes	no	no	possibly
Typic Sulfaquents	yes	no	no	very high
Aquic Quartzipsamments	no	no	yes	very low
Typic Udifluvents	no	no	no	very low*
Typic Hydraquents	yes	no	no	very high
Aquic Udifluvents	no	no	yes	possibly*
Typic Borofolists	no	no	no	essentially zero
Terric Medisaprists	no	no	no	very high**
Sapric Borohemists	no	no	no	very high**
Typic Borofibrists	no	no	no	very high**
Aeric Fragiaquepts	yes	yes	no	possibly
Typic Eutrochrepts	no	no	no	essentially zero
Mollic Epiaquepts	yes	no	no	very high
Typic Argialbolls	no	no	no	essentially zero
Vertic Natraquolls	yes	no	no	very high
Aeric Calciaquolls	yes	yes	no	possibly
Histic Epiaquods	yes	no	no	very high
Aquic Haplorthods	no	no	yes	very low
Typic Haplocryods	no	no	no	essentially zero
Typic Hapludults	no	no	no	essentially zero
Aeric Fragiaquults	yes	yes	no	possibly
Typic Umbraquults	yes	no	no	very high

* Occurs on floodplain, so some may be hydric if meeting hydric soil criteria #4.

** Organic soils don't have aquic suborder, but can be recognized by "ists" at end of classification; remember that Folists are not hydric soils.

TABLE 5.11
Number and Percent of Hydric Soils within Each Soil Order in the U.S.

Order	Total Series	Hydric Series	% Hydric
Alfisols	3527	342	9.7
Andisols	627	22	3.5
Aridisols	2407	31	1.3
Entisols	2596	290	11.2
Gelisols	5	0	0
Histosols	261	215	82.4
Inceptisols	2541	397	15.6
Mollisols	6443	691	10.7
Oxisols	40	1	2.5
Spodosols	680	63	9.3
Ultisols	1114	101	9.1
Vertisols	388	81	20.9
All Soils	20,629	2234	10.8

Source: Mike Whited, Natural Resources Conservation Service, 1998.

REFERENCES

Aandahl, A.R., S.W. Buol, D.E. Hill, and H.H. Bailey (Eds.). 1974. *Histosols: Their Characteristics, Classification, and Use.* Special Publication No. 6. Soil Science Society of America, Madison, WI.

Al-Janabi, K.Z. and D.T. Lewis. 1982. Salt-affected soils in the Platte River Valley of central Nebraska: properties and classification. *Soil Sci. Soc. Am. J.,* 46: 1037-1042.

Anderson, J.L. 1984. Part II. Soil mottling, an indication of saturation. *Soil Surv. Hor.,* Winter 1984: 13-15.

Arndt, J.L. and J.L. Richardson. 1986. The effects of groundwater hydrology on salinity in a recharge-flowthrough-discharge wetland system in North Dakota. In Proceedings 3rd Canadian Hydrogeologic Conference (April 20-23, 1986), G. van der Kamp and M. Madunicky (Eds.). Saskatchewan Res. Council, Saskatoon, Saskatchewan, Canada.

Baldwin, M., C.E. Kellogg, and J. Thorp. 1938. Soil classification. In *Soils and Men.* Yearbook of Agriculture 1938. U.S. Department of Agriculture. U.S. Government Printing Office, Washington, D.C. 979-1001.

Bell, J.C., C.A. Butler, and J.A. Thompson. 1996. Soil hydrology and morphology of three Mollisol hydrosequences in Minnesota. In Preliminary Investigations of Hydric Soil Hydrology and Morphology in the United States. J.S. Wakely, S.W. Sprecher, and W.C. Lynn (Eds.). Tech. Rep. WRP-DE-13. U.S. Army Engineer Waterways Experiment Station, Vicksburg, MS. 69-93.

Bigler, R.J. and J.L. Richardson. 1984. Classification of soils in prairie pothole wetlands with deep marsh plant species in North Dakota. *Soil Surv. Horiz.,* 25: 16-24.

Blume, H.P. and M. Helsper. 1987. Schatzung des Humusgehaltes nach der Munsell-Farbhellighkeit. *Z. Pflanzenernaehr. Bodenkd.,* 150: 354-356.

Blume, H.P. and E. Schlichting. 1985. Morphology of wetland soils. In *Wetland Soils: Characterization, Classification, and Utilization.* International Rice Research Institute, Los Baños, Laguna, Philippines. 161-176.

Boersma, L., G.H. Simonson, and D.G. Watts. 1972. Soil morphology and water table relations. I. Annual water table fluctuations. *Soil Sci. Am. Proc.,* 36: 644-649.

Botch, M.S. and V.V. Masing. 1983. Mire ecosystems in the U.S.S.R. In *Mires: Swamp, Bog, Fen and Moor.* Regional Studies. Ecosystems of the World 4B. A.J.P. Gore (Ed.). Elsevier Scientific Publishing Co., Amsterdam, The Netherlands. 95-152.

Bouma, J. 1983. Hydrology and soil genesis of soils with aquic moisture regimes. In *Pedogenesis and Soil Taxonomy. I. Concepts and Interactions*. L.P. Wilding, N.E. Smeck, and G.F. Halls (Eds.). Elsevier Science Publishers, B.V. Amsterdam, The Netherlands. 253-281.

Buurman, P. (Ed.). 1984. *Podzols*. Van Nostrand Reinhold, New York.

Buurman, P. and L.P. Van Reeuwijk. 1984. Protoimogolite and the process of podzol formation: a critical note. *J. Soil Sci.*, 35: 447-452.

Childs, C.W. 1981. Field test for ferrous iron and ferric-organic complexes (on exchange sites or in water-soluble forms) in soils. *Aust. J. Soil Res.*, 19: 175-180.

Childs, C.W. and B. Clayden. 1986. On the definition and identification of aquic soil moisture regimes. *Aust. J. Soil Res.*, 24: 311-316.

Cogger, C.G. and P.E. Kennedy. 1992. Seasonally saturated soils in the Puget Lowland. I. Saturation, reduction, and color patterns. *Soil Sci.*, 153: 421-433.

Color Communications, Inc. 1997. *EarthColors: Soil Color Book. A Guide for Soil & Earthtone Colors*. Poughkeepsie, NY.

Coombs, G.W., J.P. Gove, and R.J. Kelsea. 1985. Soil-water behavior in a drainage sequence in southeastern New Hampshire. *Soil Surv. Hor.*, 26(3): 15-20.

Conrcy, G.W. 1924. The importance of mottled colors in soils. *Am. Soc. Surv. Workers Rep.*, 5: 61-62.

Cooper, T.H. 1990. Development of students' abilities to match soil color to Munsell color chips. *J. Agron. Educ.*, 19: 141-144.

Couto, W., C. Sanzonowicz, and A.O. Barcellos. 1985. Factors affecting oxidation-reduction processes in an Oxisol with a seasonal water table. *Soil Sci. Soc. Am. J.*, 49: 1245-1248.

Coventry, R.J. and J. Williams. 1984. Quantitative relationships between morphology and current soil hydrology in some Alfisols in semiarid tropical Australia. *Geoderma*, 33: 191-218.

Cowardin, L.M., V. Carter, F.C. Golet, and E.T. LaRoe. 1979. Classification of Wetlands and Deepwater Habitats of the United States. FWS/OBS-79/31. U.S. Fish and Wildlife Service, Washington, D.C.

Craft, C.B. and C.J. Richardson. 1993. Peat accretion and N, P, and organic C accumulation in nutrient-enriched and unenriched Everglades peatlands. *Ecol. Appl.*, 3: 446-458.

Crown, P.H. and D.W. Hoffman. 1970. Relationships between water table levels and type of mottles in four Ontario gleysols. *Can. J. Soil Sci.*, 50: 453-455.

Daniels, R.B. and S.W. Buol. 1992. Water table dynamics and significance to soil genesis. In Characterization, Classification, and Utilitization of Wet Soils. J.M. Kimble (Ed.). Proc. 8th Internat. Soil Correlation Meeting (VIII ISCOM), USDA Soil Conservation Service, National Soil Survey Center, Lincoln, NE. 66-74.

Daniels, R.B. and E.E. Gamble. 1967. The edge effect in some Ultisols in the North Carolina Coastal Plain. *Geoderma*, 1: 117-124.

Daniels, R.B, E.E. Gamble, and S.W. Buol. 1973. Oxygen content in the ground water of some North Carolina aquults and udults. In Field Soil Water Regime. R.R. Bruce (Ed.). SSSA Special Publ. No. 5. Soil Sci. Soc. Am. Madison, WI. 153-166.

Daniels, R.B., E.E. Gamble, and J.G. Cady. 1971a. The relation between geomorphology and soil morphology and genesis. *Adv. Agron.*, 23. 51-88.

Daniels, R.B., E.E. Gamble, and L.A. Nelson. 1971b. Relations between soil morphology and water table levels on a dissected North Carolina Coastal Plain surface. *Soil Sci. Soc. Am. Proc.*, 35: 781-784.

Daniels, R.B., E.E. Gamble, and C.S. Holzhey. 1975. Thick Bh horizons in the North Carolina Coastal Plain: I. Morphology and relation to texture and soil ground water. *Soil Sci. Soc. Am. Proc.*, 39: 1177-1181.

Daniels, R.B., E.E. Gamble, and C.S. Holzhey. 1976. Humate, Bh soil horizons, in wet sands of the North Carolina Coastal Plain. *Southeastern Geol.*, 18: 61-81.

Daniels, R.B., G.H. Simonson, and R.L Handy. 1961. Ferrous iron content and color of sediments. *Soil Sci.*, 91: 378-382.

Darmody, R.G. and J.E. Foss. 1978. Tidal Marsh Soils of Maryland. Maryland Agricultural Expt. Station Publication MP 930. University of Maryland, College Park.

DeLaune, R.D., R.H. Baumann, and J.G. Gosselink. 1983. Relationships among vertical accretion, coastal submergence, and erosion in a Louisiana Gulf Coast marsh. *J. Sediment. Petrol.*, 53: 147-157.

DeLaune, R.D., W.H. Patrick, Jr., and S.R. Pezeshki. 1987. Foreseeable flooding and death of coastal wetland forests. *Environ. Conserv.*, 14: 129-133.

DeLaune, R.D., C.N. Reddy, and W.H. Patrick, Jr. 1981. Accumulation of plant nutrients and heavy metals through sedimentation processes and accretion in a Louisiana salt marsh. *Estuaries*, 4: 328-334.

Demas, G.P. 1998. First subaqueous soil survey completed. *National Cooperative Soil Survey Newsletter* 3: 8.

Diers, R. and J.L. Anderson. 1984. Part I. Development of soil mottling. *Soil Surv. Hor.*, Winter 1984: 9-12.

Environmental Laboratory. 1987. Corps of Engineers Wetlands Delineation Manual. Tech. Rep. Y-87-1. U.S. Army Engineer Waterways Expt. Station, Vicksburg, MS.

Eswaran, H. 1968. Point-count analysis as applied to soil micromorphology. *Pedologie*, 18: 238-252.

Eswaran, H. and T. Cook. 1985. Classification of wetland soils in soil taxonomy. In *Wetland Soils: Characterization, Classification, and Utilization*. International Rice Research Institute, Los Baños, Laguna, Philippines. 375-390.

Evans, C.V. and D.P. Franzmeier. 1986. Saturation, aeration, and color patterns in a toposequence of soils in north-central Indiana. *Soil Sci. Soc. Am. J.*, 50: 975-980.

Fanning, D.S., M.C. Rabenhorst, and J.M. Bigham. 1993. Color of acid sulfate soils. In *Soil Color*. J.M. Bigham and E.J. Ciolkosz (Eds.). Special Publication No. 31. Soil Science Society of America, Inc., Madison, WI. 91-108.

Fanning, D.S. and W.U. Reybold, III. 1968. Water Table Fluctuations in Poorly Drained Coastal Plain Soils. MD Agricultural Expt. Station Contribution No. 3987. Agriculture Expt. Station, University of Maryland, College Park.

Fanning, D.S., R.L. Hall, and J.E. Foss. 1973. Soil morphology, water tables and iron relationships in soils of the Sassafras drainage catena in Maryland. In *Pseudogley and Gley*. E. Schlicting and U. Schwertmann (Eds.). Verlag Chemie, Weinheim, Germany. 71-79.

Faulkner, S.P. and W.H. Patrick, Jr. 1992. Redox processes and diagnostic wetland soil indicators in bottomland hardwood forests. *Soil Sci. Soc. Am. J.*, 56: 856-865.

Faulkner, S.P., W.H. Patrick, Jr., and R.P. Gambrell. 1989. Field techniques for measuring wetland soil parameters. *Soil Sci. Soc. Am. J.*, 53: 883-890.

Federal Interagency Committee for Wetland Delineation. 1989. Federal Manual for Identifying and Delineating Jurisdictional Wetlands. Cooperative technical publication. U.S. Army Corps of Engineers, U.S Environmental Protection Agency, U.S. Fish and Wildlife Service, and USDA Soil Conservation Service, U.S. Government Printing Office, Washington, D.C.

Flach, K.W. 1983. List of soils with actual or high potential for hydric condition. National Bulletin No. 430-3-10. USDA Soil Conservation Service, Washington, D.C.

Florida Department of Environmental Protection, South Florida Water Management District, St. John's Water Management District, Suwannee Water Management District, Southwest Florida Water Management District, and Northwest Water Management District. 1995. The Florida Wetlands Delineation Manual. Tallahassee.

Florida Soil Survey Staff. 1992. Soil and Water Relationships of Florida's Ecological Communities. USDA Natural Resources Conservation Service, Gainesville, FL.

Franzmeier, D.P. 1993. Relation of organic matter content to texture and color of Indiana soils. *Proc. Indiana Acad. Sci.*, 98.

Franzmeier, D.P., J.E. Yahner, G.C. Steinhardt, and H.R. Sinclair, Jr. 1983. Color patterns and water table levels in some Indiana soils. *Soil Sci. Soc. Am. J.*, 47: 1196-1202.

Gambrell, R.P., R.D. DeLaune, and W.H. Patrick, Jr. 1991. Redox processes in soils following oxygen depletion. In *Plant Life Under Oxygen Deprivation*. M.B. Jackson, D.D. Davies, and H. Lambers (Eds.). SPB Academic Publishing, BV, The Hague, The Netherlands. 101-117.

Gambrell, R.P. and W.H. Patrick, Jr. 1978. Chemical and microbial properties of anaerobic soils and sediments. In *Plant Life in Anaerobic Environments*. D.D. Hook and R.M.M. Crawford (Eds.). Ann Arbor Science Publishers, Ann Arbor, MI. 375-423.

Gile, L.H. 1958. Fragipan and water table relationships of some Brown Podzolic and Low Humic-Gley soils. *Soil Sci. Soc. Am. Proc.*, 22: 560-565.

Godsalk, G.L. 1993. Synthesis of literature on the use of water-stained leaves in the delineation of wetlands. Wetland Research Program Note HY-DE-2.1. U.S. Army Engineer Waterways Expt. Station, Vicksburg, MS.

Greenland, D.J. 1962. Denitrification in some tropical soils. *J. Agric. Sci.*, 58: 227-233.

Greenwood, D.J. 1961. The effect of oxygen concentration on the decomposition of organic materials in soil. *Plant Soil*, 14: 360-376.

Griffin, R.W, L.P. Wilding, W.L. Miller, G.W. Crenwelge, R.J. Tucker, L.R. Drees, and W.C. Lynn. 1996. Preliminary investigations of hydric soil hydrology and morphology on the Texas Gulf Coast Prairie. In Preliminary Investigations of Hydric Soil Hydrology and Morphology in the United States. J.S. Wakely, S.W. Sprecher, and W.C. Lynn (Eds.). Tech. Rep. WRP-DE-13. U.S. Army Engineer Waterways Experiment Station, Vicksburg, MS. 9-30.

Guthrie, R.L. and B.F. Hajek. 1979. Morphology and water regime of a Dothan soil. *Soil Sci. Soc. Am. J.,* 43: 142-144.

Haswell, J.R. 1938. Drainage in the humid regions. In *Soils and Men*. Yearbook of Agriculture 1938. U.S. Department of Agriculture. U.S. Government Printing Office, Washington, D.C. 723-736.

Henderson, R.E. and W.H. Patrick, Jr. 1982. Soil aeration and plant productivity. In *Handbook of Agricultural Productivity,* vol. 1. CRC Press, Boca Raton, FL. 51-69.

Holstener-Jørgensen, H. 1959. Influence of shelterwood-cutting and clear cutting on ground water table in fine textured moraine. *Det. Forstlige Forsøgsvaesen I Danmark* 25: 291-306. (English summary)

Holzhey, C.S., R.B. Daniels, and E.E. Gamble. 1975. Thick Bh horizons in the North Carolina Coastal Plain: II. Physical and chemical properties and rates of organic additions from surface sources. *Soil Sci. Soc. Am. Proc.,* 39: 1182-1187.

Huddleston, J.H. and W. Austin. 1996. Preliminary investigations of hydric soil hydrology and morphology in Oregon. In Preliminary Investigations of Hydric Soil Hydrology and Morphology in the United States. J.S. Wakely, S.W. Sprecher, and W.C. Lynn (Eds.). Tech. Rep. WRP-DE-13. U.S. Army Engineer Waterways Experiment Station, Vicksburg, MS. 127-141.

Hudnall, W.H. and A.A. Szögi. 1996. Seasonally wet soils of Louisiana. In Preliminary Investigations of Hydric Soil Hydrology and Morphology in the United States. J.S. Wakely, S.W. Sprecher, and W.C. Lynn (Eds.). Tech. Rep. WRP-DE-13. U.S. Army Engineer Waterways Experiment Station, Vicksburg, MS. 31-48.

Hurt, G.W. and R.B. Brown. 1995. Development and application of hydric soil indicators in Florida. *Wetlands,* 15: 74-81.

Hurt, G.W. and V.W. Carlisle. 1998. Boundary hydric soil field indicators and interior hydric soil field indicators. *Bull. Soc. Wetland Sci.,* 15(2): 11-12.

Hurt, G.W. and J.L. Richardson. 1998. Hydric Soils Indicators. USDA Natural Resources Conservation Service, Ft. Worth, TX. CD-ROM.

Hurt, G.W., P.M. Whited, and R.F. Pringle (Eds.). 1996. Field Indicators of Hydric Soils in the United States. A Guide for Identifying and Delineating Hydric Soils (ver. 3.2). Prepared in cooperation with the National Technical Committee for Hydric Soils. USDA Natural Resources Conservation Service, Fort Worth, TX.

Hurt, G.W., P.M. Whited, and R.F. Pringle (Eds.). 1998. Field Indicators of Hydric Soils in the United States. A Guide for Identifying and Delineating Hydric Soils (ver. 4.0). Prepared in cooperation with the National Technical Committee for Hydric Soils. USDA Natural Resources Conservation Service, Fort Worth, TX.

Hyde, A.G. and R.D. Ford. 1989. Water table fluctuation in representative Immokalee and Zolfo soils of Florida. *Soil Sci. Soc. Am. J.,* 53: 1475-1478.

Johnson, W.C., R.L. Burgess, and W.R. Keammerer. 1976. Forest overstory vegetation on the Missouri River floodplain in North Dakota. *Ecol. Monogr.,* 46: 59-84.

Knuteson, J.A., J.L. Richardson, D.D. Patterson, and L. Prunty. 1989. Pedogenic carbonates in a Calciaquoll associated with a recharge wetland. *Soil Sci. Soc. Am. J.,* 53: 495-499.

Kosters, E.C., G.L. Chmura, and A. Bailey. 1987. Sedimentary and botanical factors influencing peat accumulation in the Mississippi Delta. *J. Geol. Soc. London,* 144: 423-434.

Latshaw, G.J. and R.F. Thompson. 1968. Water table study verifies soil interpretations. *J. Soil Water Conserv.,* 23: 65-67.

Leisman, G.A. 1953. The rate of organic matter accumulation on the sedge mat zones of bogs in the Itasca State Park region of Minnesota. *Ecology,* 34: 81-101.

Leisman, G.A. 1957. Further data on the rate of organic matter accumulation in bogs. *Ecology,* 38: 361.

Lugo, A.E., S. Brown, and M.M. Brinson. 1990. Concepts in wetland ecology. chap. 4. In *Forested Wetlands. Ecosystems of the World 15*. A.E. Lugo, M. Brinson, and S. Brown (Eds.). Elsevier Science Publishers B.V., Amsterdam, The Netherlands. 53-85.

Lyford, W.H. 1964. *Water Table Fluctuations in Periodically Wet Soils of Central New England*. Harvard Forest Paper No. 8. Harvard University, Petersham, MA.

Macbeth Division. 1994. Munsell Soil Color Charts. Kollmorgen Instruments Corporation, New Windsor, NY.

MacIntosh, E.E. and J. Van der Hulst. 1978. Soil drainage classes and soil water table relations in medium and coarse textured soils in Southern Ontario. *Can. J. Soil Sci.*, 58: 287-301.

Marbut, C.F. 1921. The contribution of soil surveys to soil science. *Soc. Prom. Agr. Sci. Proc.*, 41: 116-142.

Mausbach, M.J. 1994. Classification of wetland soils for wetland identification. *Soil Surv. Hor.*, Spring 1994: 17-25.

Mausbach, M.J. 1992a. History of the Development of Hydric Soils Criteria and Definition. Memo distributed to National Research Council, Wetland Characterization Committee.

Mausbach, M.J. 1992b. Soil survey interpretations for wet soils. In Proc. of the 8th Int. Soil Correlation Meeting (VIII ISCOM): Characterization, Classification, and Utilization of Wet Soils. J.M. Kimble (Ed.). USDA Soil Conservation Service, National Soil Survey Center, Lincoln, NE. 172-178.

Mausbach, M.J. and J.L. Richardson. 1994. Biogeochemical processes in hydric soil formation. *Current Topics in Wetland Biogeochemistry, Vol. 1*: 68-127.

McKeague, J.A. 1965a. Relationship of water table and Eh to properties of three clay soils in the Ottawa valley. *Can. J. Soil Sci.*, 45: 49-62.

McKeague, J.A. 1965b. Properties and genesis of three members of the Uplands catena. *Can. J. Soil Sci.*, 45: 63-76.

Melville, M.D. and G. Atkinson. 1985. Soil color: its measurement and its designation in models of uniform color space. *J. Soil Sci.*, 36: 495-512.

Michalyna, W. and R.H. Rust. 1984. Influence of drainage regime on the chemistry and morphology of some Manitoba soils: sandy Chernozemic and gleysolic soils of the Lower Assiniboine Delta. *Can. J. Soil Sci.*, 64: 587-604.

Mid-Atlantic Hydric Soil Committee. 1998. Field Indicators of Hydric Soils in the Mid-Atlantic United States. U.S. Environmental Protection Agency, Region III, Philadelphia.

Miller, J.J., D.F. Action, and R.J. St. Arnaud. 1985. The effect of groundwater on soil formation in a morainal landscape in Saskatchewan. *Can. J. Soil Sci.*, 65: 293-307.

Moore, T.R. 1974. Gley morphology and soil water regimes in some soils in southcentral England. *Geoderma*, 11: 297-304.

Moormann, F.R. and H.T.J. van de Wetering. 1985. Problems in characterizing and classifying wetland soils. In *Wetland Soils: Characterization, Classification, and Utilization*. International Rice Research Institute, Los Baños, Laguna, Philippines. 53-68.

National Research Council. 1995. *Wetlands: Characteristics and Boundaries*. Committee on Characterization of Wetlands. National Academy Press, Washington, D.C.

Natural Resources Conservation Service (Soil Conservation Service). Sept. 27, 1993. Field Indicators of Hydric Soils in the United States. U.S. Department of Agriculture, Northeast National Technical Center, Chester, PA.

Natural Resources Conservation Service. 1994. Changes in hydric soils of the United States. *Federal Register*, 59 (133): 35680.

Natural Resources Conservation Service. 1995. Hydric soils of the United States. *Federal Register*, 60 (37): 10349.

NEIWPCC Wetlands Work Group. 1995. Field Indicators for Identifying Hydric Soils in New England. New England Interstate Water Pollution Control Commission, Wilmington, MA.

NEIWPCC Wetlands Work Group. 1998. Field Indicators for Identifying Hydric Soils in New England. 2nd ed. New England Interstate Water Pollution Control Commission, Wilmington, MA.

Oenema, O. and R.D. DeLaune. 1988. Accretion rates in salt marshes in the Eastern Scheldt, Southwest Netherlands. *Estuar. Coastal Shelf Sci.*, 26: 379-394.

Ovenden, L. 1990. Peat accumulation in northern wetlands. *Quaternary Res.*, 33: 377-386.

Parker, W.B. 1992. Rationale for hydric soils criteria. Paper presented to the National Technical Committee for Hydric Soils, August 17-20, 1992. Fargo, ND.

Parker, W.B., S.P. Faulkner, and W.H. Patrick. 1984. Soil wetness and aeration in selected soils with aquic moisture regimes in the Mississippi and Pearl River deltas. In *Wetland Soils: Characterization, Classification, and Utilization*. International Rice Research Institute Workshop Proc. (March 26-April 5, 1984), Manila, Philippines. 91-107.

Patrick, W.H., R.P. Gambrell, and S.P. Faulkner. 1996. Redox measurements of soils, chap. 42. In *Methods of Soil Analysis, Part 3. Chemical Methods*. Book Series No. 5. Soil Science Society of America, Madison, WI. 1255-1273.

Patrick, W.H., Jr. and R.D. DeLaune. 1990. Subsidence, accretion, and sea level rise in south San Francisco Bay marshes. *Limnol. Oceanogr.*, 35: 1389-1395.

Pickering, E.W. and P.L.M. Veneman. 1984. Moisture regimes and morphological characteristics in a hydrosequence in central Massachusetts. *Soil Sci. Soc. Am. J.*, 48: 113-118.

Pilgrim, S. 1998. Personal communication.

Ping, C.L., M. Clark, and G. Michaelson. 1996. Preliminary investigations of hydric soil hydrology and morphology in Alaska. In Preliminary Investigations of Hydric Soil Hydrology and Morphology in the United States. J.S. Wakely, S.W. Sprecher, and W.C. Lynn (Eds.). Tech. Rep. WRP-DE-13. U.S. Army Engineer Waterways Experiment Station, Vicksburg, MS. 142-152.

Ping, C.L., J.P. Moore, and M.H. Clark. 1992. Wetland properties of permafrost soils in Alaska. In Proc. of the 8th Int. Soil Correlation Meeting (VIII ISCOM): Characterization, Classification, and Utilization of Wet Soils. J.M. Kimble (Ed.). USDA Soil Conservation Service, National Soil Survey Center, Lincoln, NE. 198-205.

Ponnamperuma, F.N. 1972. The chemistry of submerged soils. *Adv. Agron.*, 24: 29-96.

Post, D.F., S.J. Levine, R.B. Bryant, M.D. Mays, A.K. Batchily, R. Escadafal, and A.R. Huete. 1993. Correlations between field and laboratory measurements of soil color. In *Soil Color*. J.M. Bigham and E.J. Ciolkosz (Eds.). Special Publication No. 31. Soil Science Society of America, Inc., Madison, WI. 35-49.

Postma, D. 1981. Formation of siderite and vivianite and the pore-water composition of a recent bog sediment in Denmark. *Chem. Geol.*, 31: 225-244.

Prasittikhet, J. and R.P. Gambrell. 1989. Acidic sulfate soils. In *Acidic Precipitation, volume 4: Soils, Aquatic Processes, and Lake Acidification*. S.A. Norton, S.E. Lindberg, and A.L. Page (Eds.). Springer-Verlag, New York. 35-62.

Reddy, K.R., T.C. Feijtel, and W.H. Patrick, Jr. 1986. Effect of soil redox conditions on microbial oxidation of organic matter. In *The Role of Organic Matter in Modern Agriculture*. Y. Chen and Y. Avnimelech (Eds.). Martinus Nijhoff Publishers. The Netherlands. 117-156.

Richardson, J.L. and R.J. Bigler. 1984. Principal component analysis of prairie pothole soils in North Dakota. *Soil Sci. Soc. Am. J.*, 48: 1350-1355.

Richardson, J.L. and R.B. Daniels. 1993. Stratigraphic and hydraulic influences on soil color development. In *Soil Color*. J.M. Bigham and E.J. Ciolkosz (Eds.). Special Publication No. 31. Soil Science Society of America, Inc., Madison, WI. 109-125.

Richardson, J.L. and F.D. Hole. 1979. Mottling and iron distribution in a Glossoboralf-Haplaquoll hydrosequence on a glacial moraine in northeastern Wisconsin. *Soil Sci. Soc. Am. J.*, 43: 552-558.

Richardson, J.L. and D.A. Lietzke. 1983. Weathering profiles in fluvial sediments of the Middle Coastal Plain of Virginia. *Soil Sci. Soc. Am. J.*, 47: 301-304.

Richardson, J.L. and M.J. Vepraskas (Eds.). 1999. *Wetland Soils*. Ann Arbor Press, Ann Arbor, MI.

Roman, C.T., R.A. Zampella, and A.Z. Jaworski. 1985. Wetland boundaries in the New Jersey Pinelands: ecological relationships and delineation. *Water Res. Bull.*, 21: 1005-1012.

Ruhe, R.V., R.C. Prill, and F.F. Riecken. 1955. Profile characteristics of some loess-derived soils and soil aeration. *Soil Sci. Soc. Am. Proc.*, 19: 345-347.

Schulze, D.G., J.L. Nagel, G.E. Van Scoyoc, T.L. Henderson, M.F. Baumgardner, and D.E. Stott. 1993. Significance of organic matter in determining soil colors. In *Soil Color*. J.M. Bigham and E.J. Ciolkosz (Eds.). Special Publication No. 31. Soil Science Society of America, Inc., Madison, WI. 71-90.

Schwertmann, U. 1993. Relations between iron oxides, soil color, and soil formation. In *Soil Color*. J.M. Bigham and E.J. Ciolkosz (Eds.). Special Publication No. 31. Soil Science Society of America, Inc., Madison, WI. 51-69.

Schwertmann, U. and D.S. Fanning. 1976. Iron-manganese concretions in hydrosequences of soils in loess in Bavaria. *Soil Sci. Soc. Am. J.*, 40: 731-738.

Schwertmann, U. and W. Lentze. 1966. Bodnefarbe und Eisenoxidform. *Z. Pflanzernern. Düng. Bodenk.*, 115: 209-214.

Scott, M.L., W.L. Slauson, C.A. Segelquist, and G.T. Auble. 1989. Correspondence between vegetation and soils in wetlands and nearby uplands. *Wetlands*, 9: 41-60.

Segal, D.S., P.J. Latham, and G.R. Best. 1987. Delineating a wetland boundary using vegetative, soil and hydrologic characteristics: a Florida cypress dome example. *Wetlands*, 7: 51-58.

Segal, D.S., S.W. Sprecher, and F.C. Watts. 1995. Relationships Between Hydric Soil Indicators and Wetland Hydrology for Sandy Soils in Florida. Tech. Rep. WRP-DE-7. U.S. Army Engineer Waterways Experiment Station, Vicksburg, MS.

Segelquist, C.A., W.L. Slauson, M.L. Scott, and G.T. Auble. 1990. Synthesis of Soil-plant Correspondence Data from Twelve Wetland Studies Throughout the United States. Biol. Rep. 90(19). U.S. Fish and Wildlife Service, Washington, D.C.

Shaw, S. P. and C.G. Fredine. 1956. Wetlands of the United States. Their Extent and Their Value to Waterfowl and Other Wildlife. Circular 39. U.S. Fish and Wildlife Service, Washington, D.C.

Simonson, R.W. 1967. Outline of a generalized theory of soil genesis. In Selected Papers in Soil Formation and Classification. J.W. Drew (Ed.). *Soil Sci. Soc. Am. Spec. Publ. 1.*

Simonson, R.W. 1993. Soil color standards and terms for field use-history of their development. In *Soil Color.* J.M. Bigham and E.J. Ciolkosz (Eds.). Special Publication No. 31. Soil Science Society of America, Inc., Madison, WI. 1-20.

Simonson, R.W. and L. Boersma. 1972. Soil morphology and water table relations. II. Correlation between annual water table fluctuations and profile features. *Soil Sci. Soc. Am. Proc.,* 34: 649-653.

Sipple, W.S. 1988. Wetland Identification and Delineation Manual. Volume I. Rationale, Wetland Parameters, and Overview of Jurisdictional Approach. Revised Interim Final. U.S. Environmental Protection Agency, Office of Wetlands Protection, Washington, D.C.

Soil Survey Division Staff. 1993. Soil Survey Manual. USDA Handbook No. 18. U.S. Department of Agriculture, Washington, D.C.

Soil Survey Staff. 1951. Soil Survey Manual. USDA Handbook No. 18. U.S. Department of Agriculture, Washington, D.C.

Soil Survey Staff. 1975. Soil Taxonomy: A Basic System of Soil Classification for Making and Interpreting Soil Surveys. Agricultural Handbook 436. U.S. Department of Agriculture, Washington, D.C.

Soil Survey Staff. 1996. Keys to Soil Taxonomy, 7th ed. USDA Natural Resources Conservation Service, Washington, D.C.

Soil Survey Staff. 1998. Keys to Soil Taxonomy, 8th ed. USDA Natural Resources Conservation Service, Washington, D.C.

Spaeth, J.N. and C.H. Diebold. 1938. Some interrelationships between soil characteristics, water tables, soil temperature, and snow cover in the forest and adjacent open areas in south-central New York. Agric. Exp. Sta. Memoir 213. Cornell University, Ithaca, NY.

Sprecher, S.W., J.S. Wakely, and W.C. Lynn. 1996. Synthesis. In Preliminary Investigations of Hydric Soil Hydrology and Morphology in the United States. J.S. Wakely, S.W. Sprecher, and W.C. Lynn (Eds.). Tech. Rep. WRP-DE-13. U.S. Army Engineer Waterways Experiment Station, Vicksburg, MS. 153-162.

Sprecher, S. 1998. Personal communication.

Steers, J.A. 1977. Physiography, chap. 2. In *Wet Coastal Ecosystems. Ecosystems of the World 1.* V.J. Chapman (Ed.). Elsevier Scientific Publishing Co., Amsterdam, The Netherlands. 31-60.

Steinwand, A.L. and J.L. Richardson. 1989. Gypsum occurrence in soils on the margin of semipermanent prairie pothole wetlands. *Soil Sci. Soc. Am. J.,* 53: 836-842.

Stoops, G. and H. Eswaran. 1985. Morphological characteristics of wet soils. In *Wetland Soils: Characterization, Classification, and Utilization.* International Rice Research Institute, Los Baños, Laguna, Philippines. 177-189.

Siuta, J. 1967. Gleying as an indicator of the water and air regime of the soil. *Sov. Soil Sci.,* 3: 356-363.

Taylor, R.M. 1982. Colour in soils and sediments, a review. In *Developments in Sedimentology 35.* H. Van Olphen and F. Veniale (Eds.). Elsevier Publishing Co., Amsterdam, The Netherlands. 749-761.

Thien, S.J. 1979. A flow diagram for teaching texture-by-feel analysis. *J. Agron. Educ.,* 8: 54-55.

Thom, R. M. 1992. Accretion rates of low intertidal salt marshes in the Pacific Northwest. *Wetlands,* 12: 147-156.

Tiner, R.W. 1985. Wetlands of New Jersey. U.S. Fish and Wildlife Service, Newton Corner, MA.

Tiner, R.W. 1986. Hydric soils: their use in wetland identification and boundary delineation. In Proc. Natl. Wetland Assessment Symp. (Portland, ME; June 17-20, 1985). J.A. Kusler and P. Riexinger (Eds.). ASWM Rep. No. 1: 178-182. Association of State Wetland Managers, Berne, NY.

Tiner, R.W. 1991. Maine Wetlands and Their Boundaries: A Guide for Code Enforcement Officers. Maine Department of Economic and Community Development, Office of Comprehensive Planning, Augusta, ME.

Tiner, R.W. 1994. Maine Wetlands and Their Boundaries: A Guide for Code Enforcement Officers. Maine Department of Economic and Community Development, Office of Comprehensive Planning, Augusta, ME.

Tiner, R.W. 1988. Field Guide to Nontidal Wetland Identification. U.S. Fish and Wildlife Service, Newton Corner, MA and Maryland Department of Natural Resources, Annapolis, MD. Cooperative technical publication.

Tiner, R.W. 1998. *In Search of Swampland. A Wetland Sourcebook and Field Guide*. Rutgers University Press, New Brunswick, NJ.

Tiner, R.W. and W.C. Kirkham. 1984. Official list of New Jersey's hydric soils. September 26, 1984 memo from Regional Wetland Coordinator, U.S. Fish and Wildlife Service and State Soil Scientist, USDA Soil Conservation Service.

Tiner, R.W. and P. L. M. Veneman. 1987. *Hydric Soils of New England*. Bulletin C-183. University of Massachusetts Extension, Amherst.

Tiner, R.W. and P. L. M. Veneman. 1995. *Hydric Soils of New England*. Bulletin C-183R. University of Massachusetts Extension, Amherst.

USDA Soil Conservation Service. 1982. Soils — hydric soils of the United States. National Bulletin No. 430-2-7 (January 4, 1982). Department of Agriculture, Washington, D.C.

USDA Soil Conservation Service. 1985a. Hydric Soils of the United States. March 20, 1985 draft. In Cooperation with the National Technical Committee for Hydric Soils. Washington, D.C.

USDA Soil Conservation Service. 1985b. Hydric Soils of the United States. 1st ed. In Cooperation with the National Technical Committee for Hydric Soils. Washington, D.C. October 1985.

USDA Soil Conservation Service. 1987. Hydric Soils of the United States. In Cooperation with the National Technical Committee for Hydric Soils. Washington, D.C.

USDA Soil Conservation Service. 1991. Hydric Soils of the United States. In Cooperation with the National Technical Committee for Hydric Soils. Washington, D.C.

Van Breemen, N. 1988. Effects of seasonal redox processes involving iron on the chemistry of periodically reduced soils. In *Iron in soils and clay minerals*. J.W. Stucki and others (Eds.). D. Reidal Publishing Co., Boston. 797-809.

Veneman, P.L.M. and S.M. Bodine. 1982. Chemical and morphological soil characteristics in a New England drainage-toposequence. *Soil Sci. Soc. Am. J.,* 46: 359-363.

Veneman, P.L.M. and R.W. Tiner. 1990. Soil–Vegetation Correlations in the Connecticut River Floodplain of Western Massachusetts. Biol. Rep. 90(6). U.S. Fish and Wildlife Service, Washington, D.C.

Veneman, P.L.M., M.J. Vepraskas, and J. Bouma. 1976. The physical significance of soil mottling in a Wisconsin toposequence. *Geoderma,* 15: 103-118.

Veneman, P. 1998. Personal communication.

Vepraskas, M.J. 1996. Redoximorphic Features for Identifying Aquic Conditions. Technical Bulletin 301. North Carolina Agricultural Research Service, North Carolina State University, Raleigh.

Vepraskas, M.J., F.G. Baker, and J. Bouma. 1974. Soil mottling and drainage in a Mollic Hapludalf as related to suitability for septic tank construction. *Soil Sci. Soc. Am. Proc.,* 38: 497-501.

Vepraskas, M.J. and W.R. Guertal. 1992. Morphological indicators of soil wetness. In Characterization, Classification, and Utilitization of Wet Soils. J.M. Kimble (Ed.). Proc. 8th Internat. Soil Correlation Meeting (VIII ISCOM), USDA Soil Conservation Service, National Soil Survey Center, Lincoln, NE. 307-312.

Vepraskas, M.J. and S.W. Sprecher (Eds.). 1997. *Aquic Conditions and Hydric Soils: The Problem Soils*. Spec. Pub. No. 50. Soil Science Society of America, Madison, WI.

Vepraskas, M.J. and L.P. Wilding. 1983. Aquic moisture regimes in soils with and without low chroma colors. *Soil Sci. Soc. Am. J.,* 47: 280-285.

Wakely, J.S., S.W. Sprecher, and W.C. Lynn (Eds.). 1996. Preliminary Investigations of Hydric Soil Hydrology and Morphology in the United States. Tech. Rep. WRP-DE-13. U.S. Army Engineer Waterways Experiment Station, Vicksburg, MS.

Warren, R.S. and W.A. Niering. 1993. Vegetation change on a Northeast tidal marsh: interaction of sea-level rise and marsh accretion. *Ecology,* 74: 96-103.

Watts, F.C. and G.W. Hurt. 1991. Determining depths to the seasonal high water table and hydric soils in Florida. *Soil Surv. Hor.,* 32: 117-121.

Whited, M. 1998. Personal communication.

Zakharov, S.A. 1927. Achievements of Russian Science in Morphology of Soils. Russian Pedological Invest. No. 2. Academy of Sciences of the U.S.S.R., Leningrad.

Zampella, R.A. 1994. Morphologic and color pattern indicators of water table levels in sandy Pinelands soils. *Soil Sci.,* 157: 312-317.

Zobeck, T.M. and A. Richie, Jr. 1984. Analysis of soil morphology and long-term water table records from a hydrosequence of soils in central Ohio. *Soil Sci. Soc. Am. J.,* 48: 119-125.

6 Wetland Identification and Boundary Delineation Methods

INTRODUCTION

Knowledge of the location of wetlands has always been important to human civilizations. Some wetlands provided a wealth of harvestable foodstuffs (e.g., fruits and fowl) or material (e.g., peat) to heat homes. The fertile soils of other wetlands made them valuable places for agricultural development, while the unstable soils of others made them hazardous for travel or building. Shortly after this nation's settlement by Europeans, various maps were being made to chart the new world and these maps typically showed the location of large marshes and swamps as they were important natural features that affected the siting of transportation routes and cities and the development of agricultural lands. Maps prepared during the Civil War depicted wetlands for strategic military reasons (Dahl and Alford, 1996). Scientists have studied and classified wetland plant communities in the U.S. for more than a century (e.g., Shaler, 1890; Harshberger, 1900; 1909; 1916; Shreve et al., 1910). Wetland identification was an important process well before the advent of environmental protection laws in the 1970s.

While wetland identification is the simple recognition of wetlands for a multitude of purposes ranging from aesthetic appreciation to wetland inventories, two other terms are used to address the identification of wetlands on specific properties for regulatory purposes: wetland determination and wetland delineation. Wetland delineation is the detailed examination and surveying of the specific boundaries of wetlands on a particular parcel of land. Wetland determination is an intermediate step between basic wetland identification and wetland delineation. It is a process which identifies the general location of wetlands on a piece of property without identifying and marking the "exact" boundaries. Wetland determinations can aid in pre-project planning or site assessments of real estate for development potential.

Since the mid-1960s, several techniques have been developed for wetland delineation. This chapter begins with a brief discussion of the evolution and use of various methods. Most of the rest of the chapter is devoted to summarizing federal and some state methods. The following contemporary techniques are emphasized:

- The Corps manual (Environmental Laboratory, 1987)
- The EPA manual (Sipple, 1988)
- Food Security Act manual (USDA Soil Conservation Service, 1985; 1988; USDA Natural Resources Conservation Service, 1997)
- The federal interagency manual (Federal Interagency Committee for Wetland Delineation, 1989)
- The primary indicators method (Tiner, 1993)
- Selected state methods (Figure 6.1)

The discussion includes how each of the methods identifies the presence of wetlands and separates wetlands from uplands (nonwetlands). An outline of the steps involved in doing a wetland delineation is presented for the federal methods given their wide use and historic significance. Besides these summaries, some of the more significant recommendations from the National Research Council's recent scientific review of wetland delineation are given near the end of the

FIGURE 6.1   Some wetland delineation manuals developed to identify regulated wetlands.

chapter (NRC, 1995). The chapter concludes with a list of some helpful tips on performing wetland delineations and a brief discussion of best professional judgment.

EVOLUTION AND USE OF WETLAND IDENTIFICATION AND DELINEATION METHODS

Techniques to identify and delineate wetlands on-the-ground were necessitated by the passage of wetland and water resource laws. These laws established certain land use controls on both private property and public lands that required the presence of wetlands and their boundaries to be determined (see Want, 1992; Dennison and Berry, 1993; Environmental Law Institute, 1993 for reviews of these laws). After promulgating regulations to administer these laws, wetland identification and delineation were the next steps needed to protect and conserve wetland functions and values.

Until relatively recently, there were no manuals or guidebooks for wetland recognition and delineation. Without well-defined standards and trained personnel to implement them, wetland delineation is fraught with inconsistency, chaos, and uncertainty for the landowner, regulator, and general public alike. Failure to properly identify wetlands and their limits can lead to wetland destruction and loss of valued public services without appropriate administrative review by regulatory agencies.

Prior to the development of federal wetland delineation manuals in the mid- to late-1980s, wetlands were mostly identified by indicator plants or plant communities. Northeastern coastal states were the first states to regulate wetland uses. From the 1960s through the 1980s, these states passed laws to protect tidal wetlands mainly due to their importance to marine fisheries (e.g., estuaries are spawning areas and nursery grounds for many commercially and recreationally important fishes and shellfishes). These laws typically included a list of plant species or genera characteristic of these wetlands. The lists were not comprehensive, but intended to be prime examples of the types of plants indicative of these wetlands. Since most tidal wetlands are influenced by salt water, many of these wetlands have very distinctive plant communities — dominated by salt-tolerant plants called *halophytes*. Many of these species only occur in these environments and, therefore, serve as highly reliable indicators of these wetlands. States adopting this type of approach for identifying tidal wetlands included Massachusetts, Rhode Island, Connecticut, New Jersey, New York, Delaware, Maryland, and Virginia.

After passing laws to protect coastal wetlands, many northeastern states passed similar laws to protect inland or nontidal wetlands. These laws also included lists of characteristic plant species for major wetland types. While many of the listed species were viewed as highly reliable indicators, these obligate wetland species were not associated with all of these wetland types. Many freshwater wetland plant communities are mixtures of "wetland" species and "mesic upland" species. Consequently, a predominance of wetland species was then used to identify wetlands. For freshwater wetlands, when more than 50% of the plant species in a stand were hydrophytic or wetland indicator species, the plant community would be classified as a wetland. The underlying assumptions of this so-called "50% rule" were that vegetation patterns reflect significant differences in hydrology and that wetlands can be identified by recognizable plant communities.* State agencies implementing state wetland laws used this approach from the 1960s to the late 1980s when multiple-parameter/criteria methods largely replaced these techniques. Connecticut was an exception to this trend, choosing to use soils to identify nontidal wetlands (i.e., poorly drained, very poorly drained, and alluvial soils). Since soil maps had been recently completed for the state by the USDA Soil Conservation Service, the state felt that these maps would help local communities identify wetlands.

The U.S. Fish and Wildlife Service (FWS) began its National Wetlands Inventory (NWI) Project in the late 1970s. While the FWS's wetland definition included recognition of hydric soils as an indicator of wetland (Cowardin et al., 1979), the earliest NWI mapping work was largely verified by assessing the vegetation. This was necessitated by a few conditions:

- NWI staff were biologists with little or no training in soil science
- The concept of hydric soils was a new one and lists of hydric soils were in preparation
- Specific properties associated with many mineral hydric soils were not yet recorded in a form available for nonsoil scientists
- Vegetation had been the traditional method for identifying wetlands by the FWS and others.

As the NWI progressed, more information became available on hydric soils as the NWI staff soil scientist was developing a list of the nation's hydric soils. He helped train NWI regional staff in hydric soil recognition. Hydric soil properties then became used as a major feature in the recognition of wetlands on the ground. This helped improve the accuracy and comprehensiveness of the NWI over the early mapping which tended to be more conservative due to the emphasis on plant indicators (see Chapter 10).

At the same time, the EPA, the Corps, and others also began to realize that vegetation alone was insufficient for wetland identification and delineation and that soils and hydrology signs were needed under certain circumstances (Krivak, 1980; Eilers et al., 1983; Sanders et al., 1984; Roman et al., 1985; Zedler and Cox, 1985; Wentworth and Johnson, 1986). With increased federal regulation of wetlands and improved knowledge about wetland characteristics and functions, regulatory agencies (Army Corps of Engineers and Environmental Protection Agency) began to develop standardized techniques for wetland identification and delineation in order to achieve consistent and scientifically accurate results. The Corps and the EPA prepared wetland delineation manuals (Environmental Laboratory, 1987 and Sipple, 1988, respectively) for identifying the extent of jurisdictional wetlands. They both developed methods for identifying vegetated wetlands based on considering plants, soils, and signs of wetland hydrology and made use of wetland plant lists and hydric soils lists prepared to clarify the FWS's wetland definition. There were differences in how

---

* Unfortunately wetlands are more complicated than this, as factors other than soil moisture play important roles in the distribution of plants, plus many plant species either have a broad ecological tolerance (amplitude) or develop ecotypes that are well-adapted to specific environmental conditions (e.g., wetland ecotypes of normally drier-site species) (see Chapter 3).

these indicators were used to verify wetlands and delineated their boundaries (discussed later). Use of these manuals was not mandatory with the agencies. Many Corps districts continued to use their own locally created techniques or guidelines to identify jurisdictional wetlands. Adams et al. (1987) noted that in 1982 the Corps was subjectively determining wetland boundaries based on "some unstated criteria which existed in the perception of the delineators" and that such boundaries were being accepted as valid. Consequently, there was no consistency in what was identified as a federally regulated wetland across the country and in all likelihood, within districts.

Congress passed the Food Security Act of 1985 which included the Swampbuster provision designed to discourage drainage and conversion of wetlands for agricultural purposes. The USDA Natural Resources Conservation Service, formerly the Soil Conservation Service (NRCS), was charged with identifying wetlands on farmland to monitor compliance with this provision. NRCS also developed technical procedures to identify and delineate wetlands (USDA Soil Conservation Service, 1985; USDA Natural Resources Conservation Service, 1996), making use of soil survey information and aerial photographs acquired by the Department of Agriculture.

Since the four leading federal agencies in wetland regulation and conservation (Corps, EPA, FWS, and NRCS) applied different techniques to identify and delineate wetlands, considerable confusion arose over what was considered wetland and where to mark the upper boundary. This was particularly troublesome to the regulated community where the existence of two manuals (Corps and EPA) plus local Corps district approaches for delineating the boundaries of jurisdictional wetlands subject to the Clean Water Act often produced different results. Many interagency disputes centered around the presence and extent of regulated wetlands at proposed project sites.

In early 1988 after field testing their manuals, the Corps and EPA resumed previously failed discussions to attempt to resolve differences between their manuals. At that time, the FWS and NRCS were invited to participate due to their expertise in wetland vegetation and mapping, and hydric soils, respectively. At the first interagency meeting on May 19-20, 1988, after determining that the agencies' technical experts generally had the same concept of wetland, it was decided that the four agencies should attempt to develop an interagency wetland delineation manual. Eight months later, after much deliberation, the four agencies developed and adopted the *Federal Manual for Identifying and Delineating Jurisdictional Wetlands* (Federal Interagency Committee for Wetland Delineation, 1989) as the technical basis for identifying and delineating wetlands in the United States. This manual adapted existing field-tested methods employed by each of the agencies to provide a consistent wetland delineation regardless of the actual technique used. The end result, therefore, was a unified technically-based federal methodology for identification and delineation of vegetated wetlands. This manual served as the foundation for today's concept of wetland. On January 10, 1989, administrators from the four agencies adopted the manual as "the technical basis for identifying and delineating jurisdictional wetlands in the U.S."

Each agency decided how it would use the manual in its programs. On January 19, 1989, the Corps and EPA officially adopted the federal interagency manual for use by all districts and EPA regions (Page and Hanmer, 1989) which for the first time, established national consistency in section 404 wetland determinations. This manual was used from 1989 to 1991.

In the late 1980s, several states with wetland protection programs began to adopt federal methods for wetland delineation. New Jersey first used the EPA manual and subsequently replaced it with the 1989 federal interagency manual. Other states including Maryland, Pennsylvania, Vermont, and Maine also adopted the 1989 interagency manual for state use. The nation was on its way to having a single standard for wetland identification and delineation at federal, state, and local levels.

Given that there was no national consistency in identifying federally regulated wetlands prior to January 19, 1989, the mandatory use of this manual had a significant effect on the extent of federal regulation.* In many areas, the local techniques for wetland identification tended to focus on the wetter wetlands. The intention of the new manual was to identify all vegetated wetlands.

---

* Adoption of the Corps or EPA manuals would have had similar effects.

Much more "land" was identified as potentially subject to federal regulation. The use of the federal interagency manual coupled with policy changes (e.g., farmed wetlands*) greatly expanded federal jurisdiction on private land. Significant opposition to the use of the new manual arose especially in agricultural regions of the country.** Congress was brought into the process by concerned constituents, eventually passing the Johnston amendment to the 1992 Energy and Water Development Appropriations Act (the Corps funding legislation) which effectively forbade the Corps from using it or any other manual that did not fulfill the requirements of the Administrative Procedures Act (e.g., due notice and public comment). In response to this amendment, the Corps replaced the interagency manual with its 1987 Corps manual for jurisdictional purposes in late 1991 (Elmore, 1991). In January 1993, the U.S. EPA formally concurred with the Corps use of the manual for jurisdictional determinations in accordance with the Clean Water Act (U.S. Department of the Army and U.S. EPA, 1993).

Due to political pressures to decrease regulation of private wetlands (regulatory relief) and, perhaps in part, to achieve "federal consistency" (1991 abandonment of the 1989 manual due to political directives), Maine, Maryland, and Pennsylvania also replaced the federal interagency manual with the Corps manual for state regulatory uses. New Hampshire adopted the interagency manual for its wetland regulatory program after the federal government dropped its use for regulatory purposes. Other states developed their own techniques making use of elements from the federal methods and other approaches (e.g., Massachusetts, Rhode Island, and New York). Local governments also have adopted the federal interagency manual for local ordinances. For example, the Westchester County Soil and Water Conservation District (1997a,b) recommends it in their model local wetland ordinance, concluding that this manual is the most objective and comprehensive delineation system and is best suited for meeting their regulatory objectives.

In 1993, Tiner (1993) drafted the primary indicators method (PRIMET) as an alternative to three-parameter/criteria methods. This is not a new method but reflects what the FWS and others had been using for years to verify wetlands for inventory purposes. PRIMET recognizes that wetlands must possess something different than adjacent uplands and those features can be used for identify and delineate wetlands. Some state wetland manuals have adopted elements of PRIMET or something quite similar.

Over the past 3 decades, wetland identification has moved from a focus on vegetation to using a combination of vegetation, soil, and other signs in the process. Much has been learned about wetlands during this period to aid in improving the efficiency and effectiveness of wetland identification and delineation. These advances are included in many of the more recent techniques.

DIFFERING WETLAND IDENTIFICATION AND BOUNDARY DELINEATION PROTOCOLS

Plants, soils, and other features are present on the landscape that can be used to separate wetlands from uplands. Depending on an agency's philosophy and intent, these properties are interpreted in

---

* Prior to 1989, the Corps had interpreted "normal circumstances" in the federal regulatory wetland definition based on current/recent land use, so most farmed wetlands were not recognized as regulated areas. The new federal interagency manual adopted the definition of "normal circumstances" published in the Food Security Act which stated that the soils and hydrology that are normally present would be used to identify wetland, without regard to whether the vegetation was removed or not. In other words, undrained hydric soils were indicative of wetlands. By adopting the new manual without any caveats about existing farmed wetland policies, all farmed wetlands became regulated areas. Farmers, however, could continue to practice farming, but changes in land use (e.g., to houses or commercial buildings) would require federal permits.
** In 1991, the federal government proposed a revision to the federal interagency manual, which was not technically sound and received tremendous criticism from the scientific community and was never adopted (Cohen, 1992; Environmental Defense Fund and World Wildlife Fund, 1992). This controversy led to a Congressional mandate that the National Academy of Sciences conduct a review of federal wetland definitions and techniques used to identify and delineated wetlands (see National Research Council, 1995).

numerous ways to identify the presence of wetland and to delineate wetland–upland boundaries (Table 6.1). Published wetland delineation manuals provide the rules for what features constitute acceptable wetland indicators for regulatory purposes. A review of the basic protocols presented in different manuals (federal and state), plus PRIMET, is given. Readers should refer to actual manuals for details.

CORPS MANUAL

The U.S. Army Corps of Engineers (Corps) regulates wetland uses under authority of the Rivers and Harbors Act and the Clean Water Act (CWA). Private individuals, organizations, and public agencies must secure permits from the Corps prior to engaging in certain activities in wetlands (e.g., dredging and filling). The Corps' role in wetland regulation has generally increased since the mid-1970s due to court decisions and changes in public attitudes toward wetland protection. The Corps published a wetland delineation manual in 1987 (*Corps of Engineers Wetlands Delineation Manual*), but did not require its districts to use it for making wetland determinations and delineations. In late 1991, the Corps mandated nationwide use of its 1987 Corps manual for jurisdictional purposes (Elmore, 1991). Since then guidance memoranda have been published at the national and district levels to clarify points of misinterpretations in the use of the manual and to make it more technically sound (e.g., Studt, 1991; Williams, 1992; New England District, 1991; U.S. Army Corps of Engineers-Seattle District and U.S Environmental Protection Agency–Region 10, 1994).

The Corps manual utilizes a three-parameter method for identifying wetlands requiring positive indicators of hydrophytic vegetation, hydric soils, and wetland hydrology. Table 6.1 lists the acceptable indicators for each "parameter." There are exceptions to the basic rule and these are presented in two sections of the Corps manual — atypical situations (disturbed sites) and problem areas (e.g., wetland types or conditions where positive indicators may be lacking).

In practice, wetlands typically must have (1) a plant community where >50% of the dominant species have an wetland indicator status of FAC or wetter (excluding FAC–), (2) soils with certain hydric soil properties, and (3) signs of wetland hydrology. For the latter, hydrology indicators are separated into two groups — primary indicators (those referenced in the Corps manual) and secondary indicators (not listed in the manual, but added through additional guidance in Williams, by district guidance, or others that are used based on professional judgment). One primary hydrology indicator or two secondary indicators are required to verify wetland hydrology. There are two situations where soils do not need to be examined (hydric soils are assumed to exist): (1) where all the dominant plants are OBL species, and (2) where all the dominants are OBL and FACW (need at least 1 OBL) and the wetland boundary is abrupt. Exceptions to the basic rules generally require use of professional judgment which one may not choose to exercise for various reasons.

Once wetland plant communities and nonwetland plant communities are detected on a project site, the wetland–upland boundary must be determined. Wetland boundaries may be established in either of three ways.

1. Making an interpretation of significant vegetation changes between the wetland and nonwetland communities.
2. Locating the point at which there is an absence of positive indicators for one of the parameters.
3. Examining the soil for direct and indirect signs of seasonal high water tables that would indicate prolonged saturation within 12 in. of the surface.

Since guidance is not explicit, an investigator using a strict interpretation of the manual may rely on number 2 above, while a more technically accurate line may be derived by using the third way to draw the boundary.

TABLE 6.1
Use of Various Indicators by Federal Manuals

Hydrophytic Vegetation Indicators	Corps	EPA	Interagency
>50% of dominants are OBL, FACW, or FAC (excluding FAC–)	X		
>50% of dominants are OBL, FACW, or FAC			X
Presence of dominant OBL species		X	
Visual observations of plants growing in areas of prolonged inundation and/or soil saturation (>10% of the growing season)	X		
Plants with certain morphological adaptations	X	X	X
Plants with known physiological or reproductive adaptations to prolonged inundation/saturation	X	X	X
Facultative species (FACW, FAC, and FACU) when on undrained hydric soils		X	X
Plant community with a prevalence index < 3.0			X
Plant community with more coverage by OBL and FACW species than by FACU and UPL species			X
Professional judgment supported by technical literature	X		

Hydric Soils Indicators	Corps	EPA	Interagency
Organic soils (Histosols, except Folists)	X	X	X
Histic epipedons	X	X	X
Sulfidic material	X	X	X
Reducing soil conditions (observed)	X	X	X
Gleyed soils	X	X	X
Mottled soils with low chroma matrix	X	X	X
Soils with aquic or peraquic moisture regimes	X	X	X
Soil on hydric soils list[a]	X	X	X
Iron and manganese concretions	X	X	X
Oxidized root-rhizome channels along living roots	X		
High organic content in surface horizon of sandy soils	X	X	X
Organic streaking in subsurface horizons	X	X	X
Organic pans (wet Spodosols)	X	X	X

Wetland Hydrology Indicators	Corps	EPA	Interagency
Recorded data	X	X	X
Visual observations of inundation	X	X	X
Visual observations of saturation (within upper root zone – 12 in.)	X	X	X
Watermarks	X	X	X
Oxidized rhizospheres along living roots/rhizomes	A		X
Drift lines	X	X	X
Water-borne sediment deposits	X	X	X
Surface scouring		X	X
Drainage patterns within wetlands	X	X	X
Water-stained leaves	A	X	X
Bare areas (extended flooding)		X	X
Moss lines		X	
Plants with certain morphological adaptations		X	X
Hydric soil characteristics (undrained sites)			X
Local soil survey data[b]	A	X	
FAC Neutral Test	A		

[a] Both the EPA and interagency manuals required field verification of hydric soil properties, while the Corps manual emphasized field verification of typical soil profile for the series mapped.
[b] Use soil–water features table after verifying soil type (see Table 9.4 in Chapter 9.)

Note: A = added per March 1992 memorandum. (From Williams, MG A.E. 1992. U.S. Army Corps of Engineers memorandum on clarification and interpretation of the 1987 manual. March 6, 1992.) These indicators do not include indicators for problematic soils, except some sandy soils.

In general, the Corps manual provides a conservative assessment of wetlands as it will readily identify and delineate wetter wetlands, but will be less useful for recognizing drier-end wetlands. This results from several factors, including:

- Defining the significant period of wetland hydrology by growing season (e.g., does not cover all ecologically recognized wetlands or wetlands of functional significance such as certain wetlands flooded only in winter)
- The uncertainty of how to handle areas that are wet at or near the surface between 12.5 to 5% of the growing season (e.g., does an area that is saturated within 12 in. of the surface and/or flooded for ≥5% of the growing season in most years qualify as wetland?)
- Limited hydric soil indicators (i.e., can the recent list of national hydric soil indicators be used to expand the manual's list to verify hydric soils?)
- Limited wetland hydrology indicators especially for saturated wetlands
- The failure to recognize that any plant community growing on an undrained hydric soil with signs of wetland hydrology is a wetland
- The requirement that positive indicators for the three parameters be found or normally present just inside the boundary.

Proper use of the manual requires consideration of various regulatory guidance memoranda which makes its use somewhat cumbersome. While the recent regulatory guidance regarding the use of the Corps manual significantly improves its ability to identify wetlands, it still relies too heavily on the use professional judgment where clear guidance would promote more consistent and technically sound determinations. Moreover, the Corps' three-parameter approach is not supported by the science as the NRC concluded that in the absence of hydrologic modification, field-verified hydric soils or certain plant communities (e.g., OBL and FACW dominated) "can be used as strong evidence of wetland hydrology" (NRC, 1995, p. 145). The Corps manual is probably best used during the wetter part of the growing season as it is more likely that primary indicators of wetland hydrology will be found at this time, leading to a more accurate wetland determination and delineation.

EPA Manual

The U.S. Environmental Protection Agency (EPA) is responsible for administrating the Clean Water Act program. The CWA authority requires the EPA to oversee the section 404 program of the act and the Corps implementation of this program. In the mid-1980s, EPA developed a manual to reflect its definition of wetlands subject to the CWA (*Wetland Identification and Delineation Manual*; Sipple, 1988). The EPA approach to wetland identification considered three factors (vegetation, soils, and hydrology) but did not require positive evidence of hydrophytic vegetation, hydric soils, and wetland hydrology. Instead, it relied more on the strength of certain indicators and was hierarchical in nature. This resulted in differences in the extent of wetlands subject to the CWA when compared with determinations based on the Corps manual, leading to the eventual development of the federal interagency manual.

The EPA approach first considered vegetation in determining the presence or absence of wetland. Plant communities were divided into three groups:

1. Those with one or more OBL dominants.
2. Those dominated by facultative species (FACW, FAC, and FACU).
3. Those where UPL species predominated.

The first community was recognized as wetland (in the absence of significant hydrologic modification) and the latter community designated as nonwetland (upland). The intermediate community required evaluation of soils and hydrology to make a wetland determination. The presence of hydric soil indicators and signs of wetland hydrology (Table 6.1) verified that a facultative-dominated plant community was a wetland. The definition of wetland hydrology was consistent with the national hydric soil criteria — flooding, ponding, or saturation for a week or more during the growing season in most years.

Wetland boundary delineation relied more on the use of soils (hydric vs. nonhydric), "particularly if there is no evident vegetation break or when facultative species dominate two adjacent vegetation units." Changes in plants, especially shrubs and herbs in forested areas, suggest where soils should be examined. Where a distinct topographic break exists, the boundary may be determined by the presence of OBL species.

The EPA manual's approach overcomes many of the shortcomings of the Corps manual mentioned above and is an efficient method for identifying wetlands, considering the strength of individual indicators for identifying wetlands. The EPA manual has not been updated since 1988 and, therefore, does not incorporate the last decade's worth of information on wetlands (e.g., expanded field indicators of hydric soils and more detailed protocols for evaluating disturbed sites), yet this could be easily accomplished.

FOOD SECURITY ACT MANUAL

The Natural Resources Conservation Service (NRCS) developed a national list of hydric soils and field indicators to hydric soils which have greatly aided wetland identification. They are also responsible for identifying wetlands subject to the "Swampbuster" provision of the Food Security Act (FSA). This provision attempts to discourage farmers from draining and converting wetlands to produce commodity crops by removing federal subsidy payments from farmers who do so after December 23, 1985. The NRCS is responsible for identifying wetlands, farmed wetlands, and wetland pastures or haylands,* and prior converted wetlands** on participating farms. NRCS makes wetland determinations and delineations as needed.***

For offsite determinations, NRCS uses aerial photographs and soil surveys with the county list of hydric soil map units and map units with hydric inclusions as guidance for interpreting these surveys. Aerial photos and FSA compliance slides are used in combination with National Wetlands Inventory (NWI) maps and soil surveys to identify FSA wetlands.

In cases where onsite investigations are needed to determine compliance under FSA, NRCS is using hydric soil indicators listed in *Field Indicators of Hydric Soils in the United States* (Hurt et al., 1996; 1998; see Chapter 5, Table 5.9 for list). Wetland hydrology is analyzed using Corps manual indicators and guidance presented in *Hydrology Tools for Wetland Determination* (Natural Resources Conservation Service, 1997).

A majority of the FSA determinations are made using offsite procedures per the National Food Security Act Manual (USDA Soil Conservation Service, 1985; USDA Natural Resources Conservation Service, 1996). Mapping conventions have been developed for each state by interagency committees (NRCS, EPA, Corps, and FWS) and have been field tested to ensure an adequate

---

* Farmed wetlands are wetlands in cultivation (not abandoned) that are frequently flooded for at least 15 consecutive days during the growing season or for 10% of the growing season (once every other year on average) whichever is less or if the wetland is a playa, pothole, or pocosin, it must only be inundated for at least 7 consecutive days or saturated for at least 14 consecutive days during the growing season. It is, therefore, important to know the latter types from the former types, and unfortunately specific definitions are lacking. Farmed wetland pastures also have the same hydrologic requirements as potholes.

** A prior converted wetland is a wetland that has been drained and converted to the produce commodity crops (e.g., corn, soybeans, and rice) before the effective date of the Food Security Act, December 23, 1985.

*** In 1994, the NRCS was required to perform wetland delineations on agricultural lands for purposes of the Clean Water Act (Lyons et al., 1994). For this work, the NRCS is using the Corps wetlands delineation manual for delineating narrow bands and small pockets of wetlands in agricultural fields and for evaluating nonagricultural land (undisturbed native vegetation) on farm properties.

TABLE 6.2

Examples of Mapping Conventions for Offsite Wetland Determinations for the Food Security Act (Swampbuster Provision) in Virginia

Wetland or Former Wetland Type	Convention
Wetland	Woods + hydric soil + one of the following: NWI wetland, stream, wet photo-signature, not effectively drained, tidal connection, wet symbol on topo map or soil survey, stream gauge data, or previous onsite knowledge
Wetland	Beaver or other impoundment + any soil + present for 5 years
Artificial wetland	Pond on nonhydric soil or PC that was not abandoned prior to pond construction
Farmed wetland	Cropland + depressional position + hydric soil (with ponding for long duration or more) + wet signature + not abandoned
Farmed wetland (pocosin)	Cropland in southeastern Virginia + histosol (or mineral with histic epipedon) + wet signature + not abandoned
Farmed wet pasture	Pasture or hayland + hydric soils or wet signature or NWI wetland
Possible wetland	Woods + hydric soil (needs onsite evaluation)
Possible wetland	NWI wetland only
Prior converted (PC) (wetland)	Cropland + wet signature (due to saturation) + not abandoned
Prior Converted (PC) (former wetland)	Cropland + hydric soil + no wet signature + not abandoned
Converted wetland	Any wetland converted for crop production between December 23, 1985 and November 28, 1990
Nonwetland	Cropland, hayland, or pasture + no hydric soil + no wet signature + no NWI wetland
Nonwetland	Natural vegetation + no hydric soil + no wet signature + no NWI wetland

Note: Reference to NWI wetland is a wetland appearing on the existing NWI map and to cropland should be for production of commodity crops such as corn, soybeans, barley, tomatoes, and melons.

Source: Mapping Conventions and Offsite Wetland Determination/Inventory Procedures for Virginia. NRCS, August 1994.

correlation between office information and field conditions. In general the conventions use aerial photographs, FSA compliance slides, National Wetlands Inventory maps, and soil surveys to make wetland determinations (see Table 6.2 for examples).

The FSA method relies heavily on existing information and air photointerpretation. Consequently, the accuracy of the determinations are based on the quality of the data, the ability of the individual to interpret the data (e.g., aerial photointerpretation skills), and the soundness of the state mapping conventions. A major drawback is the aerial photography used in the process. The photos are acquired to serve a number of purposes besides wetland detection and are typically captured during the summer. This is not an ideal time to detect most wetlands in most places (see Chapter 10). The seasonal dryness coupled with the leaf-on condition of trees and shrubs makes it extremely difficult (near impossible) to identify the drier-end wetlands and all but the wetter forested and shrub wetlands. The FSA method should be acceptable for locating the wetter wetlands, but will either overestimate or underestimate the extent of forested and shrub wetlands depending on state conventions and staff interpretations.

FEDERAL INTERAGENCY MANUAL

In 1988, the four federal agencies with principal responsibility for regulating, managing, and conserving wetlands in the U.S. joined together to develop an interagency manual. The Federal Interagency Committee for Wetland Delineation (FICWD) was composed of members from the

Corps, EPA, NRCS, and the U.S. Fish and Wildlife Service (FWS). This committee prepared the *Federal Manual for Identifying and Delineating Jurisdictional Wetlands* (FICWD, 1989) by combining existing agency methods into a single technically based document. The document was built on consensus and each agency agreed to every statement in the manual prior to its release in January 1989. Since the document was designed as a technical publication (despite its final title*), each agency had to determine how it would use it for their programs. On January 19, 1989, the Corps and EPA officially adopted it for determining the extent of wetlands regulated under the CWA (Page and Hanmer, 1989). As mentioned earlier, this action mandated that all Corps districts use the same set of technically based protocols to identify and delineate wetlands. The manual was adopted by numerous states for use in administering their own regulatory programs. The nation was on its way toward getting a unified concept of wetland implemented for federal and state regulations. Subsequent political problems stemming from the increased scope of federal regulation among other things, however, forced the Corps to abandon use of this manual and adopt its own 1987 manual (see earlier comments). Some states continue to use the federal interagency manual (e.g., New Jersey and Vermont). It also is being used for nonregulatory purposes, such as to identify the success of wetland creation and restoration projects (Atkinson et al., 1993) and for riparian habitat managment (McKee et al., 1996).

The federal interagency manual adopted a three-criteria approach, similar to that used in the Corps manual, but allowed broader use of various indicators based on the strength of certain indicators (as in the EPA manual). Three criteria must be verified to identify a wetland — hydrophytic vegetation, hydric soils, and wetland hydrology. The wetland hydrology criterion is based on the hydrology required to form hydric soils — saturation, flooding, or ponding for 1 week or more during the growing season in most years (USDA Soil Conservation Service, 1987). While all three criteria must be satisfied to identify a wetland, the manual had more indicators, making use of most indicators previously used by federal agencies (Table 6.1). Moreover, the manual recognized that certain wetlands may be lacking hydrology indicators and allowed their identification on the basis of a community dominated by FAC and wetter species (including FAC–) growing on an undrained hydric soil (no significant drainage observed or suspected to occur). It offered protocols for handling disturbed sites and problem area wetlands incorporating the best judgment of federal agency wetland scientists in a fairly explicit set of procedures. Professional judgment was limited to the most difficult delineation situations.

Once wetland and nonwetland communities are identified, wetland delineation typically requires examination of soils except where topographic changes are abrupt and vegetation changes from wetland to upland are obvious. The objective of the soil assessment is to locate the point on the landscape where the water table is at or near the surface long enough to cause reduced (anaerobic) soil conditions. Such conditions are usually reflected by hydric soil properties especially during the drier time of the year (e.g., summer in most of the conterminous U.S.).

Like the EPA manual, the federal interagency manual overcomes many of the shortcomings of the Corps manual. It produced a technically accurate delineation consistent with the EPA manual and the FWS's concept of vegetated wetlands. The main technical problem with the federal interagency manual is that it is now somewhat dated (e.g., doesn't make use of the 1996 list of hydric soil field indicators and the latest technical criteria for hydric soils) and still utilizes the concept of growing season for identifying the critical period of wetland hydrology.

---

* The original title of the document *Federal Manual for Identifying and Delineating Vegetated Wetlands* was changed at the request of the Corps members on the FICWD prior to final publication. Substitution of "vegetated" with "jurisdictional" was intended to reflect the fact that wetlands identified using this manual were potentially subject to some form of government jurisdiction through regulations or policies. Despite the fact that the manual was a technical document that could serve regulatory and nonregulatory purposes (e.g., verification of wetlands for inventories or for assessing success/failure of wetland restoration/creation projects), the title gave the distinct impression that its sole purpose was for regulatory use.

PRIMARY INDICATORS METHOD

Tiner (1993) proposed the primary indicators method as a rapid assessment method for wetland identification and delineation in areas without significant hydrologic modification (i.e., drainage). This approach recognizes that unique plant communities and/or soil types have formed in wetlands due to varied hydrologic regimes, climatic conditions, soil formation processes, and geomorphologic settings across the country. Within similar geographic areas, wetlands have developed characteristics different than adjacent uplands (nonwetlands) due to the presence of water in or on top of the soil for prolonged periods during the year. The visible expression of this wetness may be reflected by the plant community or in the underlying soil properties. Consequently, every wetland in its natural undrained condition should possess at least one distinctive feature that distinguishes it from the adjacent upland.

This approach is not really new, but is an outgrowth of traditional methods used to recognize wetlands, including the Fish and Wildlife Service's wetland classification system (Cowardin et al., 1979). It provides a quick means of assessing the presence or absence of wetlands by looking for unique wetland characteristics.

A "primary indicator" is a single vegetation characteristic or soil property that can be reliably used to indicate the presence of wetland (Table 6.3). Since each primary indicator is decision-oriented, it does not have to be used in combination with other indicators. The list includes both vegetation and soil indicators that verify the presence of wetland in the absence of significant signs of drainage. The boundary of a wetland is simply identified by the lack of any primary indicators.

The "primary indicator approach" intentionally does not include observations of water or indirect evidence of water-carried debris, water-stained leaves, or other signs of hydrology. These ephemeral signs indicate that an event is happening or has happened, but most of these features reveal little about the duration and frequency of this event. These signs may at times be observed in nonwetlands subject to rare flooding or ponding.

The only disturbance that is relevant to making a wetland determination following PRIMET is drainage (e.g., by ditches, tile drains, groundwater withdrawals, or diverted flows) as such activity may negate the interpretative values of vegetation and soils. Other disturbances are of less concern. For example, if the vegetation has been removed and the hydrology of an area has not been disturbed, the soil indicators remain valid wetland indicators. If both vegetation and soils are removed, the area's hydrology should be considered significantly altered and should warrant further assessment. Areas of extensive ditching and tile drainage pose special problems for any delineation method (see Chapter 7).

The method has advantages over other methods for sites that are not significantly hydrologically altered (e.g., drained). Kraus (1993) noted that this is the approach that most experienced wetland delineators already use to help identify the boundary under these circumstances and then they simply complete the current regulatory field forms with data on vegetation, soils, and hydrology signs to satisfy current federal regulatory procedures. The NRC (1995) found PRIMET to be conceptually sound and recommended that the federal government seriously consider this type of approach when it develops a new federal wetland delineation manual. New York has adopted most of PRIMET's vegetation and soil indicators in its freshwater wetlands manual (Browne et al., 1996), while Rhode Island also has adopted some key elements of PRIMET in its attempt to validate the presence of hydrophytic vegetation (see discussion below). The main drawback of PRIMET is that it does not provide explicit instructions on how to determine when a site is significantly hydrologically modified.

TABLE 6.3
Recommended List of Primary Indicators of Wetlands for PRIMET with Minor Revisions

Vegetation Indicators of Wetland

V1. OBL species comprise more than 50% of the abundant species of the plant community. (An abundant species is a plant species with 20% or more areal cover in the plant community.)

V2. OBL and FACW species comprise more than 50% of the abundant species of the plant community.

V3. OBL perennial species collectively represent at least 10% areal cover in the plant community and are evenly distributed throughout the community and not restricted to depressional microsites.

V4. One abundant plant species in the community has one or more of the following morphological adaptations: pneumatophores (knees), prop roots, hypertrophied lenticels, buttressed stems or trunks, and floating leaves. (Note: Some of these features may be of limited value in tropical U.S., e.g., Hawaii.)

V5. Surface encrustations of algae, usually blue-green algae, are materially present. (Note: This is particularly useful indicator of drier wetlands in arid and semiarid regions.)

V6. The presence of significant patches of peat mosses (*Sphagnum* spp.) along the Gulf and Atlantic Coastal Plain. (Note: This may be useful elsewhere in the temperate zone.)

V7. The presence of a dominant groundcover of peat mosses (*Sphagnum* spp.) in boreal and subarctic regions. (Note: Some species may not be wetland indicators; check local authorities.)

Soil Indicators of Wetlands

S1. Organic soils (except Folists) present.

S2. Histic epipedon present.

S3. Sulfidic material (hydrogen sulfide, odor of "rotten eggs") present within 12 inches of the soil surface.

S4. Gleyed[a] (low chroma) horizon or dominant ped faces (chroma 2 or less with mottles or chroma 1 or less with or without mottles) present immediately below the surface layer (A- or E-horizons) and within 18 in. of the soil surface.

S5. Nonsandy soils with a low chroma matrix (chroma of 2 or less) within 18 in. of the soil surface and one of the following present within 12 in. of the surface:
  – Iron and manganese concretions or nodules
  – Distinct or prominent oxidized rhizospheres along several living roots
  – Low chroma mottles.

S6. Sandy soils with one of the following present:
  – Thin surface layer (1 in. or greater) of peat or muck where a leaf litter surface mat is present
  – Surface layer of peat or muck of any thickness where a leaf litter surface mat is absent
  – A Surface layer (A-horizon) having a low chroma matrix (chroma 1 or less and value of 3 or less) greater than 4 in. thick
  – Vertical organic streaking or blotchiness within 12 in. of the surface
  – Easily recognized (distinct or prominent) high chroma mottles occupy at least 2% of the low chroma subsoil matrix within 12 in. of the surface
  – Organic concretions within 12 in. of the surface
  – Easily recognized (distinct or prominent) oxidized rhizospheres along living roots within 12 in. of the surface
  – A cemented layer (ortstein) within 12 in. of the soil surface

S7. Native prairie soils with a low chroma matrix (chroma of 2 or less) within 18 in. of the soils surface *and* one of the following present: (The native prairie region extends northward from Texas to the Dakotas and adjacent Canada.)
  – Thin surface layer (at least ¼ in. thick) of peat or muck
  – Accumulation of iron (high chroma mottles, especially oxidized rhizospheres) within 12 in. of the surface
  – Iron and manganese concretions within the surface layer (A-horizon, mollic epipedon)
  – Low chroma (gray-colored) matrix or mottles present immediately below the surface layer (A-horizon, mollic epipedon) and the crushed color is chroma 2 or less

TABLE 6.3 (continued)
Recommended List of Primary Indicators of Wetlands for PRIMET with Minor Revisions

S8.   Remains of aquatic invertebrates present within 12 in. of the soil surface in nontidal pothole-like depressions.

S9.   Other regionally applicable, field-verifiable soil properties associated with prolonged seasonal high water tables (e.g., Hurt et al., 1998).

[a] Gleyed colors are low chroma colors (chroma of 2 or less in aggregated soils and chroma 1 or less in soils not aggregated, plus hues bluer than 10Y) formed by excessive soil wetness; other non-gleyed low chroma soils may occur due to (1) dark-colored materials (e.g., granite and phyllites), (2) human introduction of organic materials (e.g., manure) to improve soil fertility, and (3) podzolization (natural soil leaching process in acid woodlands where a light-colored, often grayish, E-horizon or eluvial-horizon develops below the A-horizon; these uniform light gray colors are not due to wetness).

Note: The presence of any of these characteristics in an area that has not been significantly drained typically indicates wetland. The upper limit of wetland is determined by the point at which none of these indicators are observed. Exceptions may occur as they do with any method.

Source: Tiner, R.W. 1993. The primary indicators method — a practical approach to wetland recognition and delineation in the United States. *Wetlands,* 13: 50-64.

Connecticut Method

In 1972, Connecticut passed legislation to protect inland wetlands. Rather than emphasize vegetation like other states were doing at that time, Connecticut chose to use soils. The reasons for this are, at least, two-fold: (1) soil mapping was nearing completion for the state and would serve as a good tool for local conservation commissions for implementing the law, and (2) the state recognized that vegetation alone would not be adequate to identify many of the state's wetlands, for many common species like red maple (*Acer rubrum*), perhaps the most widespread and dominant inland wetland plant in the state, also grew on uplands.

For inland wetlands, the state's definition emphasizes the presence of poorly drained, very poorly drained, alluvial, and floodplain soils (1972, 1987 CT General Statutes Sections 22a-36 to 45). A list of these soils has been compiled from the National Cooperative Soils Survey and is periodically updated. Wetlands are identified and delineated by these soils as determined by certified soil scientists. Soils must be classified to the series level in most cases to verify the presence of a listed "wetland" soil. Standard soil mapping techniques are used to determine the limits of such soils that represent the boundaries of regulated inland wetlands.

Florida Manual

Florida developed a wetland delineation manual to implement Chapter 62-340, Florida Administrative Code, Delineation of the Landward Extent of Wetlands and Surface Waters (Florida Department of Environmental Protection et al., 1995). Wetlands are defined by the federal definition, but additional language is included for clarification and to exclude certain plant communities that the state is not interested in regulating.

"Those areas that are inundated or saturated by surface water or ground water at a frequency and a duration sufficient to support, and under normal circumstances do support, a prevalence of vegetation typically adapted for life in saturated soils. Soils present in wetlands generally are classified as hydric or alluvial, or possess characteristics that are associated with reducing soil conditions. The prevalent vegetation in wetlands generally consists of facultative or obligate hydrophytic macrophytes that are typically adapted to areas having soil conditions described above. These species, due to morphological, physiological, or reproductive adaptations, have the ability to grow, reproduce, or persist in aquatic environments or anaerobic soil conditions. Florida wetlands generally include swamps, marshes, bayheads, bogs, cypress domes and strands, sloughs, wet prairies, riverine swamps and marshes, hydric seepage slopes, tidal marshes, mangrove swamps, and other similar areas. Florida wetlands do not include longleaf or slash pine flatwoods with an understory dominated by saw palmetto."

Other areas excluded from the definition are wastewater treatment areas (except where wetlands are used for this purpose), stormwater treatment areas <0.5 acres, and mosquito control areas (excavated/impounded) that were upland originally. In order to qualify as a regulated wetland, an area must meet one of the following requirements: (1) be inundated ≥7 consecutive days during normal hydrologic conditions, or (2) be saturated ≥20 consecutive days during normal hydrologic conditions.

Wetland can be recognized by one or more indicators. The most obvious wetlands can simply be identified by "applying the wetland definition to an area." This would readily identify marshes, wet prairies, tidal marshes, swamps, and a host of other obvious wetland types. Other wetlands and perhaps the upper edges of the obvious types may require using a combination of plant, soil, and other hydrologic indicators for identification and wetland boundary delineation.

OBL and FACW plants are recognized as reliable wetland indicators. Exotics, vines, and aquatic plants are not included in the assessment. The latter species would be found in obvious wetlands so they are not relevant for verification of more difficult-to-identify wetlands. FAC species are not used in evaluating dominant vegetation or used as a hydrologic indicator. Dominant species are determined for the uppermost stratum of the plant community (canopy, subcanopy, and ground cover); a stratum must have ≥ 10% areal cover to be evaluated. If the uppermost stratum is dominated by FAC species, then the next stratum must be assessed, for a FAC-dominated stratum is not recognized as a valid stratum for indicating wetland or upland. The uppermost stratum also is not used when in conflict with hydrologic conditions. In these situations, "the remaining stratum most indicative of the true nature of the site should be used to make the determination;" this is the "appropriate vegetative stratum" for evaluation. The Florida manual refers to soil characteristics listed in *Soil and Water Relationships of Florida's Ecological Communities* (Florida Soil Conservation Staff, 1992) as hydric soil indicators. The list includes muck, mucky texture, gley colors, and sulfidic odor plus "indicators of saturation near the surface" such as dark surface, organic concretions, oxidized rhizospheres, polychromatic matrix, stratified layers, Fe/Mn concretions*, distinct/prominent mottles*, and marl* (* = for loamy and clayey soils only). Delineators are expected to use "reasonable scientific judgment" in interpreting various signs of wetland hydrology. The list of potential indicators includes algal mats, aquatic mosses/liverworts, aquatic plants, aufwuchs, drift lines/rafted debris, elevated lichen lines, aquatic fauna, hydrologic data, morphological plant adaptations, secondary flow channels, sediment deposition, vegetated tussocks/hummocks, and water marks (stains).

Wetland identification involves a two-step procedure. The first step identifies obvious wetlands, while the second step applies a set of tests to verify wetlands that are more difficult to recognize. Separate analysis should be done for altered sites. The steps are outlined below.

Step 1 — Identify obvious wetlands with abrupt boundaries simply by applying the regulatory definition.

Step 2 — For other areas, apply the following technical procedures.

A Test: Area is wetland where OBL species >UPL species and where hydric soil indicators or riverwash or hydrologic indicators are present.

B Test: Area is wetland where OBL and FACW species have ≥80% coverage and where hydric soil indicators or riverwash or hydrologic indicators are present.

C Test: Wetland is present when one of the following conditions is satisfied: (1) field verification of Argiaquolls, Hydraquents, Humaquepts, Sulfaquents, Umbraqualfs, and
Umbraquults, (2) saline sands in high marsh, or (3) frequently flooded and depressional map units (boundaries verified in field). This test cannot be used in pine
flatwoods.

D Test: Wetland is verified by the presence of hydric soil and one or more hydrologic
indicators. Vegetation may be used in this test in using reasonable scientific judgment.

For altered sites, delineators should answer to basic questions. What was the ecological community prior to alteration? And is the effect of the alteration on hydrology permanent or temporary?
The Florida manual does recognize the potential use of hydrologic models for evaluating whether
the current hydrology satisfies the wetland definition, but states that the use of these models may
be permitted only upon agreement by the regulator. Yet, if the regulator rejects such usage, he or
she must give a valid reason.

From the outset, it is clear that the Florida manual is intended to identify wetlands in accordance
with the regulatory definition. This definition generally excludes longleaf or slash pine flatwoods
with an understory of saw palmetto (*Serenoa repens*). The manual also recommends the use of
reasonable scientific judgment (rather than professional judgment) which is defined as "the ability
to collect and analyze information using technical knowledge, and personal skills and experience
to serve as a basis for decision making." It emphasizes that such judgment is very important in
ecotonal seasonally wet or occasionally wet areas (not wetlands according to the statutes), wetland
communities dominated by nonlisted species (e.g., hydric hammocks), altered areas with relict
wetland vegetation and/or hydric soils, and wetland ecotones (especially throughout south Florida).

Massachusetts Manual

As the first state to regulate wetlands, Massachusetts has traditionally relied on vegetation to identify
wetlands. Salt marshes are essentially identified by the presence of smooth cordgrass (*Spartina
alterniflora*) and salt hay grass (*S. patens*). Inland wetlands bordering various waterbodies, e.g.,
streams, rivers, lakes, and ponds, are regulated as bordering vegetated wetlands. They include
several types: wet meadows, marshes, swamps, and bogs. The Massachusetts Wetlands Protection
Act (M.G.L. Chapter 131, Sec. 40) lists some examples of wetland plants that typify such areas,
but does not include explicit rules on how to use plants to identify and delineate wetlands. In 1988,
the state published *Guide to Inland Vegetated Wetlands in Massachusetts* (DiPinto and McCollum,
1988) which contained a more extensive list of inland wetland indicator plants with some general
guidance on how to identify wetlands (e.g., wetlands were typically identified where 50% or more
of the species were "wetland indicators," while soils could be used for disturbed sites and for drier
wetlands lacking typical wetland plants). In 1995, Massachusetts promulgated regulations (310
CMR 10.55) that provided a more scientific-based delineation procedure for bordering vegetated
wetlands (effective on June 30, 1995). A wetland delineation manual was published to present
consistent procedure for using both vegetation and other indicators (including soils) to identify and
delineate regulated wetlands (Jackson, 1995).

In many respects, the current wetland identification procedures have adapted elements of both
the federal interagency and Corps manuals, while maintaining the focus on vegetation. The additional indicators are used to verify wetland hydrology when necessary. Typical wetland indicator
plants are those referenced in the law (which includes peat moss, *Sphagnum* spp.), plus species
listed as OBL, FACW+, FACW, FACW–, FAC+, and FAC (as classified on the federal list, Reed,
1988) and individual plants with morphological adaptations to saturated or inundated conditions.
The manual does note four cautions regarding use of the above rules: FAC– and drier species are
not uncommon in wetlands, listing white pine (*Pinus strobus*), pitch pine (*P. rigida*), and American
beech (*Fagus grandifolia*) as examples; extended droughts may change herbaceous vegetation;

winter and early spring identification will not include many ground cover species; and altered sites (e.g., violations, lawns, golf courses, and cultivated areas) where hydric soils and other indicators are more useful identifiers of disturbed wetlands. Dominant species are determined by applying the 50/20 rule of the federal interagency manual (FICWD, 1989). Given the rolling landscape of most of Massachusetts, the state considers vegetation as an accurate indicator of wetland hydrology in most cases and recommends using wetland plant indicators for wetland identification. Typically a community dominated by FACW– and wetter species is automatically considered wetland. Soils and signs of hydrology are used to identify wetlands in altered areas or where the wetland plant community is not dominated by these species. These indicators include hydric soil properties (list of examples is presented, and some problem soils are discussed), and signs of hydrology (similar to the federal manuals including plant morphological adaptations, but also including fingernail clam and aquatic snail shells and caddisfly cases). Vegetation is used for most boundary delineations, except when boundaries are diffuse, where the community is dominated by FAC+ and/or FAC species, or where the area's vegetation or hydrology is altered. In these situations, the wetland boundary is typically established between hydric and nonhydric soils.

NEW YORK MANUAL

Since passage of the Freshwater Wetlands Act in 1975, New York has been regulating activities in these wetlands. Freshwater wetlands include vegetated types and submerged lands. With the exception of waterbodies, wetlands support vegetation that "depend on permanent or seasonal flooding or sufficiently water-logged soils to give them a competitive advantage over other [vegetation]" (Browne et al., 1996). The law required that the Department of Environmental Conservation (DEC) prepare a set of maps showing the approximate boundaries of regulated wetlands ($\geq$12.4 acres and smaller wetlands either of local importance or in the Adirondack Park). Actual wetland boundaries for project sites need to be established on the ground.

The DEC has recently published a wetland delineation manual to standardize wetland identification and delineation for regulatory purposes (Browne et al., 1996). The manual has adopted perhaps the best elements of existing federal and rapid assessment wetland delineation methods in an attempt to be efficient and effective. Hydrophytic vegetation continues to be the principal indicator of regulated freshwater wetlands, as many wetlands are easily recognized by their vegetation. Signs of hydrology and the presence of hydric soils are used as supportive evidence when needed, especially for delineating the wetland boundary in low-gradient landscapes.

After identifying dominant plant species (e.g., those with $\geq$20% areal cover), the presence of hydrophytic vegetation is determined by one of the following conditions: (1) FACW or wetter species equal more than 50% of the dominants and no FACU or UPL species are dominant, (2) OBL perennial species collectively represent at least 10% areal cover and are evenly distributed throughout the community and not restricted to depressional microsites, (3) one or more dominant species has certain specified morphological adaptations indicative of prolonged flooding and saturation, and (4) the presence of unbroken expanses of peat mosses and other regionally applicable bryophytes over persistently saturated soil. These indicators are adaptations of PRIMET (Tiner, 1993). If one of the above indicators are present, the area is a wetland. If not, then the area requires further assessment.

When more than 50% of the dominants are FAC or wetter (excluding FAC–), then hydrology and soils may require examination, whereas if dry-site species predominate, the area is typically upland (unless it is a problem area wetland, e.g., hemlock or white pine swamp). For the former situation, wetland can be verified by the presence of field indicators of wetland hydrology (by one primary indicator — visual observation of inundation or soil saturation, water marks, drift lines, water-borne sediment deposits, and wetland drainage patterns, or by two secondary indicators — oxidized root zones, water-stained leaved, surface-scoured areas, and dead nonwetland vegetation), or by the occurrence of hydric soils in areas with no indication of recent significant hydrologic modification. Hydric soil field indicators are adapted from PRIMET (Tiner, 1993). If the area has signs of significant drainage, disturbed area procedures should be used for site assessment.

Wetland boundaries are established by vegetation changes where the topography changes abruptly. In other areas, the presence of hydric soils or other wetland hydrology indicators are used to identify the extent of hydrophytic vegetation, i.e., wetland.

RHODE ISLAND METHOD

Rhode Island has been regulating freshwater wetlands since 1971. Regulated wetlands are defined by different types with characteristic vegetation — swamps, bogs, and marshes. Like Massachusetts and New York, wetland identification was traditionally based on certain plant communities, but Rhode Island's interpretation of the 50% rule required that more than 50% of the plants be hydrophytic, under normal circumstances, for identifying the plant community as a wetland. In 1994, the state clarified its procedures to identify hydrophytes to make it consistent with the state of the science (Rhode Island Department of Environmental Management, 1994). Other wetland indicators are used to help verify the presence of a predominance of hydrophytes as in PRIMET (Tiner, 1993).

Wetlands are identified by a plant community having more than 50% composition by hydrophytic species. Hydrophytic vegetation is determined by the following indicators: (1) plants listed in the act, (2) plants with an indicator status of OBL on the most recent edition of federal list of wetland plants (Reed 1988) for Rhode Island, and (3) FACW, FAC, and/or FACU species where such plants have other clear hydrologic indicators of wetlands. One of the following indicators provides sufficient proof of hydrophytes: hydric soil morphologic properties, observed soil saturation within 12 in. of the surface, sulfidic materials (hydrogen sulfide odor) within 12 in., and patches of peat moss (*Sphagnum* spp.). In the absence of any of the above, two of the following also can verify hydrophytic vegetation: distinct water marks, mound and pool topography, soil evidence of recent and/or periodic flooding, visual observation of surface water, dark or water-stained leaves on the ground surface, drift or wrack lines of water-borne materials, wetland drainage features (e.g., scoured channels), morphological plant adaptations, and distinct or prominent pore linings (oxidized rhizospheres) along living roots within 12 in. of the surface. This approach is somewhat similar to the PRIMET approach, but includes additional wetland indicators that are not unique to wetlands requiring that two of them be used to establish the presence of hydrophytic vegetation.

As with most other methods, vegetation is used to determine the wetland boundary in areas where topographic changes are abrupt. In areas of more gradual relief and where vegetation is more transitional in nature, the wetland edge is determined where hydrologic indicators are no longer present (e.g., limit of hydric soil properties or where two of the lesser hydrologic indicators cannot be found). The edge of the state regulated wetland is located 50 ft from the landward edge of any bog, swamp, marsh, or pond.

WISCONSIN METHOD

Wisconsin's wetland delineation procedures involve the use of maps and aerial photographs combined with field investigation. State and local procedures closely follow those in the 1989 interagency manual (Wisconsin Coastal Management Program, 1995). Wetlands may be the same as those identified using the Corps manual, but also may be different from both the Corps and the 1989 manual. For example, Wisconsin emphasizes the presence of wetland plants and, therefore, will consider an area with somewhat poorly drained soils as wetlands even when the hydrology is altered provided it is capable of supporting wetland plants. Explicit guidance is not given on how to do this, but consultation with a wetland delineation expert is recommended. The state prepared *Basic Guide to Wisconsin's Wetlands and Their Boundaries* to help zoning staff and other government officials understand and identify wetland ecosystems (Wisconsin Coastal Management

Program, 1995). The guide offers general procedures for wetland delineation, but not explicit stepwise methods for site evaluation. Yet as stated above, the methods are similar to the procedures referenced in the 1989 manual. Wisconsin generally uses a three-parameter approach: (1) hydrophytic vegetation (e.g., do more than 50% of the dominant species have a wetland indicator status of FAC, FACW, and/or OBL, or has vegetation been altered significantly?), (2) hydric soils, and (3) wetland hydrology (i.e., is direct evidence of wetland hydrology present or in your judgment would water be above or near the surface long enough during the growing season to affect the plants?). If the plant community is not a clearcut wetland or upland, then consider possible effects of alterations or if the site is a "problem" wetland. The guidebook intentionally does not provide further direction, yet presumably one could refer to the federal interagency manual for evaluation procedures.

Once wetland and upland plant communities are identified, the boundary is drawn between them by considering changes in plant communities (where FACU and UPL species dominate) or soil properties (where nonhydric soil indicators such as higher chroma subsoils occur) and looking for breaks in topography. During this process, the investigator is urged to "form an idea of the source of hydrology" and use this to help explain differences in vegetation and soils. It seems that soils are used to validate observed vegetation differences as soil properties are expected to change abruptly under ideal circumstances.

QUEBEC METHOD

To aid in identifying the extent of areas subject to recent aquatic habitat protection laws, Quebec developed procedures for identifying wetlands at or below the 2-year high water mark along waterbodies (Gauthier, 1997). The method uses vegetation to identify areas along rivers, lakes, and the seacoast that are frequently flooded (once every 2 years). The limit of this wetland area is determined by (1) the upper limit of certain indicator plants — obligate hydrophytes and facultative hydrophytes, equivalent to OBL and FACW categories of the American system (Reed, 1988); (2) the lower limit of lichens and the upper limit of mosses (on the south side of the trunks); or (3) other indicators (e.g., wear marks on trees, sedimentation on trunks, debris lines, signs of erosion, bare soils, and "lower limit of underbrush"). Examples of wetland plants in fresh water include trees (e.g., *Acer saccharinum, Ulmus americana, Populus deltoides, P. balsamifera, Fraxinus pennsylvanica, F. nigra, Salix alba, S. fragilis,* and *Thuja occidentalis*), shrubs (e.g., *Alnus rugosa, Myrica gale, Spiraea latifolia, Chamaedaphne calyculata, Cornus stolonifera, Salix interior,* and *Vitis riparia*), and herbs (e.g., *Calamagrostis canadensis, Onoclea sensibilis, Phalaris arundinacea, Potentilla palustris, Cicuta bulbifera, Impatiens capensis, Laportea canadensis, Matteucia struptiopteris, Osmunda regalis, Caltha palustris,* and *Lythrum salicaria*). The wetland boundary is established by the point at which terrestrial species exceed wetland species.

WETLAND DELINEATION METHODS

Most delineation manuals use vegetation sampling techniques to evaluate project sites. A variety of plant communities are identified and then are assessed for specified wetland indicators. The manuals may either locate plant communities by a simple site walk-through or by using transects to locate different communities along topographic gradients. The simpler methods use the first approach, with transects reserved for more complicated situations. Methods used by the federal government during the past decade are summarized below. Most state manuals follow similar procedures.

CORPS MANUAL

The Corps manual offers routine and comprehensive methods. Guidance from Williams (1992) clarifies the interpretation of the manual and permits alternative approaches based on federal wetland

TABLE 6.4
Corps Manual Steps for Making Routine Wetland Delineation for Areas Generally Five Acres or Less (or Areas that are Larger but Homogeneous in Vegetation)

1. Locate project area in the field and walk the site. In doing so, (a) determine the number and location of plant communities on the site; (b) determine if there is any evidence of natural or human alteration to vegetation, soil, and/or hydrology that would make wetland determinations difficult (for these areas, special procedures for "Atypical Situations" should be followed); and (c) determine whether seasonal changes in hydrology or temperatures may pose a serious problem for making a wetland determination according to the three-parameter method (follow guidance in "Problem Areas" section).

2. Evaluate each plant community for positive indicators of hydrophytic vegetation, wetland hydrology, and hydric soils, respectively. Pick one community to begin your assessment. Select one or more observation points (as necessary) that typify the community and mark the locations on a map.

3. Characterize the plant community. Determine dominant species in each stratum (two options: 4-strata or 5-strata maximum). Record the indicator status of each dominant and determine whether hydrophytic vegetation is present (more than 50% of the dominants are OBL, FACW, and FAC, excluding FAC−). If it is not, the area is usually nonwetland, unless it is a problematic wetland for identification following the Corps manual (contact local Corps district for guidance).

4. Evaluate the site's hydrology by looking for positive indicators of wetland hydrology. Any community having a positive indicator of wetland hydrology is potentially a wetland. If no indicators are present (including those mentioned in recent Corps guidance memoranda), the area is not a regulated wetland.

5. If all dominants are OBL or, OBL or FACW and the wetland boundary is abrupt, soils do not need to be examined — hydric soils are assumed to be present and the community is wetland. Other areas require examination of the soils.

6. Determine whether hydric soils are present. Dig a hole about 1 ft wide and 2 ft deep and look for positive indicators of hydric soil immediately below the A-horizon (usually within 1 ft of the soil surface). If such indicators are present, the community is wetland. If not, it is typically a nonwetland, unless it is a problematic wetland. Remember any area presently or normally having wetland indicators of all three parameters is wetland and areas lacking one are typically nonwetlands.

7. Delineate the boundary between wetland and nonwetland plant communities. The Corps manual says to draw it on a map, but in practice, the boundary is marked on the ground by a series of numbered flags (using variously colored flagging tape). The line may follow a contour that marks the limit of where positive indicators of all three parameters are found (in the most strict interpretation) or where the hydric soil indicators end (more scientific interpretation). A survey crew later will plot all wetland boundary flag locations on a detailed map.

Note: Steps have been taken to simplify this table.

Source: Environmental Laboratory. 1987. Corps of Engineers Wetlands Delineation Manual. U.S. Army Engineer Waterways Experiment Station, Vicksburg, MS.

delineation experiences between 1987 and 1992.* The major difference between the methods is in the effort of vegetation analysis, with more measurement in the latter. Routine methods are probably used in 95% or more of the wetland delineation projects. They are outlined in Tables 6.4 and 6.5. The comprehensive method is generally reserved for cases where litigation may be forthcoming or for more difficult vegetation communities. In general, the comprehensive steps follow the transect approach of the routine method, except that sampling is supposed to be done at fixed intervals based on the length of the transect rather than within recognizable plant communities along the transect, and vegetation sampling is more rigorous. If the transect is less than 1000 ft, the sampling interval is 100 ft; if length is 1000 to 5000 ft, interval is 100 to 500 ft; if length is 5000 to 10,000 ft, interval is 500 to 1000 ft; if transect length is greater, then sampling interval is 1000 ft. In practice,

* People performing wetland delineations must be familiar with this memorandum as it provides vital information on the correct interpretation of the manual and offers some more ecologically sound approaches for analyzing plant communities.

TABLE 6.5
Corps Manual Steps for Delineating Wetlands on Project Sites Larger than Five Acres (or Having Diverse Assemblages of Plant Communities)

1. Locate project area in the field and walk the site. In doing so, (a) determine the number and location of plant communities on the site; (b) determine if there is any evidence of natural or human alteration to vegetation, soil, and/or hydrology that would make wetland determinations difficult (for these areas, special procedures for "Atypical Situations" should be followed); and (c) determine whether seasonal changes in hydrology or temperatures may pose a serious problem for making a wetland determination according to the three-parameter method (follow guidance in "Problem Areas" section).

2. Establish a baseline for setting up transects. The baseline should be parallel to the major river course or perpendicular to the hydrologic gradient.

3. Determine the number and position of transects. If baseline is 1 mile long or less, locate at least 3 transects equal distance from one another. If baseline is 1 to 2 miles long, 3 to 5 transects are needed. If longer, more transects are required; transects should be no more than 0.5 mile apart. Since all communities must be included, it may be necessary to reorient a transect to ensure that all communities are on at least one transect (Figure 6.2).

4. Walk each transect, identify different plant communities along the transect as you go, and when you reach the end of the transect, turn around to return to baseline, and sample each plant community on your way back.

5. In each community, establish an observation point that best represents the plant community and evaluate each parameter in the following sequence: hydrophytic vegetation, hydric soils, and wetland hydrology.

6. Sample vegetation within a 30-ft radius plot for trees and vines and a 5-ft radius plot for herbs and saplings/shrubs. Plot size and shape may be adjusted to match the site conditions (e.g., ridges and swales or pits and mounds). Determine dominants in all strata (4- or 5-strata maximum). Record indicator status of all dominants and determine whether hydrophytic vegetation is present. If it is not, the area is usually nonwetland.

7. If all dominants are OBL or all are OBL and/or FACW (with at least one OBL dominant), hydric soils are assumed to be present. If not, then soils must be examined for hydric soil indicators usually within 1 ft from the soil surface. If such indicators are present, the community has hydric soils. If not, the area is typically nonwetland unless it is a problematic wetland for identification purposes.

8. Examine the community for positive indicators of wetland hydrology. If none are present, the community is nonwetland. If one or more indicators are found or known to normally occur, then the area is wetland.

9. Proceed to the next community along the transect and repeat steps 6 to 8, making a wetland or nonwetland determination for each community.

10. Delineate the boundary between any wetland and nonwetland along the transect.

11. Complete evaluation for all transects.

12. Connect wetland boundary lines between transects by following the contour that best reflects the wetland–nonwetland boundary and periodically evaluating all three parameters to insure that the boundary is accurately represented by this contour. Use flagging tape to mark the boundary for survey crews and others to find later.

Source: Environmental Laboratory. 1987. Corps of Engineers Wetlands Delineation Manual. U.S. Army Engineer Waterways Experiment Station, Vicksburg, MS.

however, investigators may sample at shorter intervals when a distinct plant community occurs between the prescribed sampling points.

Vegetation sampling is more quantitative than the routine methods. Trees are evaluated within a 30-ft radius plot, but the diameter at breast height (dbh) of each tree is measured and basal area (BA) calculated: $BA = \pi d^2/4$, where d = diameter at breast height. Saplings/shrubs and woody vines are sampled within a 10-ft radius plot. Height classes are established for determining dominant saplings/shrubs, while stems of woody vines are counted at the ground surface. Herbs may be evaluated in two ways — sampling one 3.28 ft × 3.28 ft plot located at the center of the fixed sampling point (identifying dominants by areal cover), or sampling within the 30-ft radius plot using multiple quadrats following the federal interagency manual procedures (identifying dominants using their mean areal cover estimates). The latter is recommended over the former for analyzing diverse or patchy herb cover (Williams, 1992). It also is possible to use the comprehensive methods

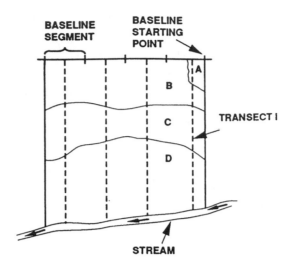

FIGURE 6.2   Establishing transects for analyzing a project site with four plant communities (A through D). Each plant community must be intersected by at least one transect. Generally transects begin at the midpoint of a baseline segment. In this case, transect 1 had to be shifted to the right to include plant community A. (Federal Interagency Committee for Wetland Delineation Manual, Washington, D.C., 1989.)

from the 1989 interagency manual or other ecologically based methods to assess dominants for comprehensive determinations (e.g., 50/20 rule and 5-strata approach).

Atypical situations procedures are included for use where positive indicators of one of the parameters are lacking due to recent human activities or natural events. Human actions include unauthorized activities (dredging, filling, removal of vegetation, or hydrologic alteration) and artificial wetlands purposefully or unintentionally created (e.g., impoundments, irrigation projects, and stream channel realignments). The manual emphasizes that it is not intended to extend federal jurisdiction to manmade wetlands that are exempted from CWA requirements by Corps regulations or policy. Natural disturbances may create new wetlands or alter existing wetlands. These actions include beaver, fire, avalanches, volcanic eruptions, and changing river courses. In considering the above, the Corps manual stresses that the approximate date of the alteration must be established to determine whether the event occurred prior to implementing Section 404 of the CWA.

The manual also includes procedures for handling "problem areas" — several types of wetlands or conditions where positive indicators of one or more parameters may not exist at certain times of the year. A list of representative examples is provided in the manual and includes wetlands on drumlins, seasonal wetlands (in arid and semiarid regions), prairie potholes, and vegetated flats. These and other examples are discussed in Chapter 7. Although presented as a list of examples, many users of the manual tend to view those listed as the only bonafide problem areas. Recent Corps headquarters guidance has affirmed that the list is a list of examples and that the situation may be applied to other wetlands (Williams, 1992), but it unfortunately did not present other examples. The procedures for wetland determinations in these situations, however, still emphasize that wetland indicators of all three parameters must normally be present during part of the growing season to be a wetland.

EPA MANUAL

The EPA manual has three basic approaches to wetland delineation — simple, detailed, and atypical and/or normally variable environmental conditions. For the latter, only general guidance is offered, relying on the delineator's professional judgment to make the decision. The simple method is recommended for sites up to 15 acres in size, noncontroversial projects, or for larger sites with

TABLE 6.6
EPA Simple Method for Delineating Wetlands

Step 1. Inspect the site and separate it into vegetation units.

Step 2. In each unit, develop species list by stratum (understory, woody vines, shrubs, saplings and trees), estimate the cover class of each (relative basal area for trees), and identify dominants (one or more of the top-ranked species cumulatively representing >50% of the stratum).

Step 3. Determine whether the unit has been hydrologically modified. If not and one or more OBL species are dominant, the unit is wetland. If not hydrologically altered and one or more UPL species are dominant, the unit is nonwetland. If FACW, FAC, and/or FACU species are the only dominants, then proceed to next step.

Step 4. Examine soils and hydrology. First verify presence or absence of hydric soils. If present, then look for hydrology indicators. The manual emphasizes that it is not necessary to directly demonstrate that wetland hydrology is present, only that the soil is saturated or inundated for a week or more during the growing season. If one or more OBL species dominate and one cannot show that the area is not significantly modified, then wetland can be verified by the presence of at least one hydrology indicator. Whereas if the vegetation unit is entirely dominated by FACW, FAC, and/or FACU species, then hydric soils and wetland hydrology indicators are needed to verify the existence of a wetland. (Note: This approach recognizes the significance of OBL species for identifying wetlands.)

Step 5. Make wetland determinations for all remaining vegetation units at the project site.

Step 6. Delineate the wetland–upland boundary. The manual suggests identifying simple boundaries based on vegetation (i.e., presence of OBL dominants) if possible, but recommends examination of soils where FACW, FAC, and/or FACU species predominate and where the boundary is gradual. In the latter case, a combination of soil properties, hydrology indicators, and vegetation are used to establish the boundary.

Note: The 15-step method is condensed to 6 steps for this table.

Source: Sipple, W.S. 1988. Wetland Identification and Delineation Manual. U.S. Environmental Protection Agency, Office of Wetlands Protection, Washington, D.C.

uniform conditions. Users are encouraged to employ this method for most projects and to use the detailed approach only where absolutely necessary. Table 6.6 presents the procedures used following the simple method.

The detailed juridictional approach is used for larger projects (>15 acres) or for controversial projects of any size where more documentation of site variables is required. This approach is designed to be undertaken by a team of specialists (minimum of 2). The main difference is in site analysis which requires employment of transect sampling procedure for evaluation of vegetation. Transects are established and vegetation units occurring along the transect evaluated. A 0.1 acre (0.04 ha) circular plot is established in each unit (for evaluating a linear plant community like a swale, a smaller plot is acceptable) and various strata are examined (understory, small plot sampling/minimum area; bryophytes, same; shrubs, cover classes; woody vines, percent cover; trees, relative basal area using Bitterlich variable plot technique and measuring diameter at breast height for all tallied individuals). The method applied the same rules regarding OBL dominants indicating wetland in the absence of hydrologic modification, etc. Soils must be examined for vegetation units dominated only by various FAC-type species. For verified hydric soil areas, hydrology indicators need to be observed to make a wetland determination. The EPA manual also has a brief section discussing atypical situations (e.g., altered sites) and normally variable environmental conditions that recommends using professial judgment to decide whether wetland indicators would normally be present during a significant part of the growing season.

FEDERAL INTERAGENCY MANUAL

The federal interagency manual combined various techniques employed by the Corps, EPA, NRCS, and the FWS into a single document. The most commonly used methods are summarized in

TABLE 6.7
Federal Interagency Manual Steps for Routine Plant Community Assessment Procedure

Step 1.   Identify plant communities on the project site.

Step 2.   For the each plant community, select one or more observation areas and visually assess the vegetation, identify dominant species (considering all strata), and determine whether the hydrophytic vegetation criterion is met. If met, the analysis continues; if not, the analysis stops and the area is nonwetland, unless it happens to be a problem area wetland.

Step 3.   Analyze soils if necessary and determine whether the hydric soil criterion is met. There are two situations where hydric soils can be assumed: (1) where all dominants are OBL and (2) all dominants are OBL and FACW *and* the wetland boundary is abrupt. In most cases, the soil is examined. If the hydric soil criterion is met, the analysis continues; if not, the analysis stops and the area is nonwetland (beware of problem soils that may not possess typical hydric soil indicators).

Step 4.   Determine whether there are appropriate signs of wetland hydrology to verify the wetland hydrology criterion. If one or more indicators are found, the area is determined to be wetland.

Step 5.   Repeat Steps 2 through 4 for all plant communities, so that each community is designated as wetland or nonwetland. After assessing all communities, go to Step 6.

Step 6.   Delineate the boundaries between wetlands and nonwetlands with flagging tape. The boundary is drawn where hydrophytic vegetation and hydric soils change to nonhydrophytic vegetation and nonhydric soils. If necessary, additional examination of soil is performed to refine the boundary. The line often follows a contour marking the upslope limit of hydric soil indicators (i.e., the limit of saturated soils indicative of seasonal high water tables and the extent to which hydrophytes occur; above this line the plants are no longer hydrophytes).

Source: Federal Interagency Committee for Wetland Delineation. 1989. Federal Manual for Identifying and Delineating Jurisdictional Wetlands. U.S. Army Corps of Engineers, EPA, FWS, USDA Soil Conservation Service, Washington, D.C.

Tables 6.7 and 6.8. Although not repeated in the procedure summaries, when using any of the methods it is neccessary to establish whether normal environmental conditions are present. Observations made at such times should find typical wetland indicators, but at other times, hydrophytic vegetation or hydrologic indicators may be absent due to seasonality (e.g., winter for vegetation, or late summer for hydrology) or to the year of observation (e.g., during long-term drought). For these latter situations, one is referred to problem area wetland procedures.

The routine methods include two approaches, hydric soil assessment procedure and plant community assessment procedure. The former is derived from field techniques employed by NRCS personnel for Swampbuster wetland determinations, while the latter originates from both the Corps and EPA manuals.

The *hydric soil assessment procedure* is, in large part, a soil scientist's view of how wetland delineation should be performed, with more focus on soils. Initially, the site's soils are generally assessed resulting in the site being divided into different soil types, separating nonhydric soil areas (nonwetlands) from potential hydric soil areas (possible wetlands). The former areas are designated as nonwetlands and the latter are more closely examined. The potential hydric soil areas are first scanned for obvious wetlands based on signs of flooding or surface saturation and then any remaining areas are examined for the presence of hydric soil (based on field indicators) and the presence of hydrophytic vegetation (by visually determining dominant species and estimating their areal cover). Where hydric soils are verified and the plant community has more coverage by OBL and FACW species than by FACU and UPL species, the area is designated as wetland. If not, then a more rigorous site assessment is required and users are advised to choose another method. This method was not as widely used as the other routine method.

The *plant community assessment procedure* reflects more of a plant ecologist's or botanist's view of wetland delineation with the project site separated into different plant communities for analysis. In theory, all plant communities at a project site should be evaluated regarding vegetation,

TABLE 6.8
Federal Interagency Manual's Quadrat Sampling Procedure of the Intermediate Methods

Step 1.   Walk the site and locating the different plant communities for evaluation.

Step 2.   Establish a baseline for setting up transects. The baseline should be perpendicular to the slope of the site; use a road if possible.

Step 3.   Locate the transects along the baseline. General guidelines are given: 3 transects if baseline length is less than 1 mile long (more if longer) and transects should be no farther than 0.5 mile apart. In practice, transects are usually closer depending on site conditions and size of parcel. Transects should run downslope which allows intersection with different plant communities along a hydrologic gradient. It is important for all plant communities to have at least one transect in them; they would not be evaluated otherwise (Figure 6.2).

Step 4.   Walk the entire length of the transect (beginning with the first transect) paying close attention to plant community changes and upon reaching the end of the transect (end of the subject property), select a location along the transect in the last plant community encountered for evaluation.

Step 5.   Analyze the vegetation and determine whether the hydrophytic vegetation criterion is met. Determine dominants by sampling vegetation within fixed plots: (1) use a 5-ft radius circular plots for herbs and (2) a 30-ft radius circular plot for trees, saplings, shrubs, and woody vines. Plot size and shape may be modified to suit site characteristics, but general area of sample should remain unchanged unless there is need for a larger plot to capture typical species (Note: This relates to minimum area concept in vegetation sampling; see Chapter 4). If the hydrophytic vegetation criterion is met, continue analysis, otherwise plant community is nonwetland (unless a problem area wetland; users need to know how to recognize such communities).

Step 6.   Analyze soils if necessary and determine whether the hydric soil criterion is met. There are two situation where hydric soils can be assumed: (1) where all dominants are OBL and (2) all dominants are OBL and FACW *and* the wetland boundary is abrupt. In most cases, the soil is examined. If the hydric soil criterion is met, the analysis continues; if not, the analysis stops and the area is nonwetland (beware of problem soils that may not possess typical hydric soil indicators).

Step 7.   Determine whether there are appropriate signs of wetland hydrology to verify the wetland hydrology criterion. If one or more indicators are found, this community on the transect is determined to be wetland.

Step 8.   Proceed to the next plant community along the transect, identify the representative sampling location, and repeat steps 4 through 7. Where wetland and nonwetland plant communities abut, a wetland boundary point has to be determined. The boundary point is typically represented by the limit of hydric soil indicators, since they reflect the degree of soil saturation (wetland hydrology). Mark the boundary with number flagging tape.

Step 9.   Repeat steps 4 through 7 until each community along the transect has been designated as wetland or nonwetland, and boundary points have been established between wetlands and nonwetlands. Go to next transect.

Step 10.   Repeat steps 4 through 9 until all transects have been evaluated, then connect the wetland boundary points between the transects to delineate the wetland boundary at the project site. This is accomplished by beginning at one of the wetland boundary points on a transect and walking in the general direction of the next transect or the end of the property, making observations regarding where the boundary is based on indicators, and flagging the boundary at appropriate places. Do this until all wetland boundary points on the transects have been connected and the wetland boundary extended out to the property line as necessary.

Source: Federal Interagency Committee for Wetland Delineation. 1989. Federal Manual for Identifying and Delineating Jurisdictional Wetlands. U.S. Army Corps of Engineers, EPA, FWS, USDA Soil Conservation Service, Washington, D.C.

soils, and signs of hydrology; yet in practice, the procedure is applied to wetlands and adjacent upland communities (i.e., those in the potential impact area), and not necessarily to all uplands on a project site. The general steps are given in Table 6.7.

The manual's intermediate-level methods offer two approaches — transect sampling method and vegetation unit analysis. The quadrat transect sampling procedure was derived from the Corps manual routine method for sites larger than five acres. Transects are established to analyze plant communities within a project site. The vegetation unit sampling procedure had its origins in the EPA manual (Sipple, 1988). This approach is similar to the routine plant community assessment procedure, with the level of vegetation analysis being more rigorous. Table 6.8 outlines steps for the quadrat transect sampling procedure which may be the more frequently used of the two approaches. The vegetation unit sampling procedure is essentially the same as the plant community assessment procedure of the routine method (see above and Table 6.7), except for vegetation sampling. Rather than making simple visual estimates of dominants in each plant community (= vegetation unit), this approach involves meandering through the vegetation unit and making frequent observations on plant species within each stratum. At each point, species are recorded by stratum and by cover class (trace = 1%, 1 = 1 to 5%, 2 = 6 to 15%, 3 = 16 to 25%, 4 = 26 to 50%, 5 = 51 to 75%, 6 = 76 to 95%, and 7 = 96 to 100%). The midpoints of each cover class are then used for calculations to determine dominant species (trace = 0, 1 = 3.0, 2 = 10.5, 3 = 20.5, 4 = 38.0, 5 = 63.0, 6 = 85.5, and 7 = 98.0).

The comprehensive methods involve making examinations of vegetation, soils, and hydrology along transects, but data collection for vegetation is more rigorous than in the intermediate method. Two procedures are offered — quadrat sampling procedure and point intercept sampling procedure. The former requires plot sampling along a transect, while the latter is an analysis of observations made at 2-ft intervals along a few transects.

The quadrat sampling procedure has two options — fixed interval transect sampling approach and plant community transect sampling approach. The former requires establishing for sampling plots at fixed intervals (every 100 to 1000 ft depending on transect length), while the latter establishes plots in different plant communities observed along the transect (similar to the approach used in the intermediate method). The former method derived from the Corps manual is not particularly efficient and probably was not used often. The latter method derived from the EPA manual was the preferred approach and, therefore, will by discussed below.

The *plant community transect sampling approach* is essentially the same as that of the intermediate method except for two points: (1) transects are supposed to be randomly selected along the baseline (but still must ensure that each community has at least one transect in it), and (2) vegetation analysis is more quantitative. All strata of the plant community are evaluated in a 30-ft radius circular plot. For herbs, the 30-ft radius plot is divided into four quadrants and a number of random samples are taken in the plot until a species-area curve for the herb stratum flattens out (see Chapter 4). This assessment, therefore, employs the minimum area concept to determine the appropriate sized area for sampling. The manual offers three choices for beginning herb sampling — eight 8 × 20 in. samples, four 20 × 20 in. samples, and four 40 × 40 in. samples, recognizing that actual number of samples will be determined by the species-area curve. Basal area is used to determine dominant trees. The diameter at breast height (dbh) is measured for each tree within the plot and the corresponding basal area is calculated. The basal areas for all trees of a given species are summed and the total basal areas for all species used to determine dominants. There also is a plotless sampling alternative for evaluating the tree stratum. This requires using a basal area factor (BAF) prism, angle-gauge (cruz-all), or similar device to identify dominant trees in a given community and then their dbh is measured and basal area calculated. The total basal area for each species is used to identify dominants. Cover classes are used for identifying dominant shrubs and woody vines (similar to vegetation unit procedure of intermediate method). Dominant saplings can be determined by either cover class estimates or basal area calculations. Again, the delineation of wetlands follows the quadrat transect sampling procedure of the intermediate methods (see above).

The *point intercept sampling procedure* is a frequency analysis of vegetation at many points along a few transects placed in an area of hydric soils (see Chapter 4 for a summary of this vegetation sampling technique). It incorporates elements of the hydric soil assessment procedure of the routine methods to first quickly cull out hydric soils in flooded or saturated condition as obvious wetlands and then perform more thorough vegetation analysis on potential hydric soil areas lacking such signs. Table 6.9 provides a summary of the steps involved in this method.

TABLE 6.9
Federal Interagency Manual's Point Intercept Sampling Procedure for Delineating Wetlands

Step 1.   After separating hydric from nonhydric soil areas and identifying obvious wetlands, the boundary of remaining hydric soil areas is refined and the starting points of three 200-ft transects are randomly selected.

Step 2.   After arriving at the starting point for a transect, the direction of the transect is determined by spinning a pencil or pen. The direction is indicated by the end of the object.

Step 3.   Lay out the 200-ft transect. This can be done by using a 200-ft tape or done as points are sampled (every 2 ft).

Step 4.   Record plants that are intersecting an imaginary line from the end of one's boot or from the point on the tape. Separate plants of the same species should be counted, so there may be more than one "hit" per species (but not from a single plant). Such observations are made every 2 feet along the transect.

Step 5.   After collecting data for an entire 200-ft transect, calculate the prevalence index (PI) for the transect (see example in Table 4.9).

Step 6.   Repeat Steps 2 through 5 for two additional transects and then calculate the mean prevalence index (PI_M) for the area and its standard error (Table 4.9). If the standard error is less than 0.20, no additional sampling is needed. If it is greater, then other transects need to be evaluated until the standard error of the mean prevalence index is acceptable (repeat Steps 2 through 6). When the PI_M is less than 3.0, the area should be a wetland, unless it is effectively drained (should have used the disturbed area procedures for the latter case). If the PI_M is 3.0 or higher, the area is usually nonwetlands, except where PI_M is between 3.0 and 3.5 and there are no signs of significant hydrologic modification (for other exceptions, users are referred to disturbed areas and problem area wetlands sections for protocols).

Source: Federal Interagency Committee for Wetland Delineation. 1989. Federal Manual for Identifying and Delineating Jurisdictional Wetlands. U.S. Army Corps of Engineers, EPA, FWS, USDA Soil Conservation Service, Washington, D.C.

The manual includes procedures for evaluating disturbed areas (i.e., areas altered by human activities or natural disturbances such as avalanches, mudslides, fire, volcanic deposition, and beaver dams). Such activities or events change the character of an area often making it difficult to make a wetland determination. This section is derived from the atypical situations section of the Corps manual with some revisions, especially concerning determination whether wetland hydrology still exists at the site.

The problem area wetlands discussion in the manual summarizes wetlands that may pose some difficulty for identification following the general procedures in the manual because they may lack certain indicators. Several examples are given.

- FACU-dominated wetlands (including certain evergreen forested wetlands)
- Wetlands on glacial till
- Highly variable seasonal wetlands
- Interdunal swale wetlands
- Vegetated river bars and adjacent flats
- Vegetated flats
- Caprock limestone wetlands
- Newly created wetlands
- Wetlands with problematic soils (Entisols, red parent material soils, Spodosols, and Mollisols)

Most of these were identified in the Corps and EPA manuals (see Chapter 7 for discussion of these wetlands).

NATIONAL RESEARCH COUNCIL FINDINGS
AND RECOMMENDATIONS

The National Research Council (NRC) conducted a 2-year study of wetland delineation and presented its findings in a 306-page report entitled *Wetlands Characteristics and Boundaries* (NRC, 1995). Their general conclusions were that "the federal regulatory system for protection of wetlands is scientifically sound and effective in most respects, but it can be more efficient, more uniform, more credible with regulated entities, and more accurate in a technical or scientific sense through constructive reforms of the type suggested in this report" (*p. 12*). Some of the more pertinent findings are outlined below:

1. Recommend a reference definition of wetlands based on the concept of an ecosystem with either constant or recurrent, shallow inundation or saturation near the surface of the substrate (*p. 59* in the NRC report; see Chapter 1 for discussion of this definition).
2. Wetland delineations established by office methods are susceptible to errors that do not affect field delineations (*p. 88*).
3. In a three-parameter method, limiting the number of hydrology indicators can greatly affect the amount of wetland acreage (*p. 88*).
4. Less area would generally be delineated as wetland following the Corps manual than with the 1989 manual due to the broader and more flexible array of indicators in the latter (*p. 89*).
5. In the absence of hydrologic alteration and evidence to the contrary, the presence of field-verified hydric soils, a predominance of OBL and FACW species, or a combination of OBL, FACW, and FAC species can be used as "strong evidence of wetland hydrology" (*p. 144-145*).
6. Procedures to evaluate hydrologically altered sites or sites where vegetation, soils, and other important indicators of site hydrology need to be developed (*p. 146*).
7. Identified problems with the concept of growing season as applied to assessing wetland hydrology (*p. 146*; see Chapter 2 for discussion).
8. Minimum wetness for wetlands should be considered as saturation within 1 ft of the surface for 14 consecutive days during the growing season in most years, until proven otherwise by regional studies (*p. 146*; see Chapter 2 for discussion).
9. The hydric soils list is useful and should be continued to be developed by NRCS; such list should consider assigning a fidelity category to designated soils (*p. 146*).
10. Regional hydric soil committees should be established (*p. 147*).
11. Field indicators of hydric soils should be based on soil color and morphology, and such indicators should be evaluated for use in delineation (*p. 146*).
12. The absence of hydric soils does not always indicate upland; analysis of hydrology and biota are needed for such lands (*p. 147*).
13. The national hydrophyte list is technically sound; its continued improvement and revision should be supported by the FWS (*p. 147*).
14. Vegetation dominated by facultative or facultative-upland species will satisfy the biologic criterion for wetland if it occurs on field-verified hydric soils with strong morphological indicators (*p. 147*).
15. Boundary determinations involving vegetation analysis should be confirmed by analysis of the soil or substrate (*p. 148*).
16. If hydrologic information is unavailable and if the hydrology has not been altered, the presence of wetland hydrology can be evaluated from information on substrate (if definitive, presence of hydric soils), or from vegetation (when hydrophytic vegetation is unequivocal), or from other indicators that have a strong relationship to wetland hydrology.

If vegetation or substrate (soils) does not provide clear evidence, then hydrologic data will be required (*p. 148*).

17. The PRIMET method is conceptually sound and should be studied for use in wetland delineation (*p. 148*).

18. Reference wetlands should be identified for long-term study of the relationships between water, substrate, and biota (*p. 148*).

19. A single federal manual derived from the existing manuals should be drafted; this manual should be supported by the development of regional supplements to provided detailed criteria and a list of indicators consistent with the federal manual (*p. 144*).

Recommendations regarding problematic wetlands are listed at the end of Chapter 7.

WETLAND DELINEATION TIPS

Wetland identification and delineation are performed in the field by personnel trained in botany and soil science. Observations of vegetation, soils, and signs of flooding or soil saturation provide important clues to site wetness. Such observations must be sufficient to allow an investigator to accurately locate and mark the wetland–nonwetland boundary. Depending on site characteristics, wetland recognition may be easy or difficult. The wetter the wetland, the easier it is to identify since there is often strong evidence of wetness (e.g., surface water or soils wet underfoot), plus conspicuous and highly wetland-specific vegetation (OBL species either dominant or common in association with FACW species). The average citizen can readily identify these areas as wetlands. Yet, the majority of wetlands have fluctuating water levels and may or may not be inundated for brief periods. The absence of standing water and saturated soils often makes wetland recognition more difficult, since the corresponding vegetation is usually not unique to wetlands. Periodically anaerobic soils typically develop specific and readily observable properties that reflect site wetness. Consequently, in areas not subject to significant hydrologic modification, hydric soil properties are invaluable for identifying the presence of wetland.

Preparation for the Field

Given that wetland identification and delineation require knowledge of plants and soils, you should read the variety of wetland books that describe wetland plant communities and their soils for the area in which you work. Such information is vital to improving your understanding of wetland systems and learning what to expect out in the field. These readings also will provide a list of plants that are typically found in the region's wetlands. Chapter 9 presents fairly extensive information on U.S. wetland plant communities. If you learn how to identify most of the listed plants, you should have little difficulty recognizing wetland plant communities across the country.

Before going into the field for a site assessment, you also should assemble and review various data sources. These sources include National Wetlands Inventory maps, soil survey reports (with accompanying maps), and aerial photographs (including Food Security Act compliance slides that may be available for examination at local NRCS offices) (see Chapter 10). These materials will give you invaluable background information on the plant communities, soils, hydrology, and topographic features that you are likely to encounter in the field. Before going out in the field, examine topographic maps to see if there are any upstream dams that may have altered the hydrology of downstream floodplain wetlands. If such dams are found, contact the appropriate facility or operator to see if additional information on the river's hydrology is available. Another useful step before heading to the field would be to contact appropriate regulatory agencies (federal, state, or local). Ask them for guidance on delineating small wetlands, linear wetlands, and complex terrains with mosaics of small wetlands interspersed among uplands (see Chapter 7 for some suggestions).

Field Tips

The following tips are offered to aid in evaluating a project site and performing a wetland delineation.

Tip 1. For sites that may present difficulty in verifying wetland hydrology, visit the project site more than once, preferably at the beginning or the wettest part of the growing season (usually to assess actual site wetness for positive wetland hydrology indicators) and later during the peak of the growing season (to assess vegetation, especially the herbaceous species which are frequently better indicators of site wetness than the woody species). If only one site visit is possible, go in the early spring as this will provide the most useful information for assessing the area's hydrology.

Tip 2. Learn to recognize certain landscape positions in the field where wetlands are most likely to form. The following landscape positions tend to support wetlands: low-lying areas subject to periodic flooding (e.g., along rivers and estuaries); gentle slopes in areas of groundwater discharge (seepage slopes) or surface water runoff (drainageways); breaks in slopes (e.g., flats on hillslopes); isolated depressions surrounded by uplands (e.g., ponds, lakes, kettle holes, potholes, playas, and vernal pools); and broad, relatively flat areas lacking drainage outlets (e.g., interstream divides).

Tip 3. Before doing any delineation at a site, simply walk the site and become familiar with its plant communities, soils, and topography. Use an auger or soil probe to get a general sense of soil changes associated with different vegetation patterns.

Tip 4. When conducting field work, avoid periods of extreme flooding or periods immediately following heavy rains because low-lying uplands may be inundated at these times.

Tip 5. Early spring and mid to late fall inspections have an advantage over other seasons in that in most temperate regions, the foliage is absent from the deciduous woody plants in forests and thickets. The lack of leaves on otherwise thick shrub understory makes it much easier to observe subtle changes in topography that typically reflect wetland–nonwetland boundary areas. This also allows one to mark the boundary with less flags, thereby making follow-up surveying by site engineers more efficient and less costly. Moreover, noxious insects are often less abundant at these times.

Tip 6. For grassland regions, field inspections during mid- to late-growing season are best for identifying the plants. This period, however, may not coincide with the optimal period for making hydrologic observations (e.g., spring in many places and winter in areas with Mediterranean climates).

Tip 7. If forced to do a winter determination in northern snow-covered regions, the results should be viewed as preliminary and should be later refined by a follow-up inspection during the growing season.

Tip 8. A tract of land often has more than one plant community and may have wetlands scattered throughout a largely upland area. Each plant community regardless of size should be treated separately for wetland determination purposes. Homogeneous stands in similar landscape positions (e.g., depressions, flats, slopes, and ridges) should be identified for evaluation. An initial walk through the project site should reveal these different communities.

Tip 9. If using a transect approach for evaluating plant communities, the number of transects used should be determined by site characteristics. If landscape is simple (i.e., area is grading downslope in one direction), two transects may be sufficient. More transects are used if the vegetation is difficult to traverse, e.g., common greenbrier (*Smilax rotundifolia*) thickets, to make more efficient use of time and trouble. These transects should be located within visible distance of each other and field work performed during leaf-off period to maximize assessment of topography and visibility.

Tip 10. If an extensive network of drainage ditches or other significant hydrologic alteration is observed or known to occur that greatly affects the site's hydrology, the hydrology will require a detailed examination (see Chapter 7). In many such cases, it may be necessary to consult an expert to determine the current hydrology or become fluent in hydrologic assessment methodologies.

Tip 11. In many locales topographic relief is more gradual and obvious differences in vegetation patterns are lacking. FAC species may dominate both wetlands and contiguous uplands. This happens in low-lying coastal plains, floodplains of major river valleys, and on gentle slopes in areas of groundwater discharge (seepage slopes). In these and similar situations including the upper edges of many wetlands, a mixed plant community of FACW, FAC, and FACU species often forms what some people call the *transition zone*. Signs of wetland hydrology may be difficult to find in some places, since flooding may be brief or soils may be saturated near the surface only during the early part of the growing season. Many of these wetlands, especially in the eastern U.S. and Alaska, are wetter longer during the nongrowing season. In these cases where hydrology is not apparent or weakly expressed and the vegetation is inconclusive, the soil properties often reveal much about site wetness. Look for subtle changes in topography and vegetation and examine the soil. Dig several holes within a few feet of each other and locate the subtle differences in soil morphology that reflect hydric soil conditions. It is widely acknowledged that soils reflect the long-term hydrology and, therefore, hydric soil properties are highly reliable indicators of wetland in the absence of significant hydrologic modification (Tiner and Veneman, 1987; Sipple, 1988; Federal Interagency Committee for Wetland Delineation, 1989; National Research Council, 1995).

Tip 12. A low-lying wetland dominated by OBL, FACW, and/or FAC species may gradually change to a community dominated by FAC, FACU, and UPL species on adjacent gentle slopes. When the latter community is dominated by nonhydrophytic vegetation (e.g., FACU and UPL species), the community is usually not wetland unless it happens to be one of the problematic wetlands (see Chapter 7 for discussion).

Tip 13. Wetland boundaries are usually marked by attaching colored flagging tape to existing vegetation. In some cases, wooden stakes or flags on metal pins may be used (e.g., in grazed meadows or mowed fields). The boundary should be marked in enough places to best reflect the configuration of the boundary and to be able to clearly see from one flag to another. In wetlands with dense tall deciduous shrub cover, much more flagging will be needed in summer than in the fall, winter, and spring when leaves are off the shrubs. Surveying of wetland boundaries should be done shortly after delineation is performed so that surveyors can easily locate flagging. All flags should be uniquely numbered. Be sure to mark the sites of your data collection/analysis with a different colored flag and label it appropriately.

Tip 14. After delineating and flagging the wetland boundary, walk the boundary and see if the boundary features are consistent. This will serve as a good quality control step. In the process, check the flag numbers and prepare a general sketch map of the wetland showing the points of the numbered flags.

Tip 15. Take good notes. Record particularly interesting and relevant observations on a set of field notes. Don't simply rely on the delineation forms to determine what is vital information. Wildlife observations and the presence of OBL species in few numbers also are important. Be sure to record all direct and indirect signs of wetland hydrology observed in each plant community. Do not limit observations of wetland hydrology to the sample plot.

Tip 16. For difficult field delineations, consider working with another delineator, preferably one with a different strength than yours. Two delineators working in tandem, one evaluating one point and the other evaluating a neighboring point, can use their skills independently to determine the wetland boundary at a given point, and then meet to discuss the matter together before further delineation. This strategy should result in a consistent, well-reasoned solution to the problem site.

Tip 17. Inexperienced delineators should work with seasoned delineators perhaps following the approach in Tip 16. This is an effective training procedure.

PROFESSIONAL JUDGMENT

A discussion of professional judgment seems necessary because it has been abused in the process of making wetland determinations and subsequently in defending one's actions. Professional judgment should mean that when all the literature and scientific concepts on wetlands are considered coupled with knowledge gained from your experiences, the evidence would support the claim that an area is or is not a wetland. It is something akin to bringing together the country's top 10 wetland experts and using their consensus opinion for making the decision. Unfortunately, in some cases, professional judgment has been interpreted to mean that "I am a wetland professional and in my opinion, this area is…" This interpretation is not in the true sense of the meaning of best professional judgment. Granted there are varying opinions on what should be regulated and what should not, and what is fair to the average landowner, and these opinions do play a role in litigation and to some extent in the interpretation of what is a wetland according to the manual. Extremists on both sides of the debate may misuse the concept of best professional judgment to support either wetland destruction by claiming the area is not wetland or protection of marginal lands by attempting to include areas that are not really wet enough to be wetlands. Currently, such decisions are in the hands of individual regulatory agents who make such decisions every week. It might be worthwhile to consider forming regional teams of experts represented by the Corps, EPA, FWS, NRCS, National Marine Fisheries Service, and state agencies to help in deciding difficult delineations. The team would, in effect, represent the government's best professional judgment and might strengthen the government's technical position in the more difficult cases.

REFERENCES

Adams, D.A., M.A. Buford, and D.M. Dumond. 1987. In search of the wetland boundary. *Wetlands,* 7: 59-70.

Atkinson, R.B., J.E. Perry, E. Smith, and J. Cairns, Jr. 1993. Use of created wetland delineation and weighted averages as a component of assessment. *Wetlands,* 13: 185-193.

Browne, S., S. Crocoll, D. Goetke, N. Heaslip, T. Kerpez, K. Koget, S. Sanford, and D. Spada. 1996. Freshwater Wetlands Delineation Manual. New York State Department of Environmental Conservation, Division of Fish and Wildlife, Albany.

Cohen, J.P. 1992. How wet must a wetland be? *Govern. Exec.,* March 1992: 20-25.

Cowardin, L.M., V. Carter, F.C. Golet, and E. T. LaRoe. 1979. Classification of Wetlands and Deepwater Habitats of the United States. Publ. No. FWS/OBS-79/31. U.S. Fish and Wildlife Service, Washington, D.C.

Dahl, T.E. and G.J. Alford. 1996. History of wetlands in the conterminous United States. In National Water Summary on Wetland Resources. J.D. Fretwell, J.S. Williams, and P.J. Redman (Comp.). Water-Supply Paper 2425. U.S. Geological Survey, Reston, VA. 19-26.

Dennison, M.S. and J.F. Berry (Eds.). 1993. *Wetlands: Guide to Science, Law, and Technology.* Noyes Publications, Park Ridge, NJ.

DiPinto, M.A. and R. McCollum (Comp.). 1988. Guide to Inland Vegetated Wetlands in Massachusetts. Department of Environmental Quality Engineering (now Environmental Protection), Pub. No. 15,190-84-1,000-2-88-C.R. Division of Wetlands and Waterways, Boston.

Eilers, H.P., A. Taylor, and W. Sanville. 1983. Vegetative delineation of coastal salt marsh boundaries. *Environ. Manage.,* 7: 443-452.

Elmore, J. 1991. U. S. Army Corps of Engineers memorandum on wetlands delineation and the 1992 Energy and Water Development Appropriations Act. August 23, 1991.

Environmental Defense Fund and World Wildlife Fund. 1992. How Wet is a Wetland? The Impacts of the Proposed Revisions to the Federal Wetlands Delineation Manual. EDF, New York, and WWF, Washington, D.C.

Environmental Laboratory. 1987. Corps of Engineers Wetlands Delineation Manual. Tech. Rpt. Y-87-1. U.S. Army Engineer Waterways Experiment Station, Vicksburg, MS.

Environmental Law Institute. 1993. *Wetlands Deskbook*. The Environmental Law Reporter, Washington, D.C.

Federal Interagency Committee for Wetland Delineation. 1989. Federal Manual for Identifying and Delineating Jurisdictional Wetlands. Cooperative technical publication. U.S. Army Corps of Engineers, U.S. Environmental Protection Agency, U.S. Fish and Wildlife Service, and USDA Soil Conservation Service, Washington, D.C.

Florida Department of Environmental Protection, South Florida Water Management District, St. John's River Water Managment District, Suwannee River Water Management District, Southwest Florida Water Managment District, and Northwest Florida Water Management District. 1995. The Florida Wetlands Delineation Manual. Florida Department of Environmental Protection, Wetland Evaluation and Delineation Section, Tallahassee.

Florida Soil Conservation Staff. 1992. Soil and Water Relationships of Florida's Ecological Communities. USDA Soil Conservation Service, Gainesville, FL.

Gauthier, B. 1997. Politique de Protection des Rives, du Littoral et des Plaines Inondables. Ministère de l'Environnement et de la Faune, Direction de la Conservation et du Patrimoine Écologique, Quebec.

Harshberger, J. W. 1900. An ecological study of the New Jersey strand flora. Proc. of the Academy of National Sciences of Philadelphia (1900). 623-671.

Harshberger, J.W. 1909. The vegetation of the salt marshes and of the salt and fresh water ponds of northern coastal New Jersey. Proc. of the Academy of National Sciences of Philadelphia (August 1909). 373-400.

Harshberger, J.W. 1916. *The Vegetation of New Jersey Pine Barrens*. 1970 reprint by Dover Publications, New York.

Hurt, G.W., P. M. Whited, and R.F. Pringle (Eds.). 1996. Field Indicators of Hydric Soils in the United States, ver. 3.2. USDA Natural Resources Conservation Service, Fort Worth, TX.

Hurt, G.W., P. M. Whited, and R.F. Pringle (Eds.). 1998. Field Indicators of Hydric Soils in the United States, ver. 4.0. USDA Natural Resources Conservation Service, Fort Worth, TX.

Jackson, S. 1995. Delineating Bordering Vegetated Wetlands Under the Massachusetts Wetlands Protection Act. Massachusetts Department of Environmental Protection, Division of Wetlands and Waterways, Boston.

Kraus, M. 1993. Wetland identification and delineation manuals: where are we? Environmental Concern, Inc., St. Michaels, MD. *Wetland J.*, 5: 8,13.

Krivak, J.A. 1980. EPA rationale for identifying wetlands under the 404 program. Memorandum dated October 9, 1980. U.S. EPA, Criteria and Standards Division, Washington, D.C.

Lyons, J.R., G.T. Frampton, R. Perciasepe, and G.E. Dickey. 1994. Memorandum of agreement among the Department of Agriculture, the Environmental Protection Agency, the Department of the Interior, and the Department of the Army concerning the delineation of wetlands for purposes of Section 404 of the Clean Water Act and Subtitle B of the Food Security Act. January 6, 1994. U.S. Departments of Agriculture, Army, and Interior, and the U.S. Environmental Protection Agency, Washington, D.C.

Mader, S.F. 1991. Forested Wetlands Classification and Mapping: A Literature Review. NCASI Tech. Bull. No. 606. National Council of the Paper Industry for Air and Stream Improvement, Inc., New York.

McKee, A., S. Gregory, and L. Ashkenas. 1996. Riparian Management Reference. Draft. U.S. Dept. of Agriculture, Chippewa National Forest, Cass Lake, MN.

National Research Council. 1995. *Wetlands Characteristics and Boundaries*. National Academy Press, Washington, D.C.

Natural Resources Conservation Service. 1994. Mapping Conventions and Off-Site Wetland Determination/Inventory Procedures for Virginia. U.S. Department of Agriculture, Richmond, VA.

Natural Resources Conservation Service. 1997. Hydrology Tools for Wetland Determination. Part 650. Engineering Field Handbook. U.S. Department of Agriculture, Washington, D.C.

New England District. 1991. Guidance for the interpretation of wetland boundaries using the 1987 Corps manual in the six New England states. U.S. Army Corps of Engineers, Waltham, MA. Regulatory guidance letter: September 9, 1991.

Page, R.W. and R.W. Hanmer. 1989. Memorandum of agreement between the Department of the Army and the Environmental Protection Agency concerning the determination of the geographic jurisdiction of the Section 404 program and the application of the exemptions under Section 404 (f) of the Clean Water Act. January 19, 1989. U.S. Department of the Army and U.S. EPA, Washington, D.C.

Reed, P.B., Jr. 1988. National List of Plant Species that Occur in Wetlands: 1988 National Summary. Biol. Rpt. 88(24). U.S. Fish and Wildlife Service, Washington D.C.

Rhode Island Department of Environmental Management. 1994. Rules and Regulations Governing the Administration and Enforcement of the Freshwater Wetlands Act. Providence, RI.

Roman, C.T., R.A. Zampella, and A.Z. Jaworski. 1985. Wetland boundaries in the New Jersey Pinelands: ecological relationships and delineation. *Water Res. Bull.,* 5: 1005-1012.

Sanders, D.R., Sr., W.B. Parker, S.W. Forsythe, and R.T. Huffman. 1984. Wetlands Delineation Manual, Parts I-IV. U.S. Army Engineer Waterways Experiment Station, Vicksburg, MS.

Shaler, N.S. 1890. General account of the fresh-water morasses of the United States with a description of the Dismal Swamp District. 10th Annual Report 1888-1889. 255-339. U.S. Geological Survey, Washington, D.C.

Shreve, F., M.A. Chrysler, F.H. Blogett, and F.W. Besley. 1910. *The Plant Life of Maryland.* Special Publication Vol. III. The Johns Hopkins Press, Baltimore, MD.

Sipple, W.S. 1988. Wetland Identification and Delineation Manual. Volume I. Rationale, Wetland Parameters, and Overview of Jurisdictional Approach. Volume II. Field Methodology. U.S. Environmental Protection Agency, Office of Wetlands Protection, Washington, D.C.

Soil Survey Staff. 1996. Keys to Soil Taxonomy, 7th ed. USDA Natural Resources Conservation Service, Washington, D.C.

Studt, J. F. 1991. U. S. Army Corps of Engineers memorandum on questions and answers on 1987 manual. October 7, 1991.

Tiner, R. W. 1993. The primary indicators method — a practical approach to wetland recognition and delineation in the United States. *Wetlands,* 13: 50-64.

Tiner, R. W. 1991. The concept of a hydrophyte for wetland identification. *BioScience,* 41: 236-247.

U.S. Army Corps of Engineers, Seattle District and U.S. Environmental Protection Agency, Region 10. 1994. Memorandum: Washington regional guidance on the 1987 wetland delineation manual. May 23, 1994. Seattle, WA.

USDA Natural Resources Conservation Service. 1996. National Food Security Act Manual, 3rd ed. Part 513 — Preparing to Make Wetland Determinations or Delineations. Washington, D.C.

USDA Soil Conservation Service. 1985. National Food Security Act Manual. Washington, D.C.

USDA Soil Conservation Service. 1987. Hydric Soils of the United States. In cooperation with the National Technical Committee for Hydric Soils. U.S. Department of Agriculture, Washington, D.C.

USDA Soil Conservation Service. 1988. National Food Security Act Manual. U.S. Department of Agriculture, Washington, D.C.

U.S. Department of the Army and U.S. Environmental Protection Agency. 1993. Amendment to the January 19, 1989 Department of the Army/Environmental Protection Agency memorandum of agreement concerning the determination of the geographic jurisdiction of the Section 404 program and the application of the exemptions under Section 404(f) of the Clean Water Act. January 4, 1993.

Want, W.L. 1992. *Law of Wetlands Regulation.* Clark Boardman Callaghan, Deerfield, IL.

Wentworth, T.R. and G.P. Johnson. 1986. Use of Vegetation in the Designation of Wetlands. North Carolina State University, School of Agriculture and Life Sciences, Raleigh. Report for the U.S. Fish and Wildlife Service's National Wetlands Inventory, St. Petersburg, FL.

Westchester County Soil and Water Conservation District. 1997a. Wetland Protection in Westchester County: A Survey of Municipal Wetland Ordinances. White Plains, NY.

Westchester County Soil and Water Conservation District. 1997b. A Model Ordinance for Wetland Protection. White Plains, NY.

Williams, MG A.E. 1992. U. S. Army Corps of Engineers memorandum on clarification and interpretation of the 1987 manual. March 6, 1992.

Wisconsin Coastal Management Program. 1995. Basic Guide to Wisconsin's Wetlands and Their Boundaries. State of Wisconsin, Department of Administration, Madison.

Zedler, J.B. and G.W. Cox. 1985. Characterizing wetland boundaries: a Pacific coast example. *Wetlands,* 4: 43-55.

7 Problem Wetlands and Field Situations

INTRODUCTION

Depending on the criteria established for wetland recognition, there may be a few or many exceptions to the "rule." Using the Corps wetland delineation manual, wetlands subject to regulation under the Clean Water Act (CWA) must typically meet a three-parameter test; they must have positive indicators of three parameters (vegetation, soils, and hydrology) that are associated with wetlands, with a few noted exceptions (Environmental Laboratory, 1987). To be designated as a potentially regulated wetland, an area must possess at the time of inspection or be expected to normally have hydrophytic vegetation, hydric soils, and certain signs of wetland hydrology (see Chapter 6 for details). Using this rule, a number of typical wetlands fail the three-parameter test and must be considered exceptions. For example, hemlock swamps usually do not have vegetation that meets the Corps basic rule for hydrophytic vegetation. Some wetlands may possess soils that do not exhibit hydric soil properties. Other wetlands may not be wet long enough during the currently specified period (i.e., growing season) or wet frequently enough (i.e., every other year on average) to qualify as a regulated wetland, despite providing critical habitat for rare wetland-dependent plants and animals (e.g., West Coast vernal pools) or other wetland functions during the nongrowing season. The above wetlands are considered "problem wetlands" that may or not be regulated. These wetlands are not necessarily less valuable or less wetland than other types, but simply present difficulties for identification when strictly applying the Corps' three-parameter test or other approaches.

The purpose of this chapter is to make readers aware of these types of wetlands and to provide some insight into how such areas may be recognized and delineated. The Corps manual listed five types and two situations as examples: wetlands on drumlins, seasonal wetlands, prairie potholes, vegetated flats, man-induced wetlands, unauthorized activities (illegal deposition of fill material), and natural events (e.g., beaver dams, fires, avalanches, volcanic activity, mudslides, and changing river courses) (Environmental Laboratory, 1987). It also mentioned some problem soils such as sandy soils, red parent material soils, and Spodosols. The federal interagency manual added a few more problem wetlands and soils (Federal Interagency Committee for Wetland Delineation, 1989):

- FACU-dominated wetlands
- Evergreen forested wetlands
- Wetlands on glacial till
- Highly variable seasonal wetlands
- Interdunal swale wetlands (sandy soils)
- Vegetated river bars and adjacent flats (western states)
- Caprock limestone wetlands (Florida Everglades)
- Entisols (floodplain soils)
- Mollisols (prairie soils)

Later guidance on the use of the Corps manual (Williams, 1992) identified some of these as additional problem areas. Most of these problem wetlands plus some others are discussed in this chapter, along with situations encountered in the field that may make wetland delineation more difficult. Since some of these problem wetlands may be contentious from the regulatory perspective, it is emphasized that the chapter is intended primarily for discussion purposes and that the ultimate decision on whether such areas should be regulated rests with individual Corps districts and EPA regions at the federal level and applicable state agencies and local governments. Questions about the regulatory status of these wetlands can only be answered by these agencies.

PROBLEMATIC WETLAND PLANT COMMUNITIES

Most so-called "problematic wetland plant communities" are simply problems arising from the definition of hydrophytic vegetation in the federal wetland delineation manuals. This is especially evident in a strict interpretation of the Corps three-parameter approach to wetland identification. For example, certain wetlands fail to meet the Corps manual's basic rule for hydrophytic vegetation because they are dominated by FAC– and/or FACU species either year-round, seasonally (usually during late summer or the driest season), or during extremely dry years. Others may lack plant cover at certain times that also makes verification of hydrophytic vegetation difficult. The Corps manual does acknowledge some of these situations as "problem areas" permitting their identification as regulated wetlands, but many cases require the use of professional judgment for accurate wetland identification. Considering a combination of landscape position and landform (e.g., depressions, sloughs, floodplains, drainageways, broad flat plains, and slopes below springs or groundwater seepage sites), soil characteristics, and signs of wetland hydrology can significantly aid in the identification of these plant communities as wetlands. A listing of the more frequently occurring of these wetland plant communities would be useful guidance, but has not been published by regulatory agencies.

WETLANDS DOMINATED BY OFTEN DRY-SITE SPECIES

While communities dominated by FACU species typify many nonwetlands, they are sometimes found in wetlands, often along wetland boundaries and on floodplains. Yet one type of FACU-dominated wetland is not restricted to drier conditions, but often grows on peaty soils or typical hydric mineral soils in depressional areas or broad flats (Plate 34). Certain evergreen tree species that dominate wetlands are more prevalent in upland forests and have been assigned an indicator status of FACU. Examples of these plants include the following: red spruce (*Picea rubens*), white spruce (*P. glauca*), eastern white pine (*Pinus strobus*), pitch pine (*P. rigida*), jack pine (*P. banksiana*), and eastern hemlock (*Tsuga canadensis*) in the Northeast and Midwest; longleaf pine (*Pinus palustris*) in the Southeast; and Engelmann spruce (*Picea engelmannii*), Sitka spruce (*P. sitchensis*), ponderosa pine (*Pinus ponderosa*), lodgepole pine (*P. contorta*), western hemlock (*Tsuga heterophylla*), Pacific silver fir (*Abies amabilis*), white fir (*A. concolor*), and subalpine fir (*A. lasiocarpa*) in the West and Alaska (see tables in Chapter 9 for sample communities). These species have a broad ecological amplitude. Hahn et al. (1920) found hypertrophied lenticels on roots of many of these species when growing in saturated soils. The individuals growing in wetlands are clearly adapted in some way for life in periodically anaerobic soils and are considered hydrophytes. As noted in Chapter 3, the species level in plant taxonomy is not sufficient for identifying these plants as hydrophytes, but consideration of possible wetland ecotypes is necessary (Tiner, 1991). They may be recognized as hydrophytes by considering associated vegetation (sometimes OBL species are co-dominant), landscape position (stream borders, drainageways, depressions, or seepage flats), the presence of hydric soils, and signs of hydrology (especially for periodically inundated wetlands).

Weinmann and Kunz (1995) emphasized that although some plant communities do not satisfy the basic hydrophytic vegetation parameter of the Corps manual, there is usually sufficient field evidence to designate such areas as wetlands, mainly by the presence of hydric soils and signs of wetland hydrology. They gave examples of two such communities.

1. A western hemlock forested community (dominants were trees = western hemlock — *Tsuga heterophylla*, FACU; shrubs = *T. heterophylla*, FACU, and salal — *Gaultheria shallon*, FACU; herbs = *T. heterophylla*, FACU, skunk cabbage — *Lysichitum americanum*, OBL, and peat moss — *Sphagnum* sp., a nonvascular plant hence no indicator status).

2. A herbaceous community dominated by creeping thistle (*Circium arvense*; FACU), quack grass (*Agropyron repens*; FAC–), velvet grass (*Holcus lanatus*; FAC), and giant horsetail (*Equisetum telmateia*; FACW).

Strict application of the Corps manual's hydrophytic vegetation standard would fail to identify these communities as hydrophytic, despite their occurrence on undrained hydric soils (with the former on peat). The Corps and EPA have provided additional guidance to help identify these types of areas as wetlands in Washington state (U.S. Army Corps of Engineers — Seattle District and U.S. EPA, Region 10, 1994). Similar guidance may be available from other districts. Where such guidance is lacking, wetland delineators and regulators alike will need to use professional judgment to evaluate these stands (e.g., undrained hydric soils with wetland hydrology indicators and favorable wetland landscape position/landform = wetland). Regulatory agencies should develop appropriate guidance to ensure consistent and accurate wetland delineations.

Examples of FAC– and FACU species common or dominant in wetlands in the northeastern U.S. are shown in Table 7.1. For the Pacific Northwest, the following species are some examples: quackgrass, creeping thistle, brittle fern (*Cystopteris fragilis*), catchweed bedstraw (*Galium aparine*), sweet scent bedstraw (*Galium triflorum*), fool's huckleberry (*Menziesia ferruginea*), indian plum (*Oemlaria cerasiformis*), timothy (*Phleum pratense*), Canada bluegrass (*Poa compressa*), bracken fern (*Pteridium aquilinum*), Himalaya blackberry (*Rubus discolor*), dewberry (*Rubus ursinus*), and red huckleberry (*Vaccinium parvifolium*) (Fred Weinmann, personal communication, 1998).

Human intervention may have caused an increase in the abundance of some FACU species. Drier-site species are often common in temporarily flooded wetlands or wetlands that are partly drained. These species may include, among others, sweet vernal grass (*Anthoxanthum odoratum*), clasping-leaf dogbane (*Apocynum cannabinum*), japanese barberry (*Berberis thunbergii*), velvet grass, quaking aspen (*Populus tremuloides*), black cherry (*Prunus serotina*), Allegheny blackberry (*Rubus allegheniensis*), common red raspberry (*Rubus idaeus*), and Canada goldenrod (*Solidago canadensis*) (Fisher et al., 1996; personal observations). In other cases, FACU species may be planted in wet meadows as forage for livestock, e.g., perennial ryegrass (*Lolium perenne*), timothy, and Kentucky bluegrass (*Poa pratensis*). Other drier-site species that may be found in wet meadows include common yarrow (*Achillea millefolium*), orchard grass (*Dactylis glomerata*), Queen Anne's lace (*Daucus carota*), red clover (*Trifolium pratense*), white clover (*T. repens*), daisy fleabane (*Erigeron annuus*), cocklebur (*Xanthium* sp.), thistle (*Cirsium* sp.), and heal-all (*Prunella vulgaris*) (Tiner and Burke, 1995). While some of these are native species, others are introduced, e.g., orchard grass and clovers. Many of the FACU species in wetlands may be non-native species that have become established near urban or agricultural areas. Some Northeast examples include quackgrass, indian mock-strawberry (*Duchesnea indica*), field garlic (*Allium vinale*), garlic mustard (*Alliaria petiolata*), japanese bamboo or knotweed (*Polygonum cuspidatum*), tartarian honeysuckle (*Lonicera tatarica*), privets (*Ligustrum* spp.), multiflora rose (*Rosa multiflora*), japanese barberry, and common buckthorn (*Rhamnus cathartica*).

Kindscher et al. (1998) reported significant vegetation changes in agriculturally altered playa wetlands in Kansas. Undisturbed playas had mean prevalence index values of 1.82 whereas playas with a disturbance history had a mean value of 3.51 (e.g., where OBL = 1.0, FAC = 3.0, and UPL = 4.0). Annual species like kochia (*Kochia scopia*) and goosefoot (*Chenopodium berlandieri*) were dominants in disturbed playas, while perennials, especially spikerushes (*Eleocharis* spp.), predominated at undisturbed sites. The authors concluded that past agricultural use of playas has altered

TABLE 7.1

Examples of Common FACU Species Found in Wetlands in the Northeast

Species	Wetland Types	References
Trees		
Picea rubens (Red Spruce)	Bogs in glaciated regions and in the Appalachians	Brooks et al., 1987; Tiner, 1988; Calhoun et al., 1994
Pinus rigida (Pitch Pine)	Lowlands in the New Jersey Pine Barrens; bogs	Little, 1959; Ledig and Little, 1979; Damman and French, 1989
Pinus strobus (Eastern White Pine)	Sandy forested wetlands in eastern and central U.S.; bogs	Huenneke, 1982; Tiner, 1988; Golet et al., 1993
Tsuga canadensis (Eastern Hemlock)	Mucky swamps in glaciated areas	Niering, 1953; Huenneke, 1982; Tiner, 1989
Fagus grandifolia (American Beech)	Temporarily flooded forested wetlands, chiefly along Atlantic coastal plain	Tiner, 1988; Golet et al., 1993; Tiner and Burke, 1995
Fraxinus americana (White Ash)	Forested wetlands	Magee, 1981; Tiner, 1985a; Golet et al., 1993
Prunus serotina (Black Cherry)	Temporarily flooded forested wetlands along floodplains	Tiner, 1985a, 1988; Golet et al., 1993
Liriodendron tulipifera (Tulip Poplar)	Forested wetlands along Atlantic coastal plain	Niering, 1953; Tiner, 1985a, 1985b, 1988
Juglans nigra (Black Walnut)	Temporarily flooded forested wetlands along Piedmont rivers and streams	Tiner and Burke, 1995
Quercus alba (White Oak)	Forested wetlands along Atlantic coastal plain	Tiner and Burke, 1995
Ilex opaca (American Holly)	Forested wetlands along Atlantic coastal plain	Golet et al., 1993; Tiner, 1988
Shrubs		
Rosa multiflora (Multifora Rose)	Deciduous forested wetlands and and wet meadows	Tiner and Burke, 1995
Viburnum prunifolium (Black Haw)	Deciduous forested wetlands in Mid-Atlantic states	Tiner, 1985a; Tiner and Burke, 1995
Vaccinium angustifolium (Lowbush Blueberry)	Northern forested wetlands and bogs	Calhoun et al., 1994
Gaylussacia baccata (Black Huckleberry)	Bogs	Damman and French, 1989; Calhoun et al., 1994
Gaylussacia dumosa (Dwarf Huckleberry)	Bogs	Tiner, 1985a; Calhoun et al., 1994
Kalmia latifolia (Mountain Laurel)	Acidic forested wetlands, especially in the Pine Barrens	Tiner, 1985a; Tiner and Burke, 1995
Asimina triloba (Pawpaw)	Deciduous forested wetlands on the Atlantic coastal plain	Tiner and Burke, 1995
Gaultheria procumbens (Wintergreen)	Acidic forested wetlands, occasionally bogs	Tiner, 1988
Herbs		
Aralia nudicaulis (Wild Sarsaparilla)	Temporarily flooded red maple swamps in New England	Golet et al., 1993
Mitchella repens (Partridgeberry)	Temporarily flooded or seasonally saturated forested wetlands (mostly evergreen)	Tiner, 1988
Claytonia virginica (Spring Beauty)	Deciduous forested wetlands, including floodplain forests	Tiner, 1985a; Tiner and Burke, 1995

TABLE 7.1 (continued)
Examples of Common FACU Species Found in Wetlands in the Northeast

Species	Wetland Types	References
Aster acuminatus[a] (Whorled or Mountain Aster)	Deciduous forested wetlands	Tiner and Burke, 1995
Duchesnea indica (Indian Mock-Strawberry)	Temporarily flooded floodplain, forested wetlands	Tiner and Burke, 1995
Geum canadense (White Avens)	Forested wetlands	Golet et al., 1993; Tiner and Burke, 1995
Anemone quinquefolia (Wood Anemone)	Deciduous frested wetlands	Golet et al., 1993
Allium vinale (Field Garlic)	Temporarily flooded forested wetlands	Tiner, 1985a
Woody Vines		
Parthenocissus quinquefolia (Virginia Creeper)	Shrub and forested wetlands	Tiner, 1988; Tiner and Burke, 1995
Smilax glauca (Cat Greenbrier)	Temporarily flooded red maple swamps and pitch pine lowlands	Tiner, 1985a; Golet et al., 1993

[a] Not on the 1988 wetland plant list, but common in some wetlands.

Note: Many are dominant species in certain wetland types.

the vegetation pattern from obligate wetland species toward a community dominated by facultative upland and upland species. Tillage destroyed the root mass of the spikerushes, creating open ground suitable for annual species, and once established, the annuals seem to persist. In these situations, dominant vegetation is not a reliable feature for classifying an area as wetland or nonwetland. The mere presence of playa lake indicator species like spikerushes, toothcups (*Ammania* spp.), umbrella sedge (*Cyperus acuminatus*) and western water clover (*Marsilea vestita*) may be useful indicators of wetland in disturbed playas. In the absence of drainage structures, the presence of hydric soils also will better reflect the presence of wetland than the dominant vegetation.

SEASONALLY VARIABLE WETLAND PLANT COMMUNITIES

All federal wetland delineation manuals acknowledged the existence of wetlands whose vegetation varies seasonally during the year. Many so-called "seasonal wetlands" have perennial OBL and FACW species dominant during the wetter part of the growing season, while UPL annuals may dominate during the drier season (Environmental Laboratory, 1987). For seasonal wetlands where UPL annuals may predominate the exposed shores of lakes or ponds during summer drawdown, considering the landscape position (lakeside or pondside) and landform (fringe) and examination of soils (for hydric soil properties) will aid in wetland identification.

Some of the wettest wetlands lack vegetation during certain times of the year, typically in winter. These wetlands, which include regularly flooded fresh tidal marshes and exposed river bars, typically are colonized by OBL species in spring and summer, but appear as nonvegetated wetlands at other times. The Corps and federal interagency manuals refer to these areas as *vegetated flats* (Environmental Laboratory, 1987; Federal Interagency Committee for Wetland Delineation, 1989). The dominant plants are mostly nonpersistent emergents — annual plants and fleshy-leaved perennials that dieback in the fall.

Tidal fresh marshes in the eastern U.S. illustrate the drastic change from nonvegetated flat to vegetated marsh during the year. In winter and early spring, these areas appear as mudflats at low tide. In early spring, seedlings of many species predominate. By late spring and early summer, the flats are covered with vegetation including sweet flag (*Acorus calamus*), arrow arum (*Peltandra virginica*), and spatterdock (*Nuphar luteum*). By mid-summer, these and other plants are common including wild rice (*Zizania aquatica*), smartweeds and tearthumbs (*Polygonum* spp.), rice cutgrass (*Leersia oryzoides*), jewelweed (*Impatiens capensis*), water hemp (*Amaranthus cannabinus*), and soft-stemmed bulrush (*Scirpus tabernaemontani*, formerly *S. validus*). In late summer, the yellow flowers of bur-marigold (*Bidens laevis*) may be visually dominant (Tiner, 1985a; Tiner and Burke, 1995). Inspections made during the growing season will detect and confirm the presence of hydrophytic vegetation, while observations at other times will not (unless one looks for overwintering perennial plant parts such as underground tubers, for example). Their wetness is obvious at all times of the year. Knowing the ecology of tidal flats in freshwater will help identify these areas as vegetated wetlands (e.g., regulated wetlands) during the nongrowing season.

CYCLICAL WETLANDS IN ARID TO SUBHUMID REGIONS

While many wetlands may exhibit a seasonal shift in the herbaceous stratum from wetter species to drier-site species, other wetlands undergo profound shifts in vegetation patterns lasting several to many years. In arid to subhumid regions, these changes are due to hydrologic fluctuations associated with the longer-term hydrologic cycle.

Wetlands whose vegetation varies greatly over time due to wide fluctuations in annual water availability typical of the regional climate may be referred to as *cyclical wetlands*. Prairie potholes, playas, and West Coast vernal pools are examples. In arid to subhumid regions, evapotranspiration demands exceed annual precipitation creating a water budget deficit. During wet periods of the natural hydrologic cycle, these wetlands possess positive indicators of hydrophytic vegetation, hydric soils, and wetland hydrology. Yet prolonged droughts typify these regions. During extended droughts, perennial upland plants (UPL) can colonize and eventually dominate certain wetlands in these regions. In agricultural areas such as the pothole region of the Upper Midwest, many wetlands are tilled and planted with crops during dry periods (Kantrud et al., 1989). Uncultivated wetland basins may be colonized by weedy FACU and UPL annuals. Consequently, the vegetation is significantly disturbed and not useful in verifying hydrophytic vegetation. This situation applies equally to playa wetlands in the Southwest and to groundwater wetlands on the Cimmaron Terrace of Oklahoma and Kansas (Taylor et al., 1984). Hydric soil properties reflect the long-term hydrology and are most useful indicators of wetlands under these circumstances, provided the area has not been effectively drained (Tiner and Veneman, 1995; Tiner, 1993).

West Coast vernal pools are cyclical wetlands whose vegetation also changes markedly within and between years in response to fluctuating hydrologic conditions. Winter rains may fill isolated depressions with water from southern Oregon to northern Baja Mexico for weeks or months in some years (Baskin, 1994). Pools range in size from 10 acres to 10 ft wide. Vernal pools achieve maximum size during unusually wet years when individual pools merge to form large pool complexes (Zedler, 1987). The isolated nature and unpredictable flooding of pools promotes endemism in plants and animals, making them especially valuable sites for conservation of biodiversity (Baskin, 1994). Underlying impermeable layers of clay or hardpans of cobbles cemented together by iron and silica create perched water tables that permit water to pond at the surface. Winter rains inundate the pools, while increased evapotranspiration in spring removes the surface water. The typical cycle of vernal pool development may be represented by four phases: wetting, aquatic, drying, and drought. Hydrophytic vegetation is evident during the first two phases, but upland species may invade during the last two phases. Some typical aquatic phase plants in southern California are quillworts (*Isoetes* spp.), water fern (*Pilularia americana*), water fern (*Marsilea vestita*), water starworts (*Callitriche* spp.), flowering quillwort (*Lilaea scilloides*), spikerushes

FIGURE 7.1 Aerial photograph showing West Coast vernal pools. They are the many whitish, irregularly shaped patches and small white rounded areas in the large field left of the airport. (Courtesy of U.S. Fish and Wildlife Service.)

(*Eleocharis* spp.), short-seed waterwort (*Elatine brachysperma*), and water pygmyweed (*Crassula aquatica*) (Zedler, 1987). Many invertebrates occupy the pools in this phase, including tadpole shrimp, fairy shrimps, clam shrimp, water fleas, copepods, ostracods, snails, worms, and insects, with crustaceans being most abundant (Baskin, 1994). During the drying phase, plants like San Diego mesa mint (*Pogogyne abramsii*), coyote thistle (*Eryngium aristulatum*), dowingia (*Dowingia cuspidata* and *D. concolor*), woolly heads (*Psilocarphus* spp.), and hairgrass (*Deschampsia danthonioides*) are among the typical species. The drought phase of these wetlands may have dove weed (*Eremocarpus setigerus*) and vinegar weed (*Trichostema lanceolatum*) that grow during the hottest part of the summer and persist well after other annuals have died back (Zedler, 1987). In the Jepson Prairie (near Sacramento), typical vernal pool species include goldfields (*Lasthenia chryostoma*), allocaryas (*Allocarya* spp.), dowingia (*D. insignis* and *D. humilis*), Boggs lake dodder (*Cuscuta howelliana*), pepper grass (*Lepidium latipes*), foxtail grass (*Setaria* spp.), salt grass (*Distichlis*), Colusa grass (*Neostapfia colusana*), and Solano grass (*Tuctoria mucronata*) (Morris, 1988). The pit and mound or ridge and swale topography is a good landscape clue to the possible occurrence of vernal pool. These features are readily observed on aerial photographs (Figure 7.1). During drawdown, some of the following signs may be useful field indicators of vernal pools:

- Algal mats/encrustations (Figure 7.2a/b)
- Matted vegetation (due to prolonged inundation; Plate 29)
- Remains of aquatic invertebrates in the upper level of the soil (Plate 15)
- Cemented hardpan
- Confining clay layer in the subsoil
- Distinct or prominent redox concentrations in a layer 2 in. thick within the upper 6 in. of the soil (national hydric soil field indicator F8; Hurt et al., 1998)
- Depleted matrix layer at least 2 in. thick within the upper 6 in. of the soil (national indicator F9)

For other indicators and details on how to perform wetland delineations in these complex landscapes, contact the appropriate regulatory agencies.

The Corps of Engineers has been conducting detailed wetland inventories on various military installations in the western U.S., mapping both vegetated wetlands and nonvegetated wetlands (e.g.,

FIGURE 7.2 Hydrologic indicators of nonvegetated wetlands: (a) multi-layered mud cracks and (b) algal crusts. (Photo "a" is courtesy of the U.S. Army Corps of Engineers.)

playas, clay pans, and intermittent streambeds or dry washes). The latter types are officially designated as "waters of the U.S." and not regulated wetlands, since they are below the ordinary high water mark but lack vegetation (Lichvar and Sprecher, 1996).* Besides conventional indicators of surface water (e.g., ponded water, drift lines, and wetland drainage patterns), Lichvar and Sprecher used some new indicators for verifying hydrology that should prove useful for identifying nonvegetated wetlands throughout the arid region of the country — mud cracks, surface staining, algal flakes, and salt crusts (Figure 7.2; Lichvar and Sprecher, 1996; Lichvar et al., 1996). A few types of mud cracks (polygonal, layered, or linear) were observed, but only the multiple-layered type was used as a positive indicator of hydrology sufficient to qualify as a water of the U.S. because it resulted from many flooding events. The shape of the cracks also were diagnostic, with large domed surface mud cracks being positive indicators on small playas and clay pans. Such cracks often had multiple layers as well. Faint to dark reddish-brown stains were used as an indicator of ponded water; these features visible on aerial photographs. Several types of algal flakes were observed, yet not all were considered positive indicators. Apparently, the researchers felt that some algal flakes were not indicative of standing water as they were formed on top of very hard clay pans. Salt crusts were considered secondary indicators of hydrology, but were not relied upon (Lichvar et al., 1996).

* These areas are considered nonvegetated wetlands according to the U.S. Fish and Wildlife Service's wetland classification system (Cowardin et al., 1979).

The variable hydrologic conditions in arid to subhumid regions noted above pose an interesting question for regulatory agencies and the scientific community alike. The concept of average hydrologic conditions is not applicable to these wetlands because they may be naturally dry for years (National Research Council, 1995; see Chapter 2). What frequency and duration of inundation and/or soil saturation is sufficient to separate wetlands from nonwetlands in these situations? Should an area that is extremely wet and dominated by OBL species for 5 years out of 20 years be considered wetland? If so, what about an similar area that is wet for 5 years out of 30 years, and so on? Eventually, a point is reached where the frequency of the event may not be sufficient to consider the area a wetland at least from the current regulatory perspective (e.g., Clean Water Act). From a scientific standpoint, this type of cyclical wetland may be recognized as an ephemeral wetland even if it functions as a wetland just a few times in a century.

CYCLICAL WETLANDS IN PERMAFROST REGIONS

Alaska's permafrost wetlands may be considered cyclical wetlands dependent on the permafrost barrier near the surface. In the subarctic where permafrost is discontinuous, permafrost aggradation is especially important to wetland hydrology of forests (National Research Council, 1995). Vegetation succession from hardwoods to spruce leads to an accumulation of organic matter that over time lowers soil temperature. A permafrost layer can develop through vegetative successsion in 200 years on Alaskan floodplains (Viereck, 1970).

The permafrost layer is greatly affected by recurring wildfires, especially in the discontinuous permafrost region (National Research Council, 1995). Severe fires destroy the confining permafrost layer which is principally responsible for creating wetland hydrology (perched water tables). Loss of the shading from the spruce forest and loss of the insulating moss layer allow the soils to warm, thereby melting the frozen layer. This can change poorly drained soils to well drained soils because the confining permafrost layer is no longer there to perch and retain surface water. Over time vegetation regrowth and the accompanying peat buildup provides the insulation required for permafrost reformation. The permafrost layer may be completely restored in a period ranging from 50 to 200 years after severe burns (Ping et al., 1992). With the restoration of permafrost, the area regains its wetland condition and functions. Mechanical alteration of the vegetation and soils can have similar effects. The thawed permafrost soil retains its low chroma mottles and still has hydric properties, but the area no longer supports hydrophytic vegetation or has wetland hydrology.

Just how permafrost wetlands in the "recovery phase" should be treated from the Clean Water Act regulatory perspective is an interesting question for the Corps and EPA to answer, since regulatory decisions are usually based on the current condition of the area whether or not it was or will at some time in the future be wetland. The National Research Council (1995) recommends further study of these wetlands, but states that permafrost wetlands should not be treated separately from other U.S. wetlands for regulatory purposes.

TROPICAL WETLANDS

These wetlands have received little attention compared to temperate U.S. wetlands. Wakely et al. (1996) applied conventional wetland delineation techniques (Corps manual) to some Hawaiian rain forests to determine whether these forests exhibited widely recognized wetland indicators. In their review of the literature, these researchers found that most of the plant species listed as wet forests in Hawaii (Char and Lamorureux, 1985; Jacobi, 1989) were FAC, FACU, and UPL species on the national wetland plant list (Reed, 1988). They reported that two situations made wetland determinations in rain forests particularly difficult — lack of consistent hydric soil morphology and lack of wetland hydrology indicators except for direct observation of inundation and near surface saturation. Soils were a special problem given that the study area was dominated by Folists (Lithic Tropofolists), Andisols (Hydric Lithic Dystrandepts), and lava flows. Andisols have inherited gray colors from volcanic ash, so low chroma colors are not reliable signs of hydric soils (see later discussion on soils with dark-colored parent materials). Sapric material and a positive α, α'-dipyridyl test proved to be the most reliable hydric soil indicators for wetlands in the study area as many soils with low chroma colors failed to exhibit signs of wetland hydrology (nonwetlands). The researchers did find a better correlation between hydrophytic vegetation defined by a prevalence index derived from plot data and hydric soils (81% agreement) than between hydrophytic vegetation based on dominant species and hydric soils (65%). They also reported that many low-lying wetland depressions may lack hydrophytic plant associations and/or hydric soil properties.

Soils of tropical wetlands also were problematic for other researchers. Buol and Camargo (1992) reported both low and high chroma colors for wet Oxisols. The low chroma colors were found in soils in direct contact with the regional water table. The abundance of hematite in tropical soils may contribute to the problem. Macedo (1986) and Macedo and Bryant (1987; 1989) found reduction of hematite in a saturated Brazilian Oxisol produced hues of 7.5YR and 10YR and lower chroma, but not chroma 2 or less. Mejia (1975) also observed soils in areas of prolonged saturation lacking low chroma colors. Clearly, more studies need to be undertaken in tropical regions of the U.S. to locate reliable indicators of wetlands and improve the list of wetland plant and hydric soil indicators, especially to help separate Folists from hydric organic soils.

PROBLEMATIC SOILS

While soils of most wetlands tend to exhibit diagnostic properties reflecting hydric soil conditions (e.g., accumulation of organic matter, gleyed matrices, and iron-depleted matrices with redox concentrations), a fair number of hydric soils are not easily recognized. Part of the problem lies in the fact that the need for this information is relatively recent. Only in the past decade has an extensive field-based effort been placed on identifying and describing finite properties associated with hydric soils for use in wetland boundary delineation. Previously, soil properties used to classify soils with an aquic moisture regime served as the primary hydric soil indicators as the concept of hydric soils was evolving in the late 1970s and 1980s (Tiner, 1986). The bulk of the problem, however, may lie in the attempt to locate a specific boundary point in a transitional zone along the soil moisture gradient.

The Corps manual mentioned problematic hydric soils including red parent material soils and sandy soils (e.g., accreting sandbars) (Environmental Laboratory, 1987). The federal interagency wetland delineation manual and recent Corps headquarters guidance listed others — Entisols (floodplain and sandy soils), Spodosols (evergreen forest soils), and Mollisols (prairie and steppe soils) (FICWD, 1989; Williams, 1992). In addition, for Vertisols with high clay content and shrink and swell potential and soils derived from dark-colored parent material (e.g., Andisols), separating hydric from nonhydric soils may be difficult.

Problem soils include a wide range of soil conditions. In general, landscape position or landform reveals much information on the likelihood of finding hydric soils and wetlands. The existence of highly reliable vegetation indicators of wetlands, especially OBL species, in suitable landscapes further supports a wetland determination, despite the presence of soils that do not possess typical hydric soil properties. Signs of wetland hydrology (e.g., watermarks, silt deposits, and water-stained leaves) yield even more support. If questions arise in the field on whether the soil exhibits hydric soil properties, it may be advisable to examine the site during the wettest time of the year to assess site wetness relative to the wetland hydrology criterion.

This section provides an overview of the topic, including the more frequently encountered problematic soils. It is intended to give readers an understanding of some of the problems using soils, some guidance on how to evaluate these situations, and some references for more information. The Soil Science Society of America has published a book on problem hydric soils entitled *Aquic Conditions and Hydric Soils: The Problem Soils* (Vepraskas and Sprecher, 1997). This book includes extensive treatments of the more problematic soils and has been used in preparing the following summaries. This year, a book on hydric soils is scheduled for publication which should provide additional worthwhile information (Richardson and Vepraskas, 1999). Areas with rocky soils are discussed under a following section dealing with problematic field conditions.

SANDY HYDRIC ENTISOLS

Sandy Entisols develop in coastal regions, along floodplains, in glacial outwash deposits, in arid regions, and in regions with significant sandy deposits (e.g., Florida, Wisconsin, Nebraska, Georgia, and Michigan). About 1.4 million ha of wet sandy soils (Psammaquents) have been mapped in the U.S., with about 40% of these located in Florida (Kuehl et al., 1997).

Sandy soils have large pores that tend to promote good internal drainage. In some places, external drainage is impeded by relatively impermeable layers (e.g., clays) in the subsoil or by landforms that prevent drainage (e.g., berms and dune ridges). In other places, poor drainage may result from naturally high water tables as local water tables may be interconnected to and directly influenced by regional aquifers in sandy soils (e.g., Cape Cod, MA and the Pine Barrens of southern New Jersey). While many hydric sandy soils have evidence of organic accumulation that is readily associated with wetlands, many sandy hydric soils do not because they often have little organic matter and are less reduced when flooded than finer-textured soils (Daniels et al., 1973). Since the silt and clay materials that best reflect gleization are virtually absent, many hydric sandy soils do not display strong evidence of redox depletions.

Current guidelines require finding evidence of prolonged saturation and reduction within 6 in. of the surface in sandy soils because of their rapid permeability (National Resources Conservation Service, 1995; Hurt et al., 1998).* This may actually be too restrictive as Segal et al. (1995) seemed to find hydric soil indicators within 12 in. in "dry" sandy Florida soils that had water tables within 12 in. for more than 10% of the monthly observations during the 2- to 5-year study interval. Monthly monitoring was not sufficient to accurately detect whether these soils experienced wetland hydrology. In any event, the large pores and the corresponding rapid infiltration might be responsible for such indicators being found below 6 in. even when the water table is above this point for significant periods. For example, Segal et al. found the presence of hydrogen sulfide odor within the upper 12 in. (30 cm) to be one of the best predictors of wetland hydrology. This suggests that the critical depth for locating hydric soil indicators in sandy soils might better be set at 12 in. (same as for nonsandy soils). Data from Comerford et al. (1996) for Florida flatwoods provide further support for requiring a somewhat deeper water depth requirement for sandy hydric soils. In plotting redox potential against water table depth, they found a natural break at 8 in. and concluded that such water tables are sufficient to produce reduced conditions at the surface that are likely to affect plant growth. If the plants in these sandy soils have most of their roots within 6 in., a water table at 10 to 12 in. should result in reducing conditions within the root zone. More work needs to be done on this as the implications are potentially significant.

Some sandy hydric soils may be easily recognized with noticeable accumulations of muck at the surface (including a histic epipedon), high organic matter content in the surface horizon, organic bodies along roots, a hydrogen sulfide odor, blotchy-colored sands, or vertical streaking in the subsoil. Most of these features have been discussed in Chapter 5 (see Table 5.9). Segal et al. (1995) believed that a histic epipedon or a sandy mucky mineral layer (at least 2 in. thick) beginning within 6 in. of the soil surface was not reliable for separating hydric from nonhydric sandy soils in their Florida study, but recognized that it was the best indicator "in low-lying [sandy] flatwoods just outside of cypress domes where water flows from one dome to another." Their difficulty with this indicator may stem from an overestimate of the water depths in the study soils which they readily

---

* The majority of Florida's flatwood soils were on the 1985 hydric soils list (USDA Soil Conservation Service, 1985; Parker, 1985). Most of them were removed from the list when the hydric soil criteria were revised in 1990 by adding criterion 2b(1) for sandy soils, which required a seasonal water table within 0.5 ft to be hydric. This change deleted 38 Florida soils accounting for over 2 million acres from the list (Reed, 1990).

acknowledged (e.g., annual rainfall deficit was 28% below normal at some sites). Areas with these indicators might be drier-end wetlands upon closer inspection. Kuehl et al. (1997) concluded that the presence of muck was the most important hydric soil indicator, along with some others related to organic matter accumulation (mucky mineral layer ≥5 cm and a thick dark surface layer ≥10 cm). They noted that a polychromatic matrix or oxidized rhizospheres also were frequently observed with these indicators.

The presence of a thin layer of mucky sand (2 in. or thicker) on top of a sandy soil is also evidence of prolonged wetness and a hydric sand throughout the country, except Alaska (national hydric soil indicator S1; Hurt et al., 1996). An inch of muck or even less on sand is probably a reliable indicator in the conterminous U.S. (Plate 11), although this has not been recognized nationally (Tiner, 1998). Hurt et al. (1998) list the presence of muck within 6 in. and a 0.5 inch layer of muck as hydric soil indicators (A8 and A9) for tropical and subtropical regions (e.g., south Florida) and southeastern and western regions, respectively. A 0.75 in. muck layer is acceptable for the lower Midwest (indicator A10) and is being tested for use in other regions. Interestingly, Segal et al. (1995) recommended using "an organic surface horizon at least 1 cm thick if a root or leaf mat is present, or any thickness in the absence of a root or leaf mat." This was found to be a reliable indicator in their examination of Florida sandy soils. Although this may work well in southern areas in general, this indicator may not be reliable for northern areas especially boreal and mountainous regions. In these regions, low evapotranspiration and moderate rainfall may produce thin organic layers on moist (not hydric) sands. This problem may be further compounded by the difficulty of separating muck and mucky materials originating from nonwetland conditions (Peter Veneman, personal communication, 1998). Additional study is needed on the issue of organic accumulation in northern climates, in general.

A thick dark A-horizon is commonly recognized as a hydric soil indicator for sandy soils (S7 — dark surface; Hurt et al., 1998; Plate 35). The guidance on this indicator is that this "salt and pepper" colored layer have at least 70% "pepper" (black-coated sand grains) vs. the "salt" (uncoated sand grains). This may be somewhat difficult for soil scientists and others to estimate as both Segal et al. (1995) and Kuehl et al. (1997) noted that soil scientists often disagree on the percentage estimates of the same soil. They also found this indicator in "dry" sandy soils, yet admitted their results were conservative due to uncertainty about water table depths observed vs. normal conditions. They urged caution in using this indicator.

Other sandy hydric soils may show evidence of vertical streaking by organic matter below the surface layer (S8 — polyvalue below surface; Hurt et al., 1998), although Segal et al. suggested that this may not be highly reliable in Florida based on admittedly limited hydrologic data. The black or dark gray streaks on a medium or light gray matrix are easily detected and seem to be useful indicators based on observations in the Northeast (Plates 35-37; Tiner and Veneman, 1995). Some hydric sandy soils may exhibit both low chroma colors (matrix chroma 2 or less) with 2% or more redox concentrations within 6 in. (S5; Hurt et al., 1998; Plate 36). The latter may occur as soft masses or pore linings. It must be recognized that many wet sandy soils exhibit more than one indicator and often a combination of thick dark A-horizon with organic streaking in the subsoil can be seen. Such coupling suggests wetter conditions than if only one of these indicators was present.

Slightly more difficult to recognize as hydric are blotchy-colored sands. These soils have a blotchy-colored subsoil (mixtures of chroma 1, 2, and/or 3 colors) due to variable organic coatings around some of the sandy grains. This condition has been called *polychromatic matrix*, but is now referred to as *stripped matrix* (S6; Hurt et al., 1998; Plate 36). The presence of organic materials in blotchy sands is determined by a simple rubbing test. When rubbed gently on the palm of the hand, organic-coated sand grains from the darker blotches leave a dark-colored, often blackish, stain on the skin. Rubbing the uncoated sand grains from light blotches leaves almost no stain.

Hydric Entisols with loamy fine sand and coarser textures within 20 in. of the surface may be recognized by a matrix chroma of 3 or less provided the hue is 2.5Y or yellower and distinct or prominent redox concentrations are present (Soil Survey Staff, 1996). Soils with these features within 6 in. are hydric. In studying Indiana soils, Franzmeier et al. (1983) found that 3 chroma colors were better predictors of seasonal high water tables than 2 chroma colors in soils on sandy outwash deposits. Soils with a chroma 3 matrix are often Aeric subgroups of aquic suborders which are drier in the upper part. Consequently, the Corps may no longer recognize such conditions as a valid indicator of hydric soil or wetlands (Steven Sprecher, personal communication, 1998).

Recently deposited sandy soils, such as sand bars along rivers, may not possess any typical hydric soil properties, but they can be recognized as hydric soils by their landscape position, associated vegetation, and signs of flooding. Hydric sandy soils associated with tidal marshes can be easily recognized by the obvious nature of this wetland type with OBL hydrophytes and by the smell of rotten eggs (hydrogen sulfide).

Nonsandy Hydric Entisols

Many of the nation's wetlands occur on floodplains where they are annually or biannually flooded for significant periods. Sediments eroded from upland areas are deposited on the alluvial plains, forming loamy or fine-textured soils. Many of the finer textured hydric Entisols possess typical hydric soil properties (see Chapter 5).

As depositional environments, floodplain soils are frequently being buried by new materials brought in by flood waters. These Entisols are young soils and often have little or no evidence of soil weathering including the presence of redoximorphic features. Floodplain (alluvial) soils typically have buried surface layers (A-horizons) at various depths. All floodplain soils are not hydric, since many are only infrequently flooded. Of those that are hydric, some possess typical hydric soil properties, while others do not. Lindbo (1997) reviewed soil data for 128 Fluvents and 42 Fluvaquents from across the country. He found that the introduction of soil from various parent materials presented problems for hydric soil recognition. In particular, some parent materials were low chroma soils due to low iron concentrations, while other floodplain soils were derived from red parent materials and failed to express their wetness in their morphology. Low organic matter was also commonplace in floodplain soils as was the likelihood for saturation with aerated water which prevents reduction.

In many areas, widespread land clearing in the last century resulted in tremendous erosion. The eroded soils buried soils with more typical hydric properties. The deposited soils, however, may not have had time to develop soil properties despite still having wetland hydrology as reported for the southern Appalachian region by Wakely (1994). Such areas should still support hydrophytic vegetation and show signs of inundation (e.g., silt marks and water-deposited debris). In spite of their lack of hydric soil morphology, the soils are, in fact, hydric by virtue of their frequency and duration of flooding (hydric soil criterion 4). A buried soil with typical hydric soil properties might be found deeper in the profile, as seen in a streamside forested wetland in a narrow drainage in North Carolina (author's observations). The depositional layer may or may not have hydric properties despite being sufficiently wet. In these cases, the occurrence of such horizon coupled with a favorable landscape position (streamside) and the presence of reliable hydrophytic species should provide ample evidence to support a wetland determination. Clearcutting of forests on steep slopes has produced similar effects (e.g., buried hydric soils) elsewhere (Mike Whited, personal communication, 1998). Veneman and Tiner (1990) found evidence of prominent high and low chroma mottles within 3 years of deposition in floodplain soils along the Connecticut River in western Massachusetts.

An additional hydric soil indicator is used in the Southeast — for loamy floodplain soils with a layer having 40% or more chroma 2, the presence of 2% or more black (manganese) or reddish brown/orange (iron) mottles occurring as soft masses (F12; Hurt et al., 1998; Plate 22). Another possible indicator is acceptable in the eastern U.S. and the upper Midwest (including North Dakota and northern Montana): the presence of stratified layers within the upper 6 in. (15 cm), with one or more layers of value 3 or less and chroma 1 or less, and/or having a muck, mucky peat, peat, or mucky modified mineral texture; the remaining layers must have values of 4 or more and chromas of 2 or less (A5; Hurt et al., 1998). In sandy floodplain soils, the presence of a thin stratum of loamy or clayey material with redoximorphic features may indicate hydric soils and wetland conditions (Mike Whited, personal communication, 1998). Some floodplain soils are predominantly red-colored due to the deposition of soil derived from red parent material (see discussion on red parent material soils).

It is important to remember that, in general, landscape position, vegetation, and evidence of flooding are useful for distinguishing the hydric floodplain soils from the nonhydric ones. For larger floodplains, hydrologic data may be available from the U.S. Geological Survey, U.S. Army Corps of Engineers, or other sources to aid in predicting the frequency and duration of flooding of various elevations on the floodplain.

HYDRIC SPODOSOLS

Spodosols are associated with evergreen forests typically on sandy soils, but also on loamy soils in northern areas. They are common in northern temperate and boreal regions of the U.S. and along the Atlantic coastal plain from New Jersey through Florida. They develop under wet or dry conditions in evergreen forests of hemlocks, spruces, and pines and also in larch and oak forests (Soil Survey Staff, 1975; Buol et al., 1980). The breakdown of the organic leaf litter from these forests creates organic acids that leach downward through the soil, stripping clean the sand grains in the subsoil layer just below the soil surface. The process is called *podzolization*. This leached layer (the E-horizon, eluvial) is often gray-colored, since it is usually free of organic matter, aluminum, and iron oxides in well-drained conditions. The latter materials are deposited in an underlying horizon called the spodic horizon (Bh-, Bs-, or Bhs-horizon). Whether wet or dry, most Spodosols have a characteristic gray E-horizon overlying a diagnostic spodic horizon of accumulated organic matter and aluminum with or without iron. Plates 5 and 38-40 show the variability in morphology among hydric Spodosols.

Wet Spodosols develop at sites of poor drainage, especially on the broad flat areas (e.g., interfluves) on the coastal plain. An estimated 6.7 million ha of wet Spodosols (Aquods) may exist in the U.S., with over 4 million ha mapped (Kuehl et al., 1997). These soils are most abundant in Florida (2.6 million ha), Alaska (419,944 ha), North Carolina (207,173 ha), Michigan (195,529 ha), and Georgia (135,578 ha).

The underlying spodic horizon was once believed to be an indication of the "most commonly occurring depth to the water table" as "organic pans" (within 12 to 30 in. of the surface) were listed as hydric soil indicators in the Corps manual (Environmental Laboratory, 1987). Studies by Evans and Mokma (1996) demonstrated that the spodic horizon was strongly stabilized and not responsive to saturation and water table fluctuations. Segal et al. (1995) concluded that the indicators listed in both the Corps and federal interagency manuals for wet Spodosols (i.e., a spodic horizon from 12 to 30 in., a thick dark surface horizon and a dull gray E-horizon) were poor predictors of hydric Spodosols as these features were found at both wet and "dry" sites in Florida (based on limited hydrologic evidence). Pilgrim and Harter (1977) did not find reliable morphologic evidence of wet Spodosols in New Hampshire and suggested using water table depths to separate wet ones (Aquods) from dry ones. The problem with forming redoximorphic features in hydric Spodosols may be caused by their low iron availability, as any iron introduced into the wet environment is readily mobilized (Smith, 1986; Rourke et al., 1988; Nichols et al., 1990).

Kuehl et al. (1997) presented a good review of the Spodosol problem in Florida. They concluded that the most important hydric soil indicator was the presence of muck. Other indicators were a mucky mineral layer at least 2 in. thick, and a thick dark surface (at least 4 in. thick). They, like Segal et al. (1996), had trouble identifying the latter indicator since determining whether 70% of the sand grains were organic-coated (black) was not easy.

TABLE 7.2
Possible Indicators of Hydric Spodosols

 1. A thin surface layer of peat or muck (not leaf litter).*
 2. A mucky mineral surface layer.*
 3. A histic epipedon.*
 4. A thick black-colored sandy surface layer (70% black grains "pepper" and 30% white grains "salt").*
 5. Streaks of organic matter in the E-horizon.*
 6. Masses of sand grains coated with organic material giving the horizon a blotchy appearance.*
 7. High chroma mottles and/or oxidized rhizospheres (at least 5% or more) within the E-horizon and usually within 12 in. of the soil surface.
 8. Iron concretions within the spodic horizon (usually within 12 in. of the soil surface).
 9. A partly or wholly cemented spodic horizon (ortstein) near the surface.
10. Mottling within the spodic horizon.
11. A 2 in.-thick (or more) dark spodic horizon (value and chroma of 3 or less) within 12 in. of the surface and with 5% or more redoximorphic features in the underlying horizon.
12. A spodic horizon within 10 in. of the surface and an E-horizon with 5% or more redoximorphic features.

Note: Those marked with an asterisk are best used with the sandy Spodosols and are also useful indicators of sandy hydric soils.

Sources: Zampella, 1994; Segal et al., 1995; Tiner and Veneman, 1995; NEIWPCC Wetlands Work Group, 1995; Kuehl et al., 1997.

In studying pitch pine-dominated lowlands in southern New Jersey, Zampella (1994) found that the wetter Spodosols had thick O- and A-horizons and lacked mottling and that the frequency of black to very dark A-horizons generally decreased from wet to dry areas. Bh-horizons with values and chromas of 2 or 3 were characteristic of soils with water tables from 6 to 24 in. (15 to 60 cm) below the surface, whereas thin, lighter colored spodic horizons (values and chromas >3) were typical of better drained soils. Despite these relationships, there were exceptions which led him to conclude that soils alone should not be used as indicators of soil saturation in somewhat poorly drained to very poorly drained soils of the Lakewood catena. He recommended considering topography, soils, vegetation, and fire effects when interpreting water tables levels in such soils. One feature that was not evaluated in his study was organic streaking (blotchiness in the E-horizon) which Peter Veneman and the author found useful for separating hydric from nonhydric soils in this region. Blotchiness is not, however, a useful indicator when the E-horizon is thin.

In some places, wet Spodosols may only form on certain types of parent material. If this is true, then being able to recognize the parent material can assist with hydric Spodosol identification. For example, the New Hampshire soil survey program found that wet Spodosols did not develop on glacial till, with one possible exception, the Westbury series (Pilgrim, 1996). Wet tills produced Aquepts and not Aquods.

Some potentially useful indicators of hydric soils are listed in Table 7.2. Given that the morphology of Spodosols is variable, the more of these indicators possessed the more likely the soil may be a hydric Spodosol. For questionable sites, the best course of action may be to make observations during the wet part of the growing season and use visible evidence of site wetness plus considerations of landscape position, hydrophytic vegetation indicators, and other signs of wetland hydrology to identify wet Spodosols (wetlands). Some other useful references on Spodosols include Mokma (1993) and Mokma and Sprecher (1994).

HYDRIC MOLLISOLS

In arid and subhumid interior regions of the country, precipitation is not sufficient to support vast forests, so grasslands (native prairies), and grasslands mixed with shrubby vegetation (steppes) predominate. Here forests tend to be limited to higher elevations in the mountains and to floodplains (riparian corridors). Grasslands characterize the middle of the country from eastern Illinois to Montana, south to Texas. Native grassland soils called Mollisols have formed in these areas. Mollisols typically have thick, dark-colored surface layers due to the accumulation of organic matter from the breakdown of extensive root systems associated with prairie and steppe grasses (Soil Survey Staff, 1975; Buol et al., 1980; Plate 41). These dark colors may suggest hydric soils, but are attributed to root decomposition and mixing of this organic matter with the soil by burrowing animals (e.g, earthworms, ants, moles, and prairie dogs) and high calcium content. In typical wetland landforms (e.g., depressions and sloughs), hydric Mollisols develop under conditions of prolonged anaerobiosis and reduction. The problem in using soil morphology to distinguish between hydric and nonhydric Mollisols is that in most cases the distinguishing features are located at considerable depth, well below the surface (Bell and Richardson, 1997). Further complications are brought about by the landscape in which wet Mollisols occur. They are in depressional areas often surrounded by cropland. Consequently, they have collected much eroded material in the last century, burying the original surface horizon. The top layer may not exhibit redoximorphic features. While the structure and color of the deposited material is similar, the buried layer can be detected by examining the carbon content where an increase in organic matter content should be observed at the depth of the original surface layer. There also may be a textural change at this depth (Bell and Richardson, 1997).

Several hydric soil field indicators that can be used to identify hydric Mollisols are listed in Hurt et al. (1998, Table 5.9) — a thin layer (at least 0.25 in. thick) of peat or muck (A9), depleted below dark surface (F4), a thick dark surface (F5), high chroma mottles (including oxidized rhizospheres, soft masses, and concretions) within 12 in. of the soil surface (redox dark surface; F6), and redox depressions (F8). Test indicators TF4 (2.5Y/5Y below dark surface) and TF7 (thick dark surface 2/1) also are other potentially useful indicators under consideration. Other possible hydric Mollisols indicators include a mollic epipedon with the remains of aquatic invertebrates within 12 in. of the surface in pothole-like depressions (Tiner, 1993), black colors (N 2/0) in the upper 12 in. with a chroma of 1 or less in the remainder of the mollic epipedon (Bell and Richardson, 1997), and mottles in the lower part of the mollic epipedon in association with a high value, chroma 1 or less matrix or a chroma 2 matrix with mottles below the mollic epipedon (Mike Whited, personal communication, 1998). Hydric Mollisols also may have yellower hues (2.5Y and 5Y) than nonhydric ones (Mike Whited, personal communication, 1998). In many cases, the color of the soil below the thick mollic epipedon must be examined. The hydric Mollisols should have gray colors beneath, while the nonhydric Mollisols typically have more reddish or yellowish colors (Mike Whited). Bell and Richardson also noted the significance of a depressional landscape position, the presence of hydrophytic vegetation and other signs of hydrology, and a profile darkness index as other possible indicators for hydric Mollisols. The latter is an index relating thickness and darkness of surface horizons to wetness. In other words, the wetter soils should have thicker and darker epipedons. Examples of hydric Mollisols are presented in Plates 14, 41, and 42.

HYDRIC VERTISOLS

In the Southwest from Texas to southern California, soils with high shrink and swell potential due to high content of montmorillonite clays have developed. These soils called Vertisols create a gilgai microtopography of low mounds and shallow depressions. Ponding may occur in the depressions during rainy periods. The high clay content in these soils causes them to swell upon wetting, creating a seal that ponds water. When drying, large cracks form as the soil shrinks. Soil from the surface then falls into the cracks, and when the soil is rewetted during the rainy season, the soil expands. Consequently, the surface layer is often very thick and dark-colored due to this mixing of organic-enriched soil near the surface. These colors may suggest hydric soil properties, but they are not necessarily produced by wetland hydrologic conditions; the high clay content and regional climate create these properties in both well drained and poorly drained soils. In Venezuela, hydric Vertisols occur on flat or concave backswamp deposits that are ponded for long periods (Comerma et al., 1978). Jacob et al. (1997) addressed the hydrologic and morphologic problems associated with Vertisols, noting that "any Vertisol that is potentially a hydric soil is a problem hydric soil" and that "there are no easily defined hydric Vertisols." The taxonomy of these soils has been revised to include an aquic suborder — the Aquerts (Soil Survey Staff, 1996). Previously, the Chrom and the Pell subgroups contained both hydric and nonhydric members and there was considerable controversy over the groupings of these soils internationally (Comerma 1985).

Jacob et al. (1997) considered depressional landscape positions (where ponding occurs) to be the best feature for separating hydric from nonhydric Vertisols as long as soil morphology and plant indicators did not suggest otherwise. They noted a problem with this feature in that depressions are not always easy to discern in the flat coastal plain. From the morphology standpoint, they reported that iron pore and ped linings or coatings were the most common redoximorphic features associated with hydric Vertisols. Other hydric properties may include iron and/or manganese nodules or concretions (sometimes resembling buckshot pellets) in a thick, blackish colored surface layer (A-horizon), gray-colored (low chroma) mottles within the A-horizon; and dominant low chroma colors immediately below the A-horizon and within 20 in. of the soil surface. Some other possible indicators are hues of 2.5Y and 5Y in combinations with reddish or high chroma mottles (Ahmad, 1983). Tensiometers may need to be employed to determine soil moisture during the wetter part of the growing season in the most complicated situations.

WET ARIDISOLS

Aridisols are soils mostly associated with deserts or steppes (dry grasslands), although some occur along the coast in tidal areas (Boettinger, 1997). Wet Aridisols were formerly designated by the aquic subgroup (e.g., Aquic Natrargids, Aquic Durorthids, and Aquentic Durorthids) or as Salorthids, but now are recognized as Aquisalids (Soil Survey Staff, 1996). They form in drainageways, depressions, floodplains, playas, and nearly level former stream or lake terrraces (Witty and Fenwick, 1985). They are poorly drained to somewhat poorly drained soils.

Boettinger (1997) provides an extensive review of Aquisalids that is briefly summarized below. High evapotranspiration in these regions promotes saturation to the surface in these soils through capillary action. This process brings salty water to the surface, leading to an accumulation of carbonates, gypsum, and other salts. Redoximorphic features are not well expressed in these soils due to several factors: high pH (demands low redox potentials to reduce iron and manganese), low availability of iron and manganese, and low microbial populations (due to high salt content and low organic matter). If reduction does not occur, saturated soils may be oxyaquic and possibly aerobic wetlands (see later discussion on these types). The absence or weak expression of redoximorphic features makes hydric soil identification difficult.

Twenty-four series of Aquisalids have been mapped in 11 western states from Texas to California and north to Montana and South Dakota. They are largely colonized by halophytic species (*Distichlis spicata, Allenrolfea occidentalis, Sporobolus airodes, Atriplex* sp., and *Sarcobatus vermiculatus*). Along the Texas coast, some Aquisalids are tidally flooded. Their sparse vegetation includes Gulf cordgrass (*Spartina spartinae*) and sea ox-eye (*Borrichia frutescens*).

Typical properties of inland Aquisalids are (1) distinct white salt crusts, (2) an organic-enriched A-horizon at vegetated sites, (3) 10YR and 2.5Y hues commonly with a chroma 2 matrix, and (4) horizons with calcium carbonates that effervesce with 10% hydrogen chloride. Many Aquisalids (18 of 24 series) have a chroma <2 with redox concentrations below 12 in. (30 cm). Only 6 of the Aquisalid series had a low chroma matrix and high chroma mottles within 12 in. Coastal series have better expression of redox depletions. The matrix color of some Aquisalids changes color when exposed to air; some of this may be due to precipitation of salts or simply that the salts are more visible upon drying.

Boettinger (1997) proposed using the presence of a salt crust (brittle crust at ponded sites vs. fluffy snowlike salt crusts where saturated by the capillary fringe) as a field indicator for Aquisalids. She also recognized that vegetation and hydrology also may be used to identify a hydric soil where salt crusts and redoximorphic features are lacking.

RED PARENT MATERIAL SOILS

Soils derived from red parent materials (e. g., strongly weathered clays and exposed Triassic and Jurassic sandstones and shales) occur in many places in the country. Some of these soils formed in place due to weathering over millions of years, while others have been deposited by various processes (fluvial, deltaic, aeolian, and glacial) (Blodgett et al., 1993). Other red soils are buried beneath sedimentary deposits. Red soils are common in the Southeast (e.g., some Ultisols of the Piedmont), but they also are frequent in the Midwest, Southwest, and West (e.g., Permian red beds in Kansas and Jurassic sediments of the eastern Rockies), in the tropics (Oxisols), and in glaciated landscapes where red sandstone and shale formations (ancient landforms) are exposed and eroding (e.g, central Connecticut). Glacial lake sediments and tills associated with Lake Superior and Lake Erie have red pigments (Johnson, 1983; Blodgett et al., 1993). In the Northeast, red parent materials can be found in the central part of southern New England, central New Jersey, and in the Hampshire and Mauch Chunk formations in western Maryland, West Virginia, and Pennsylvania (Mid-Atlantic Hydric Soil Committee, 1998). Tropical regions of the country also may have red-colored soils (see discussion of tropical wetlands in previous section). The red colors are attributed to the dominance of the iron mineral — hematite. These colors are redder than 10YR (on the Munsell charts) and obscure low chroma colors that normally develop under anaerobic, reducing conditions. Some hydric soils with red colors may have low chroma mottles present within 1.5 ft of the soil surface, but many do not.

Vegetation, other signs of wetland hydrology, and landscape position usually provide the best clues for recognizing these areas as wetlands. Wet season field inspections may be required to verify the hydrology parameter. Hurt et al. (1998) included indicator TF2 for red parent material soils: a layer within 12 in. of the soil surface that is at least 4 in. thick with a matrix hue of 7.5YR or redder and chroma 3 or less with 2% or more redox depletions and/or redox concentrations as soft masses and/or pore linings. Another possible indicator is under consideration for floodplains in the Piedmont region of the eastern U.S.: redox concentrations within 6 in. and a depleted matrix within 12 in. This indicator may be called *Piedmont depleted* (Wade Hurt, personal communication, 1998).

SATURATED SOILS NOT FORMING HYDRIC PROPERTIES

While most soils subjected to recurrent prolonged saturation develop typical hydric soil properties, some do not. In order for reduction to occur with the resultant hydric soil morphology, there must be a source of organic matter, the temperature must be high enough to support microbial activity, and free oxygen must be absent (Diers and Anderson 1984). There are certain situations where reduction does not occur despite the soil being saturated repeatedly for long periods. Moormann and van de Wetering (1985) listed four conditions where saturated soils may not be reduced.

1. Cold climates with average temperature less than 1°C.
2. Very saline waterlogged soils of deserts where high salinity restricts growth of reducing microbes.
3. Areas with little or no organic matter and moderate to high amounts of calcium carbonate limit reduction in arid and semiarid regions (e.g., irrigated rice basins in northwest India lack a low chroma matrix).
4. Groundwater discharge areas where considerable dissolved oxygen is present (e.g., areas of moderate relief and lateral water movement and soils on the edges of valleys).

FIGURE 7.3 Possible aerobic wetland on gravel bar along a cold mountain stream.

Saturated coarse sands and gravel beds along mountain streams also may be another example (Figure 7.3). McKeague (1965) reported that oxidizing conditions occurred well below the water table in some sandy soils and in one sequence of clayey soils, except when the water table was at the surface. He attributed this to a lack of organic matter and slow microbial activity. When the water table reached the surface, organic matter became available to initiate reduction. Soils that are seasonally saturated yet not reduced do not develop typical hydric soil properties (Daniels et al., 1973; Couto et al., 1985; Vepraskas, 1996) and some of them may be wet enough to qualify as wetlands. Vepraskas also noted that even when saturated and reduced, high pH soils and soils lacking iron oxide minerals may not form mottles or low chroma colors.

Vepraskas and Wilding (1983) described soils with aquic moisture regimes (i.e., reduced and saturated to the surface during some time of the year) in Texas that were wetter longer than they were reduced. They point to a problem with the definition of this moisture regime; the requirement for reduction, while morphological features (e.g., low chroma colors) used to indicate this moisture regime demand that iron, not oxygen, be reduced. Therefore, the threshold for reduction is higher in terms of the morphological indicators in contrast to the low oxygen condition of the initial phase of reduction. In their observations, Vepraskas and Wilding noted that periods of saturation and reduction do not coincide; some soils were saturated longer than reduced, while others were reduced longer than saturated due to differences in water table recharge. The latter soils showed typical hydric soil properties, whereas the former did not. The researchers recommended that the aquic moisture regime include soils with dominant chromas of 3 on ped surfaces when accompanied by a mottled ped interior with a matrix of chroma 4 or more.

In these types of situations, one must rely more on the landscape position and associated vegetation. In some places, there may be mucky mineral material on the surface or within thick A-horizons overlying a brighter subsoil (chroma 3 or 4). This indicator coupled with some positive hydrophytic indicators, e.g., OBL species like skunk cabbage or the presence of peat mosses on the coastal plain, suggest hydric soil conditions despite the brighter subsoil. In some cases, hydro-logic studies will be required to make a wetland determination.

DARK-COLORED PARENT MATERIAL SOILS

These lithochromic soils are derived from dark parent materials often with little or no iron or manganese (Moormann and van de Wetering, 1985). They occur where the parent materials are carboniferous (i.e., containing fossilized organic matter such as coal) as in many coal mining regions across the country (Peter Veneman, personal communication, 1998). Other dark-colored parent materials include phyllite, gray sandstones, conglomerates, and argillite such as found in eastern

Rhode Island (Kolesinskas and Sautter, 1990). Nonhydric soils from these and similar materials will possess blackish or grayish colors typically ascribed to hydric soils. The landscape position of these soils as well as their vegetation should be obvious clues to their nonhydric status. In addition, they should lack redox concentrations typically found in hydric soils subject to alternate wetting and drying.

Andic soils (volcanic ash) are gray-colored soils derived from volcanic ash. Even the wet ones are gray more to parent material than to iron reduction (Moormann and van de Wetering, 1985). These soils are most abundant in Alaska and Oregon, but also are found in California, Idaho, and Washington. McDaniels et al. (1997) have described conditions associated with Aquands (wet Andisols) which are summarized as follows:

Some of these soils have typical hydric soil properties such as a histic epipedon or a positive reducing test using α,α'-dipyridyl. Kyuma and Mitsuchi (1985) also recognized these features, plus typical hydric mineral soil mottling patterns. The gray colors inherited from the volcanic ash naturally produce low chroma colors that complicates recognition of hydric Andisols. Reduction of iron is further limited by the formation of high chroma colors (e.g., ferrihydrites). McDaniels et al. (1997) offer the following indicators of Aquands: accumulation of organic carbon in soil, surface layers of peat or muck, and dark-colored mineral surface horizons. They admit that such indicators are very conservative and suggest that until more studies are conducted that greater weight should be placed on hydrology and vegetation indicators for identifying hydric Andisols and associated wetlands.

CRYIC SOILS

These cold climate soils in areas with short growing seasons have not received much attention in regard to hydric soil indicators. Most of the work done in the U.S. to date is for Alaskan soils and not for areas in the lower 48 states. Studies by Clark and Ping (1997) and Ping et al. (1992) have demonstrated that reduction occurs whenever the soil is above 0°C and have recommended that the growing season based on biologic zero be redefined. They also indicate the fluctuating nature of the soils naturally changing from a poorly drained permafrost state to a well-drained permafrost-free condition induced by wildfires (see Chapter 2 for discussion on cyclical wetlands). These conditions lead to the presence of hydric soil and redoximorphic indicators on well drained soils (uplands) that were once poorly drained (wetlands). Some Alaskan soils also developed from low chroma parent materials. Problems with locating reliable hydric soil indicators due to these conditions led Ping et al. to conclude that, for these areas, wetlands cannot be characterized by any single criterion, and that all three criteria (vegetation, soils, and hydrology) must be examined. They note that hydrology appears to be the more consistent indicator of wetland conditions, yet little data exist and collection of such is expensive and time-consuming.

Cryic soils also occur at higher elevations and in northern areas of the conterminous U.S. Little or no data exist for hydric soils under these conditions. Some soil scientists feel that the hydric soils may not have developed typical hydric properties due to a lack of organic matter and slow microbial action (Sid Pilgrim, personal communication, 1998). Until sufficient studies are completed, it may be advisable to consider the combination of landscape position, the presence of hydrophytic vegetation, and signs of wetland hydrology as indicators of wetlands, in the absence of reliable hydric soil indicators. Wetland delineations may be necessarily conservative in such areas.

NEWLY FORMED SOILS

Wetlands may be intentionally created (for wildlife habitat improvement, rice culture, cranberry cultivation, or mitigation projects) or accidentally formed by undersized road culverts, impoundment seepage, irrigation drainage water, or other reasons. Abandoned sand and gravel pits often contain wet areas as water tables may have precluded further excavation. Stormwater basins (e.g., sumps) plus abandoned settling ponds also may become vegetated wetlands. Newly created wetlands also may form behind beaver dams.

Whenever a nonhydric soil becomes flooded for more than 1 week during the growing season in most years, it is considered hydric by hydric soil criteria 3 and 4. Of course, the permanency of the beaver dam or human action (e.g., road impoundments) must be considered before calling any flooded area a hydric soil or wetland. For example, if a beaver dams a road culvert blocking flow and flooding low-lying nonhydric soils and it is likely that someone will remove the blockage and attempt to prevent further beaver activity, the action is temporary and the area should not be considered to have newly created hydric soil or wetland. If the flooded condition has lasted for some time and wetland vegetation (e.g., cattails and water lilies) is established and upland plants are dying or dead, then the area should be considered to have newly created hydric soil and to be wetland. Given the recent flooding in these cases, the soil properties will not be typical of hydric soils, but will retain their characteristic upland properties. It usually takes decades or longer for a soil to develop the gleying and mottling patterns associated with most hydric mineral soils.

Soil properties can change within a relatively short period. Under ideal circumstances, redox depletions and pore linings may begin to develop after 1 week of flooding (Vepraskas et al., 1995, as reported in Lindbo, 1997). Mitsuchi (1992) reported that for originally well-drained alluvial soils subjected to flooding for rice cultivation, the upper soil surface became gleyed (bluish gray) a week after flooding. Amatekpor (1992) investigated changes in soil properties associated with flooding brought about by the construction of a dam in Ghana. By examining and describing soil profiles in a zone scheduled for periodic flooding by the dam, he was able to record any changes in soil properties. After 5 years of seasonal flooding (in the drawdown area), one soil type (an Ultic Paleustalf) experienced significant changes while the other soil type (a Typic Pellustert) did not. Color changes in the former included the development of a low chroma matrix (10YR 7/2) with redox concentrations below 20 in. (50 cm) which had been 7.5 YR 5/6 matrix with no mottling before flooding. Redox concentrations also formed in the upper part to 8 in. (20 cm). These changes were great enough to alter the classification of the soil to an Aquic Kandiustalf.

Somasiri and Deturck (1992) reported that after 80 years an irrigated rice paddy soil (Hapludults) from Sri Lanka developed a gleyed horizon (5G 5/1) over a brighter 10YR 4/6 subsoil. They mentioned that the degree of reduction and development of gleyed colors, or the shift from redder hues to less red hues, seemed to depend on the length of saturation, the amount of organic matter present, and the duration of wetland paddy farming. Mitsuchi (1992) stated that in a decade or so of rice cultivation in Japan, well-drained rice soils acquire soil properties associated with periodic flooding such as surface gleying and subsurface grayization.

In newly created hydric soils, organic matter also may accumulate at the soil surface in detectable amounts. So, depending on the circumstances, some signs of hydric conditions (redoximorphic features and organic matter buildup) may be visible within the first year, yet the soil will not likely exhibit hydric soil properties in the short term. Consequently, vegetation and signs of hydrology are often the best clues for identifying these wetlands. Oxidized rhizospheres in the upper 2 to 3 in., water-stained leaves, and algal mats have been observed in sparsely vegetated wet "soils" of abandoned sand and gravel pits (personal observations).

Anthraquic Soils

Certain soils are intentionally flooded to produce crops. Rice, sugar cane, mints, and cranberries are some common examples. In the southern U.S., rice fields may be constructed in both wetlands and uplands with artificial water control (Figure 7.4). Rice paddy soils clearly meet the definition of hydric soil since they are inundated for significant periods during the growing season. When rice fields are built on hydric soils, they possess typical hydric soil properties. When the paddies are constructed on nonhydric soils, their soil properties may vary from the original nonhydric soil depending on the length of time that rice cultivation has been practiced. Soil properties for older

FIGURE 7.4   Louisiana rice field.

permanent paddies may appear like an "inverted" hydric soil (or "inverted gley") (Moormann and van de Wetering, 1985), with a gleyed layer at the surface overlying nongley colors of the original nonhydric soil. The wetland identification problem associated with rice culture in Louisiana and elsewhere in the Southeast is due to the crop rotation history of these areas. In some years, they exhibit wetland characteristics due to rice production, while in other years, other crops may be planted. The water management practice makes these wetlands ephemeral features on the landscape. Regulatory agencies must be consulted for guidance on how to handle these "wetlands" from the regulatory perspective.

Cranberry bogs are typically constructed in existing wetlands, yet bogs have been created in former upland sandy soils by excavating down to the water table. These "upland" bogs are often built adjacent to existing "wetland" bogs. The cultivation of cranberries does not involve crop rotation, so once a bog is constructed it is maintained as a wetland until abandoned. When abandoned, even the "upland" bogs may retain wetland characteristics as long as the water table remains at or near the surface for sufficient periods.

RELICT HYDRIC SOILS

Due to dramatic shifts in climate over thousands or more years, the hydrology of many areas have changed around the world. For example, former river valley and lake basin soils in arid and semiarid regions may have developed hydric properties during former times when a humid climate occurred. Some low-lying areas near the coast in glaciated regions in North America and elsewhere in the Northern Hemisphere were once submerged by marine waters during the recession of the Wisconsin glacier 10,000 or more years ago. These lands had been compressed by the weight of the glacier and as the glacier receded, their surfaces began to rebound. At first, their surfaces were still low enough to be flooded by coastal waters, but in time, they rose above the tidal zone and either became freshwater wetlands or nonwetlands, depending on elevations and local hydrologic conditions. In the eastern U.S., this occurred north of Boston, mostly along the New Hampshire and Maine coasts. These soils developed hydric soil properties, yet such horizons underlie contemporary soil horizons. Soils that formed in former lake basins (e.g., extinct glacial lakes) also may have typical hydric soil properties at some depth below the current soil surface. In many of these areas, the hydrology is still wet enough to support wetlands today, while in other areas (e.g., more elevated, convex landforms) current hydrology is not that of wetlands and the iron-depleted layer is often below a noticeably well-drained layer. Where wetland vegetation and other signs of wetland hydrology are lacking, it is most likely that the hydric properties deeper in the profile of such soils reflect past conditions.

FIGURE 7.5   Roots exposed well above the ground suggest subsidence due to drainage.

Other situations may be responsible for the presence of relict hydric soils, such as fluvial dynamics and tectonic processes that reshape the landscape. Major rivers like the Mississippi formed extensive valleys as the river course meandered across the flat landscape over time. Today, the river may be tens of miles or more from its former course. Hydric soils formed by alluvial processes may no longer be flooded often or long enough to qualify as wetlands. In some cases, the change in hydrology is due to natural changes in river dynamics, but in many other cases, the hydrology has been significantly altered by human alterations at the landscape level (e.g., freshwater diversion, leveeing, and channelization). In other regions, mudslides, volcanic activity, earthquakes, and other tectonic processes reshape the land surface thereby affecting surface and ground-water conditions. Former wetlands may now have sufficient drainage to prohibit the retention of water that created them and resulting in better drained conditions today. Geologic processes such as stream down-cutting also may have similar consequences (Vepraskas, 1996). Some studies that reference relict hydric soils include Ruhe et al. (1955), Schelling (1960), Van Hessen (1970), Daniels et al. (1973), Van Wallenburg (1973), Vepraskas and Bouma (1976), Bouma (1983), Coventry et al. (1983), Franzmeier et al. (1983), Vepraskas and Wilding (1983), Coventry and Williams (1984), Childs and Clayden (1986), and Richardson and Daniels (1993).

All the above changes complicate interpretation of soil properties. Again, the emphasis on contemporary vegetation and signs of hydrology (e.g., oxidized rhizospheres and water-stained leaves) and observations of likely hydrologic alterations are most useful for evaluating current wetness. Along major rivers, recorded hydrologic data may be available to assess current hydrology relative to historic levels. If necessary, visit the site during the beginning of the wetter part of the growing season and determine if wetland hydrology exists.

DRAINED HYDRIC SOILS (PARTLY DRAINED WETLANDS)

Hydrologically altered wetland systems are difficult to interpret. In these cases, the hydrology has been significantly changed by drainage ditches, tile drains, groundwater withdrawals, regulated riverflows, surface water diversions, and similar actions. Hydric soils usually retain their hydric properties when drained. Only in the coarser-textured soils may properties change over a few decades (Moormann and van de Wetering, 1985). Woody plants also may be tolerant of the altered conditions and continue to predominate. This essentially negates the interpretive value of soils and woody vegetation for determining the presence of wetlands, forcing one to use other features to verify wetlands.

In organic soils, the organic matter oxidizes upon drainage exposing roots of trees leaving them well above the ground surface (Figure 7.5). While this should usually be clear evidence of subsidence,

TABLE 7.3
Actions for Determining Whether an Area Still Has Wetland Hydrology

1. Review existing hydrologic information, such as stream gauge data or groundwater well data.
2. Examine wet season (growing season) aerial photography for signs of inundation or soil saturation. Multi-year photos need to be assessed. (If area looks wet in more years than dry, it remains wetland.)
3. Conduct onsite inspection and look for wetland hydrology field indicators or assess vegetation to determine if drainage has significantly and adversely affected hydrophytic vegetation.
4. Compare vegetation of altered wetland vs. similar, neighboring wetland (same original vegetation, soils, hydrology, and topography [landscape position] as subject wetland prior to alteration). (For example, if FACU and UPL species predominate, then area is effectively drained. If vegetation remains similar, at least in terms of species of similar wetland indicator status, area is still wetland.)
5. Determine the "zone of influence' of the ditch or drainage structure by using existing soil drainage guides. (Can also use other "scope and effect" equations and computer programs like Drainmod.)
6. Conduct detailed groundwater studies. (To maximize field effort, collect data during the wetter part of the growing season.)

Note: The sequence is recommended considering the level of effort or expertise required to undertake the particular action.

Source: Federal Interagency Committee for Wetland Delineation, 1989.

wetland hydrology may likely still exist since the soil may have reached an equilibrium with the groundwater levels. Field verification of wetland hydrology in the wetter part of the growing season is recommended.

Wetlands can be drained in a number of ways. Ditches and channelization are methods that are easily recognized by the open ditch or channel. The effect of tile drains is not as easily observed. An important point to remember is that one shallow ditch is usually not sufficient to effectively drain a large wetland. A network of ditches, removal of vegetation, and land leveling are the usual indicators of effectively drained sites (former wetlands). Land leveling is required to eliminate depressions that would hold water long enough to maintain the wetland condition. Consequently, a ditch through a forested wetland usually will only alter the hydrology to the point that the area is usually considered a partly drained wetland.

Water also can be diverted from wetlands in other ways such as diversion of surface water by upstream dams or levees, and withdrawal of groundwater by private or public wells or pump irrigation projects. Areas with extensive channelization (e.g., Yazoo basin in Mississippi) are particularly challenging sites for wetland delineation as the hydrology has been greatly altered by government-supported flood control or drainage projects. Many relict hydric soils and former wetlands can be found in these situations.

The various federal manuals provide general guidance on how to evaluate significantly drained sites, but more specific procedures are needed from the regulatory agencies. The federal interagency manual proposed a sequence of actions to evaluate hydrologically altered sites (Table 7.3; Federal Interagency Committee for Wetland Delineation, 1989). The sequence is not mandatory, but recommended. Moreover, this ordering does not preclude onsite inspections of shallow groundwater wells which would likely provide the most accurate results. Onsite investigations may reveal signs of ponding (e.g., water-stained leaves, oxidized rhizospheres along living roots, and the presence of OBL herbaceous species; see Chapter 5) that may help verify the presence of wetlands.

If a ditch or channel is present but the area is not cultivated (i.e., still in natural vegetation), an assessment of the vegetation may provide the answer to the question — is the site effectively drained? If OBL species are present, the area is likely to still be a wetland, unless they are restricted to microsites or are long-lived woody plants that can persist under mesic conditions (e.g., bald cypress). The herbaceous stratum often contains the most useful indicators as this stratum tends to be more dynamic as wetland plants are typically shallow-rooted and respond more quickly to hydrologic changes. Where these species are lacking, the presence of seedlings and saplings of drier sites species like black cherry (*Prunus serotina*) would suggest that the area is drier. Yet this species also occurs in wetlands, so its presence is not a highly reliable indicator of an effectively drained hydric soil (former wetland). If, however, UPL seedlings and saplings appear to becoming common or dominant species in the community, it is a fairly reliable sign that the area is effectively drained. Besides some rather simple clues from conspicuous vegetation patterns, the evaluation of drainage will be more complicated.

Perhaps the most practical technique would be to visit the area during the wet season (e.g., growing season for regulatory determinations) and look for direct signs of wetness either surface water or high ground water levels (within 1 ft of the surface). For the best results, install shallow ground water wells (e.g., within 3 ft and above any restrictive layer) and make weekly observations during the wet part of the growing season. Supplement the well records with observations of standing water and the depth to saturation (in an unlined borehole) during the weekly visits. The latter observations are needed to document site inundation and saturation due to the capillary fringe as the wells will only yield free water level information. In limited cases, a single site visit may provide reliable results. For example, if the site is dry during the normally wet period in a "wet year," the site is likely to be effectively drained. Also, if wet during the typically wet period in a "dry year" or during the dry period of a normal rainfall year, the site should still possess wetland hydrology. While these single site visits may be quite reliable, weekly observations during the wet part of the growing season (e.g., spring in the Northeast and Midwest) are typically preferred as the single visit may reflect an aberration. Ideally, hydrologic observations should be made at the site in question, plus a reference wetland (e.g., similar type and landscape position, but unaltered, and within the same watershed).

Once weekly data for a month or two have been gathered, construct a hydrograph for each site. Compare the hydrographs and if the graphs are more or less similar (within an acceptable range, contact regulatory agency for guidance), then the altered site probably is still a wetland. But if the graphs are significantly different, then the area is effectively drained and no longer wetland. It may be advantageous in many circumstances to make these types of observations over a few consecutive years. While the hydrograph of the reference wetland should satisfy the wetland hydrology parameter or criterion, it may not during drier years. Also remember that our knowledge of wetland hydrology for altered systems and drier-end wetlands is still in its infancy and certain wetlands (e.g., temporarily flooded wetlands) may not fit the typical model despite being ecologically and functionally wetland. Professional judgment must be used in these confounding circumstances. The development of hydrographs for reference wetlands and their use for comparison with hydrographs from altered wetlands should provide the most accurate assessment of the extent of drainage.

In other cases, the use of soil drainage equations or computerized hydrologic models (e.g., Drainmod) might be warranted (USDA Natural Resources Conservation Service, 1994; Skaggs, 1991). These techniques can be used to predict the lateral effect (zone of influence) of a ditch or tile drain by taking into account such factors as soil type and ditch depth. These equations and models are based on certain assumptions. One important assumption is that there is no flooding or ponding at the site. If there is and these sites are inundated for more than 1 week during the growing season in most years, these sites with surface water remain wetlands, while areas around them may be effectively drained by the drainage structure.

Other methods to evaluate wetland hydrology include examination of aerial photos or satellite imagery, determination of runoff volumes, interpreting tide, stream, or lake gauge data, and using local NRCS soil drainage guides. These and many of the other methods discussed above are addressed in considerable detail in *Hydrology Tools for Wetland Determination* (Natural Resources Conservation Service, 1997).

FIGURE 7.6 Partly drained hydric soils in the Prairie Pothole region. Some areas may be effectively drained.

Farmed Wetlands

Nearly 20 million acres of farmed wetlands (including pastured wetlands) may exist in the U.S. (National Research Council, 1995). Agricultural conversion of wetland has been a common practice for the past 200 years as many wetland soils are suitable for crop production upon drainage (Figure 7.6). Lyford (1960) in studying drained hydric soils in New Hampshire reported that "cultivated land can be worked best when the water table is below 20 in." Traction is a problem with shallower water tables. In the Scantic and Swanton soils, Lyford found that a 200-ft-wide bed was too wide for efficient drainage even on these coarse-textured soils, as local ponding occurred. A 100-ft-wide bed was best as these soils are saturated for a significant part of the year, but may be dry and droughty by mid-summer or during long dry periods.

The presence of certain crops may provide an indication of the extent of drainage, especially in humid regions where droughts are uncommon. The situation is more complicated in arid and subhumid regions due to the frequency of droughts and the recognition that during such times, many wetlands can be and are cultivated without drainage.

For humid regions, in all likelihood, if crops like corn, cotton, or wheat are growing well in the area (not stressed), the soils are effectively drained. Flooding for 1 or 2 days can reduce corn yields by 20 to 30% (e.g., Howell and Hiller, 1974; Bhan, 1977). The presence of soybeans makes for a more difficult decision as this plant is highly adapted to wetland conditions, as is lowland rice. Soybeans can be cultivated in "farmed wetlands" after spring wetness abates, so a more detailed analysis of the hydrology is usually required. When lowland rice is present, the area is likely to remain a wetland when grown on hydric soils. Yet ricefields also are constructed on nonhydric soils where rice is part of a rotational cropping cycle that varies from wetland to nonwetland. In this situation, the growth of rice does not indicate a wetland and the ricefield is probably not a regulated area (consult the appropriate regulatory agency).

Pump-drained mucklands (organic soils) to produce crops like cabbage, lettuce, onions, and carrots are likely to still retain wetland hydrology. These soils will subside rapidly upon drainage, so good soil management requires maintaining water tables as high as possible relative to the requirements necessary for good crop yields (Harris et al., 1962).

If stressed crops (e.g., corn) are observed in depressional areas within an agricultural field, a closer look at the site might find other evidence of surface water (e.g., matted vegetation or algae). Also if the crops in such depressions are more robust than other crops during a dry year or yellow-colored (chlorotic) and significantly reduced in height during a wet or normal year, the area may still be wetland. In some parts of the country (e.g., Southeast), farmed wetlands have an abundance of crayfish mounds. A farmfield with numerous crayfish burrows is likely to retain its wetland status.

HYDROLOGICALLY PROBLEMATIC WETLANDS

Since water is not everpresent in many wetlands, surrogates must be used to confirm wetland hydrology in the three-parameter test (e.g., Corps manual or federal interagency manual), whereas in other delineation approaches (e.g., PRIMET), it may not be important provided there is no evidence of significant hydrologic modification (see Chapter 6). A strict interpretation of the Corps' three-parameter test makes the list of positive indicators of wetland hydrology critically important for ensuring accurate wetland identification and delineation. A conservative listing would lead to misidentification of some wetlands as nonwetlands based on the lack of any authorized wetland hydrology indicators. Note that partly drained wetlands (including farmed wetlands) are discussed in the soils section above.

To qualify as a wetland for regulatory purposes at the federal level, an area must be inundated or saturated (at or near the surface) for at least 5% of the growing season. Problems with the conservative definition of growing season leads to many wetlands failing to be wet enough during the specified period to qualify as regulated wetlands (see Chapter 2). In other cases, some wetland types may fail to meet the frequency test (i.e., every other year on average). The wetness threshold of the wetland hydrology parameter of the Corps manual was derived from studies of southeastern bottomland hardwood swamps. A more scientifically sound threshold would consider the variety of wetland types in the nation to develop a wetland hydrology criterion that includes all wetlands. This type of analysis would produce either a single wetness minimum based on the driest wetland type or a set minimum thresholds for a combination of wetland types and regions. As noted by the National Research Council (1995), sufficient hydrological data do not exist to accomplish this. Based on their review, the NRC proposed a minimum threshold for wetland hydrology in the U.S. — 2 consecutive weeks of saturation within a foot of the surface during the growing season in most years. They also noted problems with intermittently flooded wetlands in arid regions where annual rainfall is highly variable in amount and timing (see discussion on cyclical wetlands earlier in this chapter). Some wetlands that satisfy the regulatory criterion for wetland hydrology may lack indicators of wetland hydrology during much of the growing season. The above wetlands are problem wetlands for wetland identification and delineation as they either fail to be sufficiently wet (frequency or duration) or fail to exhibit the "acceptable" wetland hydrology indicators during dry periods.

Groundwater-Driven Wetlands

Given that most wetlands are not permanently wet but have fluctuating water tables, the wetland hydrology parameter (criterion) is usually the most difficult feature to verify during a single site inspection. This problem is exacerbated by groundwater-driven wetlands where signs of wetland hydrology other than hydric soil properties and OBL and FACW plants may be nonexistent during the dry season (usually summer). The list of wetland hydrology indicators used to identify federally regulated wetlands mostly includes signs of surface water (e.g., water marks, sediment deposits, and water-stained leaves — see Table 6.1). For groundwater wetlands, one primary indicator (observed soil saturation during the growing season) and three possible secondary indicators of wetland hydrology (i.e., oxidized rhizospheres, FAC neutral test, and use of local soil survey data — reference to the soil–water features table) may be used for identifying federally regulated wetlands following the Corps manual (Environmental Laboratory, 1987; Williams, 1992). Since the use of FAC neutral test is at the discretion of the local Corps district and is not used in some districts, many groundwater wetlands may only have one secondary indicator (usually the local soil survey data) during the dry season, and, therefore, may not qualify as a regulated wetland, unless one uses professional judgment. To avoid misclassifying these wetlands as nonregulated wetlands (or non-wetlands), either field work should be done in the earliest part of the growing season when the soils are saturated at or near the surface for extended periods, or at other times; professional judgment (i.e., undrained hydric soil = wetland) must be used. These wetlands were not "problem wetlands" using the federal interagency manual (Federal Interagency Committee for Wetland Delineation, 1989) since it recognized areas with hydrophytic vegetation growing on undrained field-verified hydric soils as wetlands in the basic evaluation process.

DRIER-END WETLANDS

The following wetland types may be difficult to verify as wetlands using guidance in most wetland delineation manuals: seasonally saturated flatwoods, saturated wet meadows, temporarily flooded floodplain wetlands, riparian wetlands along western rivers, and other temporarily flooded and/or seasonally saturated wooded swamps. Many of these wetlands fail to be wet enough to satisfy the wetland hydrology requirement in the Corps manual, yet most are either wet for extended periods during the "nongrowing season" or not inundated or saturated to the surface for sufficient periods.

Flatwoods

Flatwoods are common habitats along the Atlantic–Gulf coastal plain from southern New Jersey to Florida and west into eastern Texas (see Figure 2.5). They are perhaps the most difficult to identify wetlands since they generally have imperfectly drained soils that range from hydric poorly drained to somewhat poorly drained nonhydric soils (Plate 43). They may be the archetypal transitional wetland. The mosaic pattern of wetter areas intermixed with drier areas with little discernable change in vegetation further complicates wetland delineation. The vegetation may be hardwoods, pines, or mixtures. While many flatwoods possess plant communities that satisfy the hydrophytic vegetation criterion and hydric soil field indicators, if their hydrology is measured, they may not be wet enough during the growing season as currently defined for federal regulatory purposes. Many of these areas are wet for extended periods during the nongrowing season and have been locally referred to as "winter wet woods." For any Corps district using the 32°F temperature to approximate growing season for assessing wetland hydrology, many of the wet flatwoods will fail to meet the wetland hydrology requirement when measured. More research is needed in flatwoods to examine hydrologic regimes relative to the performance of wetland functions. The significance of this issue is reflected by the extent of area covered by these plant communities, their importance as forestry resources, and the increasing pressure to develop such areas for real estate.

Floodplain Wetlands

Forested wetlands are abundant on floodplains of rivers across the country, especially in the humid temperate region. While many of these wetlands are wet enough to qualify as wetlands, others are flooded either mostly during the nongrowing season and for short periods (e.g., a few days) on several occasions during the growing season. Like the flatwoods, more work is needed to determine whether the wetland hydrology criterion used for regulatory purposes should be revised to consider nongrowing season hydrology, especially where the soils are not frozen for significant periods.

Western Riparian Habitat

Riparian habitat includes wetlands and nonwetlands whose phreatophytic vegetation depends on river- or lake-influenced groundwater for growth and reproduction (Figure 7.7). In arid and semiarid regions, these areas are critical wildlife habitat for local species and migrants. Species in the adjacent deserts depend on these areas. For example, the white-winged dove, an effective pollinator of the saguaro cactus, needs riparian habitats (Crosswhite and Crosswhite, 1989).

FIGURE 7.7   Riparian cottonwood forest.

Some riparian habitats along western streams are wet enough to qualify as wetlands both technically and from the regulatory standpoint. Even so, they may lack certain hydrology indicators or hydric soil properties (e.g., cottonwoods growing on sandy or rocky soils) that may preclude their designation as regulated wetlands. Most riparian habitats in the arid region of the country fail to meet the hydrology parameter because they are not flooded frequently enough for sufficient duration, given regional rainfall patterns. Yet when rainfall is significant, these areas provide many functions that wetlands normally do. The riparian area includes the zone where the river course tends to migrate back and forth over time, making the zone itself vitally important to maintaining the integrity of the river system.

PROBLEMATIC FIELD CONDITIONS

In the field, an investigator may encounter one of several situations that complicates wetland identification and delineation. Complex landscapes with high interspersion of wetlands and non-wetlands, rocky areas, and significantly altered wetland systems may be particularly difficult to interpret. The Corps manual offers guidance on how to handle some of the following circumstances, but interpretations may vary among Corps districts and EPA regions, so consult with the appropriate offices for specific protocols to employ in the field.

COMPLEX LANDSCAPES

Certain landforms such as ridge and swale topography, pit and mound relief, and linear drainage-ways, force the question: how small or narrow an area should be delineated for regulatory purposes? Complex landscapes pockmarked with small wetlands and small uplands (e.g., pit and mound terrain or gilgai topography) make it practically impossible to separate wetlands from drylands. In these cases, what ratio of wetlands to nonwetlands (drylands) should be used to regulate the entire mosaic parcel — 51:49 (wetland:upland), 40:60, or some other ratio? What procedures should be used to determine these ratios in the field — line intercept transects, point intercept transects, belt transects, plot samples, or other methods? These significant questions can only be answered by the regulatory agencies seeking to protect certain wetland functions and values.

The point intercept method is applicable for determining the percentage of a pit and mound relief area that is wetland, where the pits are hydric soils and the mounds are nonhydric soils.* After examining the soils at a number of pits and mounds, three randomly selected transects are

---

* Realize that in some cases, the whole pit and mound area is upland, while in others it is entirely wetland.

evaluated. In this application, at each sample point, a decision is made a wetland (e.g., a pit) or upland (e.g., a mound). After making point obse along three 200-ft transects, a total of 300 data points will be identified as and the percentage of wetland can be calculated (e.g., 120 out of 300 poi 300 = 60%).

ROCKY LANDSCAPES

Rocky areas (e.g., wetlands on glacial till, caprock limestone wetlands in wetlands on rocky terrain) lack significant amounts of soil, making it indicators of hydric soils. Reliance on other indicators (e.g., hydropl hydrology signs, and landscape positions such as drainageways and dry positive wetland identification. If it is critical to extract a soil sample, t devices available (see Jurgensen et al., 1977; Ponder and Alley, 1997;

IRRIGATED LANDS

In the arid and semiarid regions (west Texas north to the western Dak and eastern Oregon and Washington, except mountains), irrigation is ne (Lemly, 1994). During the past 90 years, such intensive irrigation has be U.S., leading to many adverse effects on wetlands. In most cases, irrigat and conversion of wetlands to cropland or to the drying up of remaining of water. In other cases, wetlands may have been created where wate runoff. For the latter areas, wetland vegetation and signs of hydrology s of wetland.

NATIONAL RESEARCH COUNCIL RECOMMENDATIONS

The NRC offered the following suggestions for addressing the issue wetlands" including:

1. Identify permafrost wetlands, "intermittently wet" wetlands, a following the same rules and techniques used to identify other
2. Initiate studies of permafrost wetlands, especially soil–vegetat
3. For altered systems, inference of wetland features that have b d should be permitted in the wetland identification and delineati
4. The wetland delineation methods should be tested on transiti lands in all regions.

OTHER RECOMMENDATIONS

Based on the research involved in preparing this book, some addi offered:

1. For imperfectly drained landscapes such as flatwoods, gilg mound relief in woodlands, where wetlands are intersper resources managers (including wetland regulatory agencies) s managing these areas as systems not just isolated areas (wetl biota associated with the broader ecosystem are dependent o of wetlands, transitional areas, and uplands. More study nee the functions of these systems from a water quality stand ecological perspective.

still be wetland. In some parts of the country (e.g., Southeast), farmed wetlands have an abundance of crayfish mounds. A farmfield with numerous crayfish burrows is likely to retain its wetland status.

HYDROLOGICALLY PROBLEMATIC WETLANDS

Since water is not everpresent in many wetlands, surrogates must be used to confirm wetland hydrology in the three-parameter test (e.g., Corps manual or federal interagency manual), whereas in other delineation approaches (e.g., PRIMET), it may not be important provided there is no evidence of significant hydrologic modification (see Chapter 6). A strict interpretation of the Corps' three-parameter test makes the list of positive indicators of wetland hydrology critically important for ensuring accurate wetland identification and delineation. A conservative listing would lead to misidentification of some wetlands as nonwetlands based on the lack of any authorized wetland hydrology indicators. Note that partly drained wetlands (including farmed wetlands) are discussed in the soils section above.

To qualify as a wetland for regulatory purposes at the federal level, an area must be inundated or saturated (at or near the surface) for at least 5% of the growing season. Problems with the conservative definition of growing season leads to many wetlands failing to be wet enough during the specified period to qualify as regulated wetlands (see Chapter 2). In other cases, some wetland types may fail to meet the frequency test (i.e., every other year on average). The wetness threshold of the wetland hydrology parameter of the Corps manual was derived from studies of southeastern bottomland hardwood swamps. A more scientifically sound threshold would consider the variety of wetland types in the nation to develop a wetland hydrology criterion that includes all wetlands. This type of analysis would produce either a single wetness minimum based on the driest wetland type or a set minimum thresholds for a combination of wetland types and regions. As noted by the National Research Council (1995), sufficient hydrological data do not exist to accomplish this. Based on their review, the NRC proposed a minimum threshold for wetland hydrology in the U.S. — 2 consecutive weeks of saturation within a foot of the surface during the growing season in most years. They also noted problems with intermittently flooded wetlands in arid regions where annual rainfall is highly variable in amount and timing (see discussion on cyclical wetlands earlier in this chapter). Some wetlands that satisfy the regulatory criterion for wetland hydrology may lack indicators of wetland hydrology during much of the growing season. The above wetlands are problem wetlands for wetland identification and delineation as they either fail to be sufficiently wet (frequency or duration) or fail to exhibit the "acceptable" wetland hydrology indicators during dry periods.

GROUNDWATER-DRIVEN WETLANDS

Given that most wetlands are not permanently wet but have fluctuating water tables, the wetland hydrology parameter (criterion) is usually the most difficult feature to verify during a single site inspection. This problem is exacerbated by groundwater-driven wetlands where signs of wetland hydrology other than hydric soil properties and OBL and FACW plants may be nonexistent during the dry season (usually summer). The list of wetland hydrology indicators used to identify federally regulated wetlands mostly includes signs of surface water (e.g., water marks, sediment deposits, and water-stained leaves — see Table 6.1). For groundwater wetlands, one primary indicator (observed soil saturation during the growing season) and three possible secondary indicators of wetland hydrology (i.e., oxidized rhizospheres, FAC neutral test, and use of local soil survey data — reference to the soil–water features table) may be used for identifying federally regulated wetlands following the Corps manual (Environmental Laboratory, 1987; Williams, 1992). Since the use of FAC neutral test is at the discretion of the local Corps district and is not used in some districts, many groundwater wetlands may only have one secondary indicator (usually the local soil survey data) during the dry season, and, therefore, may not qualify as a regulated wetland, unless one uses professional judgment. To avoid misclassifying these wetlands as nonregulated wetlands (or non-wetlands), either field work should be done in the earliest part of the growing season when the soils are saturated at or near the surface for extended periods, or at other times; professional judgment (i.e., undrained hydric soil = wetland) must be used. These wetlands were not "problem wetlands" using the federal interagency manual (Federal Interagency Committee for Wetland Delineation, 1989) since it recognized areas with hydrophytic vegetation growing on undrained field-verified hydric soils as wetlands in the basic evaluation process.

Drier-end Wetlands

The following wetland types may be difficult to verify as wetlands using guidance in most wetland delineation manuals: seasonally saturated flatwoods, saturated wet meadows, temporarily flooded floodplain wetlands, riparian wetlands along western rivers, and other temporarily flooded and/or seasonally saturated wooded swamps. Many of these wetlands fail to be wet enough to satisfy the wetland hydrology requirement in the Corps manual, yet most are either wet for extended periods during the "nongrowing season" or not inundated or saturated to the surface for sufficient periods.

Flatwoods

Flatwoods are common habitats along the Atlantic–Gulf coastal plain from southern New Jersey to Florida and west into eastern Texas (see Figure 2.5). They are perhaps the most difficult to identify wetlands since they generally have imperfectly drained soils that range from hydric poorly drained to somewhat poorly drained nonhydric soils (Plate 43). They may be the archetypal transitional wetland. The mosaic pattern of wetter areas intermixed with drier areas with little discernable change in vegetation further complicates wetland delineation. The vegetation may be hardwoods, pines, or mixtures. While many flatwoods possess plant communities that satisfy the hydrophytic vegetation criterion and hydric soil field indicators, if their hydrology is measured, they may not be wet enough during the growing season as currently defined for federal regulatory purposes. Many of these areas are wet for extended periods during the nongrowing season and have been locally referred to as "winter wet woods." For any Corps district using the 32°F temperature to approximate growing season for assessing wetland hydrology, many of the wet flatwoods will fail to meet the wetland hydrology requirement when measured. More research is needed in flatwoods to examine hydrologic regimes relative to the performance of wetland functions. The significance of this issue is reflected by the extent of area covered by these plant communities, their importance as forestry resources, and the increasing pressure to develop such areas for real estate.

Floodplain Wetlands

Forested wetlands are abundant on floodplains of rivers across the country, especially in the humid temperate region. While many of these wetlands are wet enough to qualify as wetlands, others are flooded either mostly during the nongrowing season and for short periods (e.g., a few days) on several occasions during the growing season. Like the flatwoods, more work is needed to determine whether the wetland hydrology criterion used for regulatory purposes should be revised to consider nongrowing season hydrology, especially where the soils are not frozen for significant periods.

Western Riparian Habitat

Riparian habitat includes wetlands and nonwetlands whose phreatophytic vegetation depends on river- or lake-influenced groundwater for growth and reproduction (Figure 7.7). In arid and semiarid regions, these areas are critical wildlife habitat for local species and migrants. Species in the adjacent deserts depend on these areas. For example, the white-winged dove, an effective pollinator of the saguaro cactus, needs riparian habitats (Crosswhite and Crosswhite, 1989).

FIGURE 7.7   Riparian cottonwood forest.

Some riparian habitats along western streams are wet enough to qualify as wetlands both technically and from the regulatory standpoint. Even so, they may lack certain hydrology indicators or hydric soil properties (e.g., cottonwoods growing on sandy or rocky soils) that may preclude their designation as regulated wetlands. Most riparian habitats in the arid region of the country fail to meet the hydrology parameter because they are not flooded frequently enough for sufficient duration, given regional rainfall patterns. Yet when rainfall is significant, these areas provide many functions that wetlands normally do. The riparian area includes the zone where the river course tends to migrate back and forth over time, making the zone itself vitally important to maintaining the integrity of the river system.

PROBLEMATIC FIELD CONDITIONS

In the field, an investigator may encounter one of several situations that complicates wetland identification and delineation. Complex landscapes with high interspersion of wetlands and non-wetlands, rocky areas, and significantly altered wetland systems may be particularly difficult to interpret. The Corps manual offers guidance on how to handle some of the following circumstances, but interpretations may vary among Corps districts and EPA regions, so consult with the appropriate offices for specific protocols to employ in the field.

COMPLEX LANDSCAPES

Certain landforms such as ridge and swale topography, pit and mound relief, and linear drainage-ways, force the question: how small or narrow an area should be delineated for regulatory purposes? Complex landscapes pockmarked with small wetlands and small uplands (e.g., pit and mound terrain or gilgai topography) make it practically impossible to separate wetlands from drylands. In these cases, what ratio of wetlands to nonwetlands (drylands) should be used to regulate the entire mosaic parcel — 51:49 (wetland:upland), 40:60, or some other ratio? What procedures should be used to determine these ratios in the field — line intercept transects, point intercept transects, belt transects, plot samples, or other methods? These significant questions can only be answered by the regulatory agencies seeking to protect certain wetland functions and values.

The point intercept method is applicable for determining the percentage of a pit and mound relief area that is wetland, where the pits are hydric soils and the mounds are nonhydric soils.* After examining the soils at a number of pits and mounds, three randomly selected transects are

---

* Realize that in some cases, the whole pit and mound area is upland, while in others it is entirely wetland.

evaluated. In this application, at each sample point, a decision is made as to whether the point is wetland (e.g., a pit) or upland (e.g., a mound). After making point observations at 2-ft intervals along three 200-ft transects, a total of 300 data points will be identified as either wetland or upland and the percentage of wetland can be calculated (e.g., 120 out of 300 points = 40% or 180 out of 300 = 60%).

Rocky Landscapes

Rocky areas (e.g., wetlands on glacial till, caprock limestone wetlands in south Florida, streambed wetlands on rocky terrain) lack significant amounts of soil, making it impossible to find positive indicators of hydric soils. Reliance on other indicators (e.g., hydrophytic vegetation, wetland hydrology signs, and landscape positions such as drainageways and dry arroyos) is necessary for positive wetland identification. If it is critical to extract a soil sample, there are various sampling devices available (see Jurgensen et al., 1977; Ponder and Alley, 1997; Tuttle et al., 1984).

Irrigated Lands

In the arid and semiarid regions (west Texas north to the western Dakotas to southern California and eastern Oregon and Washington, except mountains), irrigation is needed to support agriculture (Lemly, 1994). During the past 90 years, such intensive irrigation has been practiced in the western U.S., leading to many adverse effects on wetlands. In most cases, irrigation has led to the drainage and conversion of wetlands to cropland or to the drying up of remaining wetlands that were deprived of water. In other cases, wetlands may have been created where water collected from irrigation runoff. For the latter areas, wetland vegetation and signs of hydrology should be the best indicators of wetland.

NATIONAL RESEARCH COUNCIL RECOMMENDATIONS

The NRC offered the following suggestions for addressing the issue of "especially controversial wetlands" including:

1. Identify permafrost wetlands, "intermittently wet" wetlands, and wetlands on farmland following the same rules and techniques used to identify other wetlands.
2. Initiate studies of permafrost wetlands, especially soil–vegetation–hydrology studies.
3. For altered systems, inference of wetland features that have been removed or changed should be permitted in the wetland identification and delineation process.
4. The wetland delineation methods should be tested on transitional and marginally wet lands in all regions.

OTHER RECOMMENDATIONS

Based on the research involved in preparing this book, some additional recommendations are offered:

1. For imperfectly drained landscapes such as flatwoods, gilgai topography, or pit and mound relief in woodlands, where wetlands are interspersed with uplands, natural resources managers (including wetland regulatory agencies) should begin thinking about managing these areas as systems not just isolated areas (wetlands). In all likelihood, the biota associated with the broader ecosystem are dependent on this complex association of wetlands, transitional areas, and uplands. More study needs to be given to ascertain the functions of these systems from a water quality standpoint as well as from an ecological perspective.

2. More study of the wetland–upland edge is needed since this is perhaps the most dynamic and hydrologically active portion of the landscape (Bell and Richardson, 1997), with important implications for water quality renovation and, in some areas, groundwater recharge. Studies of soil–water relations and wetland functions would be extremely worthwhile in improving our understanding of these systems and our ability to identify them.

3. Regulatory agencies should establish a wetland hydrology monitoring program to develop and maintain a set of hydrographs for the variety of wetlands that occur in their jurisdiction, emphasizing the seasonally flooded, temporarily flooded, and seasonally saturated types. This information could become the basis for evaluating the extent of drainage at altered wetlands.

REFERENCES

Ahmad, N. 1983. Vertisols. In *Pedogenesis and Soil Taxonomy. II. Soil Orders. Development in Soil Science.* L.P. Wilding, N. Smeck, and G. Hall (Eds.). Elsevier Publishing, Amsterdam.

Amatekpor, J.K. 1992. Changes in some seasonally flooded soils in Ghana: Volta Lake drawdown area. In Wetland Soils: Characterization, Classification, and Utilization of Wet Soils. J.M. Kimble (Ed.). Proc. of the 8th Int. Soil Correlation Meeting. USDA Soil Conservation Service, National Soil Survey Center, Lincoln, NE. 9-15.

Baskin, Y. 1994. California's ephemeral vernal pools may be a good model for speciation. *BioScience,* 44: 384-388.

Bell, J.C. and J.L. Richardson. 1997. Aquic conditions and hydric soil indicators for Aquolls and Albolls. In *Aquic Conditions and Hydric Soils: The Problem Soils.* M.J. Vepraskas and S.W. Sprecher (Eds.). Pub. No. 50. Soil Society of America, Inc., Madison, WI. 23-40.

Bhan, S. 1977. Effect of waterlogging on maize. *Indian J. Agric. Res.,* 2: 147-150.

Blodgett, R.H., J.P. Crabaugh, and E.F. McBride. 1993. The color of red beds — a geologic perspective. In *Soil Color.* J.M. Bigham and E.J. Ciolkosz (Eds.). Spec. Publ. No. 31. Soil Science Society of America, Madison, WI. 127-159.

Boettinger, J.L. 1997. Aquisalids (Salorthids) and other wet saline and alkaline soils: problems identifying aquic conditions and hydric soils. In *Aquic Conditions and Hydric Soils: The Problem Soils.* M.J. Vepraskas and S.W. Sprecher (Eds.). Spec. Pub. No. 50. Soil Science Society of America, Inc., Madison, WI. 79-97.

Bouma, J. 1983. Hydrology and genesis of soils with aquic moisture regimes. In *Pedogenesis and Soil Taxonomy. I. Concepts and Interactions.* L.P. Wilding, N.E. Smeck, and G.F. Hall (Eds.). Elsevier Publishing Company, Amsterdam, The Netherlands. 253-281.

Brooks, R.P., D.E. Arnold, and E.D. Bellis. 1987. Wildlife and Plant Communities of Selected Wetlands: Pocono Region of Pennsylvania. NWRC Open File Report 87-02. U.S. Fish and Wildlife Service, Washington, D.C.

Buol, S.W. and M.N. Camargo. 1992. Wet Oxisols. In Wetland Soils: Characterization, Classification, and Utilization of Wet Soils. J.M. Kimble (Ed.). Proc. of the 8th International Soil Correlation Meeting. USDA Soil Conservation Service, National Soil Survey Center, Lincoln, NE. 41-49.

Buol, S. W., F. D. Hole, and R. J. McCracken. 1980. *Soil Genesis and Classification.* Iowa State University Press, Ames.

Calhoun, A.J.K., J.E. Cormier, R.B. Owen, Jr., A.F. O'Connell, Jr., C.T. Roman, and R.W. Tiner, Jr. 1994. The Wetlands of Acadia National Park and Vicinity. Misc. Publ. No. 721. Maine Agricultural and Forest Experiment Station, Orono, ME.

Char, W.P. and C.H. Lamoureux. 1985. Puna Geothermal Area Biotic Assessment, Puna District, County of Hawaii. Final Report to the Hawaii State Department of Planning and Economic Development, Honolulu.

Childs, C.W. and B. Clayden. 1986. On the definition and identification of aquic soil moisture regimes. *Aust. J. Soil Res.,* 24: 311-316.

Clark, M.H. and C-L. Ping. 1997. Hydrology, morphology, and redox potentials in four soils of south central Alaska. In *Aquic Conditions and Hydric Soils: The Problem Soils.* M.J. Vepraskas and S.W. Sprecher (Eds.). Pub. No. 50. Soil Society of America, Inc., Madison, WI. 113-131.

Comerford, N.B., A. Jerez, A.A. Freitas, and J. Montgomery. 1996. Soil water table, reducing conditions, and hydologic regimes in a Florida flatwood landscape. *Soil Sci.,* 161: 194-199.

Comerma, J.A. 1985. Hydromorphic vertisols. In *Wetland Soils: Characterization, Classification, and Utilization.* International Rice Research Institute, Los Baños, Laguna, Philippines. 407-420.

Comerma, J.A., O. Luque, and R. Paredes. 1978. *El Drenaje como criterio de clasificacion Taxonomica en Vertisoles de Venezuela.* 8 Cong. Suelos. Barquuisimeto, Venezuela. 1-35.

Couto, W., C. Sanzonowicz, and A. De O. Barcellos. 1985. Factors affecting oxidation-reduction processes in a Oxisol with a seasonal water table. *Soil Sci. Soc. Am. J.,* 49: 1245-1248.

Coventry, R.J., R.M. Taylor, and R.W. Fitzpatrick. 1983. Pedological significance of the gravels in some red and grey earths of central north Queensland. *Aust. J. Soil Res.,* 21: 219-240.

Coventry, R.J. and J. Williams. 1984. Quantitative relationships between morphology and current soil hydrology in some Alfisols in semiarid tropical Australia. *Geoderma,* 33: 191-218.

Cowardin, L.M., V. Carter, F.C. Golet, and E.T. LaRoe. 1979. Classification of Wetlands and Deepwater Habitats of the United States. FWS/OBS-79/31. U.S. Fish and Wildlife Service, Washington, D.C.

Crosswhite, F.S. and C.D. Crosswhite. 1989. Value of riparian habitat within the desert. *Desert Plants,* 9 (3-4): back cover.

Damman, A.W. H. and T.W. French. 1987. The Ecology Peat Bogs of the Glaciated Northeastern United States: A Community Profile. Biol. Rep. 85 (7.16). U.S. Fish and Wildlife Service, Washington, D.C.

Daniels, R.B., E.E. Gamble, and S.W. Buol. 1973. Oxygen content in the ground water of some North Carolina Aquults and Udults. In *Field Soil Water Regime.* R.R. Bruce, K.W. Flach, and H.M. Taylor (Eds.). Spec. Publ. No. 5. Soil Science Society of America, Madison, WI. 153-166.

Diers, R. and J.L. Anderson. 1984. Part I. Development of soil mottling. *Soil Surv. Hor.,* Winter 1984: 9-12.

Environmental Laboratory. 1987. Corps of Engineers Wetlands Delineation Manual. Tech. Rep. Y-87-1. U.S. Army Engineer Waterways Expt. Station, Vicksburg, MS.

Evans, C.V. and D.L. Mokma. 1996. Sandy wet Spodosols: water tables, chemistry, and pedon partioning. *Soil Sci. Soc. Am. J.,* 60: 1495-1501.

Federal Interagency Committee for Wetland Delineation. 1989. Federal Manual for Identifying and Delineating Jurisdictional Wetlands. Cooperative technical publication. U.S. Army Corps of Engineers, U.S. Environmental Protection Agency, U.S. Fish and Wildlife Service, and USDA Soil Conservation Service, Washington, D.C.

Fisher, A.S., G.S. Podniesinski, and D.J. Leopold. 1996. Effects of drainage ditches on vegetation patterns in abandoned agricultural peatlands in central New York. *Wetlands,* 16: 397-409.

Franzmeier, D.P., J.E. Yahner, G.C. Steinhardt, and H.R. Sinclair. 1983. Color patterns and water table levels in some Indiana soils. *Soil Sci. Soc. Am. J.,* 47: 1196-1202.

Golet, F.C., A.J.K. Calhoun, W.R. DeRagon, D.J. Lowry, and A.J. Gold. 1993. The Ecology of Red Maple Swamps in the Glaciated Northeast: A Community Profile. Biol. Rep. 12. U.S. Fish and Wildlife Service, Washington, D.C.

Hahn, G.G., C. Hartley, and A.S. Rhoads. 1920. Hypertrophied lenticels on the roots of conifers and their relation to moisture and aeration. *J. Agric. Res.,* 20: 253-265.

Harris, C.I., H.T. Erickson, N.K. Ellis, and J.E. Larson. 1962. Water-level control in organic soil, as related to subsidence rate, crop yield, and response to nitrogen. *Soil Science,* 94: 158-161.

Howell, T.A. and E.A. Hiller. 1974. Effects of inundation period on seedling growth. Transactions of the American Society of Agricultural Engineers, 17: 286-288, 294.

Huenneke, L.F. 1982. Wetland forests of Tompkins County, New York. *Bull. Torrey Bot. Club,* 109: 51-63.

Hurt, G.W., P.M. Whited, and R.F. Pringle (Eds.). 1996. Field Indicators of Hydric Soils in the United States. A Guide for Identifying and Delineating Hydric Soils, ver. 3.2. Prepared in cooperation with the National Technical Committee for Hydric Soils. USDA Natural Resources Conservation Service, Fort Worth, TX.

Hurt, G.W., P.M. Whited, and R.F. Pringle (Eds.). 1998. Field Indicators of Hydric Soils in the United States. A Guide for Identifying and Delineating Hydric Soils, ver. 4.0. Prepared in cooperation with the National Technical Committee for Hydric Soils. USDA Natural Resources Conservation Service, Fort Worth, TX.

Hurt, G.W. 1998. Personal communication.

Jacob, J.S., R.W. Griffin, W.L. Miller, and L.P. Wilding. 1997. Aquerts and aquertic soils: a querulous proposition. In *Aquic Conditions and Hydric Soils: The Problem Soils.* M.J. Vepraskas and S.W. Sprecher (Eds.). Pub. No. 50. Soil Society of America, Inc., Madison, WI. 61-77.

Jacobi, J.D. 1989. Vegetation Maps of the Upland Plant Communities on the Islands of Hawaii, Maui, Molokai, and Lanai. Cooperative National Park Resources Studies Unit Tech. Rep. 68. U.S. Fish and Wildlife Service, Washington, D.C.

Johnson, M.D. 1983. The origin and microfabric of Lake Superior red clay. *J. Sediment. Petro.,* 53: 859-873.

Jurgenson, M.F., M.J. Larsen, and A.E. Harvey. 1977. A soil sampler for steep, rocky sites. Res. Note INT-217. USDA Forest Service, Intermountain Forest and Range Expt. Station, Ogden, UT.

Kantrud, H. A., G. L. Krapu, and G. H. Swanson. 1989. Prairie Basin Wetlands of the Dakotas: A Community Profile. Biol. Rpt. 85(7.28). U.S. Fish and Wildlife Service, Washington, D.C.

Kindscher, K., S.L. Wilson, A. Fraser, and D.P. Hurlburt. 1998. Western Kansas Playa Lake Wetlands — Alternative Endpoints in Response to Agricultural Disturbance. Kansas Biological Survey, University of Kansas, Lawrence.

Kolesinskas, K.J. and E.H. Sautter. 1990. Notes on criteria used for identifying wet soils in Connecticut and Rhode Island. USDA Soil Conservation Service, Storrs, CT. Unpublished notes.

Kuehl, R.J., N.B. Comerford, and R.B. Brown. 1997. Aquods and Psammaquents: problems in hydric soil identification. In *Aquic Conditions and Hydric Soils: The Problem Soils.* M.J. Vepraskas and S.W. Sprecher (Eds.). Spec. Pub. No. 50. Soil Science Society of America, Inc., Madison, WI. 41-59.

Kyuma, K. and M. Mitsuchi. 1985. Wet andisols. In *Wetland Soils: Characterization, Classification, and Utilization.* International Rice Research Institute, Los Baños, Laguna, Philippines. 439-463.

Ledig, F.T. and S. Little. 1979. Pitch pine (*Pinus rigida* Mill.): ecology, physiology, and genetics. In *Pine Barrens: Ecosystem and Landscape.* R.T.T. Forman (Ed.). Academic Press, Inc., New York. 347-371.

Lemly, A.D. 1994. Agriculture and wildlife: ecological implications of subsurface irrigation drainage. *J. Arid Environ.,* 28: 85-94.

Lichvar, R. and S. Sprecher. 1996. Delineation and Characterization of "Waters of the United States" at Edwards Air Force Base, CA. U.S. Army Corps of Engineers, Waterways Expt. Station, Vicksburg, MS.

Lichvar, R., S. Sprecher, and J. Wakeley. 1996. Identification and Classification of "Waters of the United States" on one Playa at the U.S. Army Dugway Proving Ground, Dugway, Utah. U.S. Army Corps of Engineers, Waterways Expt. Station, Vicksburg, MS.

Lindbo, D.L. 1997. Entisols–Fluvents and Fluvaquents: problems recognizing aquic and hydric conditions in young, flood plain soils. In *Aquic Conditions and Hydric Soils: The Problem Soils.* M.J. Vepraskas and S.W. Sprecher (Eds.). Pub. No. 50. Soil Society of America, Inc., Madison, WI. 133-151.

Little, S. 1959. Silvical characteristics of pitch pine (*Pinus rigida*). Station Paper No. 119. USDA Forest Service, Northeastern Forest Experiment Station, Upper Darby, PA.

Lyford, W.H. 1960. Water Table Fluctuations on Bedded Scantic and Swanton Soils in Southeastern New Hampshire. Tech. Bull. 102. University of New Hampshire, Agricultural Expt. Station, Durham, NH.

Macedo, J. 1986. Morphology, Mineralogy, and Genesis of a Hydrosequence of Oxisols in Brazil. Master's thesis. Cornell University, Ithaca, NY.

Macedo, J. and R.B. Bryant. 1987. Morphology, mineralogy, and genesis of a hydrosequence of Oxisols in Brazil. *Soil Sci. Soc. Am. J.,* 51: 690-698.

Macedo, J. and R.B. Bryant. 1989. Preferential microbial reduction of hematite over goethite in a Brazilian Oxisol. *Soil Sci. Soc. Am. J.,* 53: 1114-1118.

Magee, D.W. 1981. *Freshwater Wetlands: A Guide to Common Indicator Plants of the Northeast.* University of Massachusetts Press, Amherst.

McDaniel, P.A., J.H. Huddleston, C.L. Ping, and S.L. McGeehan. 1997. Aquic conditions in Andisols of the Northwest U.S. In *Aquic Conditions and Hydric Soils: The Problem Soils.* M.J. Vepraskas and S.W. Sprecher (Eds.). Spec. Pub. No. 50. Soil Science Society of America, Inc., Madison, WI. 99-111.

McKeague, J.A. 1965. Properties and genesis of three members of the Uplands catena. *Can. J. Soil Sci.,* 45:63-76.

Mejia, L. 1975. Characteristics of a Common Soil Toposequence of the Llanos Orientales of Columbia. Master's thesis. North Carolina State University, Raleigh.

Mid-Atlantic Hydric Soil Committee. 1998. Field Indicators of Hydric Soils in the Mid-Atlantic United States. U.S. Environmental Protection Agency, Region III, Philadelphia.

Mitsuchi, M. 1992. Anthropically induced wet soils. In Wetland Soils: Characterization, Classification, and Utilization of Wet Soils. J.M. Kimble (Ed.). Proc. of the 8th Int. Soil Correlation Meeting. USDA Soil Conservation Service, National Soil Survey Center, Lincoln, NE. 179-184.

Mokma, D.L. 1993. Color and amorphous materials in Spodosols from Michigan. *Soil Sci. Soc. Am. J.,* 57: 125-138.

Mokma, D.L. and S.W. Sprecher. 1994. Water table depths and color patterns in Spodosols of two hydrosequences in northern Michigan, USA. *Catena,* 22: 275-286.

Moormann, F.R. and H.T.J. van de Wetering. 1985. Problems in characterizing and classifying wetland soils. In *Wetland Soils: Characterization, Classification, and Utilization.* International Rice Research Institute, Los Baños, Laguna, Philippines. 53-68.

Morris, K. 1988. Jepson Prairie Reserve. Division of Environmental Studies, Institute of Ecology, University of California, Davis.

National Research Council. 1995. *Wetlands: Characteristics and Boundaries. Committee on Characterization of Wetlands.* National Academy Press, Washington, D.C.

Natural Resources Conservation Service. 1997. Hydrology Tools for Wetland Determination. Part 650. Engineering Field Handbook. U.S. Department of Agriculture, Washington, D.C.

Natural Resources Conservation Service. 1995. Hydric Soils of the United States. In cooperation with the National Technical Committee for Hydric Soils. U.S. Department of Agriculture, Washington, D.C.

NEIWPCC Wetlands Working Group. 1995. Field Indicators for Identifying Hydric Soils in New England, ver. 1. New England Interstate Water Pollution Control Commission, Wilmington, MA.

Nichols, J.D., M.E. Collins, and G.W. Hurt. 1990. Role of water table in Spodosol formation. In Characterization, Classification, and Utilization of Spodosols. J.M. Kimble and R.D. York (Eds.). Proc. of the 5th Intl. Soil Correlation Meeting (ISCOM). October 1-4, 1988. USDA Soil Conservation Service, Lincoln, NE. 238-241.

Niering, W.A. 1953. The past and present vegetation of High Point State Park, NJ. *Ecol. Monogr.,* 23: 127-147.

Parker, W.B. 1985. Technical hydric soils list. U.S. Fish and Wildlife Service, National Wetlands Inventory, St. Petersburg, FL. Memorandum dated November 7, 1985.

Pilgrim, S.A.L. 1996. Estimating field soil wetness — more art than science. University of New Hampshire, Durham, NH. Unpublished paper.

Pilgrim, S.A.L. and R.D. Harter. 1977. Spodic horizon characteristics of some forest soils in the White Mountains in New Hampshire. Bull 507. New Hampshire Agricultural Expt. Station, Durham, NH.

Pilgrim, S.A.L. 1998. Personal communication.

Ping, C.L., J. P. Moore, and M. H. Clark. 1992. Wetland properties of permafrost soils in Alaska. In Wetland Soils: Characterization, Classification, and Utilization of Wet Soils. J.M. Kimble (Ed.). Proc. of the 8th Intl. Soil Correlation Meeting. USDA Soil Conservation Service, National Soil Survey Center, Lincoln, NE. 198-205.

Ponder, F., Jr. and D. E. Alley. 1997. Soil sampler for rocky soils. Res. Note NC-371. USDA Forest Service, North Central Forest Expt. Station, Jefferson City, MO.

Reed, B. 1988. National List of Plant Species that Occur in Wetlands: 1988 National Summary. Biol. Rep 88 (24). U.S. Fish and Wildlife Service, Washington, D.C.

Reed, B. 1990. Transmittal of March 1990 hydric soils list and results of August meeting of National Technical Committee for Hydric Soils. U.S. Fish and Wildlife Service, National Wetlands Inventory, St. Petersburg, FL. Memorandum dated September 18, 1990.

Richardson, J.L. and R.B. Daniels. 1993. Stratigraphic and hydraulic influences on soil color development. In *Soil Color.* J.M. Bigham and E.J. Ciolkosz (Eds.). Special Publication No. 31. Soil Science Society of America, Inc., Madison, WI. 109-125.

Richardson, J.L. and M.J. Vepraskas (Eds.). 1999. *Wetland Soils.* Ann Arbor Press, Ann Arbor, MI.

Ritter, W.F. and C.E. Beer. 1969. Yield reduction by controlled flooding of corn. Transactions of the American Society of Agricultural Engineers, 12: 46-48.

Rourke, R.V., B.R. Brasher, R.D. Yeck, and F.T. Miller. 1988. Characteristic morphology of Spodosols. *Soil Sci. Soc. Am. J.,* 52: 445-449.

Ruhe, R.V., R.C. Prill, and F.F. Riecken. 1955. Profile characteristics of some loess-derived soils and soil aeration. *Soil Sci. Soc. Am. Proc.,* 19: 345-347.

Schelling, J. 1960. New aspects of soil classification with particular reference to reclaimed hydromorphic soils. *Trans. Int. Congr. Soil Sci.,* (7th, Madison, WI), 4: 218-224.

Segal, D.S., S.W. Sprecher, and F.C. Watts. 1995. Relationships Between Hydric Soil Indicators and Wetland Hydrology for Sandy Soils in Florida. Tech. Rep. WRP-DE-7. U.S. Army Engineer Waterways Experiment Station, Vicksburg, MS.

Skaggs, R.W. 1991. Drainage, chap. 10. In *Modeling Plant and Soil Systems.* Agronomy Monograph No. 31. ASA-CSSA-SSSA, Madison, WI. 205-243.

Smith, G.D. 1986. Rationale for Concepts in Soil Taxonomy. The Guy Smith Interviews. SMSS Tech. Monogr. 11. USDA Soil Conservation Service, Washington, D.C.

Soil Survey Staff. 1975. Soil Taxonomy: A Basic System for Soil Classification. Agricultural Handbook No. 436. USDA Soil Conservation Service, Washington, D.C.

Soil Survey Staff. 1996. Keys to Soil Taxonomy, 7th ed. USDA Natural Resources Conservation Service, Washington, D.C.

Somasiri, S. and P. Deturck. 1992. Anthropic wet soils of Sri Lanka. In *Wetland Soils: Characterization, Classification, and Utilization of Wet Soils.* J.M. Kimble (Ed.). Proc. of the 8th Intl. Soil Correlation Meeting. USDA Soil Conservation Service, National Soil Survey Center, Lincoln, NE. 240-247.

Sprecher, S. 1998. Personal communication.

Taylor, T.J., N.E. Erickson, R. Tumlison, J.A. Ratzlaff, and K.D. Cunningham. 1984. Groundwater wetlands of the Cimarron Terrace, northcentral Oklahoma. Dept. of Zoology, Oklahoma State University, Stillwater.

Tiner, R.W. 1985a. Wetlands of New Jersey. U.S. Fish and Wildlife Service, Newton Corner, MA.

Tiner, R.W. 1985b. Wetlands of Delaware. U.S. Fish and Wildlife Service. Cooperative Publication. National Wetlands Inventory, Newton Corner, MA and Delaware Department of Natural Resources and Environmental Control, Wetlands Section, Dover, DE.

Tiner, R.W. 1986. Hydric soils: their use in wetland identification and boundary delineation. In Proc. of the National Wetland Assessment Symp. J.A. Kusler and P. Riexinger (Eds.). (Portland, ME; June, 1985). ASWM Rep. No. 1: 178-182. Association of State Wetland Managers, Berne, NY.

Tiner, R.W. 1988. Field Guide to Nontidal Wetland Identification. U.S. Fish and Wildlife Service. Cooperative Publication. Region 5, Newton Corner, MA and Maryland Department of Natural Resources, Annapolis, MD.

Tiner, R.W. 1989. Wetlands of Rhode Island. Cooperative Publication. U.S. Fish and Wildlife Service, Region 5, Newton Corner, MA and U.S. Environmental Protection Agency, Region 1, Boston, MA.

Tiner, R. W. 1991. The concept of a hydrophyte for wetland identification. *BioScience,* 41: 236-247.

Tiner, R. W. 1993. The primary indicators method — a practical approach to wetland recognition and delineation in the United States. *Wetlands,* 13: 50-64.

Tiner, R.W. and D.G. Burke. 1995. Wetlands of Maryland. Cooperative Technical Report. U.S. Fish and Wildlife Service, Hadley, MA and Maryland Department of Natural Resources, Annapolis, MD.

Tiner, R. W., Jr. and P. L. M. Veneman. 1995. *Hydric soils of New England.* Bull. C-183R. Cooperative Extension Service, University of Massachusetts, Amherst.

Tuttle, C.L., M.S. Golden, and D.L. Sirois. 1984. A portable tool for obtaining soil cores in clayey or rocky soils. *Soil Sci. Soc. Am. J.,* 48: 1453-1455.

U.S. Army Corps of Engineers–Seattle District and U.S. Environmental Protection Agency, Region 10. 1994. Memorandum: Washington regional guidance on the 1987 wetland delineation manual. May 23, 1994. Seattle, WA.

USDA Natural Resources Conservation Service. 1994. DRAINMOD User's Guide. U.S. Department of Agriculture, Washington, D.C.

USDA Soil Conservation Service. 1985. National List of Hydric Soils. In Cooperative with the National Technical Committee for Hydric Soils. Washington, D.C.

U. S. Environmental Protection Agency, U.S. Army Corps of Engineers, U. S. Fish and Wildlife Service, and U.S. Soil Conservation Service. 1991. Proposed revisions to Federal manual for identifying and delineating jurisdictional wetlands.

Van Heesen, H.C. 1970. Presentation of the seasonal fluctuation of the water table on soil maps. *Geoderma,* 4: 257-278.

Van Wallenburg, C. 1973. Hydromorphic soil characteristics in alluvial soils in connection with soil drainage. In *Pseudogley and Gley-Genesis and Use of Hydromorphic Soils.* E. Schlichting and U. Schwertmann (Eds.). Trans. Comm. V and VI Int. Soc. Soil Sci., Verlag Chemie, Weinheim. 393-403.

Veneman, P.L.M. and R. W. Tiner. 1990. Soil-vegetation correlations in the Connecticut River floodplain of western Massachusetts. Biol. Rpt. 90(6). U. S. Fish and Wildlife Service, Washington, D.C.

Veneman, P.L.M. 1998. Personal communication.

Vepraskas, M.J. 1996. *Redoximorphic Features for Identifying Aquic Conditions.* Tech. Bull. 301. North Carolina Agricultural Research Service, NC State University, Raleigh.

Vepraskas, M.J. and J. Bouma. 1976. Model experiments on mottle formation simulating field conditions. *Geoderma,* 15: 217-230.

Vepraskas, M.J. and S.W. Sprecher (Eds.). 1997. *Aquic Conditions and Hydric Soils: The Problem Soils.* Spec. Pub. No. 50. Soil Science Society of America, Inc., Madison, WI.

Vepraskas, M.J., S.J. Teets, J.L. Richardson, and J.P. Tandarich. 1995. *Development of Redoximorphic Features in Constructed Wetlands.* Tech. Paper No. 5. Wetlands Research, Chicago, IL.

Vepraskas, M.J. and L.P. Wilding. 1983. Aquic moisture regimes in soils with and without low chroma colors. *Soil Sci. Soc. Am. J.,* 47: 280-285.

Viereck, L.A. 1970. Forest succession and soil development adjacent to the Chena River in interior Alaska. *Arc. Alp. Res.,* 2: 1-26.

Wakely, J.S. 1994. Identification of wetlands in the southern Appalachian region and the certification of wetland delineators. *Water, Air Soil Poll.,* 77: 2217-2226.

Wakely, J.S., S.W. Sprecher, and R.W. Lichvar. 1996. Relationships among wetland indicators in Hawaiian rain forest. *Wetlands,* 16 (2): 173-184.

Weinmann, F.C. and K. Kunz. 1995. A common error in assessing hydrophytic vegetation for wetland identification. In National Workshop on Wetlands, Technology Advances for Wetland Science. April 1995, New Orleans, LA. U.S. Army Engineer Waterways Expt. Station, Vicksburg, MS. 110-115.

Weinmann, F.C. 1998. Personal communication.

Whited, M. 1998. Personal communication.

Witty, J.E. and R.W. Fenwick. 1985. Wet aridisols. In *Wetland Soils: Characterization, Classification, and Utilization.* International Rice Research Institute, Los Baños, Laguna, Philippines. 465-469.

Williams, Gen. A.E. 1992. U.S. Army Corps of Engineers memorandum on clarification and interpretation of the 1987 manual. March 6, 1992.

Zampella, R.A. 1994. Morphologic and color pattern indicators of water table levels in sandy pinelands soils. *Soil Sci.,* 157: 312-318.

Zedler, P.H. 1987. The Ecology of Southern California Vernal Pools: A Community Profile. Biol. Rpt. 85(7.11). U.S. Fish and Wildlife Service, Washington, D.C.

8 Wetland Classification

INTRODUCTION

Given the range of environmental conditions affecting wetlands, scientists have developed wetland classification systems to arrange wetlands into similar groups. This is particularly important for conducting inventories and watershed planning, assessing biodiversity, evaluating wetland functions, assessing the impacts of wetland alteration and degradation, and considering potential wetland restoration. Classification from the natural resource standpoint is the grouping of habitats or natural features into categories with similar characteristics, properties, or functions. The unifying properties vary according to the needs of the classifier. For example, wetlands may be classified biologically, physically, chemically, hydrogeomorphically, and in other ways, depending on the discipline and interests of the classifier. In the U.S., most of the wetland classification has focused on the form of the wetland type rather than on the function. Recent attention has begun to emphasize the latter due to the need to evaluate wetland functions for assessing impacts of proposed projects subject to government regulations.

Wetlands can be described in many ways for various purposes. Research studies focusing on small study areas often describe wetlands in great detail based on extensive collected data. While this approach may work well for site-specific studies, it is not suitable for inventorying wetlands at the state, regional, and national levels. There are far too many individual wetlands to reasonably undertake such an intensive effort and to do so would require expenditures of labor, time, and money well beyond the budget of any natural resource agency. To determine the extent and distribution of wetlands in large geographical areas, wetlands are systematically combined into broad categories with similar ecological characteristics.

Two fundamental designs for wetland classification systems exist: horizontal or hierarchical. Horizontal classification systems divide habitats into a series of many classes or types. A wetland, therefore, must be one of 20 types, for example, as in the 1950s to 1960s Fish and Wildlife Service system (Martin et al., 1953). In contrast, hierarchical systems provide a matrix for separating wetlands into a multitude of types with different levels of types defined. Lower levels share more generalized characteristics, such as landscape position and water source, while higher levels are based on more detailed and specific characters like vegetation life form (including dominant species), substrate characteristics, and water level fluctuations. Variations of the hierarchical approach can be seen in the current American and Canadian wetland classification systems. Hierarchical systems allow for more descriptive characterizations that have more utility than horizontal classifications.

Horizontal classifications tend to be very generalized out of necessity to limit the number of wetland types. These classifications have produced terms like marsh, swamp, and bog that are familiar to the public and useful for describing wetlands to a nontechnical audience. Unfortunately, however, many of these common terms lack of universally accepted definitions and may be inconsistently interpreted. Despite this, many scientists still use these terms when referring to certain types of wetlands as they remain useful descriptors for broad categories of wetlands. For example, a book summarizing southeastern forested wetlands organized discussion into nine forested wetland types (Messina and Conner, 1998):

- Southern deepwater swamps
- Major alluvial floodplains
- Minor alluvial floodplains

- Pocosins
- Carolina bays
- Mountain bogs
- Cypress domes
- Wet flatwoods
- Mangroves

Another book intended for a general audience cited 17 major types for the Northeast (Tiner, 1998):

- Salt marshes
- Brackish marshes
- Tidal fresh marshes
- Tidal swamps
- Tidal submerged aquatic beds
- Nontidal marshes
- Savannas (including Carolina bays)
- Wet meadows
- Fens
- Bogs (including shrub and forested types)
- Shrub swamps
- Hardwood swamps (including northern hardwood swamps and coastal plain swamps)
- Larch swamps
- Floodplain forested wetlands
- Evergreen forested swamps (including cedar swamps, hemlock swamps, and pine swamps, lowlands, and flatwoods)
- Vernal pools
- Nontidal aquatic beds

While these terms are useful for general descriptions of wetlands, they may be too broad for conducting inventories of wetland resources. A hierarchical approach will likely produce more consistent mapping at different scales and permit more detailed characterization and cover-typing within individual types at finer scales. This approach provides the flexibility necessary to describe the variety within the above types in terms of hydrology, soil characteristics, water chemistry, human impacts, and vegetation patterns, for example.

In reviewing U.S. wetland classification systems, Mader (1991) mentioned the role of classification in serving the needs of land managers. In particular, he emphasized that classification should be

1. Flexible, general, and of wide geographic applicability in order to predict many kinds of information over a range of environmental situations.
2. Professionally credible, preferably through experimental validation.
3. Formed on concepts and logic that are explainable to nontechnical people.
4. Logical, consistent, and objectively quantifiable so as to function within an empirical, computer-operated information system.
5. Designed and documented so that regular professional staff can, with nominal training, use the system to identify and map field sites.

In this chapter, emphasis is placed on wetland classification systems developed in North America, with some reference to systems developed elsewhere to highlight some interesting similarities and differences. It is an overview and not an indepth analysis of the history of wetland classification. Examples of some early classification schemes are briefly discussed for historical perspective, but the major focus is on current systems used in the U.S. and Canada. For information on other classification systems, readers should consult Gore (1983), Whigham et al. (1993), and Finlayson and van der Valk (1995).

FEATURES USED TO CLASSIFY WETLANDS

After deciding to examine and describe the diversity of wetland habitats, the next question in classification is, What are the diagnostic characteristics that should be used to develop meaningful groupings? The answer will depend on the needs to be addressed through classification. To date, a wide range of features have been used to classify wetlands. The features typically correspond with the classifier's area of expertise or particular interest, e.g., ecologists, botanists, foresters, wildlife biologists, soil scientists, hydrologists, and natural resource planners. Most of the commonly used features are briefly summarized below.

VEGETATION

Vegetation has been the focal point in most wetland classification systems because most classifiers have had an interest in plant ecology, forest management, wildlife habitat management, or natural resource conservation. Some scientists contend that since vegetation is the product of hydrologic and geomorphologic factors and that many significant wetland functions are independent of vegetation, it should not be a primary criterion for classification (Semeniuk, 1987; Brinson, 1993). Nonetheless, the ecological significance of plant communities is indisputable and vegetation differences among wetlands are readily observed.

An important consideration in contemplating a vegetation-based classification is the fact that adjacent plant communities often intergrade, with zones of intermixed species occurring wherever distinct topographic breaks or drastic changes in soil type or hydrologic condition do not exist. This condition poses a problem for classification as well as for delineation, yet it is the inherent challenge of classification as we attempt to put diverse, intergrading natural systems into relatively discrete compartments. If these zones are small, they are largely ignored, but if vast, they represent a distinct community that requires separate analysis.

Some useful vegetative features to consider in classification are life form (e.g., tree, shrub, herb, moss/lichen, and aquatic); leaf type (e.g., needle-leaved and broad-leaved); foliage persistence (i.e., deciduous or evergreen for woody plants, persistent or nonpersistent for herbs), percent cover by life forms, height classes (e.g., low shrub vs. tall shrub); and community type (e.g., habitat type or dominance type). Because of the great interest in wetlands by plant ecologists and wildlife managers, plant community types (e.g., cattail marsh, red maple swamp, cottonwood bottomland, and tussock sedge meadow) have been emphasized in many classification systems. Predominant species and plant associations (e.g., *Spartina patens-Distichlis spicata* association) have been used to characterize wetlands, especially for local or site-specific studies. Regional assessments, however, usually cannot afford to capture that level of detail and tend to focus on vegetation life form and other factors (e.g., degree of wetness and water chemistry), although conspicuous dominant species of special interest may be identified (e.g., *Chamaecyparis thyoides* or *Phragmites australis*). Common wetland types recognized in many classifications describe assemblages of plants such as salt marsh, wet meadow, shallow marsh, deep marsh, shrub swamp, shrub bog, wooded swamp, and forested wetland.

Peatland scientists have described a wide range of peatlands, especially for northern climates (see discussion under Other Wetland Classification Systems). After separating bogs from fens, these scientists often identify various communities based on a combination of species associations and the physiognomy of the vegetation. Typical physiognomic types include hummock communities (growing on mounds of bryophytes and peat); lawn communities (firm turflike vegetation, including tussock-forming species); carpet communities (softer than lawns, including quaking mats); and mud bottom communities (frequently submerged muddy areas) (Sjörs, 1983).

HYDROLOGY

Hydrologic characteristics are important descriptors of wetlands, since hydrology varies among wetlands (see Chapter 2). The first step for national and regional inventories might be to separate tidal from nontidal wetlands where both types exist. Next the frequency and duration of flooding or soil saturation may be addressed, such as water permanence (e.g., Stewart and Kantrud, 1971; Millar, 1976) or water regime (Cowardin et al., 1979). Water depth also has been used to classify basin or depressional wetlands like prairie pothole wetlands. Millar (1976) identified three depth categories: (1) <3 ft, (2) 3 to 6.6 ft, and (3) >6.6. feet (2 m). The 3-ft depth correlated with the May 1 water level in shallow marshes that are vegetatively and hydrologically stable thereafter.

Directional flow of water is an important descriptor. Terms like *inflow, outflow, throughflow*, or *isolated* are used to describe whether a wetland is a source, sink, or pass-through system. These terms are widely used by hydrologists and have become increasingly used in wetland classification (Brinson, 1993; Tiner, 1997a). Brinson further describes three types of hydrodynamics: vertical fluctuations (rise and fall of water tables, e.g., in depressional wetlands); unidirectional (horizontal surface and subsurface flows, e.g., river or groundwater flows); and bidirectional flow (horizontal surface flow, e.g., tides and lake seiches).

Given the significance of hydrology in creating wetlands, influencing their vegetation, and performing various functions, classifications systems should be fairly explicit in describing hydrology. Unfortunately, most systems offer only a few broad categories. This results from a general lack of detailed hydrologic studies in the variety of wetlands that exist across the landscape. More work in wetland hydrology is needed (National Research Council, 1995). Classifiers should seek to be as descriptive as possible in regard to hydrology. The U.S. Fish and Wildlife Service's (FWS) wetland classification system (Cowardin et al., 1979) offers a wide range of water regime descriptors, recognizing flooded and saturated conditions (see later review of this system for details). Since its development in 1979, knowledge of wetland hydrology has significantly increased, making the list of FWS water regimes somewhat incomplete. In particular, a water regime for seasonally saturated wetlands needs to be added to the classification system to more accurately describe the hydrology of many wetlands, especially wet coastal plain flatwoods.

The water source of wetlands also may be described. Wetlands derive water from several sources including precipitation, snowmelt, groundwater, and surfacewater (including river overflow and tides). To characterize the hydrology of Wisconsin wetlands, Novitzki (1982) used a combination of water source and general landscape position to identify four types: surfacewater depression wetlands, surfacewater slope wetlands, groundwater depression wetlands, and groundwater slope wetlands (Figure 8.1). Brinson (1993) presented examples of several water sources: precipitation, lateral surface or near-surface transport from overbank flow (mesic climates), groundwater discharge (mesic climates), and various combinations of the first three for arid to subhumid climates.

WATER CHEMISTRY

The chemical composition of water has a profound effect on wetland vegetation, animals, and soils everywhere. Four major areas where it is particularly useful for classifying wetlands are

1. Estuaries where ocean-derived salinity (mostly sodium chloride) affects the soil environment as well as plant and animal life.
2. Arid to subhumid regions where inland salts (calcium, magnesium, sodium, potassium, chloride, bicarbonates, and sulfates) accumulate forming alkaline wetlands (e.g., wetlands along the Great Salt Lake, UT).
3. Northern climates where nutrient-poor bogs and minerotrophic fens develop.
4. Calcareous regions (e.g., karst landscapes) where alkaline (nutrient-rich) wetlands establish.

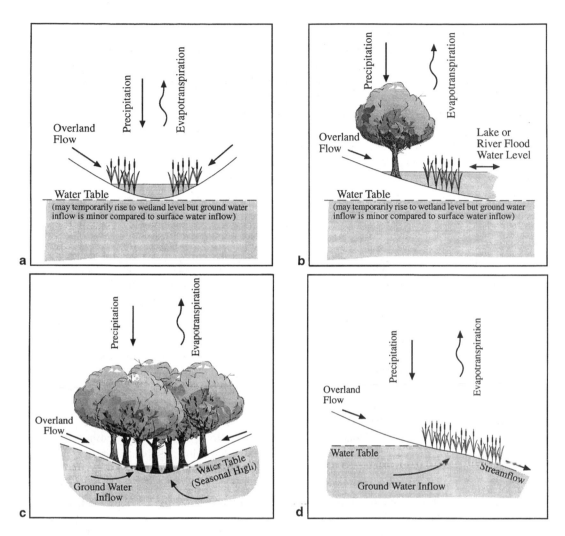

FIGURE 8.1 Hydrologic wetland types according to Novitzki (1982): (a) surfacewater depression wetland, (b) surfacewater slope wetland, (c) groundwater depression wetland, and (d) groundwater slope wetland.

Water chemistry descriptors include salinity (inland salts), halinity (coastal salts), and pH (see discussion of current FWS classification system for details). A classic ecological study by Stewart and Kantrud (1972) correlated plant species with salinity in North Dakota's prairie pothole wetlands, while Millar (1976) did this for western Canada (see also Kantrud et al., 1989). The effect of salinity on tidal wetland vegetation is well known and widely reported (e.g., Chapman, 1960; Adams, 1963; Anderson et al., 1968; Reimold and Queen, 1974; Albert, 1975; Tiner, 1987, 1993; Warren and Fell, 1995).

With more detailed information collected from field studies, terms like *ombrotrophic* (typically nutrient-poor wetland dependent on rainwater) and *minerotrophic* (nutrient-rich wetland with water enriched by ions leached from the soil) can be applied (Du Rietz, 1954). These properties have been used to separate bogs (the former) from fens (the latter). Interestingly, the effect of salt spray and atmospheric deposition in coastal areas may increase available nutrients in local ombrotrophic bogs allowing them to support some typical fen species (Damman, 1995). This led Damman to believe that the mineral content of wetland water was more important than its source for explaining vegetation patterns.

ORIGIN OF WATER

Peatland scientists have used the source of water as a descriptor for mires (peat-forming wetlands). Ombrogenous define wetlands that depended solely on precipitation (Weber, 1911; Sjörs, 1948). Topogenous and soligenous are terms introduced by von Post and Granlund (1926) and modified by Sjörs (1946) to describe ponded or saturated wetlands (stagnant water), and seepage and spring-fed wetlands, respectively. Limnogenous wetlands are those that receive flood water from rivers, streams, and lakes. Some other terms used that emphasize the origin of water are *telluric* (from groundwater), *meteoric* (from the atmosphere), and *marine* (from the ocean) (Semeniuk, 1987).

SOIL TYPES

Differences in soil types often produce different plant communities. The U.S. Fish and Wildlife Service's wetland classification system has provisions for separating wetlands on organic soils from those on mineral soils (Cowardin et al., 1979). This has been used to distinguish Atlantic white cedar swamps from other evergreen forested wetlands in the Northeast. Site-specific wetland studies make it possible to describe wetlands by soil series (or subgroups). For general wetland studies, it may be sufficient to separate organic from mineral soils and then nonsandy from sandy soils.

LANDSCAPE POSITION AND LANDFORM

Wetlands typically form in several important hydrogeologic settings: topographic depressions, slope breaks, areas of stratigraphic change, permafrost areas, and paludified landscapes (Winter and Woo, 1990; Brinson, 1993; Carter, 1996). The landscape position and landform of wetlands are geomorphologic features that significantly affect wetland functions and aid in understanding and explaining why wetlands perform certain functions and not others. Some peatland scientists have used geomorphological descriptors in their classification schemes (see discussion of one such system by Galkina [1946; 1967] in the Peatland Classifications subsection). In classifying pothole wetlands in western Canada, Millar (1976) recognized four watershed positions: isolated, overflow, channel, and terminal. Terminal wetlands are located in low topographic areas where water collects, equivalent to inflow wetlands. Lugo and Snedaker (1974) described five types of mangrove forests based largely on landscape position: basin forest, fringe forest, riverine forest, overwash forest, and dwarf forest. More recent classification systems have resurrected the significance of landscape position and landform due to increased interest in assessing wetland functions (see later discussion of Semeniuk, 1987; Brinson, 1993; Tiner, 1997a; b).

WETLAND ORIGIN

The origin of a wetland also may be a useful descriptor. Millar (1976) separated natural wetlands from altered wetlands (dugouts, borrow pits, or ditches) from those created by dams and reservoirs. He also included provisions for describing the nature of the alteration that increased or decreased the area and depth of the wetland (e.g., partial drainage, damming, dugout, road/fill/manmade structure). The current U.S. Fish and Wildlife Service wetland classification uses special modifiers to describe such impacts (Cowardin et al., 1979).

WETLAND SIZE

Wetland size also has been used as an attribute of wetlands. In classifying Canadian pothole wetlands, Millar (1976) identified nine size classes: (1) <0.25 acres, (2) 0.26 to 0.50 acres, (3) 0.51 to 1.00, (4) 1.01 to 2.50, (5) 2.51 to 5.00, (6) 5.01 to 10.00, (7) 10.01 to 20.00, (8) 20.01 to 40.00, and (9) >40 acres. Semeniuk (1987) also included size classes in his wetland classification system for western Australia (see later discussion).

ECOSYSTEM FORM/ENERGY SOURCES

A novel approach to wetland classification was proposed by Kangas (1990). He suggested classifying wetlands by energy because water-driven energy produces ecosystem characteristics. Four types of energy were described — frontal, line, sheet, and point. Frontal energy describes the energy produced by water moving parallel to the ground surface, as expressed by tides and surface water runoff. Line energy is similar to frontal energy, but line energy is focused in a small area, such as river channel flow. Sheet and point energy move perpendicular to the surface, like rainwater percolating through the soil or the water table rising during the wet season (sheet) or groundwater discharge from a spring (point). Water-driven energy then produces different landscape forms — zones (e.g., mangroves, salt marsh, and lacustrine littoral marshes); strings (oriented perpendicular to the water flow such as string bogs, oyster reefs, beaver ponds, and terraced rice paddies); islands (oriented parallel to water flow, e.g., island wetlands); and strips (e.g., floodplain and streamside wetlands) for frontal energy. For sheet energy, background (extensive landscape-level habitat types such as paludified peatlands, southern flatwoods on interfluves, and large wetland complexes such as the Dismal Swamp or Okeefenokee) and center (e.g., cypress domes, gum ponds, and prairie potholes) landscape forms are produced.

EARLY ECOLOGICAL CLASSIFICATIONS

Ecologists have traditionally characterized wetlands by their vegetation. In what is perhaps the first plant ecology textbook, *Oecology of Plants*, Warming (1909) identified two major types of wetland plant communities: saline swamps (with halophytic vegetation) and fresh-water swamps. He noted that these types were little studied. Fresh-water swamps included three types: reed-swamp, bush-swamp, and forest-swamp. The former type is dominated by herbaceous plants, whereas the latter types are characterized by shrubs and trees, respectively. The reed-swamp was mainly composed of tall perennial monocots usually growing in standing water (flowing or stagnant). Typical species that characterize plant associations included familiar North American monocots (e.g., *Phragmites australis*, *Scirpus lacustris*, *Typha* spp., *Phalaris arundinacea*, *Zizania aquatica*, *Iris pseudacorus*, *Cladium mariscoides*, *Carex aquatilis*, *C. rostrata*, *C. stricta*, *Dulichium arundinaceum*, *Eleocharis* spp., *Alisma plantago-aquatica*, *Sagittaria* spp., *Sparganium* spp., *Acorus calamus*, and *Calla palustris*), plus dicots (*Menyanthes trifoliata*, *Epilobium hirsutum*, *Lysimachia thyrisiflora*, *Sium* spp., *Cicuta* spp., *Symplocarpus foetidus*, and *Lythrum salicaria*). Bush-swamps were represented by alders (*Alnus* spp.), meadow-sweets (*Spiraea* spp.), willows (*Salix* spp.), arrowwoods (*Viburnum* spp.), glossy buckthorn (*Rhamnus frangula*), and birches (*Betula* spp. on "bog-lands"). He noted that bush-swamp and forest-swamp were extensive in the southern U.S., and listed "juniper-swamps" (Atlantic white cedar) and black gum swamps (including bald cypress) as the two main types. Other examples of these swamp types included bamboo forests, palm forests (*Nipa fruticans* on the landward side of mangrove-swamp), and spruce-birch-pine communities (*Picea excelsa-Betula pubescens-Pinus sylvestris*) of the arctic.

In an early description of the vegetation of Maryland, Shreve et al. (1910) characterized a number of wetland types including both salt marshes and several types of nontidal wetlands. In his chapter on the plant communities of the Eastern Shore, Shreve separated wetlands by vegetation life form, soil type, and/or landscape position. Forested wetlands included "clay upland swamps" (on interstream divides), "sandy loam upland swamps" (on interstream divides and including loblolly pine swamps), "flood plains," "river swamps," and "stream swamps." Herb-dominated wetlands were either "salt marshes" or "fresh marshes." In an accompanying chapter on Western Shore vegetation, Chrysler (1910) took a similar but slightly different approach to classification, emphasizing certain plant associations or zones within certain wetland types, such as the gum-pine association of lowland forests, the *Nymphaea*, *Pontederia*, *Zizania*, *Typha*, alder, and maple zones

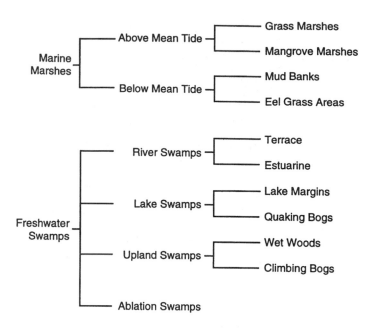

FIGURE 8.2   Shaler's wetland classification system (1890).

of fresh marshes, and different *Spartina* zones and other zones of salt marshes. He also referenced generic types like "river swamps," "cypress swamps," and "peat bogs."

WETLAND CLASSIFICATION IN THE U.S.

In the U.S., numerous classification systems have been developed over time to inventory wetlands for various purposes. The federal government's first attempts at wetland classification were largely motivated by natural resource utilization, mostly agricultural interests which sought to convert wetlands to cropland. In the late 1800s, Nathaniel Shaler, Director of the U.S. Geological Survey, prepared what is probably the first wetland classification system for the U.S. His system was developed for "morasses" east of the Cordilleras (Rockies), the region that contains most of the wetlands of the conterminous U.S. (Shaler, 1890). His classification focused on inundated lands using a tripartite approach to separate wetlands into similar types (Figure 8.2). Some other early classification systems used these and other terms to describe wetland types, such as permanent swamps, wet grazing land, periodically overflowed land, and periodically swampy land (Dachnowski, 1920; Wright, 1907). In inventorying peat deposits, Dachnowski expressed dissatisfaction by uses of certain common terms: "Progress in peat investigations has been severely checked by the widespread use of such terms as muck, overflowed land, swampy land, wetland and others." She proposed classifying peat deposits based on surface vegetation and recommended using such terms as marsh, fen, bog, heath, shrub, and forest. In 1922, the federal government conducted an inventory of wetlands based on soil surveys, topographic maps, and other sources in which the U.S. wetlands were divided into five types: tidal marsh, inland marsh, swamp and timbered overflow lands, and very deep peat (Stegman, 1976; Mader, 1991).

Later wetland classifications developed from an ecological interest, resource management use (e.g., forestry), or a need to separate wetlands and other land cover types for regional and national planning purposes (e.g., Dachnowski-Stokes, 1933; Dansereau and Segadas-Vianna, 1952; Penfound, 1952; Martin et al., 1953; Evans and Black, 1956; Curtis, 1959; Heinselman, 1963; 1970; Pestrong, 1965; Stewart and Kantrud, 1971; Chabreck, 1972; Eleuterius, 1972; Cowardin and Johnson, 1973; Chapman, 1974; Golet and Larson, 1974; Cowardin et al., 1979; Brinson, 1993;

Ferren et al., 1996; Tiner, 1997a). Most of these were regional systems, and only a few were nationally based. Mader (1991) provides an overview of forested wetland classification systems. The following discussion is a review of the current classification systems. An overview of the Martin et al. (1953) system also is included because of its historical significance and its horizontal approach to wetland classification.

1950s to 1960s Fish and Wildlife Service Classification

The system developed by Martin et al. (1953) was one of the few recent wetland classification systems developed for a national survey — to conduct an inventory of important waterfowl wetlands of the conterminous U.S. in 1954. The results of the inventory and an illustrated description of the 20 wetland types were published as U.S. Fish and Wildlife Service Circular 39 (Shaw and Fredine, 1956). This report was one of the most influential documents used in the continuous battle to preserve critically valuable, but rapidly diminishing wetlands (Stegman, 1976). This classification was used by the FWS in reporting wetland status and trends from the 1950s to the mid-1970s. The USDA Natural Resources Conservation Service also used it for administrative purposes (i.e., determining when it could provide technical assistance for drainage). The classification system was also used by state agencies to inventory wetlands in several states, and to map wetlands in Canada and Australia (Smith, 1971; Stoudt, 1971; Riggert, 1964-66).

The classification system was based on a horizontal approach containing 20 types of wetlands and associated aquatic habitats. Wetlands and aquatic habitats were described as one of the following types (see Table 8.1 for definitions):

- Seasonally flooded basins or flats
- Inland fresh meadows
- Inland shallow fresh marshes
- Inland deep marshes
- Inland open fresh water
- Shrub swamps
- Wooded swamps
- Bogs
- Inland saline flats
- Inland saline marshes
- Inland open saline water
- Coastal shallow fresh marshes
- Coastal deep fresh marshes
- Coastal open fresh water
- Coastal salt flats
- Coastal salt meadows
- Irregularly flooded salt marshes
- Regularly flooded salt marshes
- Sounds and bays
- Mangrove swamps

Most of the types are vegetated wetlands, but inland open fresh water, inland open saline water, coastal open fresh water, and sounds and bays are open water bodies.

Despite its widespread use in the 1950s and 1960s, it had some serious shortcomings, largely due to its horizontal classification approach (Leitch, 1966; Stewart and Kantrud, 1971). For example, in attempting to simplify their classification, the Martin et al.'s (1953) system not only ignored ecologically critical differences, such as the distinction between fresh and mixosaline inland wetlands, but also placed dissimilar habitats, such as forests of boreal black spruce and of southern

TABLE 8.1
Wetland Types Defined by Martin et al. (1953)

Type	Definition
Seasonally flooded basins or flats	Soil covered with water or waterlogged during variable seasonal periods; usually well drained during much of the growing season; located on river bottoms, along borders of drawn-down reservoirs, and in "dry lakes," shallow potholes, and other shallow upland depressions; vegetation includes bottomland woods and herbaceous areas.
Inland fresh meadows	Soil without standing water but waterlogged within at least a few inches of its surface during the growing season; located mostly in glaciated areas, the Nebraska sandhills, and in Florida, or may border marshes on landward side; vegetation includes grasses, sedges, rushes, and broad-leaved species.
Inland shallow fresh marshes	Soil normally waterlogged during the growing season; often covered with as much as 6 in. of water; locally mostly in glaciated areas, the Nebraska sandhills, and in Florida, or may border deep marshes on landward side, or adjoin irrigation systems; vegetation includes grasses, bulrushes, spikerushes, cattails, arrowheads, pickerelweed, smartweeds, reed, whitetop, rice cutgrass, bur-reeds, maidencane, sawgrass, and Baker cordgrass.
Inland deep marshes	Soil covered with 0.5 to 3.0 ft of water during the growing season; located mostly in glaciated areas, the Nebraska sandhills, and in Florida where they occur in and along margins of basins, potholes, limestone sinks, and sloughs; vegetation includes cattails, reed, bulrushes, spikerushes, wild rice, and various aquatic plants.
Shrub swamps	Soil normally waterlogged during the growing season, often covered with up to 6 in. of water; located mostly in the eastern U.S., also in Pacific Northwest; occur typically along sluggish streams and occasionally on floodplains; vegetation includes alders, willows, buttonbush, dogwoods, and swamp privet.
Wooded swamps	Soil waterlogged at least to within a few inches of its surface during the growing season, often covered with up to 1 ft of water; located mostly in the eastern U.S., also in the Pacific Northwest; occurs mostly along sluggish streams, on floodplains, on flat uplands, and in shallow lake basins and potholes; vegetation includes tamarack, white cedars, black spruce, balsam fir, red maple, black ash, western hemlock, red alder, and willows.
Bogs	Soil usually waterlogged, generally blanketed with a spongy covering of mosses; located in glaciated areas, western mountains, and along the Atlantic and Gulf coastal plains; mostly in shallow lake basins and potholes, on flat uplands, and along sluggish streams; vegetation includes heath shrubs (leatherleaf, labrador tea, cranberries, cyrilla, persea, and gordonia), black spruce, tamarack, pitcher plants, sphagnum moss, and sedges.
Inland saline flats	Soil without standing water, but waterlogged within at least a few inches of the surface during the growing season; located in the Great Basin, the northern Great Plains, and other arid regions; mostly in shallow lake basins; vegetation includes salt-tolerant plants such as sea blite, salt grass, Nevada bulrush, saltbush, and burro-weed.
Inland saline marshes	Soil normally waterlogged during the growing season, often flooded with as much as 2 ft of water; located in the Great Basin, the northern Great Plains, and other arid regions; mostly in shallow lake basins; vegetation includes alkali or hardstem bulrush, wigeongrass, and sago pondweed.
Coastal shallow fresh marshes	Soil always waterlogged during the growing season, may be covered at high tide with up to 6 in. of water; located along the Atlantic, Gulf, and Pacific coasts, on the landward side of deep marshes along rivers, sounds, and deltas; vegetation includes grasses (e.g., big cordgrass, reed, giant cutgrass, and maidencane), sedges (e.g., carex, spikerushes, three-squares, and sawgrass), cattails, arrowheads, smartweeds, and arrow arum.
Coastal deep fresh marshes	Soil covered at average high tide with 0.5 to 3.0 ft of water during the growing season; located along tidal rivers and elsewhere on the Atlantic and Gulf coasts; vegetation includes cattails, wild rice, pickerelweed, spatterdock, alligatorweed, water hyacinth, water lettuce, and aquatics.
Coastal salt flats	Soil almost always waterlogged during the growing season, varying from areas submerged only by occasional wind tides to others that are fairly regularly flooded with a few inches of water; located along the Atlantic, Gulf, and Pacific coasts, on the landward side of, or as islands or basins within, salt meadows and salt marshes; vegetation is sparse or patchy, including glassworts, sea blite, salt grass, saltflat grass, and saltwort.

Type	Definition
Coastal salt meadows	Soil always waterlogged during the growing season, rarely flooded; located along the Atlantic, Gulf, and Pacific coasts, on the landward side of salt marshes or bordering open water; vegetation includes saltmeadow cordgrass, salt grass, black rush, Olney three-square, salt marsh fleabane, carex, hairgrass, and jaumea.
Irregularly flooded salt meadows	Soil covered by wind tides at irregular intervals during the growing season; located on the Atlantic coast from Maryland south and on the Gulf coast, along the shores of nearly enclosed bays, sounds and river; vegetation includes black needlerush.
Regularly flooded salt marshes	Soil covered at average high tide with 0.5 ft or more of water during the growing season; located along the Atlantic, Gulf, and Pacific coasts, mostly along sounds, but also along the open ocean in places; vegetation includes salt marsh cordgrass, alkali bulrush, glassworts, and arrow-grass.
Mangrove swamps	Soil covered at average high tide with 0.5 to 2.0 ft of water during the growing season; located along the southern half of Florida; vegetation is mostly red mangrove and some black mangrove.

Source: Martin, A.C., N. Hotchkiss, F.M. Uhler, and W.S. Bourn. 1953. Classification of Wetlands of the United States. U.S. Fish and Wildlife Service, Washington, D.C.

cypress-tupelo in the same category (wooded swamp). This approach had too much lumping of diverse types with no provision for distinguishing between them. Due to the central emphasis on waterfowl habitat, far greater attention was devoted to vegetated areas of known importance to waterfowl than to nonvegetated areas. Probably the greatest single disadvantage of the Martin et al.'s system was the inadequate definition of types, which led to inconsistencies in application across the country (Cowardin et al., 1979).

CURRENT FISH AND WILDLIFE SERVICE'S WETLAND CLASSIFICATION SYSTEM

In 1974, the U.S. Fish and Wildlife Service (FWS) began planning for a comprehensive national wetlands inventory and after examination of its existing classification decided to design a new one. The decision was made that the new system should be hierarchical in structure, so that the users could select a level of detail appropriate to their needs (Sather, 1976). In addition, the new system had three primary objectives: (1) to group ecologically similar habitats so that value judgments can be made, (2) to furnish habitat units for inventory and mapping, and (3) to provide uniformity in concepts and terminology throughout the entire U.S.

The Service's wetland classification entitled *Classification of Wetlands and Deepwater Habitats of the United States* (Cowardin et al., 1979) was developed by a four-member team consisting of Lewis M. Cowardin (U.S. Fish and Wildlife Service), Virginia Carter (U.S. Geological Survey), Francis C. Golet (University of Rhode Island) and Edward T. LaRoe (National Oceanic and Atmospheric Administration), with assistance from numerous federal and state agencies, university scientists, and other interested individuals. The classification system went through three major drafts (Cowardin et al., 1976; Cowardin et al., 1977a; Cowardin et al., 1977b) and extensive field testing prior to its final publication in 1979. Since its publication, the FWS's classification system has been widely used by federal, state, and local agencies, university scientists, and private industry and nonprofit organizations for identifying and classifying wetlands. The system has provided uniformity in wetland concept and terminology (Mader, 1991). Recently, this system was adopted by the Federal Geographic Data Committee as the federal data standard for wetland classification for government data collection and reporting on wetland status and trends in the U.S.

The FWS's wetland classification system is hierarchical or vertical in nature proceeding from general to specific, as noted in Figure 8.3. In this approach, wetlands are first defined at a rather

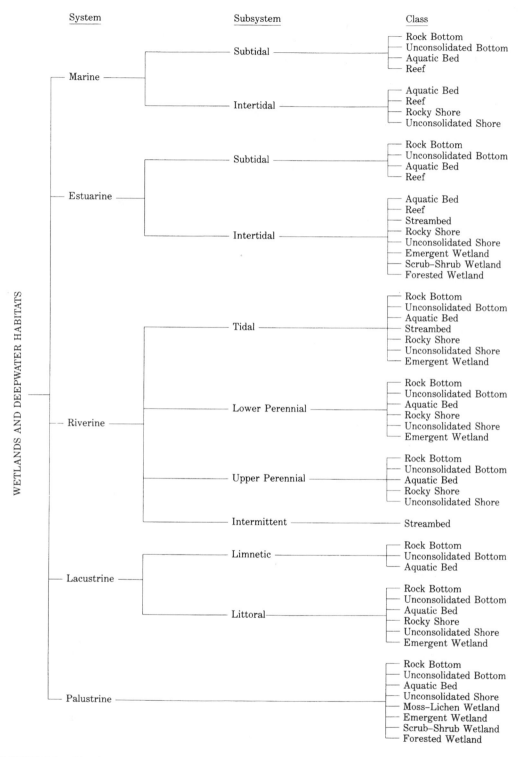

FIGURE 8.3   Classification hierarchy of wetlands and deepwater habitats (system through class) following the U.S. Fish and Wildlife Service's official classification system. (Cowardin, L.M., Carter, V., Golet, F.C., and LaRoe, E.T. 1979. *Classification of Wetlands and Deepwater Habitats of the U.S.*, U.S. Fish and Wildlife Service, Washington, D.C.)

broad level — the *system*. The term *system* represents "a complex of wetlands and deepwater habitats that share the influence of similar hydrologic, geomorphologic, chemical, or biological factors." Five systems are defined: marine, estuarine, riverine, lacustrine, and palustrine. The marine system generally consists of the open ocean and its associated high-energy coastline, while the estuarine system encompasses salt and brackish marshes, mangrove swamps, nonvegetated tidal shores, and brackish waters of coastal rivers and embayments. Freshwater wetlands and deepwater habitats fall into one of the other three systems: riverine (rivers and streams), lacustrine (lakes, reservoirs, and large ponds), or palustrine (e.g., marshes, bogs, swamps, and small shallow ponds). Thus, at the most general level, wetlands can be defined as either marine, estuarine, riverine, lacustrine, or palustrine (Figure 8.4).

Each system, with the exception of the palustrine, is further subdivided into *subsystems*. The marine and estuarine systems both have the same two subsystems, which are defined by tidal water levels: subtidal (continuously submerged areas) and intertidal (areas alternately flooded by tides and exposed to air). Similarly, the lacustrine system is separated into two systems based on water depth: littoral (wetlands extending from the lake shore to a depth of 6.6 ft or 2 m below low water or to the extent of nonpersistent emergents [e.g., arrowheads, pickerelweed, or spatterdock] if they grow beyond that depth), and limnetic (deepwater habitats lying beyond the 6.6 ft depth at low water). By contrast, the riverine system is further defined by four subsystems that represent different reaches of a flowing freshwater or lotic system: tidal (water levels subject to tidal fluctuations for at least part of the growing season); lower perennial (permanent, flowing waters with a well-developed floodplain); upper perennial (permanent, flowing water with very little or no floodplain development); and intermittent (channel containing nontidal flowing water for only part of the year).

The next level — *class* — describes the general appearance of the wetland or deepwater habitat in terms of the dominant vegetative life form or the nature and composition of the substrate, where vegetative cover is less than 30% (Table 8.2). Of the 11 classes, five refer to areas where vegetation covers 30% or more of the surface: aquatic bed, moss–lichen wetland, emergent wetland, scrub–shrub wetland, and forested wetland. The remaining six classes represent areas generally lacking vegetation, where the composition of the substrate and degree of flooding distinguish classes: rock bottom, unconsolidated bottom, reef (sedentary invertebrate colony), streambed, rocky shore, and unconsolidated shore. Permanently flooded nonvegetated areas are classified as either rock bottom or unconsolidated bottom, while exposed areas are typed as streambed, rocky shore, or unconsolidated shore. Invertebrate reefs are found in both permanently flooded and exposed areas.

Each class is further divided into *subclasses* to better define the type of substrate in nonvegetated areas (e.g., bedrock, rubble, cobble-gravel, mud, sand, and organic) or the type of dominant vegetation (e.g., persistent or nonpersistent emergents, moss, lichen, or broad-leaved deciduous, needle-leaved deciduous, broad-leaved evergreen, needle-leaved evergreen, and dead woody plants). Below the subclass level, *dominance types* can be applied to specify the predominant plant or animal in the wetland community.

To describe the hydrologic, chemical, and soil characteristics of wetlands and human impacts, the classification system contains four types of specific modifiers: water regime, water chemistry, soil, and special. These modifiers may be applied to class and lower levels of the classification hierarchy.

Water regime modifiers describe flooding or soil saturation conditions and are divided into two main groups, tidal and nontidal. Tidal water regimes are used where water level fluctuations are largely driven by oceanic tides. Tidal regimes can be subdivided into two general categories, one for salt and brackish water tidal areas and another for freshwater tidal areas. This distinction is needed because of the special importance of seasonal river overflow and groundwater inflows in freshwater tidal areas. By contrast, nontidal modifiers define conditions where surfacewater runoff, groundwater discharge, and/or wind effects (i.e., lake seiches) cause water level changes. Both tidal and nontidal water regime modifiers are presented and briefly defined in Table 8.3.

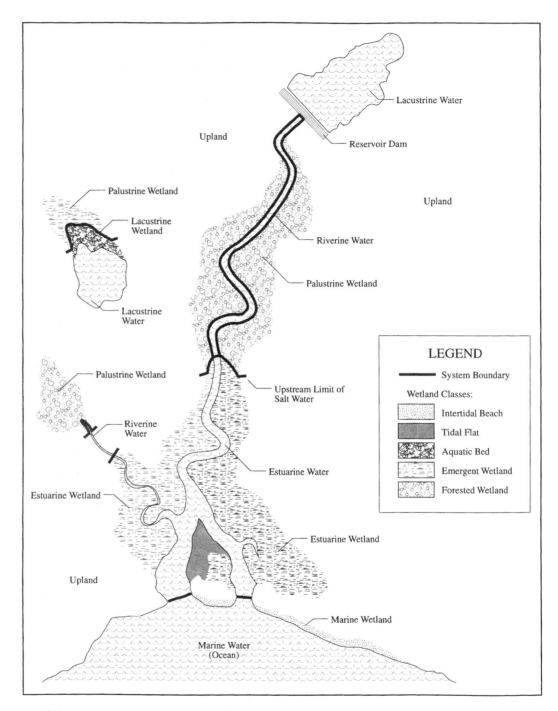

FIGURE 8.4 Diagram showing relationships between major wetland and deepwater habitat systems on the landscape. Some dominant classes are also designated. (Tiner and Burke, 1995)

Water chemistry modifiers are divided into two categories that describe the water's salinity or hydrogen ion concentration (pH), salinity modifiers and pH modifiers. Like water regimes, salinity modifiers have been further subdivided into two groups, halinity modifiers for tidal areas and salinity modifiers for nontidal areas. Estuarine and marine waters are dominated by sodium chloride, which is gradually diluted by fresh water as one moves upstream in coastal rivers. On the other hand, the

TABLE 8.2
Classes Defined by Cowardin et al. (1979)

Class	Brief Description	Subclasses
Rock bottom	Generally permanently flooded areas with bottom substrates consisting of at least 75% stones and boulders and less than 30% vegetative cover	Bedrock, rubble
Unconsolidated bottom	Generally permanently flooded areas with bottom substrates consisting of at least 25% particles smaller than stones and less than 30% vegetative cover	Cobble-gravel, sand, mud, organic
Aquatic bed	Generally permanently flooded areas vegetated by plants growing principally on or below the water surface line	Algal, aquatic moss, rooted vascular, floating vascular
Reef	Ridge-like or mound-like structures formed by the colonization and growth of sedentary invertebrates	Coral, mollusk, worm
Streambed	Channel whose bottom is completely dewatered at low water periods	Bedrock, rubble, cobble-gravel, sand, mud, organic, vegetated
Rocky shore	Wetlands characterized by bedrock, stones or boulders with areal coverage of 75% or more and with less than 30% coverage by vegetation	Bedrock, rubble
Unconsolidated shore[a]	Wetlands having unconsolidated substrates with less than 75% coverage by stone, boulders and bedrock and less than 30% vegetative cover, except by pioneer plants	Cobble-gravel, sand, mud, organic, vegetated
Moss-Lichen wetland	Wetlands dominated by mosses or lichens where other plants have less than 30% coverage	Moss, lichen
Emergent wetland	Wetlands dominated by erect, rooted, herbaceous hydrophytes	Persistent, nonpersistent
Scrub-Shrub wetland	Wetlands dominated by woody vegetation less than 20 ft (6 m) tall	Broad-leaved deciduous, needle-leaved deciduous, broad-leaved evergreen, needle-leaved evergreen, dead
Forested wetland	Wetlands dominated by woody vegetation 20 ft (6 m) or taller	Broad-leaved deciduous, needle-leaved deciduous, broad-leaved evergreen, needle-leaved evergreen, dead

[a] This class combines two classes of the 1977 operational draft system — Beach/Bar and Flat.

Source: Cowardin, L.M., V. Carter, F.C. Golet, and F.T. LaRoe. 1979. Classification of Wetlands and Deepwater Habitats of the U.S. Fish and Wildlife Service, Washington, D.C.

salinity of inland waters is dominated by four major cations (calcium, magnesium, sodium, and potassium) and three major anions (carbonate, sulfate, and chloride). Interactions between precipitation, surface runoff, groundwater flow, evaporation, and sometimes plant evapotranspiration form inland salts that are most common in arid and semiarid regions of the country. Table 8.4 shows ranges of halinity and salinity modifiers which are modifications of the Venice System (Remane and Schlieper, 1971). The other set of water chemistry modifiers are pH modifiers for identifying acid (pH<5.5), circumneutral (5.5 to 7.4) and alkaline (pH>7.4) waters. Numerous studies have shown a good correlation between plant distribution and pH levels, especially in boreal regions where pH can be used to distinguish between mineral-rich (fens) and mineral-poor wetlands (bogs) (e.g., Sjörs, 1950; Jeglum, 1971).

The third group of modifiers — soil modifiers — are presented because the nature of the soil exerts strong influences on plant growth and reproduction as well as on the animals living in it. Two soil modifiers are given, mineral and organic. In general, if a soil has 20% or more organic matter by weight in the upper 16 in., it is considered an organic soil, whereas if it has less than this amount, it is a mineral soil (see Chapter 5).

TABLE 8.3
Tidal and Nontidal Water Regimes Defined by Cowardin et al. (1979)

Group	Type of Water	Water Regime	Definition
Tidal	Saltwater and brackish areas	Subtidal	Permanently flooded tidal waters
		Irregularly exposed	Exposed less often than daily by tides
		Regularly flooded	Daily tidal flooding and exposure to air
		Irregularly flooded	Flooded less often than daily and typically exposed to air
	Freshwater	Permanently flooded–tidal	Permanently flooded by tides and river or exposed irregularly by tides
		Semipermanently flooded–tidal	Flooded for most of the growing season by river overflow but with tidal fluctuation in water levels
		Regularly flooded	Daily tidal flooding and exposure to air
		Seasonally flooded–tidal	Flooded irregularly by tides and seasonally by river overflow
		Temporarily flooded–tidal	Flooded irregularly by tides and for brief periods during growing season by river overflow
Nontidal	Inland freshwater and saline areas	Permanently flooded	Flooded throughout the year in all years
		Intermittently exposed	Flooded year-round except during extreme droughts
		Semipermanently flooded	Flooded through the growing season in most years
		Seasonally flooded	Flooded for extended periods in growing season, but surface water is usually absent by end of growing season
		Saturated	Surface water is seldom present, but substrate is saturated to the surface for most of the season
		Temporarily flooded	Flooded for only brief periods during growing season, with water table usually well below the soil surface for most of the season
		Intermittently flooded	Substrate is usually exposed and only flooded for variable periods without detectable seasonal periodicity (not always wetland; may be upland in some situations)
		Artificially flooded	Duration and amount of flooding is controlled by means of pumps or siphons in combination with dikes or dams

Source: Cowardin, L.M., V. Carter, F.C. Golet, and E.T. LaRoe. 1979. Classification of Wetlands and Deepwater Habitats of the U.S. Fish and Wildlife Service, Washington, D.C.

The final set of modifiers — special modifiers — were established to describe the activities of people or beaver affecting wetlands and deepwater habitats. These modifiers include excavated, impounded (i.e., to obstruct outflow of water), diked (i.e., to obstruct inflow of water), partly drained, farmed, and artificial (i.e., materials deposited to create or modify a wetland or deepwater habitat).

HYDROGEOMORPHIC (HGM) CLASSIFICATION

Wetland ecologist Mark Brinson (1993) developed a hydrogeomorphic-based (HGM) wetland classification to aid in performing functional assessments of wetlands, recognizing that the Cowardin system did not address certain abiotic (hydrogeomorphic) features that are directly linked to many wetland functions. The HGM wetland classification is actually more of an approach to provide a framework for wetland evaluation, rather than a classification system for mapping wetlands, as some of the types are not mutually exclusive (e.g., fringe wetlands associated with rivers and lakes vs. riverine and lacustrine wetlands; depressional wetlands [ox-bows and sloughs] within the riverine wetland type).

TABLE 8.4
Halinity and Salinity Modifiers Defined by Cowardin et al. (1979)

Coastal Modifiers[a]	Inland Modifiers[b]	Salinity (ppt)	Approximate Specific Conductance (μMhos at 25°C)
Hyperhaline	Hypersaline	>40	>60,000
Euhaline	Eusaline	30.0–40	45,000–60,000
Mixohaline (brackish)	Mixosaline[c]	0.5–30	800–45,000
Polyhaline	Polysaline	18.0–3.0	30,000–45,000
Mesohaline	Mesosaline	5.0–18	8,000–30,000
Oligohaline	Oligosaline	0.5–5	800–8,000
Fresh	Fresh	<0.5	<800

[a] Coastal modifiers are used in the marine and estuarine systems.
[b] Inland modifiers are used in the riverine, lacustrine, and palustrine systems.
[c] The term *brackish* should not be used for inland wetlands or deepwater habitats.

Source: Cowardin, L.M., V. Carter, F.C. Golet, and E.T. LaRoe. 1979. Classification of Wetlands and Deepwater Habitats of the U.S. Fish and Wildlife Service, Washington, D.C.

The HGM system identified seven basic hydromorphic classes.

1. Depressional wetlands (within topographic depressions, e.g., kettles, potholes, vernal pools, playas, and Carolina bays).
2. Organic soil flats (extensive peatlands, e.g., bogs and pocosins).
3. Mineral soil flats (broad nearly level wetlands with inorganic soils, e.g., flatwoods).
4. Riverine wetlands (along rivers and streams, e.g., floodplains and riparian areas).
5. Slope wetlands.
6. Lacustrine fringe (lakeshore wetlands).
7. Estuarine fringe (tidal wetlands).

Riverine wetlands are further divided into three gradients that reflect stream flow and fluvial processes:

- High gradient (flow likely continuous, but mostly flashy, wetland on coarse substrate maintained by upslope groundwater source; unstable substrate colonized by pioneer species; streamside vegetation contributes to allochthonous organic supply)
- Middle gradient (flow likely continuous, channel processes establish variable topography/hydroperiod/habitat interspersion on floodplain; alluvium is renewed by surface accretion and point bar deposition, interspersion of plant communities contributes to beta diversity)
- Low gradient (flow continuous with cool season flooding, high suspended sediments, flood storage, conserves groundwater discharge; major fish and wildlife habitat and biodiversity, strong biogeochemical activity and nutrient retention).

Examples of these hydrogeomorphic classes are given in Table 8.5.

The development of HGM has been explosive and a number of guidebooks have been or are being prepared. Modification of the original classification has already occurred. An example is one developed for use in Pennsylvania by Brooks et al. (1996). This system classifies wetlands by morphometry (shape of the wetland), landscape position (relationship to the contributing watershed),

TABLE 8.5
Examples of Brinson's Hydrogeomorphic Classes of Wetlands

Hydrogeomorphic Class	Examples of Wetland Types
Riverine	Floodplain wetlands, bottomland hardwood swamps, riparian wetlands, willow streambank thickets
Depressional	Bogs (small), potholes, Grady ponds, playas (small), woodland vernal pools, California vernal pools, fens, wet meadows isolated swamps, Carolina bays, farm ponds
Slope	Seepage wetlands, snow-meltwater wetlands, fens, wet meadows, avalanche chutes
Mineral soil flats	Coastal plain flatwoods, playas (large)
Organic soil flats	Bogs (extensive), pocosins
Lacustrine fringe	Great Lakes marshes, Flathead Lake marshes, Great Salt Lake marshes
Estuarine fringe	Atlantic and Gulf coast salt and brackish marshes, San Francisco Bay marshes, coastal marshes of the Yukon-Kushokwim delta

Note: Some types fall within more than one class depending on site characteristics (e.g., size or landscape position)

Source: Brinson, M.M. 1993. A Hydrogeomorphic Classification of Wetlands. U.S. Army Engineer Waterways Expt. Station, Vicksburg, MS.

water source (dominant hydrologic source supporting the wetland), and disturbance level criteria. Six morphometric types are identified: depression, slope, riverine (headwater or mainstem), fringing (water levels controlled by contiguous water body including lakes, reservoirs, and large ponds), floodplain, and impounded. For landscape position, four groups are possible: isolated, headwater (associated with first and second order streams), mainstem (third order stream and higher), and riparian (within transitional areas between uplands and edges of water bodies, may or may not be subjected to flooding). The three main water sources are groundwater, surface water (overbank, sheetflow, or impounded), and precipitation. Disturbance level criteria are defined to rank a wetland condition as pristine, moderate disturbance, or severe disturbance in terms of vegetation, water quality, and surrounding landscape condition. A pristine wetland should be pristine in at least two of the three categories with only minor disturbances in the third. This system has been used to categorize wetlands for applying HGM in Pennsylvania (Cole et al., 1997).

FWS/HGM-Type Classification Systems

With the great interest in HGM classification, the FWS recognized the need to expand its mapping and characterization of wetlands to aid map users in better understanding the functions of individual wetlands. In 1994, following the lead of Brinson's HGM classification, the FWS began creating a set of HGM-type descriptors in the Northeast. In developing the HGM system, Brinson used several key terms found in the FWS's classification system, but defined them differently. This eliminated the possibility of a simple application of HGM terminology to FWS wetlands inventory data. For example, the HGM riverine class includes both riverine and palustrine wetlands as defined by the FWS. Likewise, the HGM lacustrine class includes more than the lacustrine wetlands of the FWS (Cowardin et al., 1979). Consequently, new terminology had to be developed. By 1997, an operational draft of these descriptors was published for use in pilot studies across the country (Tiner, 1997a; b). These descriptors are designed to enhance the FWS classification system. The descriptors link the Cowardin system with the HGM system in a way that can be easily applied to existing National Wetlands Inventory (NWI) maps and databases and used for future NWI maps. This will help the FWS and others produce preliminary functional assessments for individual watersheds or other areas from NWI data.

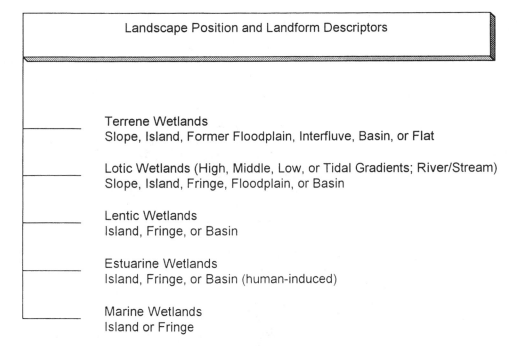

FIGURE 8.5   Wetland landscape position and landform modifiers used as a pilot basis by the U.S. Fish and Wildlife Service. These descriptors are added to the FWS's official wetland classification hierarchy to expand the amount of information presented for mapped wetlands.

Three descriptors provide additional information on each mapped wetland: landscape position, landform, and water flow path (Figure 8.5). Landscape position describes the linkage between the wetland and a water body. Five landscape position descriptors are defined: marine (ocean); estuarine (salt and brackish embayments and rivers); lotic (freshwater rivers and streams); lentic (lakes, reservoirs, and large deep ponds); and terrene (relatively isolated wetlands including shallow ponds and headwater wetlands). The first two landscape positions are equivalent to Cowardin's usage of the terms. The lotic landscape position is further divided into four gradients (high, middle, low, and tidal) and then rivers are separated from streams (simply based on width, with streams being linear map features and rivers polygonal features).

After identifying landscape position, wetlands are described by their landform and water flow path. Landform is defined by the shape of the wetland. Seven major types of landforms are identified for inland areas: slope, island, fringe, floodplain, interfluve, basin, and flat. For coastal wetlands, three landforms are defined: fringe, island, and basin. The latter type is typically formed by impoundment or some other human-built obstruction (e.g., road), whereas the former has free, uninterrupted tidal flowage including sheet flow. Each landform is further characterized by water flow path for inland wetlands (isolated, throughflow, inflow, or outflow). Some may be even further classified by the nature of the flow (what's above and below the wetland), by paludification, or by the type of island wetland (delta, river, stream, lake, pond, or floating mat). Former floodplains also may be designated. For coastal wetlands, the water flow is essentially bidirectional except where tidally restricted or controlled. Coastal landforms may be additionally defined by their position in the estuary (e.g., fringe types as barrier island, barrier beach, bay, coastal pond, river, headland, or ocean).

The system has provisions for identifying headwater wetlands, drainage-divide wetlands, plus an endless array of defined regional types of individual wetland landforms, particularly wetland basin types like prairie potholes, West Coast vernal pools, playas, woodland vernal pools, pocosins, sinkholes, cypress domes, and southern bottomlands. These descriptors coupled with the Cowardin types and information gained from HGM and previous studies of wetland functions allow for preliminary assessments of wetland functions to be performed over wide geographic areas, such as watersheds, by using existing NWI maps in digital form (Tiner, 1997b; Tiner et al., 1999). Some new modifiers have been established based on an application of the system to the Casco Bay watershed in Maine. A descriptor for fragmented wetlands has been added, since many such wetlands were observed in developed areas. A tidally restricted modifier also has been added to aid in recognizing coastal wetlands where hydrologic restoration may be needed. Although not yet implemented, the general condition of wetlands also may be realized by looking externally to the wetlands (i.e., the condition of the 300-ft buffer on the landward side and to the water quality of the adjacent waterbody when present). For example, wetlands with forested buffers may be in better condition than wetlands surrounded by urban or agricultural development, other circumstances not withstanding. Adding descriptors for external characteristics (e.g., condition of the buffer) and for potential inland wetland restoration sites will be tested in the future.

Another classification system utilizing both the FWS system and the HGM approach has been developed for central and southern California (Ferren et al., 1996). Like Tiner (1997a), it builds upon framework provided by the FWS classification system and adds components for hydrogeo-morphic units. It goes farther and presents choices for dominance types (substrate or vegetation) based on extensive knowledge of the region's wetland plant communities. Nine groups of HGM-type units are defined: (1) water bodies; (2) channels, fissures, drainages, inverts, and falls; (3) shores, beaches, banks, benches, and margins; (4) beds, bottoms, bars, and reefs; (5) flats, plains, deltas, washes, floodplains, and terraces; (6) headlands, bluffs, slopes, and fans; (7) seeps and springs; (8) pools, ponds, lakes, meadows, marshes, and swales; and (9) artificial structures. Each of these types is further categorized. For example, floodplains are divided into four types: stream, river, canyon, and montane. Nine types of seeps are recognized: drainage head, bluff and slope, canyon, stream bank/bed, river bank/bed, montane, foothill, valley and plain, and lake. This classification produces an impressive collection of wetlands. Although complex, it describes the variability among California's wetlands and will help identify biodiversity among the state's remaining wetlands and provide the information necessary to explain some important differences. The hierarchical approach to classification allows descriptions at various scales, with field studies required to make full use of this system.

WETLAND CLASSIFICATION IN CANADA

The need for a national wetland classification system in Canada evolved from a national effort to develop a systematic land classification by the National Committee on Forest Lands. The first national wetland classification for Canada was proposed by Zoltai et al. (1975). This system was later refined by Tarnocai (1980). Like the U.S., various regional classifications had been developed, including Jeglum et al. (1974) for Ontario, Couillard and Grondin (1986) for Quebec, Millar (1976) for the Prairies, and Runka and Lewis (1981) for British Columbia. In 1976, the National Wetlands Working Group of the Canada Committee on Ecological Land Classification was established to combine existing systems into a national wetland classification system, among other objectives.

The Canadian wetland classification like the FWS classification system is a hierarchical one, but the categories are quite different. The Canadian system has three levels: class, form, and type (National Wetlands Working Group, 1987). The class reflects the genetic origin of the wetland. The form distinguishes differences in surface morphology, surface pattern, water type, and underlying soil morphology. Wetland type is based on the physiognomy of the vegetation.

Five wetland classes are identified and within each class, different forms and types are recognized. The five classes are: bog, fen, marsh, swamp, and shallow water. Each class and its component forms and types are briefly discussed below; for more information consult the classification document or *Wetlands of Canada* (National Wetlands Working Group, 1988).

Bogs are acidic peatlands with water tables at or near the surface found in nutrient-poor areas. Their soils are organic (i.e., the remains of peat moss or wood). Ericaceous shrubs and/or various trees such as black spruce comprise the vegetation. Seventeen forms are recognized based on surface form, relief, or proximity to water bodies.

- Atlantic plateau bog
- Basin bog
- Blanket bog
- Collapse scar bog
- Domed bog
- Flat bog
- Floating bog
- Lowland polygon bog
- Mound bog
- Northern plateau bog
- Palsa bog
- Peat mound bog
- Peat plateau bog
- Polygonal peat plateau bog
- Shore bog
- String bog
- Veneer bog

Other peatlands called fens are associated with nutrient-rich, minerotrophic waters. Their soils are also organic, but are composed of the remains of sedges and/or brown moss. Sedges, grasses, reeds, and brown mosses predominate, with some shrubs and occasionally sparse tree cover. Seventeen forms are distinguished.

- Atlantic ribbed fen
- Basin fen
- Channel fen
- Collapse scar fen
- Feather fen
- Floating fen
- Horizontal fen
- Ladder fen
- Lowland polygon fen
- Net fen
- Northern ribbed fen
- Palsa fen
- Shore fen
- Slope fen
- Snowpatch fen
- Spring fen
- Stream fen

Marshes include peatlands and wetlands on mineral soils that are periodically inundated by standing or slow-moving, nutrient-rich water. Vegetation is often interspersed with water, with sedges, grasses, rushes, and reeds being the dominant emergent types and submerged species typifying the shallow water zone. Fifteen forms are recognized.

- Active delta marsh
- Channel marsh
- Coastal high marsh
- Coastal low marsh
- Estuarine high marsh
- Estuarine low marsh
- Floodplain marsh
- Inactive delta marsh
- Kettle marsh
- Seepage track marsh
- Shallow basin marsh
- Shore marsh
- Stream marsh
- Terminal basin marsh
- Tidal freshwater marsh

Swamps are similar to marshes by their occurrence on nutrient-rich peat and mineral soils, but are dominated by woody plants (trees and shrubs). When on peat, the material is comprised of well-decomposed wood, sometimes underlain by sedge peat. Seven forms are defined.

- Basin swamp
- Flat swamp
- Floodplain swamp
- Peat margin swamp
- Shore swamp
- Spring swamp
- Stream swamp

The fifth wetland class is shallow water that includes both intermittently and permanently flooded waters (ponds, pools, shallow lakes, oxbows, reaches, channels, and impoundments). In mid-summer, their water depth is less than 6.6 ft (2 m) and their vegetative cover is less than 25%. Thirteen forms are classified.

- Channel water
- Delta water
- Estuarine water
- Kettle water
- Nontidal water
- Oxbow water
- Shallow basin water
- Shore water
- Stream water
- Terminal basin water
- Thermokarst water
- Tidal water
- Tundra pool water

Eight general vegetation types are defined by the presence of particular vegetation or the absence of vegetation.

- Treed
- Shrub
- Forb
- Graminoid
- Moss
- Lichen
- Aquatic bed
- Nonvegetated

Specific types are identified within some of the types. Treed types may be coniferous or hardwood. Shrub types include tall, low, and mixed types. The forb type is dominated by nongrassy herbs. Graminoid types may be grass, reed, tall rush (including cattails and bulrushes), low rush, and sedge. The moss type is dominated by mosses, mainly peat mosses (*Sphagnum* spp.), feather-mosses (*Pleurozium* spp., *Hylocomium* spp., and *Ptilium* spp.), and brown mosses (*Drepanocladus* spp., *Scorpidium* spp., and *Tomenthypnum* spp.). The lichen type is characterized by reindeer lichen (*Cladonia* spp.). The aquatic bed type has two specific types, floating aquatic and submerged aquatic. The last type is nonvegetated where vegetation cover occupies less than 5% of the ground surface.

OTHER WETLAND CLASSIFICATION SYSTEMS

While the emphasis of this book is on North American wetlands, especially wetlands of U.S., some reference to wetlands in other continents will be made. Classification systems used in other countries may be of interest to some readers, so an overview of several systems will be given here. There is no particular significance to those described vs. those not described. They are presented to show some basic similarities as well as some key differences in worldwide approaches to wetland classification.

RAMSAR: A MULTINATIONAL CLASSIFICATION

In 1971, an international environmental convention held in Ramsar, Iran led to a multinational agreement to protect and preserve certain globally significant wetlands. To date over 770 sites (more than 52 million acres) have been designated as wetlands of international importance (Hunt, 1996). The Ramsar wetland classification system includes both deepwater habitats and typical wetlands (Ramsar Convention Bureau, 1998). Eleven types of marine and coastal types are identified.

- Permanent shallow marine waters (less than 6 m at low tide)
- Marine subtidal aquatic beds
- Coral reefs
- Rocky marine shores
- Sand, shingle, or pebble shores (including dune systems)
- Estuarine waters
- Intertidal mud, sand, or salt flats
- Intertidal marshes
- Intertidal forested wetlands (e.g., mangroves and tidal freshwater swamp forests)
- Coastal brackish/saline lagoons
- Coastal freshwater lagoons

Nineteen inland wetland types are listed.

- Permanent inland deltas
- Permanent rivers/streams/creeks
- Seasonal, intermittent, or irregular rivers/streams/creeks
- Permanent freshwater lakes (over 8 ha)
- Seasonal or intermittent freshwater lakes (over 8 ha)
- Seasonal or intermittent saline/brackish/alkaline lakes and flats
- Permanent saline/brackish/alkaline marshes and pools
- Seasonal or intermittent saline/brackish/alkaline marshes and pools
- Permanent freshwater marshes and pools (including ponds less than 8 ha)
- Seasonal or intermittent freshwater marshes and pools
- Nonforested peatlands (e.g., shrub bogs, open bogs, swamps, and fens)
- Alpine wetlands (including meadows and temporary snowmelt waters)
- Tundra wetlands
- Shrub-dominated wetlands
- Freshwater tree-dominated wetlands (on inorganic soils)
- Forested peatlands
- Freshwater springs and oases
- Geothermal wetlands
- Subterranean karst and cave hydrological systems

In addition to the above natural wetlands, nine manmade wetlands are cited.

- Aquaculture ponds (e.g., fish/shrimp)
- Ponds (e.g., farm ponds and stock ponds, generally less than 8 ha)
- Irrigated land (including rice fields)
- Seasonally flooded agricultural land (including wet meadows and pastures)
- Salt exploitation sites (e.g., salt pans and salinas)
- Water storage areas (including reservoirs and impoundments, generally more than 8 ha)
- Excavations (e.g., pools formed by gravel, clay, brick, or borrow pits)
- Wastewater treatment areas (including settling ponds)
- Canals and drainage channels (ditches)

PEATLAND CLASSIFICATIONS

Peatland scientists have been active in classifying the variety of peatland types. Much of this work has been done in northern Europe, Russia, and Canada where peatlands represent a significant portion of the landscape and an important forestry resource. Many terms have been used to describe differences among peatlands (bogs and fens) and most of these have already been discussed (Water Chemistry and Origin of Water subsections). Some other frequently used descriptors are eccentric, concentric (domed or raised), and blanket. The first two terms have been used to describe the surface pattern of bogs and fens. Eccentric bogs and fens are patterned peatlands marked with a series of alternating ponded and elevated sites — an irregular or regular pattern of patches of shallow open water (*flarks*, with mud bottom communities) and narrow ridges (*strings*, with typical fen or bog vegetation). They typically form on very gentle slopes, with the strings forming perpendicular to the slope (Figure 8.6). Concentric raised bogs are generally flat in the center with the margins sloping downward to give the bog an overall convex shape in cross-section or domed appearance. Plateau bogs are similar to concentric bogs, but have only one sloping front, thus appearing plateau-like in cross-section. Blanket bogs are bogs that form on slopes resulting from the process of paludification. Most seem to represent an upslope extension of a valley bog in areas

FIGURE 8.6   Patterned peatland in Maine. (Hank Tyler photo.)

of high precipitation and low evapotranspiration. These bogs have been found climbing 25% slopes in western Ireland where they have attained peat depths of more than 6.6 ft (2 m) (Moore and Bellamy, 1974). Some blanket bogs may form directly on slopes in areas with shallow impermeable soils and groundwater seepage (Sjörs, 1983).

The Finnish scientist A.K. Cajander (1913) proposed one of the earliest peatland or mire classification systems based on his previous work on vegetation classification. He defined mire as both a plant community that forms peat and the peat deposit itself. His system was instrumental in stimulating ecological work as well as having practical utility for agricultural and forestry planning (Ruuhijärvi, 1983). The current Finnish system identifies a series of mire site-types defined on the basis of three ecological gradients: wetness (wet to dry direction, with three categories: flark [hollow], lawn, and hummock [mound] vegetation); nutrient status (oligotrophic to mesotrophic to eutrophic, with increasing nutrients); and the distinction between the mire margin and the mire center (e.g., treed types from treeless types). Cajander's approach was broader than a site-type characterization and considered mire-complexes or ecosystems based on vegetation, fauna, ecology, morphology, and stratigraphy (Ruuhijärvi, 1983). Four complex types were recognized: raised bogs, Karelian mire-complexes, aapa mires, and palsa mires. A sloping mire complex was later added to this list (Auer, 1922). In his review of Finnish wetlands, Ruuhijärvi listed the five main types: plateau bogs, concentric bogs, eccentric bogs, aapa mires (minerotrophic wet mires with characteristic strings and flarks), and palsa mires (peat mounds with permafrost cores and surrounded by minerotrophic fens). Heikurainen and Pakarinen (1982) provided brief descriptions of the Finnish peatland site types.

Another example of a mire classification system incorporating much of peatland terminology is Masing's (1975) classification of Estonian mires (see Water Chemistry and Origin of Water subsections for definitions of major terms). He identified two major groupings of mires based on nutrient status: minerotrophic mires and oligotrophic mires. The former was divided into four subtypes: soligenous (woodless spring fens); topogenous (rich and poor fens); limnogenous (excluding semi-aquatic reeds, but including floating fens, floodplain fens and swamps, and wooded swamps); and transitional mires (topo-ombrogenous, limno-ombrogenous, with transitional fens and bogs). Oligotrophic mires were separated into two subtypes: moor sites (ombrogenous on thin peat overlying sand) and bogs (with bog margins distinguished from bog centers).

A different approach to mire classification focused on geomorphological principles (Galkina, 1946; 1967; Ivanov, 1975; among others as reported in Botch and Masing, 1983). Mires were classified by Galkina based on their location in "watershed areas," on slopes and terraces, or in river and lake valleys. Four watershed areas were given: on vast plains or gently sloping areas between rivers, in closed hollows (without outflow), in hollows with an outflow, and in oblong depressions (troughs) with an outflow. The latter three areas seem to represent headwater-type watersheds that may or may not be connected to a larger drainage system. Five types of slope and terrace locations were listed: round depressions with through-flow, troughs with through-flow, gentle slopes, the foot of slopes, and slopes with diverging outflow. Four geomorphical conditions were associated with river and lake valleys: on lake sides, in river flood valleys, on river sides, and in river arms.

AUSTRALIA

In an overview of Australian wetlands, Paijmans et al. (1985) presented a classification system describing a wide variety of wetland types and included a series of figures that showed the general distribution of these types. Given the aridity of much of Australia and the attendent cycling and irregularity of precipitation, these classifiers recognized the difficulty of wetland classification in arid regions. Many wetlands are continuous gradations such as lakes grading into swamps and then into periodically flooded lands.

Wetlands were classified into six major categories: lakes (open water bodies >1 m deep with little or no persistent free-standing vegetation); swamps (vegetated areas with water <1 m deep); lands subject to inundation (water not present long enough to support typical wetland species, but may be important waterbird habitat); river and creek channels; tidal flats; and coastal water bodies. Table 8.6 shows further subdivisions within each major type. The authors' concept of wetland, therefore, includes inland and coastal deep open water habitats as well as vegetated swamps and channels.

Several features of this classification are particularly noteworthy. First, the system clearly recognizes hydrologic differences among wetlands with classes designated for permanent/near-permanent, seasonal (wet and dry every year with seasonal periodicity), intermittent (wet and dry less often and at irregular intervals), and episodic (mostly dry with rare wet phases) wetness. The latter hydrologic regime is especially notable as it broadens the concept of wetland to address a significant ecological condition of arid and semiarid regions — the presence of ephemeral wetland habitats. When the rare heavy rainfall patterns create lakes in normally dry basins, these areas become important habitats for certain wetland animals, particularly waterbirds, for some time thereafter. There is probably an entire ecosystem of minute plants and animals adapted for life in such areas. Although flooded only rarely, these soils usually are saturated below a salt crust (Paijmans et al., 1985). The swamp type can be further separated into five subtypes based on predominant vegetation:

- Herbaceous swamps (dominated by graminoids like *Eleocharis*, *Paspalum*, *Leersia*, and *Oryza*, and other herbs such as *Eichhornia crassipes*)
- Sedge swamps (dominated by members of the Cyperacea including *Lepironia*, *Baumea*, *Cyperus*, *Eleocharis*, *Rhynchospora*, *Schoenus*, *Scirpus*, and *Scleria*)
- Heath and scrub swamps (with shrubs less than 8 m tall including wet heaths)
- Woodland and forest swamps (dominated by trees, mostly *Melaleuca* and also palms)
- *Sphagnum* swamps or bogs

The distinction between swamp and land subject to inundation recognizes that some wetlands do not hold water long enough or often enough to support hydrophytic vegetation, yet they are still important waterbird habitats. This aspect of the classification recognizes the significance of one wetland function — waterbird habitat — as a reason for classification and identification. These lands are subjected to variable flooding from annual to once a century. Tidal flats also include a broad range of habitats from nonvegetated mud flats to mangrove swamps to areas that are tidally inundated once a year. Some of the latter areas are flooded during the wet season by a combination of tidal action and freshwater river overflow.

TABLE 8.6
Australian Wetland Classification

I. **Lakes: areas of open water generally over 1 m deep with little or no persistent emergent vegetation**
    1. Permanent and near-permanent lakes
        • Permanent floodplain lakes including billabongs and waterholes in channels
        • Permanent lakes of coastal dunes and beach ridge plains
        • Permanent lakes in terminal drainage basins
        • Permanent lakes associated with lava flows
        • Permanent crater lakes
        • Permanent karst lakes
        • Permanent glacial lakes
        • Permanent manmade lakes
    2. Seasonal lakes: alternately wet and dry every year according to season
        • Seasonal floodplain lakes
        • Seasonal lakes in terminal drainage basins
    3. Intermittent lakes: alternately wet and dry but less frequently and regularly than seasonal lakes
        • Intermittent floodplain lakes
        • Intermittent coastal dune lakes
        • Intermittent lakes in terminal drainage depressions
        • Intermittent manmade lakes
    4. Episodic lakes: dry most of the time with rare and very irregular wet phases
        • Episodic lakes in terminal drainage depressions
        • Episodic lakes on present or former floodplains
          (Other subclasses rare or non-existent)

II. **Swamps: dominantly vegetated; where present, water generally less than 1 m deep; persistent emergent vegetation**
    1. Permanent swamps: wet most of the time
        • Permanent floodplain swamps
        • Permanent swamps of coastal dunes and beach ridge plains
        • Permanent swamps in terminal drainage depressions
        • Permanent swamps associated with lava flows
        • Permanent crater swamps
        • High-mountain permanent swamps
        • Permanent swamps fed by springs
          (These swamps can be further subdivided according to vegetation into herbaceous, sedge, heath and scrub, woodland and forest, and sphagnum types.)
    2. Seasonal swamps: seasonally wet and dry each year
        • Seasonal floodplain swamps
          (Other subclasses rare or absent)
    3. Intermittent swamps: alternately, but irregularly, wet and dry
        • Intermittent floodplain swamps
        • Intermittent swamps in terminal drainage depressions
    4. Episodic swamps: very rarely contain water and lacking swamp vegetation
          (Rare. No subclasses distinguished)

III. **Land Subject to Inundation: water not present long enough for typical wetland vegetation to develop, but may be important waterbird habitat**
    1. Seasonally inundated
        • Floodplains
        • River and creek banks
    2. Intermittently inundated
        • Floodplains
        • River and creek banks

TABLE 8.6 (continued)
Australian Wetland Classification

IV. **River and Creek Channels**
  1. Permanent and near-permanent channels
     • Rocky
     • Sandy
     • Silty/clayey
  2. Seasonal channels
     • Rocky
     • Sandy
     • Silty/clayey
  3. Intermittent channels
     • Rocky
     • Sandy
     • Silty/clayey
  4. Episodic channels
     • Rocky
     • Sandy
     • Silty/clayey

V. **Tidal Flats**
  1. Daily tidal flooding
     • Intertidal flats of open coasts
     • Intertidal estuarine flats
     • Intertidal stream banks
  2. Spring tidal and less frequent flooding
     • Supratidal surfaces
     • Supratidal stream banks
     • Saline pools
  3. Spring tidal and less frequent flooding combined with seasonal freshwater flooding
     • Supratidal flats
     • Brackish pools and billabongs

VI. **Coastal Water Bodies**
  1. Permanently open to the sea
     • Saline to brackish estuaries and inlets
  2. Intermittently open to the sea
     • Saline to brackish lagoons
  3. Rarely open to the sea
     • Brackish to fresh lagoons and lakes

Source: Paijamans, K., et al. 1985. Aspects of Australian Wetlands. Commonwealth Scientific and Industrial Research Organization, Melbourne, Australia.

WESTERN AUSTRALIA

Semeniuk (1987) summarized wetland classification systems for western Australia. One of these was the Martin et al. (1953) classification. It was used by the Department of Fisheries and Fauna of Western Australia to evaluate wetlands used by waterfowl (Riggert, 1964-66). Another wetland study of western Australia by Tingay and Tingay (1976) used classifications by Hutchinson (1957) and Bayly and Williams (1973) that first separated wetlands into lotic and lentic types. The latter were further divided into different types of lakes (tectonic, volcanic, landslide, glacial, solution, fluviatile, wind action or coastal) and shallow water bodies of various types (e.g., underground water, springs, water associated with terrestrial vegetation, puddles, rock pools, and ponds, permanent or temporary). Lotic wetlands were separated into permanent, temporary, or episodic types and also by other features including unidirectional flow, fluctuation in flow rates, and linear morphology. The Wetlands Advisory Committee (1977) established another system with main divisions being lentic, lotic, estuarine, and artificial. Other features such as size, salinity, permanence, and vegetation cover were then used to classify different types.

Semeniuk (1987) used many of the above features to devise a wetland classification system using a geomorphic framework for improving the understanding of wetlands, their distribution, and relationship to biota. The system emphasizes water and landform characteristics and probably served as a model for Brinson's HGM classification system (Brinson, 1993). The Semeniuk system has been used to inventory wetlands in several areas in the country, including Perth (Water Authority of Western Australia, 1993).

The water attributes of this system include water permanence, water salinity, consistency of water salinity, and water maintenance. Three degrees of water permanence are specified: permanent inundation, seasonal or intermittent inundation, and seasonal or intermittent waterlogging. Categories for water salinity include fresh, subsaline, hyposaline, mesosaline, hypersaline, and brine. Two categories for consistency of water salinity are poikilohaline (variable) and stasohaline (stable), and three categories of water maintenance (or origin) are specified: telluric (groundwater), meteoric (from the atmosphere, precipitation, essentially surface water), and marine (oceanic).

The landform component of the classification system includes cross-sectional shape (basin, flat, and channel) as the principal element along with others (size, plan shape, stratigraphy, and origin with the latter two of lesser significance). Plan shape includes nine forms: five for basins (linear, elongate, irregular, ovoid, and round) and four for channels (straight, sinuous, anastomosing, and irregular). Five size descriptors are given: megascale (>10 km × 10 km for basins/flats), macroscale (1 km × 1 km to 10 km × 10 km for basins/flats; >1 km wide channels), mesoscale (500 m × 500 m to 1 km × 1 km; 100 m+ wide channels), microscale (100 m × 100 m to 500 m × 500 m; 10 m+ channels), and leptoscale (channels <10 m wide).

Combining landform with water permanence yield nine possibilities with seven common types recognized: lake, sumpland (seasonally inundated basin), dampland (waterlogged basin), floodplain (seasonally inundated flat), palusplain (waterlogged flat), river, and creek (seasonally inundated channel). There seems to be no provision for identifying slope wetlands. Semeniuk recognized the merits of using vegetation descriptors such as structure, life form, species dominance, percent cover, and zonation to aid in identifying similar wetlands and to help understand ecological relationships. He suggested using vegetation as a later step in describing wetlands (as a tertiary or quaternary modifier) but offers no specific terminology or categories. He also noted that the dampland type has in the past been variously identified as wetlands by some or excluded by others, but included it due to its importance for management considerations. The hydrographs of selected damplands shown in Semeniuk (1987) suggest that most of these areas would easily qualify as wetland according to U.S. criteria since their water table is at the surface for several months.

REFERENCES

Adams, D.A. 1963. Factors influencing vascular plant zonation in North Carolina salt marshes. *Ecology,* 44: 445-456.

Albert, R. 1975. Salt regulation in halophytes. *Oecologia,* 21: 51-71.

Anderson, R.R., R.G. Brown, and R.D. Rappleye. 1968. Water quality and plant distribution along the upper Patuxent River in Maryland. *Chesapeake Sci.,* 9: 145-156.

Auer, V. 1922. Suotutkimuksia Kuusamon ja Kuolajärven vaara-alueilta. *Commun. Inst. Quaest. For. Finl.,* 6(1): 1-368.

Bayly, J.A.E. and W.D. Williams. 1973. *Inland Waters and Their Ecology.* Longman, Cheshire, England.

Botch, M.S. and V.V. Masing. 1983. Mire ecosystems in the U.S.S.R., chap. 4. In *Mires: Swamp, Bog, Fen and Moor*. A.J.P. Gore (Ed.). Regional Studies. Ecosystems of the World 4B. Elsevier Publishing Co., Amsterdam. 95-152.

Brinson, M.M. 1993. A Hydrogeomorphic Classification for Wetlands. Wetlands Research Program Tech. Rep. WRP-DE-4. U.S. Army Engineer Waterways Expt. Station, Vicksburg, MS.

Brooks, R.P., C.A. Cole, D. H. Wardrop, L. Bishel-Machung, D.J. Prosser, D.A. Campbell, and M.T. Gaudette. 1996. *Wetlands, Wildlife, and Watershed Assessment Techniques for Evaluation and Restoration*, vol. 1. Penn State Cooperative Wetlands Center, Environmental Resources Research Institute, University Park, PA.

Cajander, A.K. 1913. Studien über die Moore Finnlands. *Acta For. Fenn.*, 2(3): 1-208.

Carter, V. 1996. Wetland hydrology, water quality, and associated functions. In National Water Summary on Wetland Resources. J.D. Fretwell, J.S. Williams, and P.J. Redman (Comp.). Water-Supply Paper 2425. U.S. Geological Survey, Reston, VA.

Chabreck, R.H. 1972. *Vegetation, Water and Soil Characteristics of the Louisiana Coastal Region*. Bull. 664. Louisiana State University, Agricultural Expt. Station, Baton Rouge.

Chapman, V.J. 1960. *Salt Marshes and Salt Deserts of the World*. Interscience Publishers, Inc., New York.

Chrysler, M.A. 1910. The ecological plant geography of Maryland; Coastal Zone; Western Shore District. In *The Plant Life of Maryland*. F. Shreve, M.A. Chrysler, F.H. Blodgett, and F.W. Besley (Eds.). The John Hopkins Press, Baltimore, MD. 150-197.

Cole, A.C., R.P. Brooks, and D.H. Wardrop. 1997. Wetland hydrology as a function of hydrogeomorphic (HGM) subclass. *Wetlands,* 17: 456-467.

Couillard, L, and P. Grondin. 1986. La vegetation des milieux humides du Quebec. Les publications du Quebec. Quebec Ministry of Environment, Quebec, Canada.

Cowardin, L.M., V. Carter, F.C. Golet, and E.T. LaRoe. 1976. Interim Classification of Wetland and Aquatic Habitats of the United States. March 1, 1976. U.S. Fish and Wildlife Service, Washington, D.C.

Cowardin, L.M., V. Carter, F.C. Golet, and E.T. LaRoe. 1977a. Classification of Wetlands and Aquatic Habitats of the United States. April 1977. U.S. Fish and Wildlife Service, Washington, D.C.

Cowardin, L.M., V. Carter, F.C. Golet, and E.T. LaRoe. 1977b. Classification of Wetlands and Deep-Water Habitats of the United States (An Operational Draft). October 1977. U.S. Fish and Wildlife Service, Washington, D.C.

Cowardin, L.M., V. Carter, F.C. Golet, and E.T. LaRoe. 1979. Classification of Wetlands and Deepwater Habitats of the United States. FWS/OBS-79-31. U.S. Fish and Wildlife Service, Washington, D.C.

Cowardin, L.M. and D.H. Johnson. 1973. A Preliminary Classification of Wetland Plant Communities in Northcentral Minnesota. Special Scientific Report — Wildlife No. 168. U.S. Fish and Wildlife Service, Washington, D.C.

Curtis, J.T. 1959. *The Vegetation of Wisconsin*. University of Wisconsin Press, Madison.

Dachnowski, A.P. 1920. Peat deposits in the United States and their classification. *Soil Sci.*, 10: 453-465.

Dachnowski-Stokes, A.P. 1933. Peat deposits in the U.S.A. Their characteristic profiles and classification. In *Handbuch der Moorkunde 7: 1-140*. K.V. Bulow (Ed.). Springer-Verlag, Berlin.

Dansereau, P. and F. Segadas-Vianna. 1952. Ecological study of peat bogs of eastern North America. I. Structure and evolution of vegetation. *Can. J. Bot.*, 30: 490-518.

Damman, A.W.H. 1995. Major mire vegetation units in relation to the concepts of ombrotrophy and minerotrophy: a worldwide perspective. In Regional Variation and Conservation of Mire Ecosystems. *Grunneria*, 70: 23-34.

Du Rietz, G.E. 1954. Die Mineralbodenwasserzeigergrenze als Grundlage einer natürlichen Zweigliederung der nord- und mitteleuropäischen Moore. *Vegetatio*, 5-6: 571-585.

Eleutrius, L.M. 1972. The marshes of Mississippi. *Castanea*, 37: 153-168.

Evans, C.D. and K.E. Black. 1956. Duck Production Studies on the Prairie Potholes of South Dakota. Spec. Sci. Rep.: Wildlife No. 32. U.S. Fish and Wildlife Service, Washington, D.C.

Ferren, W.R., Jr., P.L. Fiedler, R.A. Leidy, K.D. Lafferty, and L.A.K. Mertes. 1996. Part II. Classification and description of wetlands of the central and southern California coast and coastal watersheds. *Madroño*, 43: 125-182.

Finlayson, C.M. and A.G. van der Valk (Eds.). 1995. *Classification and Inventory of the World's Wetlands*. Kluwer Academic Publishers, Dordrecht, The Netherlands.

Galkina, E.A. 1946. Bolotnye landshafty I printsipy ikh klassifikatsii. In *Sborn. Nauachn. rabot Bot. Inst. Imeni V.L. Komarova Akad. Nauk S.S.S.R., vypolnennykh v Leningrade za tri goda Velikoi Otechestvennoi voiny (1941-1943)*. Akad. Nauk S.S.S.R., Moscow, Leningrad. 139-156.

Galkina, E.A. 1967. Ispol'zovanie aerosnimkov dlya ustanovleniya zakonomernostei raspredeleniya bolotnykh urochishch po territorii lesnoi zony S.S.S.R. In *Aeros'emka I yee primenenie*. Nauka. Leningrad. 329-336.

Golet, F.C. and J.S. Larson. 1974. Classification of Freshwater Wetlands in the Glaciated Northeast. Res. Pub. 116. U.S. Fish and Wildlife Service, Washington, D.C.

Gore, A.J.P. 1983. *Mires: Swamp, Bog, Fen and Moor.* Regional Studies. Elsevier Scientific Publishing Co., Amsterdam, The Netherlands.

Heikurainen, L. and P. Pakarinen. 1982. Peatland classification. 3.1. Mire vegetation and site types. In *Peatlands and their Utilization in Finland*. J. Laine (Ed.). Finnish Peatland Society and Finnish National Committee of the International Peat Society, Helsinki, Finland. 14-23.

Heinselman, M.L. 1963. Forest sites, bog processes, and peatland types in the Glacial Lake Agassiz Region, Minnesota. *Ecol. Monogr.,* 33: 327-374.

Heinselman, M.L. 1970. Landscape evolution, peatland types, and the environment in the Lake Agassiz Peatlands Natural Area, Minnesota. *Ecol. Monogr.,* 40: 235-261.

Hunt, C. 1996. The Ramsar Convention: a global effort to protect aquatic ecosystems. Water Science and Technology Board, National Research Council, Washington, D.C. *WSTB Newsletter,* 13(3): 1-2.

Hutchinson, G.E. 1957. *A Treatise on Limnology,* vol. 1. John Wiley & Sons, New York.

Ivanov, K.E. 1975. *Vodoobmen v bolotnykh landshaftakh*. Gidrometizdat, Leningrad.

Jeglum, J.K. 1971. Plant indicators of pH and water levels in peatlands at Candle Lake, Saskatchewan. *Can. J. Bot.,* 49: 1661-1676.

Jeglum, J.K., A.N. Boissonneau, and V.F. Haavisto. 1974. Toward a Wetland Classification for Ontario. Information Report O-X-215. Canadian Forestry Service, Environment Canada, Sault Ste Marie, Ontario.

Kangas, P.C. 1990. An energy theory of landscape for classifying wetlands, chap. 2. In *Forested Wetlands*. A.E. Lugo, M.M. Brinson, and S. Brown (Eds.). Elsevier Publishers, Amsterdam. 15-22.

Kantrud, H.A., J.B. Millar, and A.G. van der Valk. 1989. Vegetation of wetlands in the Prairie Pothole Region, chap. 5. In *Northern Prairie Wetlands*. A.G. van der Valk (Ed.). Iowa State University Press, Ames. 132-187.

Leitch, W.G. 1966. Historical and ecological factors in wetland inventory. *Trans. N. Amer. Wildl. and Natur. Resources Conf.,* 31: 88-96.

Lugo, A.E. and S.C. Snedaker. 1974. The ecology of mangroves. *Annual Review of Ecology and Systematics,* 5: 39-64.

Mader, S.F. 1991. *Forested Wetlands Classification and Mapping: A Literature Review*. Technical Bulletin No. 606. National Council of the Paper Industry for Air and Stream Improvement, Inc., New York.

Martin, A.C., N. Hotchkiss, F.M. Uhler, and W.S. Bourn. 1953. Classification of Wetlands of the United States. Spec. Sci. Rep.: Wildlife No. 20. U.S. Fish and Wildlife Service, Washington, D.C.

Masing, V. 1975. Mire typology of the Estonian S S R. In *Some Aspects of Botanical Research in the Estonian S.S.R*. Tartu, Estonia. 123-133.

Messina, M.G. and W.H. Conner. 1998. *Southern Forested Wetlands. Ecology and Management*. Lewis Publishers, Boca Raton, Florida.

Millar, J.B. 1976. Wetland Classification in Western Canada: A Guide to Marshes and Shallow Open Water Wetlands in the Grasslands and Parklands of the Prairie Provinces. Report Series No. 37. Canadian Wildlife Service, Environment Canada, Ottawa, Ontario.

Moore, P.D. and D.J. Bellany. 1974. *Peatlands*. Elek Science, London.

National Research Council. 1995. *Wetlands: Characteristics and Boundaries*. National Academy Press, Washington, D.C.

National Wetlands Working Group, Canada Committee on Ecological Land Classification. 1987. Ecological Land Classification Series No. 21. The Canadian Wetland Classification System. Land Conservation Branch. Canadian Wildlife Service, Environment Canada.

National Wetlands Working Group (Eds.). 1988. Wetlands of Canada. Polyscience Publishers, Montreal, Quebec. Ecological Land Classification Series No. 24, Environment Canada, Ottawa, Ontario.

Novitzki, R.P. 1982. Hydrology of Wisconsin Wetlands. Information Circular 40. U.S. Geological Survey, Reston, VA.

Paijamans, K., R.W. Galloway, D.P. Faith, P.M. Fleming, H.A. Haantjens, P.C. Heyligers, J.D. Kalma, and E. Loffler. 1985. Aspects of Australian Wetlands. Division of Water and Land Resources Tech. Paper No. 44. Commonwealth Scientific and Industrial Research Organization, Melbourne, Australia.

Penfound, W.T. 1952. Southern swamps and marshes. *Bot. Rev.,* 18: 413-446.

Pestrong, R. 1965. The development of drainage patterns on tidal marshes. *Stanford University Publications in Geological Science,* 10: 1-87.

Ramsar Convention Bureau. 1998. Information sheet on Ramsar wetlands. Gland, Switzerland.

Reimold, R.J. and W.H. Queen (Eds.). 1974. *Ecology of Halophytes.* Academic Press, New York.

Remane, A. and C. Schlieper. 1971. *Biology of Brackish Water.* Wiley Interscience Division, John Wiley & Sons, New York.

Riggert, T.L. 1964-66. Wetlands of Western Australia. Department of Fisheries and Fauna, Western Australia.

Runka, G.G. and T. Lewis. 1981. Preliminary Wetlands Managers Manual, Cariboo Resource Management Region. Technical Paper No. 5. Assessment and Planning Division, British Columbia Ministry of Environment, Victoria, British Columbia.

Ruuhijärvi, R. 1983. The Finnish mire types and their regional distribution, chap. 2. In *Mires: Swamp, Bog, Fen and Moor. Regional Studies. Ecosystems of the World 4B.* A.J.P. Gore (Ed.). Elsevier Publishing Co., Amsterdam. pp. 47-67.

Sather, J.H. (Ed.). 1976. Proc. of the National Wetland Classification and Inventory Workshop. FWS/OBS-76/09. U.S. Fish and Wildlife Service, Washington, D.C.

Semeniuk, C.A. 1987. Wetlands of the Darling System — a geomorphic approach to habitat classification. *J. Roy. Soc. West. Aus.,* 69: 95-112.

Shaler, N.S. 1890. Fresh-water Morasses of the United States. *U.S. Geol. Survey Ann. Rep.,* 10: 261-339.

Shaw, S.P. and C.G. Fredine. 1956. Wetlands of the United States. Their Extent and Their Value to Waterfowl and Other Wildlife. Circular 39. U.S. Fish and Wildlife Service, Washington, D.C.

Shreve, F. 1910. The ecological plant geography of Maryland; Coastal Zone; Eastern Shore District. In *The Plant Life of Maryland.* F. Shreve, M.A. Chrysler, F.H. Blodgett, and F.W. Besley (Eds.). The John Hopkins Press, Baltimore, MD. 101-149.

Shreve, F., M.A. Chrysler, F.H. Blodgett, and F.W. Besley. 1910. *The Plant Life of Maryland.* The John Hopkins Press, Baltimore, MD.

Sjörs, H. 1946. The mire vegetation of Upper Långan district in Jämtland. *Ark. Bot.,* 33A(6): 1-96.

Sjörs, H. 1948. Mire vegetation in Bergslagen. *Acta Phytogeogra. Suec.,* 21: 1-299.

Sjörs, H. 1950. Regional studies in northern Swedish mire vegetation. *Bot. Not.,* 1950: 173-222.

Sjörs, H. 1983. Mires of Sweden, chap. 3. In *Mires: Swamp, Bog, Fen and Moor. Regional Studies. Ecosystems of the World 4B.* A.J.P. Gore (Ed.). Elsevier Publishing Co., Amsterdam. 69-94.

Smith, A.G. 1971. Ecological Factors Affecting Waterfowl Production in the Alberta Parklands. Resource Pub. No. 98. U.S. Fish and Wildlife Service, Washington, D.C.

Stegman, J.L. 1976. U.S. Fish and Wildlife Service. In Proc. of the National Wetland Classification and Inventory Workshop. J.H. Sather (Ed.). FWS/OBS-76/09. U.S. Fish and Wildlife Service, Washington, D.C. 102-115.

Stewart, R.E. and H.A. Kantrud. 1971. Classification of Natural Ponds and Lakes in the Glaciated Prairie Region. Resource Publication No. 92. U.S. Fish and Wildlife Service, Washington, D.C.

Stewart, R.E. and H.A. Kantrud. 1972. Vegetation of Prairie Potholes, North Dakota, in Relation to Quality of Water and Other Environmental Factors. Geol. Survey Prof. Paper 585-D. U.S. Geological Survey, Reston, VA.

Stoudt, J.H. 1971. Ecological Factors Affecting Waterfowl Production in the Saskatchewan Parklands. Resource Pub. No. 99. U.S. Fish and Wildlife Service, Washington, D.C.

Tarnocai, C. 1980. Canadian wetland registry. In Proc., Workshop on Canadian Wetlands. C.D.A. Rubec and F.C. Pollett (Eds.). Lands Directorate, Environment Canada, Ottawa, Canada. Ecological Land Classification Series No. 12.

Tiner, R.W. 1987. *A Field Guide to Coastal Wetland Plants of the Northeastern United States.* University of Massachusetts Press, Amherst.

Tiner, R.W. 1993. *Field Guide to Coastal Wetland Plants of the Southeastern United States.* University of Massachusetts Press, Amherst.

Tiner, R.W. 1997a. Keys to Landscape Position and Landform Descriptors for U.S. Wetlands (Operational Draft). U.S. Fish and Wildlife Service, National Wetlands Inventory Project, Hadley, MA.

Tiner, R.W. 1997b. Piloting a more descriptive NWI. *National Wetlands Newsletter,* 19(5): 14-16.

Tiner, R.W. 1998. *In Search of Swampland. A Wetland Sourcebook and Field Guide.* Rutgers University Press, New Brunswick, NJ.

Tiner, R.W. and D.G. Burke. 1995. Wetlands of Maryland. U.S. Fish and Wildlife Service, Region 5, Hadley, MA and the Maryland Dept. of Natural Resources, Annapolis. Cooperative publication.

Tiner. R.W., S. Schaller, D. Petersen, K. Snider, K. Ruhlman, and J. Swords. 1999. Wetland Characterization and Preliminary Assessment of Wetland Functions for the Casco Bay Watershed, Southern Maine. U.S. Fish and Wildlife Service, Hadley, MA.

Tingay, A. and S.R. Tingay. 1976. The Wetland of System Six. Bulletin 28. Dept. Conservation & Environment.

von Post, L. and E. Granlund. 1926. Södra Sveroges torvtillgångar I. *Sver. Geol. Unders.,* C335: 1-127.

Warming, E. 1909. *Oecology of Plants. An Introduction to the Study of Plant-communities.* Clarendon Press, Oxford, England. (Updated English version of a 1896 text.)

Warren, R.S. and P.E. Fell. 1995. Tidal wetland ecology of Long Island Sound. In *Tidal Marshes of Long Island Sound: Ecology, History and Restoration.* G.D. Dreyer and W.A. Niering (Eds.). Bull. No. 34. The Connecticut Arboretum, Connecticut College, New London. 22-41.

Water Authority of Western Australia. 1993. Wetlands of the Perth to Bunbury Regions. Map/poster.

Weber, C.A. 1911. Das Moor. *Hannoversche Geschichtbl.,* 14: 255-270.

Wetland Advisory Committee. 1977. The Status of Wetlands Reserves in System Six. Report of the Wetland Advisory Committee to the Environmental Protection Authority. Australia.

Whigham, D., D. Dykyjová, and S. Henjný (Eds.). 1993. *Wetlands of the World: Inventory, Ecology and Management,* vol. I. Kluwer Academic Publishers, Dordrecht, The Netherlands.

Winter, T.C. and M-K. Woo. 1990. Hydrology of lakes and wetlands. In *Surface Water Hydrology of North America.* The Geology of North America. M.G. Wolman and H.C. Riggs (Eds.). Vol. 0-1.Geological Society of America, Boulder, CO.

Wright, W.O. 1907. Swamp and Overflowed Lands in the United States. Circular 76. U.S. Department of Agriculture, Office of Experiment Stations, Washington, D.C.

Zoltai, S.C., F.C. Pollett, J.K. Jeglum, and G.D. Adams. 1975. Developing a wetland classification in Canada. In Proc., 4th N. Am. Forest Soils Conf. B. Bernier and C.H. Winget (Eds.). Laval University Press, Quebec, Canada. 497-511.

9 Wetlands of the U.S.: An Introduction, with Emphasis on Their Plant Communities

INTRODUCTION

The climatic and physiographic variations in the U.S. have produced a great diversity of wetlands, ranging from cold wet tundra and black spruce forested wetlands in Alaska to temperate hardwood swamps in the East to subhumid pothole marshes in the western interior and Midwest to desert wetlands in the arid Southwest to steamy subtropical marshes and wet hammocks in Florida and tropical rain forests in Hawaii and Puerto Rico. Within regions, local differences in water sources, soils, and geology further influence plant composition of wetlands. In addition to natural processes affecting wetlands, human actions also have a profound influence on water quality (e.g., pollutants and increased nutrients), sedimentation rates, and hydrologic regimes.

This chapter provides a general overview of wetlands in the U.S. emphasizing their vegetation. It introduces readers to the variety of plant associations which themselves often serve as useful indicators of wetland. The brief descriptions and accompanying tables and figures are intended to illustrate the major vegetated wetland types.* A listing of some publications describing representative plant communities in different parts of the country is presented at the end of the summary for further information. The discussion is organized according to wetland types as defined by the official U.S. wetland classification system (Cowardin et al., 1979). For a more extensive review of U.S. wetlands, see Fretwell et al. (1996) and for information of wetland ecology, refer to Niering (1985), Majumdar et al. (1989), Mitsch and Gosselink (1993) and Tiner (1998). An excellent treatise on Canada's wetlands has been published (National Wetlands Working Group, 1988). If interested in wetland communities in other countries, consult summaries provided in Chapman (1977), Gore (1983), Lugo et al. (1990), and Whigham et al. (1993).

WETLAND DISTRIBUTION

The U.S. possesses an estimated 282 million acres of wetlands. The distribution of wetlands is not uniform across the country due to climatic and topographic differences. One look at a map of the general distribution of wetlands in the U.S. (Figure 9.1) shows concentrations of wetlands in the following areas: (1) Alaska's arctic coastal plain, (2) Alaska's Yukon-Kushokwim Delta, (3) interior Alaska, (4) Gulf-Atlantic coastal plain (within 100 miles of the coast from New Jersey to west Texas), (5) Mississippi alluvial plain, and (6) glaciated portions of the Northeast and Midwest. Over half of the nation's wetlands (about 175 million acres) are in Alaska alone, with most of these occurring in the zone underlain by continuous permafrost. The state distribution of wetlands is outlined in Table 9.1.

---

* Scientific names in the tables typically follow the names used in the referenced sources.

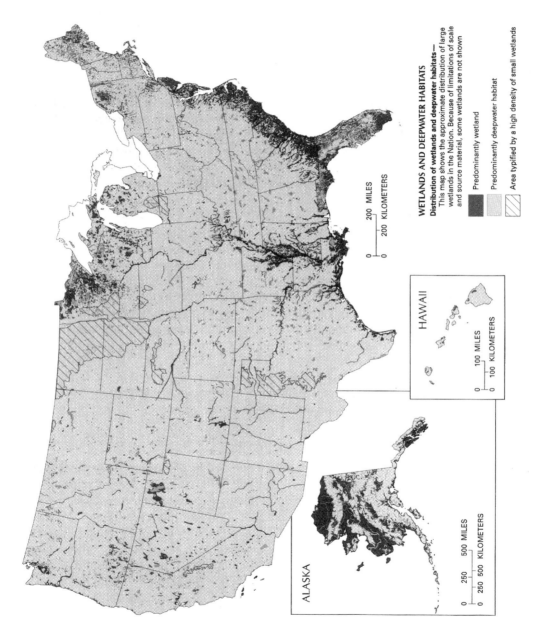

FIGURE 9.1   Map showing the general distribution wetlands and deepwater habitats of the U.S. (Fretwell et al., 1996)

TABLE 9.1
Wetland Acreage in the U.S.

State	Wetland Acreage (square miles)	% of State	Source
Alabama	2,651,000	8.2%	Hefner et al., 1994
Alaska	174,683,900	47.8%	Hall et al., 1994
Arizona	600,000	0.8%	*
Arkansas	3,573,000	10.7%	Hefner et al., 1994
California	462,163	0.5%	Frayer et al., 1989 *
Colorado	1,000,000	1.5%	Colorado Division of Parks and Outdoor Recreation, 1987
Connecticut	172,548	5.5%	Metzler and Tiner, 1992
Delaware	222,993	18.0%	Tiner, 1985b
Florida	11,403,000	32.9%	*
Georgia	6,520,000	17.5%	*
Hawaii	51,800	1.2%	*
Idaho	385,700	0.7%	*
Illinois	1,253,891	3.5%	Suloway and Hubbell, 1994
Indiana	809,565	3.5%	Indiana Dept. of Natural Resources, 1997
Iowa	691,725	1.9%	*
Kansas	435,400	0.8%	*
Kentucky	388,000	1.5%	Hefner et al., 1994
Louisiana	10,313,000	36.2%	*
Maine	5,199,200	26.2%	Widoff, 1988
Maryland	598,388	9.5%	Tiner and Burke, 1995
Massachusetts	589,871	11.8%	Tiner, 1992
Michigan	5,583,400	15.3%	*
Minnesota	10,607,000	20.8%	Minnesota Dept. of Natural Resources, 1997
Mississippi	4,365,000	14.4%	Hefner et al., 1994
Missouri	643,000	1.5%	U.S. Dept. of Agriculture, 1982
Montana	840,300	0.9%	*
Nebraska	1,905,500	3.9%	*
Nevada	457,000	0.6%	*
New Hampshire	200,000	3.5%	*
New Jersey	915,960	19.2%	Tiner, 1985a
New Mexico	481,900	0.6%	*
New York	1,025,000	3.4%	New York State Office of Parks, Reservations and Historic Preservation, 1988
North Carolina	5,048,000	16.1%	Hefner ct al., 1994
North Dakota	2,490,000	5.6%	*
Ohio	718,630	2.7%	Ohio Dept. Of Natural Resources, 1996
Oklahoma	949,700	2.2%	Brabander et al., 1985; Stinnett et al., 1997; Oklahoma Tourism and Recreation Dept., 1987
Oregon	1,393,900	2.3%	*
Pennsylvania	404,000	1.4%	Tiner, 1990
Rhode Island	65,154	9.7%	Tiner, 1989
South Carolina	4,091,400	21.2%	*
South Dakota	2,240,000	4.6%	Johnson et al., 1997 *
Tennessee	632,000	2.4%	Hefner et al., 1994
Texas	6,412,000	3.8%	Texas Parks and Wildlife Dept, 1988; U.S. Fish and Wildlife Service, 1979; Moulton et al., 1997
Utah	558,000	1.1%	Jensen, 1974; Utah Division of Parks and Recreation and Utah Dept. of Natural Resources, 1988
Vermont	243,047	4.1%	Tiner, 1987a

TABLE 9.1 (continued)
Wetland Acreage in the U.S.

State	Wetland Acreage (square miles)	% of State	Source
Virginia	1,074,613	4.2%	Tiner and Finn, 1986; Commonwealth of Virginia, 1988
Washington	938,000	2.2%	Peters, 1990
West Virginia	56,678	0.4%	Tiner, 1996
Wisconsin	5,331,392	15.3%	Wisconsin Dept. of Natural Resources, 1992
Wyoming	1,250,000	2.0%	Wyoming Game and Fish Dept., 1987
Conterminous U.S.	108,241,818	5.7%	
All 50 States	282,925,718	12.5%	

* Asterisk denotes unpublished data.

Source: Compiled by the U.S Fish and Wildlife Service's National Wetlands Inventory from various sources.

PALUSTRINE WETLANDS

Palustrine wetlands include marshes, bogs, fens, wet meadows, swamps, and seasonally wet woods (e.g., flatwoods and bottomlands) of the nation's interior. Most of these wetlands occur beyond the influence of salt-laden tides (i.e., nontidal), although those along coastal plain rivers may be affected by the tides, with their water levels rising with the incoming tide and falling on the ebb tide (with no change in salinity). The hydrology of the majority of palustrine wetlands is affected by precipitation, surface water runoff, and groundwater discharge in varying combinations. In arid to sub-humid regions, many inland wetlands may be saline due to excessive evaporation and an accumulation of inland salts (e.g., magnesium, potassium, and bicarbonates) present in the groundwater or in surface water draining from irrigated lands.

Given their broad geographic range, palustrine wetlands are much more plentiful than their coastal counterparts (estuarine wetlands). They constitute about 95% of the wetlands (or 97.8 million acres) in the conterminous U.S. and 98.8% (or 172.5 million acres) of Alaska's wetlands (Dahl and Johnson, 1991; Hall et al., 1994, respectively). As shown in Figure 9.1, their distribution and extent is not uniform. In temperate regions, they are more abundant in areas where precipitation exceeds rainfall and where the landscape is conducive to retaining water for significant periods. The relatively flat Gulf–Atlantic coastal plain, the floodplain of the Mississippi River, the prairie pothole region, and northern glaciated regions have the necessary ingredients for extensive wetland formation and states in these areas possess the bulk of the nation's wetlands in the lower 48 states. Alaska's cold climate, permafrost soils, river valleys, and vast nearly level flats on its coastal plains and river deltas have led to the development of expansive belts of wetland nearly 200 miles wide.

Four major hydrogeologic settings favor wetland creation: topographic depressions, slope breaks, areas of stratigraphic change, and permafrost areas (Figure 9.2; Carter, 1996; Winter and Woo, 1990). The first three situations pertain to all regions of the country, whereas the latter setting only occurs in Alaska.

Most wetlands in the conterminous U.S. form in topographic depressions associated with lake basins, river valleys, ponds, and upland depressions. These depressions have varied origins depending on the region of the country and its geologic history. Many in northern areas are glacially formed depressions carved out by glacial action, formed by melting ice blocks left when the Wisconsin glacier retreated about 10,000 to 12,000 years ago, dammed by morainal deposits (obstructing former drainages), or former glacial lake bottoms. Prairie potholes in the Upper

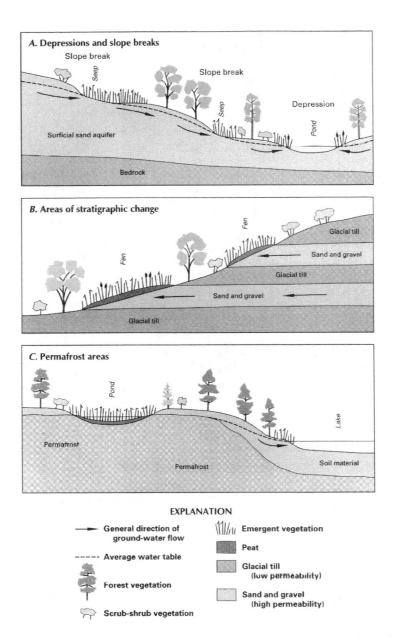

FIGURE 9.2 Typical hydrogeologic settings favoring wetland development: (a) depressions and slope breaks, (b) areas of stratigraphic change, and (c) permafrost. (Fretwell, J.D., Williams, J.S., and Redman, P.J., 1996. *National Water Summary on Wetland Resources,* U.S. Geological Survey, Reston, VA.)

Midwest and kettlehole wetlands in New England and upstate New York are well-known examples. Many wetlands have established along the shores of lakes and rivers and most lakes in northern areas were glacially formed, including the Great Lakes. Former glacial lake bottoms have fine-textured soils and relatively flat to slightly depressed surfaces that support extensive wetlands. Southern areas are typically devoid of lakes, except where karst geology (e.g., central Florida) or tectonic activity (e.g., Tennessee's Reelfoot Lake) have promoted their formation. In limestone (karst) regions, sinkhole depressions develop where underground limestone has been dissolved and the land above collapses as exemplified by central Florida. Wetlands are common in these depressions. In sand-dominated regions, wetland depressions have formed where aeolian forces (wind action) blow out sand down to the level of seasonal high water tables. These wetlands can be found in coastal dunes (wet hollows) and in the Nebraska Sandhills (the largest sand dune system in the Western Hemisphere). Shallow "lakes" also have formed in the latter area, with marshes associated with their margins. In Alaska, wetlands occupy thermokarst basins created by alternate freeze and thaw action, while in Hawaii, depressional wetlands develop in flooded collapsed calderas of inactive volcanoes.

Palustrine wetlands also form in seepage areas or below springs where the water table intersects the ground surface (Figure 9.2). This commonly occurs at the toes of slopes and in swales between hills. Other slopes may also support wetlands, such as slopes below melting glaciers. In areas of high rainfall, slopes with shallow or dense (clay) soils may be covered with wetlands. Site wetness may actually initiate the formation of the clay soils. Changes in stratigraphy (i.e., a more permeable substrate overlying a less permeable surface) may allow water movement through the more permeable layer and eventually to the soil surface. Under these circumstances, wetlands may develop on parts of slopes that are sufficiently wet to support wetland vegetation.

In the arctic and subarctic, the occurrence of permafrost (frozen soil) close to the surface creates a restrictive layer that prevents downward percolation of water. In areas with shallow permafrost, seasonal melting of permafrost and restricted drainage saturates the soil above, leading to formation of organic soils and promoting the growth of wetland species, especially peat moss. Paludification processes promote an extension of the peat bog (mire), so that the entire landscape may be completely covered by wetlands.

Palustrine wetlands may be flooded permanently, periodically, or never flooded, but saturated for extended periods during the year.* Water tables are highest in late winter (due to low evapotranspiration, little or no plant activity, and low water storage capacity associated with frozen soils), following snowmelt, or during the wet season, depending on the region (Carter, 1996). In the temperate region of the conterminous U.S., many palustrine wetlands are flooded for at least a week or two early in the growing season (but longer during the nongrowing season) (see Figure 2.3 in Chapter 2 for examples of hydrographs). Predominantly saturated wetlands occur on slopes, on broad interfluves (flatwoods), or in northern climes on peat deposits. Those associated with bogs and fens of northern climates remain saturated for most of the growing season, yet those on mineral slopes or on interfluves (coastal plains or glaciolacustrine plains) may be only seasonally wet (usually from late winter into early spring).

Palustrine wetlands are largely dominated by trees, shrubs, and persistent herbaceous plants that remain visible in wetlands through the winter and into the following spring. They consist mainly of freshwater types, although inland saline wetlands exist in arid, semiarid, and subhumid regions of the country (e.g. Southwest, West, and Upper Midwest). Three major vegetated types of palustrine wetlands exist: emergent wetland, scrub-shrub wetland, and forested wetland. Many wetlands are represented by complexes of these types. In Alaska, moss-lichen wetlands also are common where peat mosses (*Sphagnum* spp.) form saturated wetlands. Shallow open water bodies (i.e., ponds and playa lakes, less than 20 acres in size and less than 6.6 ft deep, along with their aquatic beds) and nonvegetated salt flats represent the remaining palustrine wetland types.

EMERGENT WETLANDS

Palustrine emergent wetlands are dominated by erect, herbaceous vegetation (Figure 9.3; Plates 44-45). They appear as grasslands or stands of reedy growth. These wetlands are commonly referred to by

---

* The term *flooding* used in this chapter is synonomous with inundation, referring to the condition where surface water is present whether from river overflow, tides, snowmelt, high groundwater, or other sources.

FIGURE 9.3   Cattail-tule marsh in the western U.S.

a host of terms, including marsh, wet meadow, fen, prairie pothole, playa, vernal pool, inland salt marsh, and alkali marsh, depending on the region of the country and individual characteristics (e.g., vegetation, hydrology, and water chemistry). There are approximately 25 million acres of these wetlands in the conterminous U.S. and 42 million acres in Alaska (Dahl and Johnson, 1991; Hall et al., 1994, respectively).

Marshes represent emergent wetlands that are flooded for all or most of the year. The more widespread marsh genera and species are given in Table 9.2. Cattail marshes (*Typha* spp.) are common throughout the country, while tule marshes (*Scirpus* spp.) are more abundant in the west. Sawgrass (*Cladium jamaicense*) dominates marshes in the Florida Everglades, while cattails (*Typha* spp.) have increased in abundance where water levels are stabilized due to impoundment. Prairie pothole marshes formed in glacial depressions in the Midwest are among the most familiar marshes because of their importance as primary waterfowl breeding areas.

Fens are seasonally flooded and saturated emergent wetlands in northern climates where ericaeous bogs also are common. They differ from bogs in being nutrient-rich due to groundwater discharge which brings nutrients to the wetlands. These conditions promote sedge growth over woody ericads. Fens are usually dominated by members of the sedge family (Cyperaceae), especially cotton-grasses (*Eriophorum*) and sedges (*Carex*). The nutrient-rich ones often have calciphilous species such as yellow sedge (*Carex viridula*), dioeceous sedge (*C. sterilis*), moor rush (*Juncus stygius*), marsh muhly (*Muhlenbergia glomerata*), buckbean (*Menyanthes trifoliata*), marsh cinquefoil (*Potentilla palustris*), and yellow lady's-slipper (*Cypripedium calceolus*). Calcareous fens are rare habitats in many states. Some fens may be dominated by shrubs (e.g., *Salix candida*, *Potentilla fruticosa*, *Rhamnus alnifolia*) or by trees (e.g., *Thuja occidentalis*).

While the term *fen* is usually restricted to boreal and recently glaciated regions, similar graminoid wetlands develop in more southerly locations. These wetlands may be called *sedge meadows* where sedges predominate or *wet meadows* where grasses and various other species abound. Typical wet meadow species include grasses such as bluejoint (*Calamagrostis canadensis*) and reed canary grass (*Phalaris arundinacea*), other graminoids like sedges and rushes (*Juncus*), and various forbs such as Joe-Pye-weeds (*Eupatoriadelphus*), thoroughworts (*Eupatorium*), smartweeds (*Polygonum*), goldenrods (*Euthamia* and *Solidago*), and asters (*Aster*) (see Table 9.2).

Within the expansive forests of the Atlantic–Gulf coastal plain, open wet virtually treeless, grasslands called *savannahs* can be found. Grasses and sedges usually predominate, with typical species including lowland broom-sedge (*Andropogon glomeratus*), Walter's sedge (*Carex walteriana*), pineland three-awn grass (*Aristida stricta*), switchgrass (*Panicum virgatum*), beak-rushes

TABLE 9.2
Examples of Common Wetland Herbs in Palustrine Emergent Wetlands of the Conterminous U.S. and Alaska

Eastern Species — Habitats

Acorus calamus — WM, M	*Juncus acuminatus* — WM
Alternanthera philoxeroides — M*, W*	*Juncus canadensis* — WM
Andropogon glomeratus — WM, M	*Juncus effusus* — WM, M
Aristida spp. — M*	*Juncus militaris* — M
Aster puniceus — WM	*Leersia oryzoides* — M, WM
Aster vimineus/lateriflorus — WM	*Lycopus* spp. — WM
Bidens frondosa — WM, M	*Lythrum salicaria* — WM, M
Calamagrostis canadensis — WM	*Muhlenbergia capillaris* — M*
Calla palustris — M	*Nuphar luteum* — M
Carex crinita — WM	*Nymphaea odorata* — M
Carex lacustris — M, WM	*Onoclea sensibilis* — WM
Carex lasiocarpa — WM	*Panicum hemitomom* — M*
Carex lurida — WM	*Paspalum lividum* — WM*
Carex utriculata (including *C. rostrata*) — WM	*Peltandra virginica* — M
Carex stipata — WM	*Phalaris arundinacea* — WM
Carex stricta — WM, M	*Phragmites australis* — M, WM
Carex vulpinoidea — WM	*Polygonum* spp. — M, WM
Cicuta maculata — M, WM	*Pontederia cordata* — M
Cladium jamaicense — M*	*Rhychospora* spp. — M, WM
Crinum americana — M*	*Sagittaria* spp. — M
Cyperus spp. — WM, M	*Scirpus atrovirens* — WM
Decodon verticillatus — M	*Scirpus cyperinus* — M, WM
Dulichium arundinaceum — M	*Scirpus fluviatilis* — M, WM
Echinochloa walteri — M	*Scirpus pungens* — M, WM
Eleocharis spp. — M, WM	*Scirpus tabernaemontani* (formerly *S. validus*) — M
Erianthus giganteus — M*	*Sium suave* — M
Eriocaulon spp. — M	*Sparganium* spp. — M
Eriophorum spp. — WM	*Thalia genticulata* — M*
Eupatorium perfoliatum — WM, M	*Thelypteris thelypteroides* — WM, M
Eupatoriadelphus spp. — WM	*Typha* spp. — M
Glyceria spp. — M, WM	*Utricularia* spp. — M
Hibiscus spp. — M	*Vernonia noveboracensis* — WM
Iris versicolor — M, WM	*Zizania aquatica* — M
Iris virginica — M, WM	*Zizaniopsis milicea* — M*

Western Species — Habitats

Agrostis spp. — WM	*Caltha leptosepala* — WM, S
Alopecurus spp. — WM	*Carex aquatilis* — S, WM
Alisma plantago-aquatica — M	*Carex arcta* — WM
Angelica genuflexa — M, WM	*Carex atheroides* — M
Aster simplex — WM	*Carex buxbaumi* — WM
Athyrium filix-femina — WM	*Carex canescens* — M
Beckmannia syzigachne — M	*Carex echinata* — WM
Bidens spp. — M	*Carex eurycarpa* — WM
Boltonia asteroides — WM	*Carex languinosa* — WM
Calamagrostis canadensis — WM	*Carex lasiocarpa* — WM
Calamagrostis inexpansa — WM	*Carex lenticularis* — WM
Callitriche spp. — M	*Carex leptalea* — WM
Caltha biflora — WM, S	*Carex limosa* — M

Carex luzulina — WM
Carex lyngbyei — WM
Carex microptera — WM
Carex muricata — WM
Carex nebraskensis — WM
Carex nigra — WM
Carex obnupta — WM, M
Carex retrorsa — WM
Carex sartwelli — WM
Carex scirpoidea — S
Carex scopulorum — WM, S
Carex simulata — WM
Carex sitchensis — WM, S
Carex stipata — WM
Carex utriculata (includes *C. rostrata*) — M, WM, S
Carex vesicaria — S, WM
Cicuta douglasii — WM
Deschampsia cespitosa — WM
Distichlis spicata — WM
Dodecatheon spp. — WM, S
Dulichium arundinaceum — M
Eleocharis palustris — M
Eleocharis pauciflora — WM
Eleocharis rostellata — WM
Equisetum fluviatile — M
Eriophorum chamissonis — WM
Galium trifidum — WM
Glyceria grandis — M
Glyceria striata — WM
Hordeum jubatum — WM
Hypericum anagalloides — WM
Impatiens spp. — WM
Jaumea carnosa — M
Juncus acuminatus — M, WM
Juncus articulatus — WM
Juncus drummondii — WM
Juncus effusus — WM
Juncus ensifolius — WM
Juncus mertensiana — WM
Ludwigia palustris — M
Lysichum americanus — WM
Lythrum salicaria — M, WM
Mentha spp. — WM, M
Menyanthes trifoliata — M
Mimulus guttatus — WM, M
Mitella spp. — S
Muhlenbergia filiformis — WM

Nuphar luteum — M
Nymphaea spp. — M
Parnassia fimbriata — WM, S
Pedicularis groenlandica — WM
Petasites spp. — WM, S
Phalaris arundinacea — WM
Phragmites australis — WM, M
Platanthera spp. — WM, S
Poa alpina — WM
Poa palustris — WM
Poa pratensis — WM
Polygonum bistortoides — WM
Polygonum coccineum — M
Polygonum hydropiper — M
Polygonum hydropiperoides — M
Potentilla anserina — WM
Puccinellia nuttalliana — M
Ranunculus aquatilis — M
Ranunculus flammula — WM
Rhynchospora alba — WM
Rorippa islandica — M
Sagittaria spp. — M
Salicornia rubra — M
Salicornia virginica — M, WM
Saxifraga spp. — S
Scirpus acutus — M
Scirpus atrovirens — WM
Scirpus cyperinus — WM
Scirpus fluviatilis — M
Scirpus heterochaetus — M
Scirpus maritimus — M
Scirpus microcarpus — WM
Scirpus pungens — M
Scirpus tabernaemontani — M
Scolochloa festucaea — M
Senecio cymbalarioides — WM
Sisyrinchium spp. — WM
Sium suave — M
Sparganium spp. — M
Spartina pectinata — WM
Triglochin maritimum — WM
Trollius laxus — WM
Typha spp. — M
Veratrum spp. — WM
Veronica americana — M, WM
Veronica wormskjoldii — WM

Alaskan Species — Habitats

Alopecurus alpinus — WM
Arctophila fulva — M, WM
Calamagrostis canadensis — M, WM
Callitiche spp. — M
Carex aquatilis — M, WM
Carex bigelovii — WM

Carex livida — WM
Carex lyngbyaei — WM
Carex pluriflora — WM
Carex rostrata — WM
Carex saxatilis — WM
Carex sitchensis — WM

TABLE 9.2 (continued)
Examples of Common Wetland Herbs in Palustrine Emergent Wetlands of the Conterminous U.S. and Alaska

Cicuta mackenziana — WM	*Lathyrus palustris* — WM
Deschampsia beringensis — WM	*Menyanthes trifoliata* — M, WM
Dupontia fisheri — WM	*Myriophyllum* spp. — M
Eleocharis palustris — M	*Nuphar polysepalum* — M
Equisetum fluviatile — M, WM	*Nymphaea tetragona* — M
Eriophorum angustifolium — WM	*Petasites frigidus* — WM
Eriophorum vaginatum — WM	*Polygonum amphibium* — M
Fauria crista-galli — WM	*Potentilla palustris* — M, WM
Glyceria borealis — M	*Scirpus tabernaemontani* (formerly *S. validus*) — M
Hippuris vulgaris — M	*Senecio congestus* — WM
Juncus arcticus — WM	

Note: Habitat codes — M = marsh, WM = wet meadow/fen, and S = mountain streams. Asterisk (*) denotes a strictly southern species. Plains and prairie pothole species are included on the western list.

(*Rhynchospora*), fimbrys (*Fimbristylis*), and nut-rushes (*Scleria*). A scattering of woody plants include slash pine (*Pinus elliottii*), longleaf pine (*P. palustris*), blueberries (*Vaccinium*), huckleberries (*Gaylussacia*), arrowwoods (*Viburnum*), azaleas (*Rhododendron*), wax myrtle (*Myrica cerifera*), plus members of the holly family (*Ilex*). Other herbs that can be found in these wetlands are orchids (*Habenaria*), pipeworts (*Eriocaulon*), star-grasses (*Hypoxis*), yellow-eyed grasses (*Xyris*), meadow-beauties (*Rhexia*), milkworts (*Polygala*), seedboxes (*Ludwigia*), pitcher plants (*Sarracenia*), and sundews (*Drosera*).

In the Southwest, playa wetlands are common, especially in west Texas. Due to variable annual rainfall patterns, playa basins have dynamic plant communities related to changing water levels. They may be devoid of vegetation or colonized by different species during the drawdown phase. Some dominant species include spikerush (*Eleocharis palustris*), ragweed (*Ambrosia grayi*), curly dock (*Rumex crispus*), alkaline mallow (*Malvella leprosa*), kochia (*Kochia scoparia*), barnyard grass (*Echinochloa crusgalli*), red sprangle-top (*Leptochloa filiformis*), smartweeds (*Polygonum amphibium*, *P. pensylvanicum*, *P. lapathifolium*), arrowhead (*Sagittaria longiloba*), knotgrass or jointgrass (*Paspalum distichum*), and annual salt marsh aster (*Aster subulatus*). During the open water phase, pondweeds (*Potamogeton* spp.), southern cattail (*Typha domingensis*), and bulrushes (*Scirpus* spp.) may be present.

Inland salt marshes are found in Utah, Nevada, and other arid areas in adjoining states. High evapotranspiration and low rainfall create salty soils in areas where the water table is near the surface. The shoreline of the Great Salt Lake is occupied by halophytic species. These salt-loving species includes red saltwort (*Salicornia rubra*), Utah glasswort (*S. utahensis*), iodine bush (*Allenrolfea occidentalis*), sea-blites (*Suaeda* spp.), salt grass (*Distichlis spicata*), alkali sacaton (*Sporobolus airoides*), and foxtail barley (*Hordeum jubatum*).

For more information on palustrine emergent wetlands in the U.S., consult the following:

Ashworth (1997)	Chadde et al. (1998)	Damman and French (1987)
Batten (1990)	Chapman (1960)	Davis (1943)
Batten and Murray (1982)	Conard (1935)	Davis and Ogden (1994)
Bolen et al. (1989)	Conner and Day (1987)	Dix and Smeins (1967)
Caldwell and Crow (1992)	Cooper et al. (1997)	Drew and Schomer (1984)
Calhoun et al. (1994)	Curtis (1959)	Eicher (1988)

Elliot (1981)
Farrar (1982)
Farrar and Gersib (1991)
Ferren et al. (1996)
Folkerts (1982)
Geis and Kee (1977)
Gilbert (1989)
Good and Good (1974)
Good et al. (1978)
Harshberger (1914)
Haukos et al. (1998)
Herdendorf et al. (1981)
Herdendorf et al. (1986)
Herdendorf (1987)
Hobbie (1984)
Hofstetter (1983)
Hopper (1968)
Hubbard (1988)
Hubbard et al. (1988)

Jervis (1969)
Kantrud et al. (1989)
Kindscher et al. (1996)
Kovalchik (1987)
Kunze (1994)
Kushlan (1991)
Laessle (1942)
Lodge (1994)
McPherson (1973)
Metzler and Tiner (1992)
Minc and Albert (1998)
Moore and Bellamy (1974)
Nachlinger (1988)
Nelson et al. (1983)
Nicholson (1995)
Odum et al. (1984)
Penfound (1952)
Penfound and Hathaway (1938)

Reinartz and Warne (1993)
Rowell (1971,1981)
Schomer and Drew (1982)
Seyer (1979)
Simpson et al. (1983)
Stewart and Kantrud (1971; 1972)
Steward and Ornes (1975)
Stuckey (1989)
Tiner (1985a; 1985b; 1988; 1989; 1998)
Tiner and Burke (1995)
Tyndall et al. (1990)
Ungar (1974)
van der Valk (1985; 1989)
Viereck et al. (1992)
Walker et al. (1989)
Weller (1981)
Windell et al. (1986)
Zedler (1987)

SCRUB-SHRUB WETLANDS

Inland wetlands dominated by woody vegetation less than 20 ft tall are palustrine scrub-shrub wetlands (Figures 9.4a/b/c; Plate 46). Slightly more than 15 million acres of shrub swamps are present in the conterminous U.S. (Dahl and Johnson, 1991). In Alaska, they are the dominant type occupying 114.5 million acres (Hall et al., 1994). These shrub-dominated wetlands are commonly called bogs, pocosins, shrub-carrs, or simply shrub swamps. Examples of shrub wetland communities for eastern states, western states, and Alaska are presented in Tables 9.3-9.5, respectively. *(See tables at the end of the chapter.)*

Peat bogs are particularly interesting scrub-shrub wetlands that are almost continuously saturated. Many bogs have a somewhat rolling microtopography with hummocks (mounds) and hollows (pools). Surface water may be present in the pools at times. Peat bogs are most common in northern regions where low temperatures and corresponding low evapotranspiration rates favor the formation of peaty soils. Many Alaskan bogs are underlain by permafrost (frozen soil near the surface), whereas elsewhere in the U.S., bogs are prevalent in isolated depressions, along river courses, and along the margins of lakes in northern states. Bogs are abundant in Alaska, Maine, Michigan, Minnesota, and Wisconsin. Typical northern bog plants include leatherleaf (*Chamaedaphne calyculata*), sweet gale (*Myrica gale*), cotton-grasses (*Eriophorum*), peat mosses (*Sphagnum*), bog rosemary (*Andromeda glaucophylla*), Labrador tea (*Ledum groenlandicum*), bog laurel (*Kalmia polifolia*), blueberries (*Vaccinium*), cranberries (e.g., *Vaccinum macrocarpon* and *V. oxycoccus*), as well as stunted trees of black spruce (*Picea marina*), larch (*Larix laricina*), and balsam fir (*Abies balsamea*). Alaskan bogs may have these and other species (Table 9.5).

A special type of bog has formed in depressions on broad flat imperfectly drained interfluves along the southeastern coastal plain. In this relatively flat terrain, the interfluves represent low-level plateaus positioned above neighboring streams that led to their native American name *pocosin* meaning *swamp on a hill*. Pocosins are dominated by evergreen and deciduous shrubs, especially sweet pepperbush (*Clethra alnifolia*), inkberry (*Ilex glabra*), large gallberry (*Ilex coriacea*), fetterbush (*Lyonia lucida*), dusty zenobia (*Zenobia pulverulenta*), and titi (*Cyrilla racemiflora*). Scattered trees of pond pine (*Pinus serotina*) and Atlantic white cedar (*Chamaecyparis thyoides*) may occur.

FIGURE 9.4   Shrub swamps: (a) alder swamp in the Appalachians, (b) willow riparian wetlands along a western stream, and (c) Alaskan muskeg. (Photo (c) courtesy of the U.S. Fish and Wildlife Service.)

Other important scrub-shrub wetlands in the U.S. are characterized by buttonbush (*Cephalanthus occidentalis*), alders (*Alnus*), willows (*Salix*), dogwoods (*Cornus*), meadowsweets (*Spiraea*), arrowwoods (*Viburnum*), shrubby St. John's-worts (*Hypericum*), dwarf birches in northern areas (*Betula glandulosa*, *B. nana*, and *B. pumila*), and saplings of tree species like red maple (*Acer rubrum*) and cottonwoods (*Populus* spp.). Alders, several species of willows, salt cedar (*Tamarix pentandra*), and mulefat (*Baccharis glutinosa*) can be found in and along western streams.

Some references on U.S. scrub-shrub wetlands include

Batten (1990)
Batten and Murray (1982)
Bennett and Nelson (1991)
Bray (1930)
Calhoun et al. (1994)
Chadde et al. (1998)
Conway (1949)
Cooper et al. (1997)
Crow (1969)
Crum (1988)
Curtis (1959)
Damman (1977)
Damman and French (1987)
Dansereau and Segadas-Vianna (1952)
Deevey (1958)
Drury (1962)
Elliot (1981)
Faber et al. (1989)

Ferren et al. (1996)
Folkerts (1982)
Franklin and Dyrness (1973)
Gates (1942)
Glaser (1987)
Glaser et al. (1981)
Heinselman (1963, 1965, 1970)
Hofstetter (1983)
Hopper (1968)
Janssen (1967)
Johnson (1985)
Kologiski (1977)
Kovalchik (1987)
Kunze (1994)
Larsen (1982)
Metzler and Tiner (1992)
Moore and Bellamy (1974)
Neiland (1971)
Nicholson (1995)

Osvald (1955)
Richardson (1981)
Rigg (1937)
Schomer and Drew (1982)
Schlesinger (1978b)
Seyer (1979)
Shanks (1966)
Sharitz and Gibbons (1982)
Swan and Gill (1970)
Swinehart and Starks (1994)
Szaro (1989)
Tiner (1985a, 1985b, 1988, 1989, 1998)
Tiner and Burke (1995)
Viereck et al. (1992)
Vogl and Henrickson (1971)
Windell et al. (1986)
Worley (1981)
Wright et al. (1992)

FORESTED WETLANDS

Forested wetlands are dominated by trees 20 ft or taller (Figures 9.5a/b; Plates 47-48). These wet forests cover nearly 52 million acres in the conterminous U.S., while slightly more than 13 million occur in Alaska (Dahl and Johnson, 1991; Hall et al., 1994). These wetlands are most abundant in humid and tropical regions as such conditions favor the growth of trees. They are the dominant wetland type in the eastern half of the country, whereas in arid to subhumid regions, forested wetlands are relatively uncommon, usually restricted to river valleys where water is abundant or higher elevations where more moist conditions are found. Flooding is extremely variable depending on regional climate, topographic position, and local hydrology. Among the many types of forested wetlands in the U.S. are maple swamps, floodplain or bottomland hardwood swamps (including western riparian forested wetlands), forested bogs, cypress-gum (tupelo) swamps, bay swamps, wet flatwoods, and pine swamps and lowlands. Within these forested communities, shrubs and herbs also are common. Regional differences in composition of forested wetlands are shown by examples in Tables 9.6-9.10 for various regions at the end of the chapter.

In the northern U.S., important trees of the wetter swamps include red maple, black ash (*Fraxinus nigra*), green ash (*F. pennsylvanica*), northern white cedar, black spruce, balsam fir (*Abies balsamea*), and larch. Bald cypress (*Taxodium distichum*), water tupelo (*Nyssa aquatica*), red maple, swamp black gum (*Nyssa sylvatica* var. *biflora*), Atlantic white cedar, overcup oak (*Quercus lyrata*), sweet gum (*Liquidambar styraciflua*), sweet bay (*Magnolia virginiana*), Carolina ash (*Fraxinus caroliniana*), pumpkin ash (*F. profunda*), and black willow (*Salix nigra*) are common in southern wet swamps. Florida's swamps may have these plus slash pine (*Pinus elliottii*), cabbage palm (*Sabal palmetto*), loblolly bay (*Gordonia lasianthus*), red bay (*Persea borbonia*), and Carolina willow (*S. caroliniana*). In the northwestern U.S., western hemlock (*Tsuga heterophylla*), western red cedar (*Thuja plicata*), Sitka spruce (*Picea sitchensis*), red alder (*Alnus rubra*), Oregon ash (*F. latifolia*), and willows are important species.

Drier swamps (i.e., those flooded only briefly during the growing season) are characterized by silver maple (*Acer saccharinum*), pin oak (*Quercus palustris*), box elder (*Acer negundo*), and

FIGURE 9.5 Forested wetlands: (a) red maple swamp in the Northeast and (b) southern bottomland swamp (Okeefenokee Swamp). (Photo (b) courtesy of the U.S. Fish and Wildlife Service.)

sycamore (*Platanus occidentalis*) in northern areas. In the South, these swamps are dominated by sweet gum, loblolly pine (*Pinus taeda*), slash pine, tulip poplar (*Liriodendron tulipifera*), beech (*Fagus grandifolia*), sycamore, water hickory (*Carya aquatica*), pignut hickory (*C. glabra*), and oaks (e.g., *Quercus nigra*, *Q. laurifolia*, and *Q. phellos*).

Riparian wetlands along western streams are dominated by sugarberry (*Celtis laevigata*), sweet gum, willow oak (*Quercus phellos*), water oak (*Q. nigra*), overcup oak (*Q. lyrata*), water hickory, Fremont's cottonwood (*Populus fremontii*), black cottonwood (*P. trichocarpa*), box elder (*Acer negundo*), alders, willows, green ash (*Fraxinus pennsylvanica*), willows (*Salix* spp.), and elms (*Ulmus* spp.). In the western mountains, subalpine fir (*Abies lasiocarpa*), Engelmann spruce (*Picea engelmannii*), narrowleaf cottonwood (*P. angustifolia*), and quaking aspen (*Populus tremuloides*) may frequently occur as narrow forested bands along streams. Western red cedar (*Thuja plicata*), western hemlock (*Tsuga heterophylla*), lodgepole pine (*Pinus contorta*), and Sitka spruce (*Picea sitchensis*) are other important forested wetland species in western states.

Major forested wetland species in Alaska are black spruce (*Picea mariana*), larch or tamarack (*Larix laricina*), lodgepole pine (*Pinus contorta*), balsam fir, mountain hemlock (*Tsuga mertensiana*), western hemlock, Sitka spruce (*Picea sitchensis*), and red alder (*Alnus rubra*). Paper birch (*Betula papyrifera*) also may be co-dominant in some communities.

Some significant reports on forested wetlands include the following:

Abrams (1986)
Alvarez-López (1990)
Battan (1990)
Beaven and Oosting (1939)
Brabander et al. (1985)
Bray (1930)
Brinson (1977)
Brinson et al. (1981)
Brown and Peterson (1983)
Bruner (1931)
Buell and Wistendahl (1955)
Bush and Van Auken (1984)
Calhoun et al. (1994)
Carter (1988)
Chadde et al. (1998)
Chamless and Nixon (1975)
Christensen et al. (1988)
Clark and Benforado (1981)
Clewell et al. (1982)
Cohen et al. (1984)
Conner and Day (1976; 1987)
Conner et al. (1981)
Cooper (1986)
Crites and Ebinger (1969)
Crum (1988)
Curtis (1959)
Dabel and Day (1977)
Damman and French (1987)
Devall (1990)
Dick-Peddie et al. (1987)
Drew and Schomer (1894)
Duever et al. (1984)
Dunn and Stearns (1987)
Edmisten (1963)
Epperson (1992)
Erickson and Leslie (1988)
Ewel (1991)
Ewel and Odum (1984)

Faber et al. (1989)
Ferren et al. (1996)
Ford and Auken (1982)
Frye and Quinn (1979)
Golet et al. (1993)
Hall and Penfound (1943)
Hannan and Lassetter (1982)
Harper (1914)
Harshberger (1970)
Hawk and Zobel (1974)
Heinselman (1970)
Heitmeyer et al. (1991)
Hoagland and Jones (1992)
Hofstetter (1983)
Hook and Lea (1989)
Hopper (1968)
Hosner and Minckler (1963)
Huenneke (1982)
Jahn and Anderson (1986)
Janssen (1967)
Johnson et al. (1985)
Keammerer et al. (1975)
Kearney (1901)
Kennedy and Nowacki (1997)
Kologiski (1977)
Kunze (1994)
Laderman (1987; 1989)
Laessle (1942)
Larsen (1982)
Lee (1983)
Leitman (1975)
Leitman et al. (1984)
Lugo et al. (1990)
McCormick and Somes (1982)
Meijer et al. (1981)
Messina and Conner (1998)
Metzler and Damman (1985)

Metzler and Tiner (1992)
Moizuk and Livingston (1966)
Monk (1966)
Musselman et al. (1977)
Neiland (1971)
Nicholson (1995)
Niering (1953)
Paratley and Fahey (1986)
Patrick et al. (1981)
Patterson et al. (1985)
Penfound (1952)
Post (1996)
Reiners (1972)
Rheinhardt (1991; 1992)
Rice (1965)
Richardson (1981)
Schlesinger (1978a,b)
Schomer and Drew (1982)
Shanks (1966)
Sharitz and Gibbons (1982)
Shear et al. (1997)
Shelford (1954)
Szaro (1989)
Taylor (1985)
Tiner (1985a; 1985b; 1988; 1989; 1998)
Tiner and Burke (1995)
Trettin et al. (1997)
Veneman and Tiner (1990)
Viereck et al. (1992)
Vince et al. (1989)
Wells (1928)
Wharton (1978)
Wharton et al. (1976; 1982)
Wilkinson et al. (1987)
Windell et al. (1986)
Wistendahl (1958)
Wright and Wright (1932)

Other reports that have descriptions of forested wetlands include

Alexander (1986)
Cooper et al. (1983)
Daubenmire and Daubenmire (1968)
DeVelice et al. (1984)
Hall (1973)
Hess (1981)

Hess and Wasser (1982)
Hoffman and Alexander (1980)
Hopkins (1979)
Komarkova (1984)
Kovalchik (1987)
Mauk and Henderson (1984)
Pfister et al. (1977)

Steele et al. (1981; 1983)
Steen and Dix (1974)
Vollund (1976)
Youngblood (1984)
Youngblood and Mueggler (1981)
Youngblood et al. (1985)

ESTUARINE WETLANDS

Estuarine wetlands are salt and brackish marshes associated with tidal estuaries (i.e., salt and brackish tidal waters) located along the entire U.S. coastline. They typically form behind barrier islands and beaches or along coastal rivers and semienclosed embayments protected from the full force of ocean waves. The conterminous U.S. possesses almost 5.5 million acres of these wetlands, while Alaska has 2.1 million acres (Dahl and Johnson, 1991; Hall et al., 1994). From a salinity standpoint, estuaries can be divided into three distinct reaches: polyhaline — strongly saline areas (18 to 30 parts per thousand, ppt); mesohaline — moderate salinity areas (5 to 18 ppt); and oligohaline — slightly brackish areas (0.5 to 5 ppt) (Cowardin et al., 1979). Salt marshes occupy the polyhaline reach, while brackish marshes are found in the mesohaline zone, and slightly brackish or transitional marshes in the oligohaline reach. In southern Florida, mangrove swamps dominate the polyhaline zone and protected euhaline embayments (30 to 40 ppt). Large coastal rivers become increasingly fresher upstream from the river's mouth as salt water is diluted by the river's freshwater discharge. Since river discharge varies during the year, the salinity of coastal river systems varies seasonally.

A variety of wetlands develop in estuaries largely because of differences in salinity and duration and frequency of tidal inundation. Major wetland types include: emergent wetlands, intertidal unconsolidated shores, and scrub-shrub wetlands. Other coastal wetlands include intertidal coral and mollusk reefs, rocky shores (often covered with macroalgae), streambeds, and some forested wetlands. Submerged aquatic vegetation forming dense beds (e.g., eelgrass) in permanent coastal waters are considered deepwater habitats (see Zieman, 1982; Phillips, 1984; Thayer et al., 1984; Zieman and Zieman, 1989).

ESTUARINE EMERGENT WETLANDS

Estuarine emergent wetlands are usually dominated by grass or grasslike herbaceous plants (Plate 49). These wetlands, commonly called *salt marshes* and *brackish tidal marshes,* are the dominant estuarine wetland type in the conterminous United States. They are best represented along the coastlines of the Atlantic and the Gulf of Mexico. About 4 million acres of these wetlands occur in the lower 48 states, whereas only 360,200 acres are present in Alaska (Dahl and Johnson, 1991; Hall et al., 1994).

Differences in salinity and tidal flooding within estuaries have a profound effect on the emergent vegetation (e.g., Johnson and York, 1915; Hinde, 1954; Adams, 1963; Chapman, 1974; Tiner, 1987b; 1993). Plant composition markedly changes from the more saline portions of estuaries to the slightly brackish areas. Even within areas of similar salinity, vegetation differs largely due to frequency and duration of tidal flooding. Two distinct zones can be observed based on hydrologic differences in frequency and duration of flooding: regularly flooded marsh (low marsh) and irregularly flooded (high marsh) (Figure 9.6).

The regularly flooded marsh (low marsh) is flooded and exposed at least once daily by the tides. In the Northeast, this marsh is generally limited to tidal creek banks and the shores of coastal embayments and rivers, while in the Southeast along the coasts of Georgia and South Carolina, regularly flooded salt marsh is the dominant coastal wetland type. Smooth cordgrass (*Spartina alterniflora*) dominates the regularly flooded marsh in estuaries along the coasts of the Atlantic Ocean and the Gulf of Mexico. In similar areas on the Pacific coast, California cordgrass (*Spartina foliosa*) prevails. These grasses are among the most productive marsh plants in the world. In southern California, Bigelow's glasswort (*Salicornia bigelovii*) may occur in low marshes in exposed locations, while perennial glasswort (*S. virginica*) is a colonizer of mudflats in sheltered areas in Humboldt Bay (Macdonald, 1977). In the Pacific Northwest, smooth cordgrass, an introduced species, is considered an undesirable invasive. Common glasswort (*Salicornia europaea*) and seaside arrow-grass (*Triglochin maritimum*) are characteristic of Alaskan low salt marshes (Jon Hall, personal communication, 1998). Other low marsh species in Alaska include alkali grasses

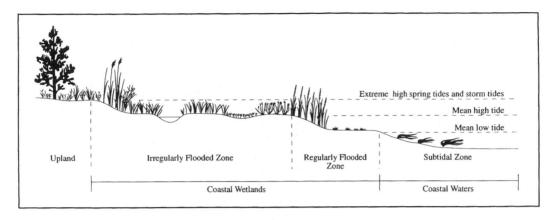

FIGURE 9.6   Diagram showing tidal marsh zones based on frequency of tidal flooding. Low marsh is flooded at least once daily and high marsh is flooded less often. (Tiner, R.W. and Burke, D.G., *Wetlands in Maryland*, U.S. Fish and Wildlife Services, Hadley, MA and Maryland Department of Natural Resources, Annapolis, MD.

(*Puccinellia* spp.), Pursh seepweed (*Suaeda depressa*), and seaside plantain (*Plantago maritima*) (Macdonald, 1977).

The irregularly flooded marsh occurs above the low marsh where it is exposed to air for long periods and flooded for shorter durations at varying intervals. Vegetation in this zone is more diverse, with the highest diversity typically found along the upper edges of the high marsh where salt stress is lowest. Characteristic plants in the Northeast include salt hay grass (*Spartina patens*), salt grass (*Distichlis spicata*), black grass (*Juncus gerardii*), alkali grasses (*Puccinellia* spp.), baltic rush (*Juncus balticus*), glassworts (*Salicornia* spp.), salt marsh aster (*Aster tenuifolius*), and salt marsh bulrush (*Scirpus robustus* and *S. maritimus*), switchgrass (*Panicum virgatum*), high-tide bush (*Iva frutescens*), and groundsel-bush (*Baccharis halimifolia*). Common reed (*Phragmites australis*) dominates many tidally restricted former salt marshes, often ideal sites for estuarine marsh restoration. In the Southeast, salt grass, glassworts, asters, high-tide bush, and groundsel-bush remain common, but other species become abundant, including black needlerush (*Juncus roemerianus*), sea ox-eye (*Borrichia frutescens*), and key grass (*Monanthochloe littoralis*). Common reed is abundant in many Louisiana tidal marshes. On the Pacific coast, common plants include California cordgrass, glassworts, salt dodder (*Cuscuta salina*), salt grass, California sea-blite (*Suaeda californica*), alkali heath (*Frankenia grandifolia*), California sea lavender (*Limonium californicum*), spreading alkaliweed (*Cressa truxillensis*), seaside plantain, spiny rush (*Juncus acutus*), Lyngbye's sedge (*Carex lyngbyei*), tufted hairgrass (*Deschampsia caespitosa*), and fleshy jaumea (*Jaumea carnosa*). Key grass is common in southern California. In Washington and Oregon, Lyngbye's sedge is often dominant on silty soils between the low marsh and higher high marsh (Macdonald, 1977). Alaskan salt marshes are vegetated by alkali grasses, tufted hairgrass, Lyngbye's sedge, sedge (*C. cryptocarpa*), Ramensk's sedge (*C. ramenskii*), Hoppner's sedge (*C. subspathacea*), seaside arrow-grass, MacKenzie water-hemlock (*Cicuta mackenziena*), vetchling peavine (*Lathyrus palustris*), seabeach sandwort (*Honckenya peploides*), seaside plantain (*Plantago maritima*), Canada sand spurrey (*Spergularia canadensis*), scurvy grass (*Cochlearia officinalis*), oysterleaf (*Mertensia maritima*), and spearscale (*Atriplex*).

The dilution of saltwater with freshwater in large coastal rivers lowers salt stress and different plant communities become established. Salinity in these brackish marshes fluctuates greatly with the tides, river flow, and the seasons. Plants found along the edges of salt marshes often grow out to the water's edge under brackish conditions. Nearest the salt marshes, plants with higher salt tolerance prevail. For example, black needlerush found along the margins of Southeast salt marshes is a dominant brackish marsh species. Big cordgrass (*Spartina cynosuroides*), salt hay grass, switchgrass, narrow-leaved cattail (*Typha angustifolia*), and bulrushes also are dominant in brackish waters on the East Coast. On the West Coast, soft-stemmed bulrush (*Scirpus validus*), California bulrush (*S. californicus*), various sedges, and narrow-leaved cattail are among the brackish marsh species. In Alaska, four-tail marestail (*Hippuris tetraphylla*) is a dominant species.

As the upstream limit of saltwater influence is approached in tidal marshes that are mostly freshwater, a highly diverse assemblage of emergent plants often becomes established. They are among the most floristically diverse plant communities in North America. Characteristic species of these slightly brackish (oligohaline) marshes on the East Coast include big cordgrass, narrow-leaved cattail, pickerelweed (*Pontederia cordata*), wild rice (*Zizania aquatica*), rose mallow (*Hibiscus moscheutos*), arrowheads (*Sagittaria* spp.), smartweeds (*Polygonum* spp.), sedges (*Carex* spp.), bulrushes, beggars-ticks (*Bidens* spp.), and common reed. Most of these plants reach their maximum abundance in palustrine wetlands.

Numerous references on salt and brackish marsh vegetation in the U.S. are available including

Adams (1963)
Akins and Jefferson (1973)
Anderson et al. (1968)
Bertness (1992)
Calhoun et al. (1994)
Carlton (1975, 1977)
Chabreck (1972)
Chapman (1938, 1940a,b, 1960, 1976a)
Conard (1924, 1935)
Conard and Galligar (1929)
Conner and Day (1987)
Cooper (1931)
Cooper and Waits (1973)
Copeland et al. (1983, 1984)
Crow (1966)
Day et al. (1973)
de la Cruz (1981)
Drew and Schomer (1984)
Dreyer and Niering (1995)
Egler (1952)
Eleuterius (1972, 1980, 1984)
Eleuterius and McDaniel (1978)
Ferren et al. (1996)
Frey and Basan (1985)
Good (1965)

Gosselink (1984)
Hackney and de la Cruz (1982)
Hanson (1951, 1953)
Harshberger (1900, 1902, 1909)
Hinde (1954)
House (1914)
Jefferson (1973, 1974)
Josselyn (1983)
Josselyn et al. (1990)
Kearney (1900)
Kurz and Wagner (1957)
Laessle (1942)
Latham et al. (1994)
Lewis and Estevez (1988)
Livingston (1984)
Macdonald (1977)
Macdonald and Barbouur (1974)
McCormick and Somes (1982)
Metzler and Tiner (1992)
Miller and Egler (1950)
Montague and Wiegert (1991)
Mudie (1970)
Nichols (1920)
Niering and Warren (1980)
Nixon (1982)
Nixon and Oviatt (1973)

Odum et al. (1984)
Penfound and Hathaway (1938)
Pomeroy and Wiegert (1981)
Purer (1942)
Redfield (1972)
Reimold (1977)
Reimold and Queen (1974)
Schomer and Drew (1982)
Seliskar and Gallagher (1983)
Shaler (1885, 1886)
Silberhorn et al. (1974)
Stephens and Billings (1967)
Stevenson and Emery (1958)
Stout (1984)
Taylor (1938)
Teal and Teal (1969)
Tiner (1977, 1985a, 1985b, 1987b, 1988, 1989, 1993, 1998)
Tiner and Burke (1995)
Viereck et al. (1992)
Vogl (1966)
Wass and Wright (1969)
White et al. (1993)
Wiegert and Freeman (1990)
Zedler and Nordby (1986)
Zedler (1982)

ESTUARINE SCRUB-SHRUB WETLANDS

Estuarine scrub-shrub wetlands are characterized by salt-tolerant woody vegetation less than 20 ft in height. Roughly 710,000 acres of estuarine shrub swamps occur in the conterminous U.S., with most of them found in Florida (668,000 acres) (Dahl and Johnson, 1991; Frayer and Hefner, 1991). Although limited in distribution, mangrove swamps represent the greatest acreage of estuarine shrub swamps. Mangroves are generally found south of the 30° N latitude (the Ponce de Leon Inlet on the East Coast), reaching their maximum abundance in Florida, Puerto Rico, and the Virgin Islands. These wetlands are dominated by two forms of mangroves: red mangrove (*Rhizophora mangle*) and black mangrove (*Avicennia germinans*) (Figure 9.7). The former tends to dominate the regularly flooded zone, while the latter species characterizes higher irregularly flooded areas. White mangrove (*Laguncularia racemosa*) may be intermixed, while buttonwood (*Conocarpus*

FIGURE 9.7   Black mangrove swamp.

erectus) often occurs at higher elevations along the upland border. Salt marshes of smooth cordgrass, black needlerush, salt grass, woody glasswort (*Salicornia perennis*) and saltwort (*Batis maritima*) may be interspersed among Florida's mangroves swamps. Other estuarine shrub swamps along the Atlantic and Gulf of Mexico coasts may be vegetated with high-tide bush, groundsel tree, and sea ox-eye. These shrubs typically occur at higher levels in the salt marshes. The former species is common along mosquito ditches where substrate material has been mounded and along the upper borders of many salt marshes.

Additional information on mangroves can be found in the following references: Chapman (1976b), Davis (1940), Egler (1952), Kuenzler (1974), Lugo and Brown (1988), Lugo and Snedaker (1974), Odum and McIvor (1991), Odum et al. (1982), Pool et al. (1977), Schomer and Drew (1982), Tiner (1993), and Twilley (1998).

ESTUARINE INTERTIDAL SHORES

Intertidal shores of unconsolidated material (sand, gravel, or mud) called *tidal flats* often lie seaward of tidal marshes and mangroves or in coastal embayments subjected to frequent prolonged tidal inundation. They also occur as barren areas within the high salt marsh in the Southeast. Although largely nonvegetated except for microscopic plants (e.g., diatoms, bluegreen algae, and dinoflagellates), some tidal flats are locally dominated by macroscopic algae such as sea lettuce (*Ulva lactuca*) and a filamentous green algae (*Enteromorpha intestinalis*) or by rockweeds (e.g., *Fucus* spp. and *Ascophyllum nodosum*) on bolder-strewn mudflats. These wetlands are particularly extensive in areas with high tidal ranges such as occur in Alaska and Maine. They are the predominant estuarine wetland type in Alaska where about 1.8 million acres exist, while only 450,000 acres of tidal flats (and associated rocky shores) occur in the lower 48 states (Hall et al., 1994; Dahl and Johnson, 1991).

Rocky shores form where the rocky headlands meet the sea. They are common in Alaska, California, Maine, Oregon, and Washington. Many of these shores are nonvegetated, but fucoid algae such as rockweeds and other algae often colonize portions that are flooded daily by the tides. Barnacles also are quite common on these shores.

TABLE 9.3
Examples of Palustrine Scrub-Shrub Wetland Plant Communities in the Eastern U.S.

Wetland Type and Location	Dominant Plants	Associated Vegetation	Source
Northern coastal raised bog; E. Maine	*Kalmia angustifolia*	*Empetrum nigrum, Sphagnum flaviocomans, S. imbricatum, Icmadophila ericetorum, Rubus chamaemorus, Picea mariana,* and *Larix laricina*	Damman, 1977
Shrub bog; southern part of glaciated N.E. U.S.	*Sphagnum centrale* and *Chamaedaphne calyculata*	*Sphagnum fallax, S. fimbriatum, Carex stricta, Spiraea* spp., *Vaccinium corymbosum,* and *Rhododendron viscosum*	Damman and French, 1987
Northern bog; N. Minnesota	*Kalmia polifolia, Andromeda glaucophylla, Ledum groelandicum, Chamaedaphne calyculata,* and *Sphagnum* spp.	*Carex oligosperma*	Glaser, 1987
Open bog; Wisconsin	*Andromeda glaucophylla, Chamaedaphne calyculata, Kalmia polifolia, Ledum groelandicum, Vaccinium macrocarpon, V. myrtilloides,* and *V. oxycoccos*	*Cornus canadensis, Carex trisperma, Drosera rotundifolia, Sarracenia purpurea, Smilacina trifolia, Vaccinium angustifolium,* and others	Curtis, 1959
Leatherleaf bog; S. New Jersey	*Chamaedaphne calyculata*	*Pinus rigida, Acer rubrum, Vaccinium corymbosum, Ilex glabra, Woodwardia virginica,* and *Sphagnum* spp.	Tiner, 1985a
Buttonbush swamp; N. New Jersey	*Cephalanthus occidentalis*	*Spiraea tomentosa, Acer rubrum, Peltandra virginica, Juncus effusus, Scirpus cyperinus, Iris versicolor, Hypericum* sp., *Carex stricta, Boehmeria cylindrica, Polygonum sagittatum* and *Lemnaceae*	Tiner, 1985a
Alder thicket; Wisconsin	*Alnus rugosa*	*Aster simplex, Calamagrostis canadensis, Campanula aparinoides, Eupatoriadelphus maculatus, Galium asprellum, Glyceria grandis, Impatiens capensis, Onoclea sensibilis, Poa palustris, Scirpus atrovirens, Solidago canadensis, Spiraea alba, Thelypteris thelypteroides,* and others	Curtis, 1959
Alder swamp; Connecticut	*Alnus rugosa*	*Acer rubrum, Spiraea latifolia, Lyonia ligustrina, Vaccinium corymbosum, Osmunda regalis, Scirpus cyperinus, Iris versicolor,* and *Thelypteris thelypteroides*	Metzler and Tiner, 1992
Alder-dogwood thicket; W. Maryland	*Alnus rugosa* and *Cornus stolonifera*	*Sambucus canadensis, Viburnum cassinoides, Rosa palustris, Calamagrostis canadensis, Solidago* sp., and *Onoclea sensibilis*	Tiner and Burke, 1995
Alder-swamp rose swamp; E. Maryland	*Alnus serrulata* and *Rosa palustris*	*Diospyros virginiana, Salix nigra, Sambucus canadensis, Cornus amomum, Typha latifolia, Aster puniceus, Eupatorium perfoliatum, Sagittaria latifolia, Impatiens capensis, Carex stricta, C.lurida, Polygonum sagittatum, P. hydropiper, Juncus effusus,* and others	Tiner and Burke, 1995

Wetland Type and Location	Dominant Plants	Associated Vegetation	Source
Rich shrub fen; N. Minnesota	*Betula pumila, Andromeda glaucophylla, Vaccinium oxycoccus,* and *Chamaedaphne calyculata*	*Potentilla fruticosa* and *Carex cephalantha*	Glaser, 1987
Shrub fen; Maine	*Rhododendron canadense, Aronia melanocarpa, Nemopanthus mucronata, Vaccinium corymbosum, Viburnum cassinoides, Spiraea tomentosa,* and *Spiraea latifolia*	*Larix laricina, Acer rubrum, Thuja occidentalis, Picea mariana, Ledum groenlandicum, Kalmia angustifolia, Myrica gale, Chamaedaphne calyculata, Carex paupercula, C. trisperma, Smilacina trifolia*	Calhoun et al., 1994
Blueberry thicket; Rhode Island	*Vaccinium corymbosum*	*Ilex verticillata, Rhododendron viscosum, Acer rubrum, Eleocharis* sp., *Sphagnum* spp., *Carex stricta, Aronia* sp., *Amelanchier* sp., *Nyssa sylvatica, Pinus strobus, Osmunda cinnamomea, Maianthemum canadense, Iris versicolor, Betula populifoila, Kalmia angustifolia,* and *Spiraea latifolia*	Tiner, 1989
Black huckleberry swamp; Connecticut	*Aronia melanocarpa*	*Kalmia angustifolia, Picea mariana, Pinus rigida, Carex* spp., *Trientalis borealis, Drosera* sp., *Sphagnum* spp., and lichens	Metzler and Tiner, 1992
Meadowsweet thicket; W. Maryland	*Spiraea alba*	*Calamagrostis canadensis, Carex* spp., *Scirpus cyperinus, Alnus* sp., and *Hypericum densiflorum*	Tiner, 1988
Pocosin; coastal North Carolina	*Pinus serotina, Cyrilla racemosa, Zenobia pulverulenta, Gordonia lasianthus,* and/or *Lyonia lucida*	*Clethra alnifolia, Kalmia angustifolia, Ilex glabra,* and *Chamaedaphne calyculata*	Christensen et al., 1981
Pocosin (Carolina bay); South Carolina	*Lyonia lucida, Leucothoe racemosa, Lyonia ligustrina, Vaccinium* spp., *Gaylussacia frondosa, Rhododendron canescens, R. viscosum, Ilex glabra, Ilex coriacea,* and *Smilax laurifolia*	*Gordonia lasianthus, Pinus serotina, Aronia arbutifolia, Zenobia pulverulenta, Kalmia angustifolia,* and *Sphagnum* spp.	Bennett and Nelson, 1991

TABLE 9.4
Examples of Palustrine Scrub-Shrub Wetland Plant Communities in the Western U.S.

Wetland Type and Location	Dominant Plants	Associated Vegetation	Source
Snow willow alpine dwarf shrub swamp; S.W. Montana	*Salix reticulata* and *Caltha leptosepala*	*Carex haydenii, Carex nova, Carex scirpoidea, Deschampsia cespitosa, Luzula spicata, Poa alpina, Salix rotundifolia, Salix arctica*, and *Silene acaulis*	Cooper et al., 1997
Planeleaf willow alpine shrub meadow; S.W. Montana	*Salix planifolia* and *Carex scopulorum*	*Deschampsia cespitosa, Trollius laxus, Senecio cymbalarioides, Veronica wormskjolddii*, and *Epilobium alpinum*	Cooper et al., 1997
Meadowsweet-willow shrub swamp; W. Oregon	*Spiraea douglasii* and *Salix densiflorus*	*Vaccinium uliginosum, Carex luzulina, Carex lenticularis, C. vesicaria, C. sitchensis, C. saxatilis, Scirpus microcarpus, Juncus balticus, Calamagrostis canadensis, Deschampsia caespitosa, Agrostis exarata, A. hallii, A. thurberiana, Epilobium glandulosum, E. glaberrinum, Hypericum anagalloides*, and *Dodecatheon* spp.	Hemstom et al., 1987
Shrub thicket; W. Oregon	*Spiraea douglasii, Vaccinium uliginosum* and *Salix* spp.	*Juncus ensifolius, Carex vesicaria, C. sitchensis, C. luzulina, C. muricata, C. obnupta, Epilobium glandulosum, Tofieldia glutinosa, Angelica genuflexa, Caltha biflora, Calamagrostis canadensis*, and *Agrostis thurberiana*	Hemstom et al., 1987
Mountain alder thicket; central Oregon	*Alnus incana*	*Glyceria elata, Carex amplifolia, Scirpus microcarpus, Equisetum arvense, Galium triflorum, Geum macrophyllum*, and *Smilacina stellata*	Kovalchik, 1987
Mountain alder thicket; central Oregon	*Alnus incana* and *Spiraea douglasii*	*Ribes lacustre, Rosa woodsii, Carex disperma, Carex eurycarpa, Achillea millefolium, Aconitum columbianum, Equisetum arvense, Geum triflorum, Senecio triangularis*, and *Smilacina stellata*	Kovalchik, 1987
Spring-fed mountain alder thicket; central Oregon	*Alnus incana*	*Ribes lacustre, Glyceria elata, Carex amplifolia, Aconitium columbianum, Circaea alpina, Galium triflorum, Geum macrophyllum, Listera* spp., *Smilacina stellata, Mimulus guttatus*, and *Polemonium occidentale*	Kovalchik, 1987
Willow thicket; central Oregon	*Salix* spp.	*Salix geyeriana, Salix lemmonii, Salix boothii, Spiraea douglasii, Calamagrostis canadensis, Carex eurycarpa, Galium trifidum*, and *Geum macrophyllum*	Kovalchik, 1987
Wet shrub meadow; central Oregon	*Salix* spp. and *Carex aquatilis*	*Salix geyeriana, S. lemmonii, S. boothii, Calamagrostis* spp., *Juncus balticus, Epilobium watsonii, Geum macrophyllum, Mimulus guttatus, Pedicularis groenlandica, Polemonium occidentale, Polygonum bistortoides, Saxifraga oregana*, and *Sidalcea oregana*	Kovalchik, 1987
Wet shrub meadow; central Oregon	*Salix* spp. and *Carex sitchensis*	*Salix geyeriana, S.lemmonii, Betula glandulosa, S. douglassi*, and *Vaccinium occidentalis*	Kovalchik, 1987
Wet shrub meadow; central Oregon	*Salix* spp. and *Carex rostrata*	*Salix boothii, S. geyeriana, Deschampsia cespitosa, Carex vesicaria, Epilobium watsonii, Polemonium occidentale, Rumex occidentalis*, and *Trifolium longipes*	

Wetland Type and Location	Dominant Plants	Associated Vegetation	Source
Wet shrub meadow; central Oregon	*Vaccinium occidentalis* and *Carex sitchensis*	*Salix eastwoodiae* and *Salix* spp.	
Mixed emergent-shrub bog; central Oregon	*Vaccinium occidentalis* and *Eleocharis pauciflora*	*Picea englemannii, Pinus contorta, Betula glandulosa, Kalmia microphylla, Salix* spp., *Deschampsia cespitosa, Carex muricata, C. rostrata, C. sitchensis, Dodecatheon* sp., *Drosera* spp. *Habenaria dilitata, Pedicularis groenlandica, Polygonum bistortoides, Sphenosciadium capitellatum, Spiranthes romanzoffiana, Tofieldia glutinosa,* and mosses	
Shrub swamp; Idaho	*Salix* spp. and *Carex vesicaria* and *C. muricata*	*Salix boothii, S. wolfii, S. drummondiana, Carex scopulorum, Agrostis alba, Carex rostrata, Calamagrostis canadensis, Betula glandulosa, Aster foliaceus, Carex simulata, Angelica arguta,* and others	Bill Sipple, pers. comm., 1998
Shrub swamp; Idaho	*Salix boothii* and *Carex aquatilis, C. lanuginosa,* and *C. rostrata*	*Pinus contorta, Lonicera involucrata, Alnus sinuata, Salix drummondiana, S. lutea, Aster modestus, Agrostis alba,* and *Calamagrostis canadensis*	Bill Sipple, pers. comm., 1998
Riparian willow thicket; New Mexico	*Salix exigua*	*Populus wislizenii, Elaeagnus angustifolia, Conyza canadensis, Juncus* spp., *Apocynum cannabinum, Agrostis stolonifera, Elymus canadensis, Muhlenbergia asperifolia, Bromus japonicus, Sporobolus contractus,* and others	Dick-Peddie et al., 1984
Riparian sandbar thicket; New Mexico	*Populus fremontii* and *Salix gooddingii*	*Salix exigua, Baccharis glutinosa, Salsola kali, Conyza canadensis, Ambrosia artemisifolia, Sporobolus* spp., and others	Dick-Peddie et al., 1987
Riparian willow thicket; Arizona and New Mexico	*Salix exigua*	*Cowania mexicana, Rhus trilobata, Carex* spp., *Conium maculatum, Equisetum laevigatum, Melilotus alba,* and *Muhlenbergia* spp.	Szaro, 1989
Riparian willow thicket; Arizona and New Mexico	*Salix irrorata*	*Salix amygdaloides, Salix lasiandra, Populus fremontii, Acer negundo, Rosa* spp., *Rubus strigosus, Bromus* spp., *Eragrostis* spp., *Fragaria ovalis, Geranium* spp., *Melilotus officinalis, Poa pratensis, Taraxacum officinale,* and *Thalictrum fendleri*	Szaro, 1989

Note: Dominant plants include shrub and emergent dominants where appropriate.

TABLE 9.5
Examples of Palustrine Scrub-Shrub Wetland Plant Communities in Alaska

Wetland Type and Location	Dominant Plants	Associated Vegetation	Source
Willow gravel bar thicket; interior Alaska	*Salix alaxensis*	*Salix richardsonii, S. reticulata, S. polaris, Poa alpina, Calamagrostis canadensis*, and *Trisetum spicatum*	Hanson, 1958 (as reported in Batten and Murray, 1982)
Raised or blanket bog; S.E. Alaska	*Tsuga mertensiana, T. heterophylla*, and *Pinus contorta*	*Sphagnum* spp., *Ledum* sp., *Empetrum* sp., *Kalmia* sp., *Carex pluriflora, Carex* spp., *Rubus chamaemorus, Vaccinium vitis-idaea*, and *Carex livida*	Neiland, 1971 (as reported in Batten and Murray, 1982)
Open black spruce dwarf tree scrub; interior, S. central and W. Alaska	*Picea mariana*	*Larix laricina* and *Betula papyrifera*	Viereck et al., 1992
Black spruce dwarf tree woodland; interior, S. central and W. Alaska	*Picea mariana*	*Betula nana, Ledum decumbens, Vaccinium uliginosum, V. vitis-idaea, Myrica gale, Eriophorum vaginatum, Carex bigelowii, Aulacomnium* spp., *Hylocomium splendens, Sphagnum* spp., lichens, *Peltigera aphthosa*, and *Cladonia* spp.	Viereck et al., 1992
Closed tall shrub swamp; interior, S. central and S.E. Alaska	*Alnus tenuifolia, Salix planifolia*, and *S. lanata*	*Chamaedaphne calyculata, Viburnum edule, Ribes* spp., *Sambucus callicarpa, Rosa acicularis, Oplopanax horridus, Calamagrostis canadensis, Equisetum* spp., *Cornus canadensis, Trientalis europaea, Potentilla palustris, Carex* spp., and *Sphagnum* spp.	Viereck et al., 1992
Open tall shrub swamp; interior and S. central Alaska	*Alnus tenuifolia, Salix planifolia*, and *S. lanata*	*Alnus crispa, A. sinuata, Myrica gale, Spiraea beauverdiana, Viburnum edule, Rosa acicularis, Ribes triste, Calamagrostis canadensis, Carex aquatilis, Equisetum arvense, E. fluviatile, Potentilla palustris, Polemonium acutiflorum, Mnium* spp., and *Sphagnum* spp.	Viereck et al., 1992
Closed low shrub birch-willow shrub; N. and W. Alaska	*Betula glandulosa, B. nana*, and *Salix planifolia*, and *S. lanata*	*Hylocomium* spp. and *Aulacomnium* spp.	Viereck et al., 1992
Open low mixes shrub-sedge tussock tundra; N. and W. Alaska	*Eriophorum vaginatum* and *Carex bigelowii*	*Betula glandulosa, B. nana, Ledum decumbens, Vaccinium vitis-idaea, V. uliginosum, Rhododendron lapponicum, Salix planifolia, S. reticulata, Arctostaphylos rubra, Rubus chamaemorus, Arctagrostis latifolia, Poa arctica, Eriophorum angustifolium, Pedicularis labradorica, Petasites frigidus, Pleurozium schreberi, Hylocomium splendens, Aulacomnium* spp., *Sphagnum* spp., and lichens	Viereck et al., 1992
Open low mixed-shrub-sedge tussock bog; interior and S. central Alaska	*Eriophorum vaginatum*	*Betula glandulosa, B. nana, Ledum decumbens, Vaccinium ulignosum, V. vitis-idaea, Chamaedaphne calyculata, Vaccinium oxycoccos, Potentilla fruiticosa, Salix planifolia, S. fuscescens, Alnus tenuifolia, Picea mariana, Rubus chamaemorus, Equisetum* spp., *Carex* spp., *Sphagnum* spp., *Pleurozium schreberi*, and *Hylocomium splendens*	Viereck et al., 1992

Wetland Type and Location	Dominant Plants	Associated Vegetation	Source
Open low shrub birch-ericaceous shrub bog; S. central, interior, W. and to some extent N. Alaska	*Betula glandulosa, B. nana, Vaccinium uliginosum, V. vitis-idaea, Ledum decumbens, Empetrum nigrum*, and *Andromeda polifolia*	*Myrica gale, Potentilla fruticosa, Salix planifolia, S. reticulata, Picea mariana, Rubus chamaemorus, Eriophorum angustifolium, Carex aquatilis, C. limosa, C. pauciflora, C. rotundata, C. magellanica, Eriophorum vaginatum, Carex bigelowii, Equisetum fluviatile, Sphagnum* spp., *Dicranum* spp., and *Polytrichum* spp.	Viereck et al., 1992
Open low ericaceous shrub bog; S.E. Alaska, gulf coast and Aleutian Islands	*Kalmia polifolia, Empetrum nigrum, Vaccinium uliginosum, V. vitis idaea, Andromeda polifolia, Vaccinium oxycoccos*, and *Ledum decumbens*	*Pinus contorta, Chamaecyparis nootkatensis, Tsuga mertensiana, Picea sitchensis, Tsuga heterophylla, Eriophorum augustifolium, Trichophorum caespitosum, Carex pluriflora, C. pauciflora, Rubus chamaemorus, Drosera* spp., and *Gentiana douglasiana*	Viereck et al., 1992
Open low willow-sedge shrub tundra; N. and W. Alaska	*Salix planifolia* and *S. lanata*	*Carex aquatilis, C. vaginata, C. bigelowii, Salix arctica, S. reticulata, mosses, Tomenthypnum nitens, Distichium capillaceum, Drepanocladus* spp., and *Campylium stellatum*	Viereck et al., 1992
Open low willow-graminoid shrub bog; interior, S.W., S. central, and S.E. Alaska	*Salix barclayi* and *S. communtata*	*Betula glandulosa, B. nana, Calamagrostis canadensis, Carex aquatilis, C. pluriflora*, and nonsphagnaceous mosses	Viereck et al., 1992
Open low sweetgale-graminoid bog; S.E. S. Central and S.W. Alaska	*Myrica gale*	*Salix fuscescens, S. barclayi, Chamaedaphne calyculata, Betula glandulosa, B. nana, Alnus tenuifolia, Picea* sp., *Calamagrostis canadensis, Carex livida, C. aquatilis, C. pluriflora, C. limosa, C. sitchensis, C. magellanica, C. canescens, C. lyngbyaei, Trichophorum caespitosum, Potentilla palustris, Menyanthes trifoliata, Equisetum* spp., *Utricularia* spp., and *Sphagnum* spp.	Viereck et al., 1992
Open low alder shrub; S.W., S. central and interior Alaska	*Alnus crispa* and *A. tenuifolia*	*Betula nana, Ledum decumbens, Empetrum nigrum, Vaccinium uliginosum, V. vitis-idaea, Arctostaphylos alpina, Carex bigelowii, Eriophorum vaginatum, Hylocomium splendens, Aulacomnium* spp., *Tomenthypnum nitens*, and *Sphagnum* spp.	Viereck et al., 1992

TABLE 9.6
Examples of Palustrine Forested Wetland Plant Communities in the Northeastern and Midwestern U.S.

Wetland Type and Location	Dominant Plants	Associated Vegetation	Source
Northern white cedar swamp; Maine	*Thuja occidentalis*	*Larix laricina, Picea rubens, Picea mariana, Acer rubrum, Ilex verticillata, Alnus rugosa, Vaccinium corymbosum, Viburnum cassinoides, Rhododendron canadense, Ribes* spp., *Spiraea latifolia, Symplocarpus foetidus, Cornus canadensis, Carex trisperma, Carex stricta, Calamagrostis canadensis, Glyceria canadensis, Trientalis borealis, Maianthemum canadense, Rubus hispidus, Vaccinium macrocarpon, Platanthera clavellata, Malaxis unifolia,* and *Listera cordata*	Calhoun et al., 1994
Forested wetland; N. Minnesota	*Thuja occidentalis, Alnus rugosa, Fraxinus nigra, Larix laricina,* and *Picea mariana*	*Sphagnum* spp., *Gaultheria hispidula, Vaccinium oxycoccus, Ledum groenlandicum, Trientalis borealis, Smilacina trifolia, Calamagrostis canadensis, Impatiens biflora, Coptis trifolia, Mitella nuda, Linnaea borealis, Cornus canadensis, Rubus pubescens,* and *Carex trisperma*	Heinselman, 1970
Forested fen; N. Minnesota	*Larix laricina* and *Picea mariana*	*Carex pseudo-cyperus, Aronia melanocarpa, Rubus pubescens,* and *Lonicera villosa*	Glaser, 1987
Forested bog; N.E. Pennsylvania	*Larix laricina* and *Picea mariana*	*Acer rubrum, Vaccinium corymbosum, Ilex verticillata, Viburnum cassinoides, Vaccinium* sp., *Chamaedaphne calyculata, Ledum groenlandicum, Andromeda glaucophylla, Kalmia polifolia, Vaccinium oxycoccus, Carex trisperma, Osmunda cinnamomea, Sphagnum* spp., and others	Brooks et al., 1987
Black spruce swamp; Maine	*Picea mariana*	*Larix laricina, Acer rubrum, Nemopanthus mucronata, Rhododendron canadense, Vaccinium corymbosum, Alnus rugosa, A. crispa, Ilex verticillata, Ledum groenlandicum, Gaylussacia baccata, Kalmia angustifolia, Myrica gale, Chamaedaphne calyculata, Vaccinium macrocarpon, V. oxycoccus, Gaultheria hispidula, Osmunda cinnamomea, O. regalis, Symplocarpus foetidus, Smilacina trifolia, Iris versicolor, Drosera rotundifolia, Carex trisperma, Trientalis borealis, Sarracenia purpurea, Coptis groenlandicum, Cornus canadensis,* and *Sphagnum* spp.	Calhoun et al., 1994
Black spruce bog; N. Minnesota	*Picea mariana*	*Larix laricina, Kalmia polifolia, Andromeda glaucophylla, Ledum groelandicum, Chamaedaphne calyculata, Gaultheria hispidula, Sphagnum* spp., *Carex trisperma, Vaccinium vitis-idaea, Smilacina trifolia, Pleurozium schreberi, Dicranum* sp., and *Polytrichum strictum*	Glaser, 1987
Black spruce swamp; Wisconsin	*Picea mariana*	*Larix laricina, Thuja occidentalis, Abies balsamea, Pinus banksiana, Carex trisperma, Ledum groelandicum, Smilacina trifolia, Gaultheria hispidula, Vaccinium myrtilloides, Alnus rugosa, Cornus canadensis, Dryopteris* sp., *Kalmia polifolia, Trientalis borealis,* and others	Curtis, 1959

Wetland Type and Location	Dominant Plants	Associated Vegetation	Source
Red maple/white pine forested wetland; Rhode Island	*Acer rubrum* and *Pinus strobus*	*Betula alleghenesis, Quercus alba, Alnus* sp., *Vaccinium corymbosum, Ilex verticillata, Clethra alnifolia, Kalmia angustifolia, Viburnum recognitum, Osmunda cinnamomea, Aster* sp., *Sphagnum* spp., and *Vitis* sp.	Tiner, 1989
Red maple swamp; Rhode Island	*Acer rubrum*	*Quercus alba, Nyssa sylvatica, Betula alleghaniensis, Clethra alnifolia, Amelanchier canadensis, Ilex verticillata, Vaccinium corymbosum, Leucothoe racemosa, Rhododendron viscosum, Smilax rotundifolia, Trientalis borealis, Maianthemum canadense, Osmunda cinnamomea, Rubus hispidus, Lycopodium obscurum, Lilium superbum, Thelypteris simulata, Symplocarpus foetidus, Sphagnum* spp., and others	Golet et al., 1993
Red maple swamp; N. New Jersey	*Acer rubrum*	*Ulmus* spp., *Fraxinus americana, Quercus bicolor, Lindera benzoin, Sambucus canadensis, Rosa multiflora, Prunus pensylvanica, Ilex verticillata, Cornus amomum, Viburnum dentatum, Impatiens capensis, Geum* sp., *Solanum dulcamara, Carex stricta, Rumex* sp., *Aster novi-belgii, Eleocharis* sp., *Epilobium* sp., *Polygonum sagittatum, P. arifolium, Leersia oryzoides, Bidens* spp., *Arisaema triphyllum, Symplocarpus foetidus, Lysimachia ciliata, Toxicodendron radicans,* and *Parthenocissus quinquefolia*	Tiner, 1985a
Coastal plain swamp; E. Maryland	*Acer rubrum, Liquidambar styraciflua,* and *Fraxinus pennsylvanica*	*Magnolia virginiana, Vaccinium corymbosum, Smilax rotundifolia, Rhododendron viscosum,* and *Symplocarpus foetidus*	Tiner, 1988
Silver maple floodplain forest; central Connecticut	*Acer saccharinum*	*Populus deltoides, Fraxinus pennsylvanica, Ulmus americana, Boehmeria cylindrica, Laportea canadensis, Onoclea sensibilis, Matteucia struthiopteris,* and *Toxicodendron radicans*	Metzler and Tiner, 1992
Silver maple floodplain swamp; Wisconsin	*Acer saccharinum*	*Carya cordiformis, C. ovata, Fraxinus pennsylvanica, Quercus bicolor, Platanus occidentalis, Ulmus americana, Salix nigra, Populus deltoides, Laportea canadensis, Toxicodendron radicans, Vitis riparia, Leersia virginica, Cryptotaenia canadensis, Impatiens capensis, Elymus virginicus, Parthenocissus quinquefolia, Pilea pumila, Boehmeria cylindrica, Amphicarpa bracteata,* and others	Curtis, 1959
Pin oak-red maple forested wetland; central New Jersey	*Quercus palustris* and *Acer rubrum*	*Quercus alba, Fraxinus americana, Carya ovata, Ulmus* sp., *Populus deltoides, Viburnum dentatum, Lonicera japonica, Ligustrum vulgare, Carex* sp., *Geum* sp., *Iris versicolor, Allium vinale, Aster* spp., *Smilax* spp., and mosses	Tiner, 1985a
Bottomland hardwood forested wetland; E. Maryland	*Liriodendron tulipifera, Betula nigra,* and *Liquidambar styraciflua*	*Acer rubrum, Carpinus caroliniana, Athyrium filix-femina, Geum canadense, Polygonum* sp., *Polystichum acrosticoides, Impatiens capensis, Onoclea sensibilis, Arisaema triphyllum, Eupatoriadelphus fistulosus, Lycopus* sp., *Eupatorium perfoliatum, Viola* sp., *Thalictrum pubescens, Panicum* spp., and *Carex* sp.	Tiner and Burke, 1995

TABLE 9.6 (continued)
Examples of Palustrine Forested Wetland Plant Communities in the Northeastern and Midwestern U.S.

Wetland Type and Location	Dominant Plants	Associated Vegetation	Source
Hemlock swamp; W. Connecticut	*Tsuga canadensis*	*Acer rubrum, Pinus strobus, Vaccinium corymbosum, Ilex verticillata, Lindera benzoin, Nemopanthus mucronata, Osmunda cinnamomea, Coptis groelandicum, Symplocarpus foetidus, Mitchella repens, Trientalis borealis,* mosses, and liverworts	Metzler and Tiner, 1992
White pine swamp; Rhode Island	*Pinus strobus*	*Acer rubrum, Quercus alba, Clethra alnifolia, Viburnum cassinoides, Vaccinium corymbosum, Kalmia angustifolia, Rhododendron viscosum, Gaultheria procumbens, Osmunda cinnamomea, Symplocarpus foetidus, Similax rotundifolia, Mitchella repens, Lycopodium obscurum, Sphagnum* sp., and other mosses	Tiner, 1989
Pitch pine lowland; S. New Jersey	*Pinus rigida*	*Sassafras albidum, Betula populifolia, Acer rubrum, Vaccinium corymbosum, Clethra alnifolia, Woodwardia virginica,* and *Smilax rotundifolia*	Tiner, 1985a
Loblolly pine forested wetland; Delaware	*Pinus taeda*	*Liquidambar styraciflua, Acer rubrum, Toxicodendron radicans, Vaccinium corymbosum, Ilex glabra, Magnolia virginiana, Parthenocissus quinquefolia, Smilax rotundifolia, Ilex opaca, Carpinus caroliniana, Nyssa sylvatica,* and *Fagus grandifolia*	Tiner, 1985b

TABLE 9.7
Examples of palustrine forested wetland plant communities in the southern U.S.

Wetland Type and Location	Dominant Plants	Associated Vegetation	Source
Bottomland hardwood forest; W. Kentucky	*Acer saccharinum*	*Fraxinus pennsylvanica, Acer rubrum, Carpinus caroliniana, Carya cordiformis, C. ovata, Quercus bicolor, Ulmus americana, Lonicera japonica, Pilea pumila,* and *Eupatorium rugosum*	Taylor et al., 1982
Bottomland hardwood forest; Reelfoot Lake area of Tennessee	*Salix nigra* and *Populus deltoides*	*Campsis radicans, Toxicodendron radicans, Vitis* sp., *Ampelopsis arborea, Ampelamus albidus, Brunnichia cirrhosa, Ipomoea lacunosa, Sambucus canadensis, Cornus drummondii, Echinochloa* sp., *Vernonia* sp., and *Rubus* sp.	Shelford, 1954
Bottomland hardwood forest; Louisiana	*Acer rubrum* var. *drummondii* and *Nyssa aquatica*	*Acer negundo, Populus heterophylla, Taxodium distichum, Cornus drummondii, Salix nigra, Ulmus americana, Carya ovata, Fraxinus tomentosa, Quercus nigra, Celtis laevigata, Diospyros virginiana, Ilex decidua, Carya cordiformis, Quercus shumardii, Liquidambar styraciflua, Forestiera acuminata, Quercus nuttallii, Persea palustris, Styra americana, Quercus laurifolia, Sambucus canadensis, Toxicodendron radicans, Gelsemium sempervirens, Smilax* spp., *Ampelopsis* spp., and *Parthenocissus quinquefolia*	Conner and Day, 1976
Bottomland hardwood forest; E. Texas	*Fraxinus pennsylvanica Ulmus crassifolia, Celtis laevigata, Quercus phellos, Q. nigra, Q. lyrata, Ulmus americana, Liquidambar styraciflua,* and *Carya aquatica*	*Morus rubra, Carpinus caroliniana, Crataegus* spp., *Diospyros virginiana, Ilex opaca, Ilex decidua, Cornus drummondii, C. foemina, Sebastiania fruticosa, Halesia diptera, Ilex vomitoria, Callicarpa americana, Toxicodendron radicans, Smilax rotundifolia, Berchemia scandens,* and *Vitis rotundifolia*	Wilkinson et al., 1987
Hardwood swamp; S. Florida	*Ilex cassine* and *Persea palustris*	*Blechnum serrulatum, Myrica cerifera, Quercus laurifolia Serenoa repens, Vitis rotundifolia,* and *Gordonia lasianthus*	Florida Dept. of Environmental Protection et al., 1995
Mixed hardwood swamp; E. Texas	*Acer negundo, Quercus phellos, Quercus nigra, Celtis laevigata,* and *Carpinus caroliniana*	*Aesculus pavia, Platanus occidentalis, Betula nigra, Fraxinus pennsylvania, Betula nigra, Ulmus alata, Arundinaria gigantea, Parthenocissus quinquefolia, Berchemia scandens, Podophyllum peltatum, Sambucus canadensis, Claytonia virginica, Viola* sp., *Ranunculus abortivus, Galium* sp., and *Cardamine pennsylvanica*	*
Willow oak swamp ("pin oak flats"); E. Texas	*Quercus phellos*	*Quercus stellata, Ulmus alata, Fraxinus pennsylvanica, Crataegus* sp., *Vaccinium* sp., *Claytonia virginica, Uniola* sp., *Viola* sp., *Carex* spp., and *Smilax bona-nox*	Christensen et al., 1981
Pine flatwoods; N. Florida	*Pinus palustris*	*Pinus serotina, Pinus australis, Serenoa repens, Ilex glabra, Myrica cerifera, Rubus betulifolius, Aronia arbutifolia,* and *Smilax laurifolia*	Laessle, 1942

TABLE 9.7 (continued)
Examples of palustrine forested wetland plant communities in the southern U.S.

Wetland Type and Location	Dominant Plants	Associated Vegetation	Source
Slash pine flatwoods; N. Florida	*Pinus elliottii*	*Andropogon glomeratus, Aronia arbutifolia, Boehmeria cylindrica, Cyrilla racemiflora, Eupatorium perfoliatum, Euthamia* sp., *Ilex glabra, Osmunda cinnamomea, Panicum dichotomum, Persea palustris, Pluchea* sp., *Rhexia nuttallii, Serenoa repens, Smilax glabra,* and *Solidago fistulosa*	Florida Dept.of Protection et al., 1995
Wet flatwoods; S. Florida	*Pinus elliottii* var. *densa, Sabal palmetto*	*Ilex cassine, Persea palustris, Myrica cerifera, Serenoa repens, Amphicarpum mulhenbergianum, Aristida* spp., *Rhynchospora* spp., *Panicum hemitomom, Scleria* spp., *Lachnanthes caroliniana, Xyris* spp., *Pontederia cordata, Aletris lutea, Drosera* spp., *Polygala* spp., *Hypericum* spp., *Sabatia* spp., and*Eriocaulon* spp.	Miller and Gunsalus, 1997
Pond pine seepage; N.E. Florida	*Pinus serotina*	*Gordonia lasianthus, Magnolia virginiana australis, Persea palustris, Pinus elliottii, Aristida stricta, Arundinaria gigantea, Gaylussacia frondosa, Ilex coriacea, Ilex glabra, Kalmia hirsuta, Lyonia lucida, Myrica cerifera, Osmunda cinnamomea, Pteridium aquilinum, Rhododendron canescens, Serenoa repens, Toxicodendron vernix, Vaccinium corymbosum,* and *Vaccinium myrsinites*	Florida Dept. of Environmental Protection et al., 1995
Bayhead; N. Florida	*Gordonia lasianthus, Persea palustris,* and *Magnolia virginiana*	*Ilex glabra, Lyonia lucida, Myrica cerifera, Smilax laurifolia, Osmunda cinnamomea, Woodwardia virginica,* and *Sphagnum* spp.	Laessle, 1942
Bayhead swamp; N.W. Florida	*Gordonia lasianthus, Magnolia virginiana, Nyssa sylvatica,* and *Taxodium ascendens*	*Cliftonia monophylla, Aronia arbutifolia, Clethra alnifolia, Cyrilla racemiflora, Gordonia lasianthus, Ilex coriacea, Ilex glabra, Lyonia ligustrina, Lyonia lucida, Lyonia mariana, Pteridium aquilinum, Rhododendron viscosum, Scleria* spp., *Serenoa repens, Smilax laurifolia, Sphagnum* spp., *Vaccinium corymbosum,* and *Vitis rotundifolia*	Florida Dept. of Environmental Protection et al., 1995
River swamp; coastal North Carolina	*Nyssa aquatica*	*Taxodium distichum, Fraxinus caroliniana, Saururus cernuus, Sagittaria* sp., *Peltandra virginica, Smilax* sp., *Ludwigia palustris, Nitella flexilis, Hydrocotyle* sp., *Fontinalis* sp., and algae	Brinson, 1977
Black gum swamp; Dismal Swamp, Virginia	*Nyssa sylvatica*	*Acer rubrum, Taxodium distichum, Nyssa aquatica, Chamaecyparis thyoides, Fraxinus caroliniana, Quercus phellos, Pinus taeda, Pinus serotina, Ilex opaca, Magnolia virginiana, Persea borbonia, Liriodendron tulipifera, Salix* sp., *Fagus grandifolia,* and *Ulmus* sp.	Whitehead, 1972
Strand swamp; N.W. Florida	*Nyssa sylvatica biflora, Acer rubrum,* and *Ulmus americana floridana*	*Magnolia virginiana australis, Taxodium ascendens, Sabal palmetto, Pinus elliottii, Amphicarpum muhlenbergianum, Campsis radicans, Carex* spp., *Cladium jamaicense, Clematis crispa, Hibiscus moscheutos, Hypericum hypericoides, Lycopus rubellus, Osmunda cinnamonea, Osmunda regalis,*	Florida Dept. of Environmental Protection et al., 1995

Wetland Type and Location	Dominant Plants	Associated Vegetation	Source
		Panicum dichotomum, Rhynchospora spp., *Sabal palmetto, Sagittaria graminea, Saururus cernuus, Smilax laurifolia, Thelypteris* spp., and *Toxicodendron radicans*	
River swamp; W. central Florida	*Acer rubrum* and *Nyssa sylvatica biflora*	*Pinus elliottii, Cephalanthus occidentalis, Magnolia virginiana, Persea palustris, Gordonia lasianthus, Fraxinus* sp., plus numerous herbs	Clewell et al., 1982
River swamp; central Florida	*Carpinus caroliniana, Liquidambar styraciflua, Pinus ellottii, Pinus serotina, Quercus laurifolia,* and *Sabal palmetto*	*Axonopus furcatus, Carex albolutescens, Centella asiatica, Chasmanthium sessiliflorum, Hypericum hypericoides, Hypoxis leptocarpa, Mitchella repens, Oxalis* sp., and *Panicum* sp.	Florida Dept. of Environmental Protection et al., 1995
Cypress swamp; Okefenokee Swamp, Georgia	*Taxodium distichum*	*Tillandsia usneoids* (epiphyte), *Lyonia lucida, Nyssa sylvatica* var. *biflora, Clethra alnifolia, Itea virginica, Leucothoe racemosa, Cyrilla racemiflora, Ilex cassine, Pieris phyllyreifolia, Decodon verticillatus, Smilax walteri,* and *Eriocaulon compressum*	Schlesinger, 1978
Cypress dome swamp; central Florida	*Taxodium ascendens*	*Agalinis* sp., *Andropogon virginicus glaucus, Bacopa caroliniana, Bigelowia nudata, Carex* spp., *Centella asiatica, Cyperus haspan, Erianthus giganteus, Eriocaulon compressum, Eriocaulon decangulare, Eupatorium leptophyllum, Gratiola* spp., *Helenium pinnatifidum, Hypericum cistifolium, Juncus repens, Juncus polycephalus, Jucus trigonocarpus, Lachinanthes caroliniana, Ludwigia virgata, Myrica cerifera, Panicum hemitomon, Panicum longifolium, Paspalum praecox, Pluchea rosea, Polygala cymosa, Polygonum hydropiperoides, Pontederia cordata, Rhexia* sp., *Rhynchospora* spp., *Sagittaria graminea, Scleria* sp., *Sphagnum* spp., *Utricularia inflata, Utricularia purpurea, Woodwardia virginica, Xyris elliotii,* and *Xyris fimbriata*	Florida Dept. of Environmental Protection et al., 1995
Cypress swamp; S. Florida	*Taxodium* spp.	*Salix caroliniana, Nyssa sylvatica* var. *biflora, Acer rubrum, Cephalanthus occidentalis, Myrsine guianensis, Lyonia lucida, Osmunda regalis, O. cinnamomea, Blechum serrulatum, Woodwardia* spp., *Thelypteris* spp., *Peltandra virginica, Saururus cernuus, Pontederia cordata,* and *Sphagnum* spp.	Miller and Gunsalus, 1997
Cypress swamp; Corkscrew Swamp, Florida	*Taxodium distichum*	*Acer rubrum, Ficus aurea, Fraxinus caroliniana, Annona glabra, Cephalanthus occidantaltis, Persea borbonia, Ilex* sp., *Myrica cerifera, Salix caroliniana,* epiphytic ferns, orchids and bromeliads	Duever et al., 1984
Atlantic white cedar swamp; Alligator River, North Carolina	*Chamaecyparis thyoides*	*Clethra alnifolia, Vaccinium corymbosum, Gaylussacia frondosa, Ilex coriacea, Ilex glabra, Viburnum nudum, Lyonia lucida, Lyonia ligustrina, Myrica heterophylla, Mitchella repens, Peltandra virginica, Woodwardia areolata,* and *Sphagnum* spp.	Moore and Carter, 1984
Pocosin; coastal North Carolina	*Pinus serotina, Taxodium distichum, Acer rubrum,* and *Nyssa sylvatica* var. *biflora*	*Cyrilla racemiflora* and *Lyonia lucida*	Christensen et al., 1981

* Author's observations.

TABLE 9.8
Examples of Palustrine Forested Wetland Plant Communities in the Western States, Except the Pacific Northwest

Wetland Types and Locations	Dominant Plants	Associated Vegetation	Source
Western red cedar forest; N. Idaho	*Thuja plicata* and *Athyrium filix-femina*	*Pinus monticola, Abies grandis, Picea engelmannii,* and *Galium trifolium*	Cooper et al., 1983; Daubermire and Daubermire, 1968
Blue spruce forest; mountains of S. Utah	*Picea pungens* and *Equisetum arvense*	*Picea engelmannii, Populus tremuloides, Geranium richardsonii, Thalictrum fendleri,* and *Osmorhiza chilensis*	Youngblood, 1984
Subalpine fir forest; mountains of N. Montana	*Abies lasiocarpa* and *Oplopanax horridum*	*Picea engelmannii, Pinus monticola, Larix occidentalis,* and *Taxus brevifolia*	Pfister et al., 1977
Subalpine fir forest; mountains of central and S. Idaho	*Abies lasiocarpa* and *Calamagrostis canadensis*	*Picea engelmanni, Pinus contorta, Populus tremuloides, Geranium triflorum, Vaccinium caespitosum, Ledum glandulosum,* and *Senecio triangularis*	Steele et al., 1981
Blue spruce forest; mountains of S. Utah	*Picea pungens* and *Equisetum arvense*	*Populus tremuloides, Pseudotsuga menziesii, Geranium richardsonii, Thalictrum fendleri,* and *Osmorhiza chilensis*	Youngblood, 1984
Engelmann spruce forest; mountains of N.W. Wyoming and E. Idaho	*Picea engelmannii* and *Equisetum arvense*	*Pinus contorta, Abies lasiocarpa, Streptopus amplexifolius, Senecio triangularis,* and *Luzula parviflora*	Steele et al., 1983
Subalpine fir forest; mountains of N. Utah	*Abies lasiocarpa* and *Calamagrostis canadensis*	*Picea engelmannii, Pinus contorta, Populus tremuloides, Linnaea borealis, Equisetum arvense,* and *Geum triflorum*	Mauk and Henderson, 1984
Alder forest; streambanks, montane zone, N. central Colorado	*Alnus tenuifolia* and *Equisetum arvense*	*Betula occidentalis, Populus tremuloides, Salix* spp., *Rosa woodsii,* and *Acer glabrum*	Hess, 1981
Cottonwood forest; streambanks, montane zone, N. central Colorado	*Populus angustifolia* and *Salix exigua*	*Populus tremuloides, Juniperus scopulorum, Salix* spp., and *Acer glabrum*	Hess, 1981
Quaking aspen forest; mountains of N.W. Colorado	*Populus tremuloides* and *Veratrum tenuipetalum*	*Mertensia ciliata, Ligusticum porteri,* and *Bromus ciliatus*	Hoffman and Alexander, 1980
Blue spruce forest; mountains of S. Utah	*Picea pungens* and *Equisetum arvense*	*Pica engelmannii, Pseudotsuga menziesii, Geranium richardsonii, Thalictrum fendleri,* and *Osmorhiza chilensis*	Youngblood, 1984
Subalpine fir forest; mountains of central and S. Idaho, W. Wyoming N. Utah, and N. central Colorado	*Abies lasiocarpa* and *Calamagrostis canadensis* (*Picea engelmanni* and *C. canadensis*)	*Pinus contorta, Picea engelmannii, Picea menziessi, Picea pungens, Senecio triangularis,* and *Galium triflorum*	Hess, 1981; Mauk and Henderson, 1984; Steele et al., 1981

Wetland Types and Locations	Dominant Plants	Associated Vegetation	Source
Fir-spruce forest; mountains of central and S. Colorado	*Abies lasiocarpa-Picea engelmannii* and *Cardamine cordifolia* (*A. lasiocarpa* and *Mertensia ciliata*)	*Picea engelmannii*, *Mertensia ciliata*, *Mitella pentandra*, and *Carex bella*	Delvice et al., 1984; Steen and Dix, 1974
Alder forest; streambanks, montane zone, N. central Colorado	*Alnus tenuifolia* and *Equisetum arvense*	*Betula occidentalis*, *Picea pungens*, *Salix* spp., *Rosa woodsii*, and *Acer glabrum*	Hess, 1981
Cottonwood forest; streambanks, montane zone, N. central Colorado	*Populus angustifolia* and *Salix exigua*	*Juniperus scopulorum*, *Picea pungens*, *Salix* spp., and *Acer glabrum*	Hess, 1981
Subalpine fir forest; high mountains of Colorado	*Abies lasiocarpa* and *Salix pseudolapponum* (*A. lasiocarpa-Picea engelmannii* and *S. glauca*)	*Picea engelmanni*, *Pinus flexilis*, *Vaccinium myrtillus*, *Polemonium pulcherrimum*, and *Acomastylis rossii*	Hess, 1981; Hess and Wasser, 1982; Komarkova, 1984
Subalpine fir forest; mountains of central Montana, Idaho, N.W. Wyoming, N. Utah, and central Colorado	*Abies lasiocarpa* and *Calamagrostis canadensis* (*Picea engelmannii* and *C. canadensis*)	*Picea engelmanni*, *Populus tremuloides*, *Picea pungens*, and *Pseudotsuga menziessi Vaccinium caespitosum,Ledum glandulosum*, *Senecio triangularis*, and *Galium triflorum*	Cooper et al., 1983; Hess, 1981; Mauk and Henderson, 1984; Pfister et al., 1977; Steele et al., 1981; 1983
Subalpine fir forest; mountains of central Idaho	*Abies lasiocarpa* and *Caltha biflora*	*Picea engelmannii*, *Lonicera involucrata*, *Pedicularis bracteosa*, and *Dodecatheon jeffreyi*	Steele et al., 1981; 1983
Engelmann spruce forest; mountains of Montana, central Idaho, N.W. Wyoming, N. Utah, central and S. Colorado	*Picea engelmannii* and *Equisetum arvense*	*Abies lasiocarpa*, *Picea pungens*, *Equisetum scirpoides*, *Streptopus amplexifolium*, *Senecio triangularis*, and *Luzula parviflora*	Mauk and Henderson, 1984; Pfister et al., 1977; Steele et al., 1983
Quaking aspen forest; mountains of W. Wyoming	*Populus tremuloides* and *Equisetum arvense*	*Picea engelmannii*, *Abies lasiocarpa*, *Elymus glaucus*, and *Thalictrum fendleri*	Youngblood and Mueggler, 1981
Engelmann spruce forest; mountains of Montana, central Idaho, N.W. Wyoming, and N. Utah	*Picea engelmannii* and *Equisetum arvense*	*Abies lasiocarpa*, *Pinus contorta*, *Picea pungens*, *Equisetum scirpoides*, *Streptopus amplexifolius*, *Senecio triangularis*, and *Luzula parviflora*	Mauk and Henderson, 1984; Pfister et al., 1977; Steele et al., 1981; 1983
Subalpine fir forest; mountains of N. Montana	*Abies lasiocarpa* and *Oplopanax horridum*	*Pinus monticola*, *Pseudotsuga menziesii*,*Larix occidentalis*, and *Taxus brevifolia*	Pfister et al., 1977
Subalpine fir forest; high mountains of Colorado	*Abies lasiocarpa* and *Salix pseudolapponum* (*A. lasiocarpa-Picea engelmannii* and *S. glauca*)	*Pinus contorta*, *Pinus flexilis*, *Vaccinium myrtillus*, *Polemonium pulcherrimum*, and *Acomastylis rossill*	Hess, 1981; Hess and Wasser, 1982; Komarkova, 1984
Subalpine fir forest; mountains of central Montana, Idaho, N.W. Wyoming, N. Utah, N. central and W. Colorado	*Abies lasiocarpa* and *Calamagrostis canadensis;* *A. lasiocarpa-Picea engelmannii* and *C. canadensis* (*P. engelmannii* and *C. canadensis*)	*Pinus contorta*, *Populus tremuloides*, *Pseudotsuga menziesii*, *Picea pungens*, *Vaccinium caespitosum*, *Ledum glandulosum*, *Senecio triangularis*, and *Galium triflorum*	Cooper et al., 1983; Hess, 1981; Komarkova, 1984; Mauk and Henderson, 1984; Pfister et al., 1977; Steele et al., 1981; 1983

TABLE 9.8 (continued)
Examples of Palustrine Forested Wetland Plant Communities in the Western States, Except the Pacific Northwest

Wetland Types and Locations	Dominant Plants	Associated Vegetation	Source
Subalpine fir forest; mountains of central Idaho	*Abies lasiocarpa* and *Caltha biflora*	*Pinus contorta, Lonicera involucrata, Pedicularis bracteosa*, and *Dodecatheon jeffreyi*	Steele et al., 1981
Fir-spruce forest; mountains of central and S. Colorado	*Abies lasiocarpa-Picea engelmannii* and *Cardamine cordifolia* (*A. lasiocarpa* and *Mertensia ciliata*)	*Populus tremuloides, Cardamine cordifolia, Mitella pentandra*, and *Carex bella*	DeVelice et al., 1984; Steen and Dix, 1974
Fir-spruce forest; mountains of central and W. Colorado	*Abies lasiocarpa* and *Senecio triangularis, Picea engelmannii* and *S. triangularis* (*P. engelmannii* and *S. triangularis*)	*Mertensia ciliata, Cardamine cordifolia*, and *Equisetum arvense*	Hess, 1981; Komarkova, 1984
Blue spruce forest; mountains of S. Utah	*Picea pungens* and *Equisetum arvense*	*P. menziesii, Populus tremuloides, Geranium richardsonii, Thalictrum fendleri*, and *Osmorhiza chilensis*	Youngblood, 1984
Quaking aspen forest; mountains of W. Wyoming	*Populus tremuloides* and *Equisetum arvense*	*Pinus contorta, Elymus glaucus*, and *Thalictrum fendleri*	Youngblood and Mueggler, 1981
Quaking aspen forest; mountains of W. Wyoming	*Populus tremuloides* and *Ranunculus alismaefolius*	*Abies lasiocarpa, Carex microptera*, and *Trifolium longipes*	Youngblood and Mueggler, 1981
Subalpine fir forest; mountains of N. Montana	*Abies lasiocarpa* and *Oplopanax horridum*	*Picea engelmannii, Pinus monticola, Larix occidentalis*, and *Taxus brevifolia*	Pfister et al., 1977
Subalpine fir forest; high mountains of Colorado	*Abies lasiocarpa* and *Salix pseudolapponum, A. lasiocarpa-Picea engelmannii* and *S. glauca*	*Pinus contorta, Pinus flexilis, Vaccinium myrtillus, Polemonium pulcherrimum* and *Acomastylis rossii*	Hess, 1981; Hess and Wasser, 1982; Komarkova, 1984
Fir-spruce forest; mountains central Montana, Idaho, N.W. Wyoming, N. Utah N. central and W. Colorado	*Abies lasiocarpa* and *Calamagrostis canadensis; A. lasiocarpa, Picea engelmannii*, and *C. canadensis* (*P. engelmannii* and *C. canadensis*)	*Picea contorta, Populus tremuloides, Pseudotsuga menziessii (ID), Picea pungens (UT), Galium triflorum, Vaccinium caespitosum, Ledum glandulosum*, and *Senecio triangularis*	Cooper et al., 1983; Hess, 1981; Komarkova, 1984; Mauk and Henderson, 1984; Pfister et al., 1977; Steele et al., 1981; 1983
Subalpine fir forest; mountains of central Idaho	*Abies lasiocarpa* and *Caltha biflora*	*Lonicera involucrata, Pedicularis bracteosa*, and *Dodecatheon jeffreyi*	Steele et al., 1981
Fir-spruce forest; mountains of central and S. Colorado	*Abies lasiocarpa-Picea engelmannii* and *Cardamine cordifolia* (*A. lasiocarpa* and *Mertensia ciliata*)	*Mitella pentandra* and *Carex bella*	DeVelice et al., 1984; Steen and Dix, 1974

Wetland Types and Locations	Dominant Plants	Associated Vegetation	Source
Fir-spruce forest; mountains of N. central and W. Colorado	*Abies lasiocarpa* and *Senecio triangularis, A. lasiocarpa-Picea engelmannii* and *S. triangularis* (*P. engelmannii* and *S. triangularis*)	*Cardamine cordifolia, Equisetum arvense,* and *Mertensia ciliata*	Hess, 1981; Komarkova, 1984
Subalpine fir forest; mountains of Idaho and N.W. Utah	*Abies lasiocarpa* and *Streptopus amplexifolius*	*Picea engelmannii* and *Pinus contorta*	Cooper et al., 1983; Steele et al., 1981; 1983
Engelmann spruce forest; mountains of S. Montana, N.W. Wyoming, central Idaho	*Picea engelmanni* and *Equisetum arvense*	*Pinus contorta, Picea pungens, Streptopus amplexifolius, Senecio triangularis,* and *Luzula parviflora*	Mauk and Henderson, 1984; Steele et al., 1981; 1983
Western red cedar forest; mountains of N.W. Montana	*Thuja plicata* and *Oplopanax horridum*	*Tsuga heterophylla, Tsuga mertensiana, Pinus monticola, Athyrium filix-femina,* and *Gymnocarpium dryopteris*	Pfister et al., 1977
Quaking aspen forest; Wyoming and Wind River Mountains, W. Wyoming	*Populus tremuloides* and *Equisetum arvense*	*Picea engelmannii, Abies lasiocarpa, Pinus contorta, Equisetum arvense, Elymus glaucus,* and *Thalictrum fendleri*	Youngblood and Mueggler, 1981
Quaking aspen forest; Wyoming Mountains, W. Wyoming	*Populus tremuloides* and *Ranunculus alismaefolius*	*Abies lasiocarpa, Picea engelmannii, R. alismaefolius, Carex* spp., *Claytonia lanceolata,* and *Trifolium longipes*	Youngblood and Mueggler, 1981
Engelmann spruce forest; Absaroka and Wind River Mountains, N.W. Wyoming	*Picea engelmannii* and *Caltha leptosepala*	*Abies lasiocarpa* (minor climax) *Pinus contorta, Caltha leptosepala, Mitella pentandra, Parnassia fimbriata, S. arguta, Senecio triangularis,* and *Trollius laxus*	Steele et al., 1983
Engelmann spruce forest; Absaroka and Wind River Mountains, N.W. Wyoming	*Picea engelmannii* and *Equisetum arvense*	*Picea pungens, Pinus contorta, Abies lasiocarpa, Luzula parviflora, Carex* spp., *Juncus* spp., *Parnassia fimbriata, Senecio triangularis,* and *Streptopus amplexifolius*	Steele et al., 1983
Subalpine fir forest; Yellowstone Plateau, N.W. Wyoming	*Abies lasiocarpa* and *Calamagrostis canadensis* (*Ledum glandulosum* phase and *Vaccinium caespitosum* phase)	*Picea engelmannii, Pinus contorta,* and *Carex* spp.	Steele et al., 1983
Cottonwood-willow forest; floodplains of Snake, Wind, Green and Shoshone Rivers, N.W. Wyoming	*Populus* spp. and *Salix* spp.	*Picea engelmannii, Picea pungens, Betula occidentalis, Crateagus douglasii,* and *Elaeagnus commutata*	Steele et al., 1983
Balsam poplar forest; Boise River, Idaho	*Populus balsamifera* var. *trichocarpa* and *Equisetum arvense-Agropyron repens*	*Acer saccharinum, Rosa woodsii, Salix lutea, Salix* spp., *Alnus sinuata, Poa pratensis, Dacylis glomerata, Agrostis stolonifera, Solidago canadensis, Asclepias syriaca,* and others	Bill Sipple, pers. comm., 1998

TABLE 9.8 (continued)
Examples of Palustrine Forested Wetland Plant Communities in the Western States, Except the Pacific Northwest

Wetland Types and Locations	Dominant Plants	Associated Vegetation	Source
Alder forest; San Jose Creek, California	*Alnus rhombifolia*	*Salix laevigata, Salix lasiolepis, Rubus ursinus, Toxicodendron diversilobum, Cyperinus alternifolius,* and *Pennisetum claudestinum*	Faber et al., 1989
Riparian box elder forest; Arizona and New Mexico	*Acer negundo*	*Amorpha fruticosa, Cornus stolonifera, Rhus radicans, Rosa* spp., *Salix irrorata, Clematis ligusticifolia, Parthenocissus inserta, Vitis arizonica, Agropyron* spp., *Aster* spp., *Bromus* spp., *Carex* spp., *Equisetem* spp., *Geranium* spp., *Marrubium vulgare, Melilotus alba, Mentha* spp., *Poa pratensis, Taraxacum officinale,* and *Verbascum thapsus*	Szaro, 1989
Riparian alder forest; Arizona and New Mexico	*Alnus oblongifolia*	*Acer negundo, Fraxinus velutina, Parthenocissus inserta, Vitis arizonica, Aquilegia chrysantha, Bromus* spp., *Carex* spp., *Cucurbita foetidissima, Equisetum* spp., *Fragaria ovalis, Galium triflorum, Geranium* spp., *Meliotus officinalis, Mimulus guttatus, Pteridium aquilinum,* and *Thalictrum fendleri*	Szaro, 1989
Riparian ash forest; Arizona and New Mexico	*Fraxinus pennsylvanica*	*Salix gooddingii, Salix bonplandiana,* and *Platanus wrightii*	Szaro, 1989
Riparian walnut-sycamore forest; Arizona and New Mexico	*Juglans major-Platanus wrightii*	*Alnus oblongifolia, Populus angustifolia, Cupressus arizonica, Celtis reticulata, Rhus radicans, Vitis arizonica, Aquilegia chrysantha, Carex* spp., *Cynodon dactylon, Elymus glaucus, Leptochloa dugia, Melilotus alba, Mimulus guttatus, Rudbeckia lacinata, Rumex acetosella,* and *Vicia americana*	Szaro, 1989
Riparian cottonwood forest; Arizona and New Mexico	*Populus fremontii*	*Bouteloua curtipendula, Cynodon dactylon, Melilotus officinalis, Fraxinus pennsylvanica, Salix bonplandiana, Salix gooddingii,* and *Tamarix pentandra*	Szaro, 1989

Wetland Types and Locations	Dominant Plants	Associated Vegetation	Source
Riparian cottonwood-willow forest; Arizona and New Mexico	*Populus fremontii-Salix gooddingii*	*Baccharis salicifolia, Prosopis velutina,* and *Cynodon dactylon*	Szaro, 1989
Riparian willow forest; Arizona and New Mexico	*Salix bonplandiana*	*Baccharis salicifolia, Fraxinus pennsylvanica, Populus fremontii,* and *Vitis arizonica*	Szaro, 1989
Riparian willow forest; Arizona and New Mexico	*Salix gooddingii*	*Populus fremontii, Prosopis velutina, Tamarix pentandra, Baccharis salicifolia, Cynodon dactylon,* and *Rorippa nasturtium-aquaticum*	Szaro, 1989
Riparian salt cedar forest; Arizona and New Mexico	*Tamarix pentandra*		Szaro, 1989
Riparian desert palm forest; Arizona and California	*Washingtonia filifera*	*Cynodon dactylon, Franseria tenuifolia, Penstemon* spp., and *Sphaeralcea* spp.	Szaro, 1989
Riparian alder forest; Arizona and New Mexico	*Alnus tenuifolia*	*Rosa* spp., *Achillea lanulosa, Carex* spp., *Equisetum* spp., *Helianthus annuus, Juncus* spp., *Poa pratensis, Taraxacum officinale, Picea pungens, Salix irrorata, Salix lasiolepis,* and *Salix lutea*	Szaro, 1989

Note: Dominant plants include overstory and understory dominants. Although all communities were designated as "wet" forests by the various sources, some nonwetland riparian forests may be included in the list.

TABLE 9.9
Examples of Palustrine Forested Wetland Communities in the Pacific Northwest

Wetland Type and Location	Dominant plants	Associated Vegetation	Source
Engelmann spruce forest; N. central Washington	*Picea engelmannii* and *Equisetum arvense*	*Populus tremuloides, Ribes lacustre, Alnus sinuata, Cornus canadensis, Linnaea borealis, Smilacina stellata, Galium triflorum, Carex* spp., *Thalictrum occidentale, Mitella* sp., *Fragaria* sp., *Viola glabrella*, and *Equisetum scirpoides*	Williams and Lillybridge, 1983
Subalpine fir forest; N. central Washington	*Abies lasiocarpa*	*Ledum glandulosum, Carex* spp., and mosses	Williams and Lillybridge, 1983
Lodgepole pine forest; central Oregon	*Pinus contorta, Spiraea douglasii*, and *Carex eurycarpa*	*Rosa woodsii, Salix boothii, S. geyeriana S. lemmonii, Elymus glaucus, Fragaria virginiana, Polemonium occidentale, Trifolium longipes, Smilacina stellata, Senecio pseudaureus*, and *Vicia americana*	Kovalchik, 1987
Lodgepole pine forest; central Oregon	*Pinus contorta, Vaccinium occidentalis*, and *Carex eurycarpa*	*Salix* spp., *Spiraea douglassi, Vaccinium caespitosum, Fragaria virginiana, Ribes lacustre, Lonicera caerulea, Carex* sp., *Calamagrostis canadensis, Equisetum arvense, Sidaleca oregana Mimulus guttatus*, and *Sphagnum* sp.	Kovalchik, 1987
Lodgepole pine forest; central Oregon	*Pinus contorta* and *Carex eurycarpa*	*Salix geyeriana, S. douglasii, Poa pratensis, Calamagrostis* sp., *Juncus balticus, Achillea millefolium, Fragaria virginiana, Geum macrophyllumn, Epilobium watsonii, Poa palustris, Elymus glaucus, Mimulus guttatus, Trifolium longipes*, and *Smilacina stellata*	Kovalchik, 1987
Lodgepole pine forest; central Oregon	*Pinus contorta* and *Carex aquatilis*	*Elymus glaucus, Hordeum brachyantherum, Poa pratensis, Achillea millefolium, Geum macrophyllum, Ligusticum grayii, Potentilla gracilis, Trifolium longipes, Ribes lacustre, Deschampsia cespitosa, Carex vesicaria, C. lasiocarpa*, and *Equisetum arvense*	Kovalchick, 1987
Pine-spruce forest; *central Oregon	*Pinus contorta, Picea engelmannii*, and *Eleocharis pauciflora*	*Betula glandulosa, Salix* spp., *Vaccinium* spp., *Deschampsia cespitosa, Carex aquatilis, Carex muricata, Aster foliaceus, Fragaria virginiana, Habenaria dilitata, Pedicularis groenlandica, Saxifraga oregana, Trifolium longipes*, and mosses	Kovalchick, 1987
Engelmann spruce forest; *central Oregon	*Picea engelmannii, Vaccinium occidentalis*, and *Carex eurycarpa*	*Pinus contorta, Salix* spp., *Spirea douglasii, Calamagrostis canadensis, Carex scopulorum, Carex jonesii, Dodecatheon* spp., *Equisetum arvense*, and *Habenaria dilitata*	Kovalchick, 1987
Engelmann spruce forest; central Oregon	*Picea engelmannii* and *Carex eurycarpa*	*Pinus contorta, Vaccinium occidentalis, Calamagrostis* spp., *Glyceria elata, Equisetum arvense, Fragaria virginiana, Gaultheria* spp., and *Viola* spp.	Kovalchick, 1987
Engelmann spruce forest; central Oregon	*Picea engelmannii* and *Equisetum arvense*	*Abies lasiocarpa, Abies concolor, Scirpus microcarpus, Acontium columbianum, Mitella pentandra, Galium triflorum, Gaultheria* spp., *Senecio triangularis*, and *Streptopus amplexifolius*	Kovalchick, 1987

Wetland Type and Location	Dominant plants	Associated Vegetation	Source
Quaking aspen forest; central Oregon	*Populus tremuloides* and *Pinus contorta*	*Spiraea douglasii, Salix* sp. *Rosa woodsii, Elymus glaucus, Carex eurycarpa, Poa pratensis, Fragaris virginiana,* and *Smilacina stellata*	Kovalchick, 1987
Silver fir forest; Olympic National Forest, Washington	*Abies amabilis* and *Oplopanax horridum*	*Tsuga heterophylla, Thuja plicata, Pseudotsuga menziesii, Taxus brevifolia, Tsuga mertensiana, Vaccinium alaskaense, Tiarella trifoliata, Rubus spectabilis, Athyrium filix-femina, Blechnum spicant, Polystichum munitum, Clintonia uniflora, Vaccinium parviflorum, Achlys triphylla, Streptopus amplexifolius, Rubus pedatus,* and *Trillium ovatum*	Henderson et al., 1989
Silver fir forest; Olympic National Forest, Washington	*Abies amabilis* and *Lysichitum americanum*	*Tsuga heteropylla, Thuja plicata, Alnus rubus, Vaccinium, alaskaense, Gaultheria shallon, Vaccinium ovalifolium, Menziesia ferruginea, Coptis laciniata, Blechnum spicant, Maianthemum dilatatum, Rubus pedatus, Cornus canadensis, Streptopus roseus, Clintonia uniflora, Linnaea borealis, Polypodium glycyrrhiza, Selaginella oregana,* and *Streptopus amplexifloius*	Henderson et al., 1989
Western hemlock forest; Olympic National Forest, Washington	*Tsuga heterophylla* and *Oplopanax horridum*	*Pseudotsuga menziesii, Thuja plicata, Acer macrophyllum, Alnus rubra, Polystichum munitum, Tiarella trifoliata, Athyrium filix-femina, Vaccinium parviflorum, Achlys triphylla, Acer circinatum, Smilacina stellata, Gymnocarpium dryopteris, Galium triflorum, Gaultheria shallon, Rubus ursinus Dryopteris austriaca, Trientalis latifolia,* and *Trillium ovatum*	Henderson et al., 1989
Western hemlock swamp; Olympic National Forest, Washington	*Tsuga heterophylla* and *Lysichitum americanum*	*Thuja plicata, Rhamnus purshiana, Rubus spectabilis, Boykinia major, Vaccinium alaskaense, Acer circinatum, Tiarella trifoliata, Maianthemum dilatatum, Menziesia ferruginea, Rubus pedatus, Vaccinium parviflorum, Gaultheria shallon,* and *Polystichum munitum*	Henderson et al., 1989
Hardwood forested wetland; W. Washington	*Fraxinus latifolia, Populus balsamifera,* and *Alnus rubra*	*Cardamine oligospermum, Carex obnupta, C. stipata, Athyrium filix-femina, Cornus sericea, Equisetum telmateia, Juncus effusus, J. ensifolius, Lonciera involucrata, Lysichtum americanum, Malus fusca, Oenanthe sarmentosa, Phalaris arundinacea, Ranunculus repens, Ribes divericatum, Rubus spectabilis, R. ursinus, Salix sitchensis, S. lasiandra, Scirpus microcarpus, Spiraea douglasii,* and *Utrica dioica*	Fred Weinmann, pers. comm., 1998
Western hemlock swamp; Spring Lake, W. Washington	*Tsuga heterophylla* and *Lysichitum americanum*	*Gaultheria shallon, Sphagnum* spp., *Athyrium filiz-femina, Blechunum spicant, Claytonia sibirica, Claystopteris fragilis, Dryopteris intermedia, Galium triflorum, Luzula parviflora, Menziesia ferruginea, Oemlaria cerasiformis, Pinus contorta, Pteridium aquilinum, Rubus spectabilis, R. ursinus, Sambucus racemosa, Tolmiea menziesii,* and *Vaccinium parvifolium*	Weinmann and Kunz, 1995; F. Weinmann, pers. comm., 1998

TABLE 9.9 (continued)
Examples of Palustrine Forested Wetland Communities in the Pacific Northwest

Wetland Type and Location	Dominant plants	Associated Vegetation	Source
Red alder swamp; low elevations, W. Washington	*Alnus rubra* and *Lysichitum americanum*	*Athyrium filix-femina, Carex obnupta, Cornus stolonifera, Blechum spicant, Impatiens capensis, Lonciera involucrata, Maianthemum dilatatum, Oenanthe sarmentosa, Picea sitchensis, Pteridium aquilinum, Pyrus fusca, Rubus spectabilis, Thuja plicata, Tiarella trifoliata, Tolmiea menxiessii, Tsuga heterophylla*, and *Vaccinium parvifolium*	Kunze, 1994
Red alder swamp; low elevations, W. Washington	*Alnus rubra* and *Rubus spectabilis*	*Athyrium filix-femina, Equisetum hyemale, Galium trifidum, Lonicera involucrata, Maianthemum dilatatum, Rhamnus purshiana, Ribes* sp., *Rubus ursinus, Sambucus racemosa, Thuja plicata, Tiarella trifoliata, Tolmiea menziesii*, and *Tsuga heterophylla*	Kunze, 1994
Oregon ash swamp; low elevations, W. Washington	*Fraxinus latifolia* and *Carex obnupta*	*Alnus rubra, Maianthemum dilatatum, Oenanthe sarmentosa, Pyrus fusca, Rhamnus purshiana, Rubus spectabilis, Salix* spp., *Sambucus racemosa, Sphagnum* spp., and *Spiraea douglasii*	Kunze, 1994
Oregon ash swamp; low elevations, W. Washington	*Fraxinus latifolia* and *Symphoricarpos albus*	*Carex deweyana, C. obnupta, Montia sibirica, Rubus ursinus, Elymus glaucus, Eleocharis acicularis, Galium aparine, G. trifidum, Lonicera involucrata, Luzula parviflora, Maianthemum dilatatum, Myosotis laxa, Oemlaria cerasiformis, Oenanthe sarmentosa, Polystichum munitum, Populus tremuloides, P. trichocarpa, Pyrus fusca, Quercus garryana, Rosa pisocarpa, Sambucus racemosa, Spiraea douglasii, Stachys cooleyae, Stellaria calycantha, Tellima grandiflora*, and *Veronica scutellata*	Kunze, 1994

Note: Dominant plants include both overstory and understory dominants. Although all communities were designated as "wet" forests by the various sources, some nonwetland riparian forests may be included in the list.

* Some of these forested types may be open forests mixed with shrubs or emergents.

TABLE 9.10
Examples of Palustrine Forested Wetland Plant Communities in Alaska

Wetland Type and Location	Dominant Plants	Associated Vegetation	Source
Closed mountain hemlock forest; S.W. Alaska	*Tsuga mertensiana*	*Cornus canadensis, Rubus pedatus, Coptis aspleniifolia, Blechnum spicant,* and *Fauria crista-galli*	Viereck et al., 1992
Closed black spruce forest; interior and S. central Alaska	*Picea mariana*	*Alnus crispa, Rosa acicularis, Salix* spp., *Ledum groenlandicum, Vaccinium uliginosum, V. vitis-idaea, Linnaea borealis, Ledum decumbens,* and *Empetrum nigrum*	Viereck et al., 1992
Open W. hemlock-sitka spruce forest;	*Tsuga heterophylla* and *Picea sitchensis*	*Oplopanax horridus, Vaccinium* spp., *Menziesia ferruginea, Rubus spectabilis, Lysichiton americanum, Rubus pedatus* and *Athyrium filix-femina*	Viereck et al., 1992
Open mountain hemlock forest; S. central and S.E. Alaska	*Tsuga mertensiana*	*Vaccinium alaskaense, V. ovalifolium Menziesia ferruginea, Cladothamnus Cladothamnus pyrolaeflorus, Cassiope, mertensiana, C. stelleriana, Phyllodoce aleutica* spp. *glanduliflora,* and *Luetkea pectinata*	Viereck et al., 1992
Open black spruce forest; interior and S. central Alaska	*Picea mariana*	*Vaccinium uliginosum, V. vitis-idaea, Ledum groenlandicum, Rosa acicularis, Potentilla fruticosa, Empetrum nigrum, Ledum decumbens, Alnus crispa, Betula glandulosa, Salix* spp., *Calamagrostis* spp., *Equisetum sylvaticum, Rubus chamaemorus, Eriophorum vaginatum, Carex bigelowii, Pleurozium schreberi, Hylocomium splendens, Polytrichum* spp., *Sphagnum* spp., and fruiticose and foliose lichens	Viereck et al., 1992
Lodgepole pine woodland; S.E. Alaska	*Pinus contorta, Thuja plicata,* and *Tsuga mertensiana*	*Menziesia ferruginea, Vaccinium alaskaense, V. ovalifolium, Empetrum nigrum, Vaccinium caespitosum, V. uliginosum, V. vitis-idaea, Kalmia polifolia, Andromeda polifolia, Ledum groenlandicum, Vaccinium oxycoccos, Fauria crista-galli, Trichophorum caespitosum, Eriophorum angustifolium, Cornus canadensis,* and *Sphagnum* spp.	Viereck et al., 1992
Black spruce woodland; interior, W. and S. central Alaska	*Picea mariana, Betula papyrifera,* and *Larix laricina*	*Betula glandulosa, Salix lanata, S. planifolia, S. glauca, Vaccinium uliginosum, V. vitis-idaea, Ledum, decumbens, L. groenlandicum, Empetrum nigrum, Carex* spp., *Eriophorum vaginatum, Calamagrostis canadensis, Rubus chamaemorus, Geocaulon lividum, Hylocomium splendens, Pleurozium schreberi,* and *Sphagnum* spp.	Viereck et al., 1992
Closed red alder forest; S.E. Alaska	*Alnus rubra*	*Carex macrochaeta, Calamagrostis nutkaensis, Carex lyngbyei,* and *Potentilla palustris*	Viereck et al., 1992
Closed spruce-paper birch forest; interior and S. central Alaska	*Betula papyrifera,* and *Picea glauca* or *Picea mariana*	*Alnus crispa, Salix bebbiana, S. glauca, S. scouleriana, Rosa acicularis, Vaccinium vitis-idaea, V. uliginosum, Ledum groenlandicum, Empetrum nigrum, Calamagrostis canadensis, Pleurozium schreberi,* and *Hylocomium splendens*	Viereck et al., 1992

REFERENCES

Abrams, M.D. 1986. Historical development of gallery forests in northeast Kansas. *Vegetatio,* 65: 29-37.

Adams, D. A. 1963. Factors influencing vascular plant zonation in North Carolina salt marshes. *Ecology,* 44: 445-456.

Akins, G.J. and C.A. Jefferson. 1973. Coastal Wetlands of Oregon. Oregon Coastal Conservation and Development Commission, Florence, OR.

Alexander, R.R. 1986. Major Habitat Types, Community Types, and Plant Communities in the Rocky Mountains. USDA Forest Service, Rocky Mountain Forest and Range Expt. Stat., Ft. Collins, CO. Gen. Tech. Rept. RM-123.

Alvarez-López, M. 1990. Ecology of *Pterocarpus officinalis* forested wetlands of Puerto Rico. In *Forested Wetlands,* vol. 15. Ecosystems of the World. A.E. Lugo, M.M. Brinson, and S. Brown (Eds.). Elsevier, Amsterdam, The Netherlands. 213-225.

Anderson, R. R., R.G. Brown, and R.D. Rappleye. 1968. Water quality and plant distribution along the upper Patuxent River, Maryland. *Chesapeake Sci.,* 9: 145-156.

Ashworth, S.M. 1997. Comparison between restored and reference sedge meadow wetlands in southcentral Wisconsin. *Wetlands,* 17: 518-527.

Batten, A.R. 1990. A synopsis of Alaska wetland vegetation. In *Alaska — Regional Wetland Functions.* Proceedings of a workshop, May 28-29, 1986, Anchorage, AK. University of Massachusetts, Amherst, MA. Environmental Institute Publication 90-1. 23-44.

Batten, A. R. and D. F. Murray. 1982. A Literature Survey on the Wetland Vegetation of Alaska. Technical Report Y-82-2. U.S. Army Corps of Engineers, Wetland Research Program, Vicksburg, MS.

Beaven, G.F. and H.J. Oosting. 1939. Pocomoke Swamp: a study of a cypress swamp on the Eastern Shore of Maryland. *Bull. Torrey Bot. Club,* 66: 367-389.

Bennett, S.H. and J.B. Nelson. 1991. Distribution and Status of Carolina Bays in South Carolina. South Carolina Wildlife and Marine Resources Dept., Columbia.

Bertness, M.D. 1992. The ecology of a New England salt marsh. *Amer. Sci.,* 80: 260-268.

Bolen, E.G., L.M. Smith, and H.L. Schramm, Jr. 1989. Playa lakes: prairie wetlands of the Southern High Plains. *BioScience,* 39: 615-623

Brabander, J. J., R. E. Masters, and R. M. Short. 1985. Bottomland Hardwoods of Eastern Oklahoma. U.S. Fish and Wildlife Service, Washington, D.C.

Bray, W.L. 1930. The Development of the Vegetation of New York State. Tech. Pub. No. 29. New York State College of Forestry, Syracuse University, Syracuse, NY.

Brinson, M.M. 1977. Decomposition and nutrient exchange of litter in an alluvial swamp forest. *Ecology,* 58: 601-609.

Brinson, M. M., B.L. Swift, R.C. Plantico, and J.S. Barclay. 1981. Riparian Ecosystems: Their Ecology and Status. FWS/OBS-81/17. U.S. Fish and Wildlife Service, Washington, D.C.

Brooks, R. P., D.E. Arnold, and E.D. Bellis. 1987. Wildlife and Plant Communities of Selected Wetlands: Pocono Region of Pennsylvania. NWRC Open File Report 87-02. U.S. Fish and Wildlife Service, Washington, D.C.

Brown, S. and D.L. Peterson. 1983. Structural characteristics and biomass productivity of two Illinois bottomland forests. *Am. Midl. Nat.,* 110: 107-117.

Bruner, W.E. 1931. The vegetation of Oklahoma. *Ecol. Monogr.,* 1: 99-188.

Buell, M.F. and W.A. Wistendahl. 1955. Flood plain forests of the Raritan River. *Bull. Torrey Bot. Club,* 82: 463-472.

Bush, J.K. and O.W. Van Auken. 1984. Woody-species composition of the upper San Antonio River gallery forest. *Texas J. Sci.,* 36: 139-148.

Caldwell, F.A. and G.E. Crow. 1992. A floristic and vegetation analysis of a freshwater tidal marsh on the Merrimack River, West Newbury, MA. *Rhodora,* 94: 63-97.

Calhoun, A.J.K., J.E. Cormier, R.B. Owen, Jr., A.F. O'Connell, Jr., C.T. Roman, and R.W. Tiner, Jr. 1994. The Wetlands of Acadia National Park and Vicinity. Misc. Pub. No. 721. Maine Agricultural and Forest Experiment Station, Orono, ME.

Carlton, J.M. 1975. A Guide to Comon Florida Salt Marsh and Mangrove Vegetation. Florida Dept. Of Natural Resources, Marine Research Laboratory, St. Petersburg, FL. Mar. Res. Publ. No. 6.

Carlton, J. M. 1977. A survey of selected coastal vegetation communities of Florida. Florida Department of Natural Resources, Marine Research Laboratory, St. Petersburg, FL. Mar. Res. Publ. No. 30.

Carter, V.C. 1988. The Relation of Hydrogeology, Soils and Vegetation on the Wetland-to-Upland Transition Zone of the Great Dismal Swamp of Virginia and North Carolina. Ph. D. dissertation. George Washington University, Washington, D.C.

Carter, V.C. 1996. Wetland hydrology, water quality, and associated functions. In National Water Summary on Wetland Resources. J.D. Fretwell, J.S. Williams, and P.J. Redman (Comp.). Water-Supply Paper 2425. U.S. Geological Survey, Reston, VA. 35-48.

Chabreck, R. H. 1972. Vegetation, water and soil characteristics of the Louisiana coastal region. Bull. 664. Louisiana State University Agricultural Experiment Station, Baton Rouge, LA.

Chadde, S.W., J.S. Shelly, R.J. Bursik, R.K. Moseley, A.G. Evenden, M. Mantas, F. Rabe, and B. Heidel. 1998. Peatlands on National Forests of the Northern Rocky Mountains: Ecology and Conservation. USDA Forest Service, Rocky Mountain Research Station, Ogden, UT. General Tech. Rep. RMRS-GTR-11.

Chamless, L.F. and E.S. Nixon. 1975. Woody vegetation-soil relations in a bottomland forest in east Texas. *Texas J. Sci.*, 36: 407-416.

Chapman, V. J. 1938. Studies in salt marsh ecology I-III. *J. Ecology*, 26: 144-221.

Chapman, V.J. 1940a. Succession on the New England salt marshes. *Ecology*, 21: 279-282.

Chapman, V. J. 1940b. Studies in salt marsh ecology VI-VII. *J. Ecology*, 28: 118-179.

Chapman, V. J. 1960. *Salt Marshes and Salt Deserts of the World*. Interscience Publishers, Inc., New York.

Chapman, V. J. 1974. *Salt Marshes and Salt Deserts of the World. 2nd supplemented ed*. J. Cramer, Leuterhausen, Vaduz, Germany.

Chapman, V. J. 1976a. *Coastal Vegetation. 2nd ed*. Pergamon Press, Oxford, England.

Chapman, V. J. 1976b. *Mangrove Vegetation*. J. Cramer, Leuterhausen, Vaduz, Germany.

Chapman, V.J. (Ed.) 1977. *Wet Coastal Ecosystems*. Elsevier Scientific Publishing Company, Amsterdam, The Netherlands.

Christensen, N. L., R.B. Burchele, A. Liggett, and E.L. Simms. 1981. The structure and development of pocosin vegetation. In *Pocosin Wetlands*. C. J. Richardson (Ed.). Hutchinson Ross Publishing Co., Stroudsburg, PA. 43-61.

Christensen, N.L., R.B. Wilbur, and J.S. McLean. 1988. Soil-Vegetation Correlations in the Pocosins of Croatan National Forest, North Carolina. Biol. Rep. 88(28). U.S. Fish and Wildlife Service, Washington, D.C.

Clark, J. R. and J. Benforado (Eds.). 1981. *Wetlands of Bottomland Hardwood Forests*. Elsevier Scientific Publishing Co., New York.

Clewell, A.F., J.A. Goolsby, and A.G. Shuey. 1982. Riverine forests of the South Prong Alafia River system, Florida. *Wetlands*, 2: 21-72.

Cohen, A.D., D.J. Casagrande, M.J. Andrejko, and G.R. Best (Eds.). 1984. *The Okefenokee Swamp: Its Natural History, Geology, and Geochemistry*. Wetlands Surveys, Los Alamos, NM.

Colorado Division of Parks and Outdoor Recreation. 1987. Statewide Comprehensive Outdoor Recreation Plan — Wetlands Amendment. Denver, CO.

Commonwealth of Virginia. 1988. The Virginia Outdoors Plan — Wetlands. Virginia Department of Conservation and Recreation, Division of Planning and Recreational Resources, Richmond, VA.

Conard, H.S. 1924. Second survey of vegetation of a Long Island salt marsh. *Ecology*, 5: 379-388.

Conard, H.S. 1935. The plant associations on central Long Island: a study in descriptive sociology. *Am. Midl. Nat.*, 16: 433-516.

Conard, H.S. and G.C. Galligar. 1929. Third survey of a Long Island salt marsh. *Ecology*, 10: 326-336.

Conner, W. H. and J.W. Day, Jr. 1976. Productivity and composition of a bald cypress — water tupelo site and a bottomland hardwood site in Louisiana swamp. *Am. J. Bot.*, 63: 1354-1364.

Conner, W. H. and J.W. Day, Jr. (Eds.). 1987. The Ecology of Barataria Basin, Louisiana: An Estuarine Profile. Biological Report 85(7.13). U.S. Fish and Wildlife Service, Washington, D.C.

Conner, W.H., J.G. Gosselink, and R.T. Parrondo. 1981. Comparison of the vegetation of three Louisiana swamp sites with different flooding regimes. *Am. J. Bot.*, 63: 320-331.

Conway, V. M. 1949. The bogs of central Minnesota. *Ecol. Monogr.*, 19: 173-206.

Cooper, A.W. and E.D. Waits. 1973. Vegetation types in an irregularly flooded salt marsh on the North Carolina Outer Banks. *J. Mitchell Soc.*, 89: 78-91.

Cooper, D.J. 1986. *Ecological Studies in Wetland Vegetation, Cross Creek Valley, Holy Cross Wilderness, Sawatch Range, Colorado*. Technical Report. Holy Cross Wilderness Defense Fund, Boulder, CO.

Cooper, S., S. Neiman, and R. Steele. 1983. Forest Types of Northern Idaho. USDA Forest Service, Inter-mountain Forest and Range Expt. Stat., Ogden, UT.

Cooper, S.V., P Lesica, and D. Page-Dumroese. 1997. Plant Community Classification for Alpine Vegetation on the Beaverhead National Forest, Montana. Gen. Tech. Rep. INT-GTR-362. USDA Forest Service, Rocky Mountain Research Station, Ogden, UT.

Cooper, W.S. 1931. A third expedition to Glacier Bay, Alaska. *Ecology,* 12: 61-95.

Copeland, B.J., R.G. Hodson, S.R. Riggs, and J.E. Easley, Jr. 1983. The Ecology of Albemarle Sound, NC: An Estuarine Profile. FWS/OBS-83/01. U.S. Fish and Wildlife Service, Washington, D.C.

Copeland, B. J., R.G. Hodson, and S.R. Riggs. 1984. The Ecology of the Pamlico River, NC: An Estuarine Profile. FWS/OBS-82/06. U.S. Fish and Wildlife Service, Washington, D.C.

Cowardin, L. M., V. Carter, F.C. Golet, and E.T. LaRoe. 1979. Classification of Wetlands and Deepwater Habitats of the United States. FWS/OBS-79/31. U.S. Fish and Wildlife Service, Washington, D.C.

Crites, R.W. and J.E. Ebinger. 1969. Vegetation survey of floodplain forests in east-central Illinois. *Trans. Ill. Acad. Sci.,* 62: 316-330.

Crow, G.E. 1969. A phytogeographical analysis of a southern Michigan bog. *Mich. Bot.,* 8: 11-27.

Crow, J.H. 1966. Plant Ecology of the Copper River Delta, Alaska. MS thesis. Washington State University, Seattle.

Crum, H. 1988. *A Focus on Peatlands and Peat Mosses.* University of Michigan Press, Ann Arbor, MI.

Curtis, J. T. 1959. *The Vegetation of Wisconsin.* The University of Wisconsin Press, Madison, WI.

Dabel, C. V. and Day, F. P., Jr. 1977. Structural comparison of four plant communities in the Great Dismal Swamp, Virginia. *Bull. Torrey Bot. Club,* 104: 352-360.

Dahl, T.E. and C.E. Johnson. 1991. Wetlands Status and Trends in the Conterminous United States Mid-1970's to Mid-1980's. U.S. Fish and Wildlife Service, Washington, D.C.

Damman, A. W. H. 1977. Geographic changes in the vegetation patterns of raised bogs in the Bay of Fundy Region of Maine and New Brunswick. *Vegetatio,* 35: 137-151.

Damman, A. W. H. and T.W. French. 1987. The Ecology of Peat Bogs of the Glaciated Northwestern United States: A Community Profile. Biol. Rep. 85 (7.16). U.S. Fish and Wildlife Service, Washington, D.C.

Dansereau, P. and F. Segadas-Vianna. 1952. Ecological study of peat bogs of eastern North America. I. Structure and evolution of vegetation. *Can. J. Bot.,* 30: 490-518.

Daubenmire, R. and J.B. Daubenmire. 1968. Forest Vegetation of Eastern Washington and Northern Idaho. Tech. Bull. 60. Washington Agr. Expt. Stat., Washington State University, Pullman, WA.

Davis, J.H., Jr. 1940. The ecology and geologic role of mangroves in Florida. Carnegie Inst., Wash. Publ. No. 517. *Pap. Tortugas Lab.* 32: 304-412.

Davis, J.H. Jr. 1943. The Natural Features of Southern Florida, Especially the Vegetation and the Everglades. *FL. Geolo. Surv. Bull.,* No. 25.

Davis, S.M. and J.C. Ogden (Eds.). 1994. *Everglades: The Ecosystem and its Restoration.* St. Lucie Press, Boca Raton, FL.

Day, J. W., Jr., W.G. Smith, P.R. Wagner, and W.C. Stowe. 1973. Community Structure and Carbon Budget of a Salt Marsh and Shallow Bay Estuarine System in Louisiana. Publication No. LSU-SG-72-04. Center for Wetland Resources, Louisiana State University, Baton Rouge.

de la Cruz, A. A. 1981. Difference between South Atlantic and Gulf Coast marshes. In Proc. of the U.S. Fish and Wildlife Service Workshop on Coastal Ecosystems of the Southeastern United States. R.C. Carey, P.S. Markovits, and J.B. Kirkwood (Eds.). FWS/OBS-80/59. U.S. Fish and Wildlife Service, Washington, D.C. 10-20.

Deevey, E.S., Jr. 1958. Bogs. *Sci. Amer.,* (Oct. 1958): 1-9.

Devall, M.S. 1990. Cat Island Swamp: window to a fading Louisiana ecology. *For. Ecol. Manage.,* 33/34: 303-314.

DeVelice, R.L., J.A. Ludwig, W.H. Moir, and F. Ronco, Jr. 1984. A Classification of Forest Habitat Types in Northern New Mexico and Southern Colorado. USDA Forest Service, Rocky Mountain Forest and Range Expt. Stat., Ft. Collins, CO.

Dick-Peddie, W. A., J.K. Meents, and R. Spellenberg. 1984. Vegetation Resources Analysis for the Velarde Community Ditch Project, Rio Arriba and Santa Fe Counties, New Mexico. U.S. Bureau of Reclamation, Amarillo, TX.

Dick-Peddie, W.A., J.V. Hardesty, E. Muldavin, and B. Sallach. 1987. Soil-Vegetation Correlations in the Riparian Zones of the Gila and San Francisco Rivers in New Mexico. Biol. Rep. 87(9). U.S. Fish and Wildlife Service, Washington, D.C.

Dix, R.L. and F.E. Smeins. 1967. The prairie, meadow and marsh vegetation of Nelson County, North Dakota. *Can. J. Bot.*, 45: 21-58.

Drew, R. D. and N.S. Schomer. 1984. An Ecological Characterization of the Caloosahatchee River/Big Cypress Watershed. FWS/OBS-82/58.2. U.S. Fish and Wildlife Service, Washington, D.C.

Dreyer, G.D. and W.A. Niering (Eds.). 1995. *Tidal Marshes of Long Island Sound: Ecology, History and Restoration.* Bull. No. 34. The Connecticut College Arboretum, New London, CT.

Drury, W. H., Jr. 1962. Patterned Ground and Vegetation on Southern Bylot Island, Northwest Territories, Canada. Gray Herbarium Contributions 190. Harvard University, Cambridge, MA.

Duever, M. J., J.E. Carlson, and L.A. Riopelle, L. A. 1984 Corkscrew Swamp: a virgin cypress strand. In *Cypress Swamps*. K.C. Ewel and H.T. Odum (Eds.). University of Florida Press, Gainesville, FL. 334-348.

Dunn, C.P. and F. Stearns. 1987. Relationship of vegetation layers to soils in southeast Wisconsin forested wetlands. *Am. Midl. Nat.*, 118: 366-374.

Edmisten, J.E. 1963. The Ecology of the Florida Pine Flatwoods. Ph. D. dissertation. University of Florida, Gaineville, FL.

Eicher, A. L. 1988. Soil-Vegetation Correlations in Wetland and Adjacent Uplands of the San Francisco Bay Estuary, California. Biological Report 88(21). U.S. Fish and Wildlife Service, Washington, D.C.

Egler, F.E. 1952. Southeast saline Everglades vegetation, Florida and its management. *Vegetatio,* 3: 213-265.

Eleuterius, L. M. 1972. The marshes of Mississippi. *Castanea,* 37: 153-168.

Eleuterius, L. M. 1980. An Illustrated Guide to Tidal Marsh Plants of Mississippi and Adjacent States. Publication No. MASGP-77-039. Mississippi-Alabama Sea Grant Consortium, Ocean Springs, MS.

Eleuterius, L.M. 1984. Autecology of the black needlerush *Juncus roemerianus. Gulf Res. Rep.,* 7: 339-350.

Eleuterius, L.M. and S. McDaniel. 1978. The salt marsh flora of Mississippi. *Castanea,* 43: 86-95.

Elliot, M.E. 1981. Wetlands and wetland vegetation of the Hawaiian Islands. M.A. thesis. University of Hawaii, Honolulu.

Epperson, J.E. 1992. Missouri Wetlands — A Vanishing Resource. Water Resources Report No. 39. Missouri Division Of Geology and Land Survey.

Erickson, N. E. and D.M. Leslie, Jr. 1988. Soil–Vegetation Correlations in Coastal Mississippi Wetlands. Biol. Rep. 89(3). U.S. Fish and Wildlife Service, Washington, D.C.

Ewel, K.C. 1991. Swamps, chap. 9. In *Ecosystems of Florida.* R.L. Meyers and J.J. Ewel (Eds.). University of Central Florida Press, Orlando, FL. 281-323.

Ewel, K.C. and H.T. Odum (Eds.). 1984. *Cypress Swamps.* University of Florida Press, Gainesville, FL.

Faber, P. M., Keller, E., Sands, A. and Massey, B. M. 1989. The Ecology of Riparian Habitats of the Southern California Coastal Region: A Community Profile. Biol. Rep. 85(7.27). U.S. Fish and Wildlife Service, Washington DC.

Farrar, J. 1982. The Rainwater Basin: Nebraska's Vanishing Wetlands. Nebraska Game and Parks Commission, Lincoln, NE.

Farrar, J. and R. Gersib. 1991. Nebraska's Salt Marshes — Last of the Least. Nebraska Game and Parks Commission, Lincoln, NE.

Ferren, W.R., Jr., P.L. Fiedler, R.A. Leidy, K.D. Lafferty, and L.A.K. Mertes. 1996. Part III. Key to and catalogue of wetlands of the central and southern California coast and coastal watersheds. *Madroño,* 43: 183 233.

Florida Department of Environmental Protection, South Florida Water Management District, St. John's River Water Managment District, Suwannee River Water Management District, Southwest Florida Water Management District, and Northwest Florida Water Management District. 1995. The Florida Wetlands Delineation Manual. Florida Department of Environmental Protection, Wetland Evaluation and Delineation Section, Tallahassee, FL.

Folkerts, G.W. 1982. The Gulf Coast pitcher plant bogs. *Am. Sci.,* 70: 260-267.

Ford, A.L. and O.W. Van Auken. 1982. The distribution of wood species in the Guadalupe River floodplain forest in the Edwards Plateau of Texas. *Southwest. Nat.,* 27: 383-392.

Franklin, J.F. and C.T. Dyrness. 1973. Natural Vegetation of Oregon and Washington. Gen. Tech. Rept. PNW-8. USDA Forest Service, Pacific Northwest Forest and Range Expt. Stat., Portland, OR.

Frayer, W.E. and J.M. Hefner. 1991. Florida Wetlands Status and Trends, 1970s to 1980s. U.S. Fish and Wildlife Service, Southeast Region, Atlanta, GA.

Frayer, W.E., D.D. Peters, and H.R. Pywell. 1989. Wetlands of the California Central Valley Status and Trends. U.S. Fish and Wildlife Service, Region 1, Portland, OR.

Fretwell, J.D., J.S. Williams, and P.J. Redman (Comp.). 1996. National Water Summary on Wetland Resources. Water-Supply Paper 2425. U.S. Geological Survey, Reston, VA.

Frey, R.W. and P.B. Basan. 1985. Coastal salt marshes. In *Coastal Sedimentary Environments*. R.A. Davis, Jr. (Ed.). Springer Verlag, New York. 225-301.

Frye, R.J., II and J.A. Quinn. 1979. Forest development in relation to topography and soils on a floodplain of the Raritan River, New Jersey. *Bull. Torrey Bot. Club,* 106: 334-345.

Gates, F.C. 1942. The bogs of northern lower Michigan. *Ecol. Monogr.,* 12: 213-254.

Geis, J.W., and J.L. Kee. 1977. *Coastal Wetlands along Lake Ontario and St. Lawrence River in Jefferson County, New York.* State University of New York, Institute of Environmental Program Affairs, Syracuse, NY.

Gilbert, M.C. 1989. Ordination and Mapping of Wetland Communities in Nebraska's Rainwater Basin Region. CEMRO Environmental Report 89-1. U.S. Army Corps of Engineers, Omaha, NE.

Glaser, P. H. 1987. The Ecology of Patterned Boreal Peatlands of Northern Minnesota: A Community Profile. Biol. Rep. 85(7.14). U.S. Fish and Wildlife Service, Washington, D.C.

Glaser, P.H., G.A. Wheeler, E. Gorham, and H.E. Wright, Jr. 1981. The patterned mires of the Red Lake peatland, northern Minnesota — vegetation, water chemistry, and landforms. *J. Ecol.,* 69: 575-599.

Golet, F.C., A.J.K. Calhoun, W.R. DeRagon, D.J. Lowry, and A.J. Gold. 1993. The Ecology of Red Maple Swamps in the Glaciated Northeast: A Community Profile. Biol. Rep. 12. U.S. Fish and Wildlife Service, Washington, D.C.

Good, R.E. 1965. Salt marsh vegetation, Cape May, New Jersey. *N.J. Acad. Sci. Bull.,* 10: 1-11.

Good, R.E. and N.F. Good. 1974. Vegetation and production of the Woodbury Creek-Hessian Run freshwater tidal marshes. *Bartonia,* 43: 38-45.

Good, R. E., D.F. Whigham, and R.L. Simpson (Eds.). 1978. *Freshwater Wetlands, Ecological Processes and Management Potential.* Academic Press, Inc. New York.

Gore, A.J.P. (Ed.). 1983. *Mires: Swamp, Bog, Fen and Moor. 4B. Regional Studies.* Elsevier, Amsterdam, The Netherlands.

Gosselink, J. G. 1984. The Ecology of Delta Marshes of Coastal Louisiana: A Community Profile. FWS/OBS-84/09. U.S. Fish and Wildlife Service, Washington, D.C.

Hackney, C. T. and A.A. de la Cruz. 1982. The structure and function of brackish marshes in the north central Gulf of Mexico: a 10-year case study. In *Wetlands: Ecology and Management.* B. Gopal, R.E. Turner, R.G. Wetzel, and D.F. Whigham (Eds.). International Scientific Publishers, Jaipur, India. 89-107.

Hall, F.C. 1973. Plant Communities of the Blue Mountains in Eastern Oregon and Southeastern Washington. Area Guide 3-1. USDA Forest Service, Pacific Northwest Region, Portland, OR.

Hall, J.V., W.E. Frayer, and B.O. Wilen. 1994. Status of Alaska Wetlands. U.S. Fish and Wildlife Service, Alaska Region, Anchorage, AK.

Hall, T. F. and W. T. Penfound. 1943. Cypress-gum communities in the Blue Girth Swamp near Selma, Alabama. *Ecology,* 24: 208-217.

Hannan, R.R. and J.S. Lassetter. 1982. The vascular flora of the Brodhead Swamp Forest, Rockcastle County, Kentucky. *Trans. Kent. Acad. Sci.,* 43: 43-49.

Hanson, H.C. 1951. Characteristics of some grassland, marsh, and other plant communities in western Alaska. *Ecol. Monogr.,* 21: 317-375.

Hanson, H.C. 1953. Vegetation types in northwestern Alaska and comparisons with communities in other Arctic regions. *Ecology,* 34: 111-140.

Hanson, H. C. 1958. Analysis of Nelchina Caribou Range. Federal Aid in Wildlife Job Completion Report, Job 6, W-3-k-12, U.S. Fish and Wildlife Service, Juneau, AK.

Harper, R.M. 1914. Geography and vegetation of northern Florida. Florida Geological Survey, 6th Annu. Rep. 163-451.

Harshberger, J.W. 1900. An ecological study of the New Jersey strand flora. *Proc. Acad. Nat. Sci. Phila.,* 52: 623-671.

Harshberger, J.W. 1902. Additional observations on the strand flora of New Jersey. *Proc. Nat. Sci. Phila.,* 54: 642-669.

Harshberger, J.W. 1909. The vegetation of the salt marshes and of the salt and freshwater ponds of northern coastal New Jersey. *Proc. Acad. Nat. Sci. Phila.,* 1909: 373-400.

Harshberger, J.W. 1914. The vegetation of south Florida. *Trans. Wagner Free Inst. Sci. Philos.,* 3: 51-189.

Harshberger, J.W. 1970. *The Vegetation of the New Jersey Pine-Barrens. An Ecologic Investigation.* Dover Publications, Inc., New York. (Reprint of a 1916 publication by Christopher Sower Co., Philadelphia.)

Haukos, D.A., H.Z. Sun, D.B. Wester, and L.M. Smith. 1998. Sample size, power, and analytical considerations for vertical structure data from profile boards in wetland vegetation. *Wetlands,* 18: 203-215.

Hawk, G.M. and D.B. Zobel. 1974. Forest succession on alluvial landforms of the McKenzie River valley, Oregon. *Northwest. Sci.,* 43: 245-265.

Hefner, J.M., B.O. Wilen, T.E. Dahl, and W.E. Frayer. 1994. Southeast Wetlands: Status and Trends, Mid-1970's to Mid-1980's. U.S. Fish and Wildlife Service, Region 4, Atlanta, GA.

Heinselman, M. L. 1963. Forest sites, bog processes, and peatland types in the Glacial Lake Agassiz Region, Minnesota. *Ecol. Monogr.,* 33: 327-374.

Heinselman, M.L. 1965. String bogs and other patterned organic terrain near Seney, Upper Michigan. *Ecology,* 46: 185-188.

Heinselman, M. L. 1970. Landscape evolution, peatland types, and the environment in the Lake Agassiz Peatlands Natural Area, Minnesota. *Ecol. Monogr.,* 40: 235-261.

Heitmeyer, M.E., L.H. Frederickson, and G.F. Krause. 1991. Water relationships among wetland habitat types in the Mingo Swamp, Missouri. *Wetlands,* 11: 55-66.

Hemstom, M.A., S.E. Logan, and W. Pavlat. 1987. Plant Association and Management Guide. Willamette National Forest. R6-Ecol257-B-86. USDA Forest Service, Pacific Northwest Region, Portland, OR.

Henderson, J.A., D.H. Peter, R.D. Lesher, and D.C. Shaw. 1989. Forested Plant Associations of the Olympic National Forest. R6-Ecol Tech. Paper 001-88. USDA Forest Service, Pacific Northwest Region, Portland, OR.

Herdendorf, C. E., S.M. Hartley, and M.D. Barnes (Eds.). 1981. Fish and Wildlife Resources of the Great Lakes Coastal Wetlands within the United States. Volume I: Overview. FWS/OBS-81/02. U.S. Fish and Wildlife Service, Washington, D.C.

Herdendorf, C.E., C.N. Raphael, and E. Jaworski. 1986. The Ecology of Lake St. Clair Wetlands: A Community Profile. Biol. Rep. 85(7.7). U.S. Fish and Wildlife Service, Washington, D.C.

Herdendorf, C. E. 1987. The Ecology of the Coastal Marshes of Western Lake Erie: A Community Profile. Biol. Rep. 85(7.9). U.S. Fish and Wildlife Service, Washington, D.C.

Hess, K. 1981. Phyto-edaphic Study of Habitat Types of the Arapaho and Roosevelt National Forests. Ph.D. dissertation. Colorado State University, Ft. Collins, CO.

Hess, K. and C.H. Wasser. 1982. Grassland, Shrubland and Forestland Habitat Types of the White River-Arapaho National Forest. USDA Forest Service, Rocky Mountain Forest and Range Expt. Stat., Ft. Collins, CO.

Hinde, H.P. 1954. Vertical distribution of salt marsh phanerogams in relation to tide levels. *Ecol. Monogr.,* 24: 209-225.

Hoagland, B.W. and R.L. Jones. 1992. Wetland and riparian flora of the upper Green River basin, south-central Kentucky. *Trans. Kent. Acad. Sci.,* 53: 141-153.

Hobbie, J. E. 1984. The Ecology of Tundra Ponds of the Arctic Coastal Plain: A Community Profile. FWS/OBS-83/25. U.S. Fish and Wildlife Service. Washington, D.C.

Hoffman, G.R. and R.R. Alexander. 1980. Forest Vegetation of the Routt National Forest in Northwestern Colorado: A Habitat Type Classification. Res. Paper RM-221. USDA Forest Service, Rocky Mountain Forest and Range Expt. Stat., Ft. Collins, CO.

Hofstetter, R.H. 1983. Wetlands in the United States, chap. 7. In *Mires: Swamp, Bog, Fen and Moor. 4B Regional Studies.* A.J.P. Gore (Ed.). Elsevier, Amsterdam, The Netherlands. 201-244.

Hook, D. D. and R. Lea (Eds.). 1989. Proc. of the Symposium — The Forested Wetlands of the Southern United States. Gen. Tech. Rep. SE-50. U.S. Forest Service, Southeastern Forest Experiment Station, Washington, D.C.

Hopkins, W.A. 1979. Plant Associations of South Chiloquin and Klamath Ranger Districts, Winema National Forest. R6-ECOL-79-005. USDA Forest Service, Pacific Northwest Region, Portland, OR.

Hopper, R.M. 1968. Wetlands of Colorado. Tech. Publ. No. 22. Colorado Dept. Of Game, Fish, and Parks.

Hosner, J.F. and L.S. Minckler. 1963. Bottomland hardwood forests of southern Illinois — regeneration and succession. *Ecology,* 44: 29-41.

House, H.D. 1914. Vegetation of the Coos Bay Region, Oregon. *Muhlenbergia,* 9: 81-100.

Hubbard, D. E. 1988. Glaciated Prairie Wetland Functions and Values: A Synthesis of the Literature. Biol. Rep. 88(43). U.S. Fish and Wildlife Service, Washington, D.C.

Hubbard, D. E., D.D. Millar, D. D. Malo, D.D. and K.F. Higgins. 1988. Soil-Vegetation Correlations in Prairie Potholes of Beadle and Deuel Counties, South Dakota. Biol. Rep. 88(22). U.S. Fish and Wildlife Service, Washington, D.C.

Huenneke, L.F. 1982. Wetland forests of Tompkins County, New York. *Bull. Torrey Bot. Club,* 109: 51-63.

Indiana Department of Natural Resources. 1997. Acreage Summary of Wetlands and Deepwater Habitats in Indiana; compiled from NWI Digital Data. Terra Haute, IN.

Jahn, L. A. and R.V. Anderson. 1986. The Ecology of Pools 19 and 20, Upper Mississippi River: A Community Profile. Biol. Rep. 85(7.6). U.S. Fish and Wildlife Service, Washington, D.C.

Janssen, C.R. 1967. A floristic study of forests and bog vegetation, northwestern Minnesota. *Ecology,* 48: 751-765.

Jefferson, C.A. 1973. Salt marsh mapping and description. In Coastal Wetlands of Oregon. G.J. Akins and C.A. Jefferson (Eds.). Oregon Coastal Conservation and Development Commission, Florence, OR.

Jefferson, C.A. 1974. Plant Communities and Succession in Oregon Coastal Salt Marshes. Oregon State University, Corvallis, OR. Master's Thesis.

Jensen, C.F. 1974. Evaluation of Existing Wetland Habitat in Utah. Pub. No. 74-17. Utah Department of Natural Resources, Division of Wildlife Resources, Salt Lake City, UT.

Jervis, R. A. 1969. Primary production in the freshwater marsh ecosystem of Troy Meadows, New Jersey. *Bull. Torrey Bot. Club,* 96: 209-231.

Johnson, C. W. 1985. *Bogs of the Northeast.* University Press of New England, Hanover, NH.

Johnson, D.S. and H.H. York. 1915. The relation of plants to tide-levels. Pub. No. 206: 1-162. Carnegie Institute, Washington, D.C.

Johnson, R.R., K.F. Higgins, M.L. Kjellsen, and C.R. Elliott. 1997. Eastern South Dakota Wetlands. South Dakota State University, Brookings, SD.

Josselyn, M. 1983. The Ecology of San Francisco Bay Tidal Marshes: A Community Profile. FWS/OBS-83/23. U.S. Fish and Wildlife Service, Washington, D.C.

Josselyn, M.N., S.P. Faulkner, and W.H. Patrick, Jr. 1990. Relationships between seasonally wet soils and occurrence of wetland plants in California. *Wetlands,* 10: 7-26.

Kantrud, H. A., G.L. Krapu, and G.A. Swanson. 1989. Prairie Basin Wetlands of the Dakotas: A Community Profile. Biol. Rep. 85(7.28). U.S. Fish and Wildlife Service, Washington, D.C.

Keammerer, W.R., W.C. Johnson, and R.L. Burgess. 1975. Floristic analysis of the Missouri River bottomland forests in North Dakota. *Can. Field-Nat.,* 89: 5-19.

Kearney, T.H. 1900. The plant covering of Okracoke Island: a study in the ecology of the North Carolina strand vegetation. *U.S. Natl. Herb.,* 5: 261-319.

Kearney, T. H. 1901. Report on a botanical survey of the Dismal Swamp Region. U.S. Department of Agriculture, Washington, D.C. Contributions to the *U.S. Natl. Herb.,* 5: 321-550.

Kennedy, H.E. and G.J. Nowacki. 1997. An Old-Growth Definition for Seasonally Wet Oak-Hardwood Woodlands. Gen. Tech. Rep. SRS-8. USDA Forest Service, Southern Research Station, Asheville, NC.

Kindscher, K., S. Wilson, A. Fraser, and C. Lauver. 1996. Vegetation of Western Kansas Playa Lakes — 1993-1995. Kansas Biological Survey Report No. 70. University of Kansas, Lawrence.

Kologiski, R. L. 1977. The Phytosociology of the Green Swamp, North Carolina. Tech. Bull. 250. North Carolina Agricultural Experiment Station, Raleigh.

Komarkova, V. 1984. Habitat Types on Selected Parts of the Gunnison and Uncompahgre National Forests. USDA Forest Service, Rocky Mountain Forest and Range Expt. Stat., Ft. Collins, CO.

Kovalchik, B.L. 1987. Riparian Zone Associations. Deschutes, Ochoco, Fremont, and Winema National Forests. Region 6 Ecology Tech. Paper 279-87. USDA Forest Service, Pacific Northwest Region, Portland, OR.

Kuenzler, E. J. 1974. Mangrove swamp systems. In *Coastal Ecological Systems of the United States.* vol. 1. H.T. Odum, B.J. Copeland, and E.A. McMahan (Eds.). The Conservation Foundation, Washington, D.C. 346-371.

Kunze, L. 1994. Preliminary Classification of Native, Low Elevation, Freshwater Wetland Vegetation in Western Washington. Washington State Department of Natural Resources, Natural Heritage Program, Olympia, WA.

Kurz, H. and K. Wagner, K. 1957. *Tidal Marshes of the Gulf and Atlantic Coasts of Northern Florida and Charleston, South Carolina.* FSU Studies No. 24. Florida State University, Tallahassee.

Kushlan, J.A. 1991. Freshwater marshes, chap. 10. In *Ecosystems of Florida*. R.L. Meyers and J.J. Ewel (Eds.). University of Central Florida Press, Orlando. 324-363.

Laderman, A.D. (Ed.). 1987. *Atlantic White Cedar Wetlands*. Westview Press, Boulder, CO.

Laderman, A.D. 1989. The Ecology of Atlantic White Cedar Wetlands: A Community Profile. Biological Report 85(7.21). U.S. Fish and Wildlife Service, Washington, D.C.

Laessle, A.M. 1942. *The Plant Communities of the Welaka Area with Special Reference to Correlation between Soils and Vegetational Succession*. Biological Science Series IV(1). University of Florida, Gainesville.

Larsen, J.A. 1982. *Ecology of the Northern Lowland Bogs and Conifer Forests*. Academic Press, New York.

Latham, P.J., L.G. Pearlstine, and W.M. Kitchens. 1994. Species association changes across a gradient of freshwater, oligohaline, and mesohaline tidal marshes along the lower Savannah River. *Wetlands,* 14: 174-183.

Lee, L.C. 1983. The Floodplain and Wetland Vegetation of Two Pacific Northwest River Ecosystems. Ph. D. Dissertation. University of Washington, Seattle.

Leitman, H.M. 1975. Correlation of Apalachicola River floodplain tree communities with water levels, elevation, and soils. Master's thesis. Florida State University, Tallahassee.

Leitman, H.M., J.E. Sohm, and M.A. Franklin. 1984. Wetland Hydrology and Tree Distribution of the Appalachicola River Flood Plain, Florida. U.S. Geological Survey Water-Supply Paper 2196-A.

Lewis, R. R. III and E.D. Estevez. 1988. The Ecology of Tampa Bay, Florida: An Estuarine Profile. Biol. Rep. 85(7.18). U.S. Fish and Wildlife Service, Washington, D.C.

Livingston, R. J. 1984. The Ecology of the Appalachicola Bay System: An Estuarine Profile. FWS/OBS-82/05. U.S. Fish and Wildlife Service, Washington, D.C.

Lodge, T.E. 1994. *The Everglades Handbook: Understanding the Ecosystem*. St. Lucie Press, Boca Raton, FL.

Lugo, A.E. and S. Brown. 1988. The wetlands of the Caribbean Islands. *Acta Cientifica,* 2: 48-61.

Lugo, A.E. and S.C. Snedaker. 1974. The ecology of mangroves. *Annu. Rev. Ecol. Syst.,* 5: 39-64.

Lugo, A.E., M.M. Brinson, and S.L. Brown (Eds.). 1990. *Forested Wetlands, vol. 15. Ecosystems of the World*. Elsevier, Amsterdam, The Netherlands.

Macdonald, K.B. 1977. Plant and animal communities of Pacific North American salt marshes. In *Wet Coastal Ecosystems*. V.J. Chapman (Ed.). Elsevier Scientific Publishing Co., Amsterdam, The Netherlands. 167-191.

Macdonald, K.B. and M.G. Barbour. 1974. Beach and salt marsh vegetation of the North American Pacific Coast. In *Ecology of Halophytes*. R.J. Reimold and W.H. Queen (Eds.). Academic Press, New York. 175-233.

Majumdar, S.K., R.P. Brooks, F.J. Brenner, and R.W. Tiner, Jr. (Eds.). 1989. *Wetlands Ecology and Conservation: Emphasis in Pennsylvania*. The Pennsylvania Academic of Science, Lafayette College, Easton, PA.

Mauk, D.L. and J.A. Henderson. 1984. Forest Habitat Types of Northern Utah. Gen. Tech. Rept. INT-170. USDA Forest Service, Intermountain Forest and Range Expt. Stat., Ogden, UT.

McCormick, J. and H. A. Somes, Jr. 1982. The Coastal Wetlands of Maryland. Maryland Department of Natural Resources, Coastal Zone Management Program. Annapolis, MD.

McPherson, B.F. 1973. Vegetation in Relation to Water Depth in Conservation Area 3, Florida. Open-File Report 73-0173. U.S. Geological Survey, Reston, VA.

Meijer, W., J.J.N. Campbell, H.Setser, and L.E. Meade. 1981. Swamp forests on high terrace deposits in the Bluegrass and Knobs Regions of Kentucky. *Castanea,* 46: 122-135.

Messina, M.G. and W.H. Conner (Eds.) 1998. *Southern Forested Wetlands. Ecology and Management.* Lewis Publishers, Boca Raton, FL.

Metzler, K. J. and A. W. H. Damman. 1985. Vegetation patterns in the Connecticut River floodplain in relation to frequency and duration of flooding. *Le Naturaliste Canadien,* 112: 535-549.

Metzler, K.J. and R.W. Tiner. 1992. Wetlands of Connecticut. State Geological and Natural History Survey of Connecticut, Rept. of Investigations No. 13. Department of Environmental Protection, Hartford, CT.

Miller, R.E., Jr. and B.E. Gunsalus. 1997. Wetland Rapid Assessment Procedure. Tech. Pub. REG-001. Natural Resource Management Division, Regulation Department, South Florida Water Management District, West Palm Beach, FL.

Miller, W.R. and F.E. Egler. 1950. Vegetation of the Wequetequock-Pawcatuck tidal-marshes, Connecticut. *Ecol. Monogr.,* 20: 143-172.

Minc, L.D. and D.A. Albert. 1998. Great Lakes Coastal Wetlands: Abiotic and Floristic Characterization. Michigan Natural Features Inventory report.

Minnesota Department of Natural Resources. 1997. Minnesota Wetlands and Surface Water Resources. Poster. Minneapolis.

Mitsch, W. J. and J.G. Gosselink, J. G. 1993. *Wetlands, 2nd ed.* Van Nostrand Reinhold Co., New York.

Moizuk, G.A. and R.B. Livingston. 1966. Ecology of red maple (*Acer rubrum* L.) in a Massachusetts upland bog. *Ecology,* 47: 942-950.

Monk, C.D. 1966. An ecological study of hardwood swamps in northcentral Florida. *Ecology,* 47: 649-654.

Montague, C.L. and R.G. Wiegert. 1991. Salt marshes, chap. 14. In *Ecosystems of Florida.* R.L. Meyers and J.J. Ewel (Eds.). University of Central Florida Press, Orlando. 481-516.

Moore, J.H. and J.H. Carter, III. 1984. Habitats of white cedar in North Carolina. In *Atlantic White Cedar Wetlands.* A.D. Laderman (Ed.). Westview Press, Boulder, CO. 177-190.

Moore, P.D. and D.J. Bellamy. 1974. *Peatlands.* Springer-Verlag, New York.

Moulton, D.W., T.E. Dahl, and D.M. Dall. 1997. Texas Coastal Wetlands: Status and Trends, Mid-1950's to Early 1990's. U.S. Fish and Wildlife Service, Region 2, Albuquerque, NM.

Mudie, P.J. 1970. A survey of the coastal wetland vegetation of San Diego Bay. *Calif. Dep. Fish Game Contr.* W26. D25-51.

Musselman, L.J., O.L. Nickrent, and G.F. Levy. 1977. A contribution toward a vascular flora of the Great Dismal Swamp. *Rhodora,* 79: 240-268.

Nachlinger, J.L. 1988. Soil-vegetation Correlations in Riparian and Emergent Wetlands, Lyon County, Nevada. Biol. Rep. 88(17). U.S. Fish and Wildlife Service, Washington, D.C.

National Wetlands Working Group (Eds.). 1988. Wetlands of Canada. Polyscience Publishers, Montreal, Quebec. Ecological Land Classification Series No. 24. Environment Canada, Ottawa.

Neiland, B.J. 1971. The forest-bog complex of Southeast Alaska. *Vegetatio,* 22: 1-64.

Nelson, R.W., W.J. Logan, and E.C. Weller. 1983. Playa Wetlands and Wildlife on the Southern Great Plains: A Characterization of Habitat. FWS/OBS-83/28. U.S. Fish and Wildlife Service, Washington, D.C.

New York State Office of Parks, Recreation and Historic Preservation. 1988. Wetlands Protection in New York State. Albany, NY.

Nichols, G.E. 1920. The vegetation of Connecticut. VII. The plant associations of depositing areas along the coast. *Bull. Torrey Bot. Club,* 47: 47: 511-548.

Nicholson, B.J. 1995. The wetlands of Elk Island National Park: vegetation, classification, water chemistry, and hydrotopographic relationships. *Wetlands,* 15: 119-133.

Niering, W.A. 1953. The past and present vegetation of High Point State Park, New Jersey. *Ecol. Monogr.,* 23: 127-147.

Niering, W.A. 1985. *Wetlands.* Alfred A. Knopf, Inc., New York. (The Audubon Society Nature Guides.)

Niering, W.A. and R.S. Warren. 1980. Vegetation patterns and processes in New England salt marshes. *BioScience,* 30: 301-307.

Nixon, S.W. 1982. The Ecology of New England High Salt Marshes: A Community Profile. FWS/OBS-81/55. U.S. Fish and Wildlife Service, Washington, D.C.

Nixon, S.W. and C.A. Oviatt. 1973. Ecology of a New England salt marsh. *Ecol. Monogr.,* 43: 463-498.

Odum, W.E. and C.C. McIvor. 1991. Mangroves, chap. 15. In Ecosystems of Florida. R.L. Meyers and J.J. Ewel (Eds.). University of Central Florida Press, Orlando. 517-548.

Odum, W.E., C.D. McIvor, and T.J. Smith III. 1982. The Ecology of the Mangroves of South Florida: A Community Profile. FWS/OBS-81/24. U.S. Fish and Wildlife Service, Washington, D.C.

Odum, W.E., T.J. Smith III, J.K. Hoover, and C.C. McIvor. 1984. The Ecology of Tidal Freshwater Marshes of the United States East Coast: A Community Profile. FWS/OBS-83/17. U.S. Fish and Wildlife Service, Washington, D.C.

Ohio Department of Natural Resources. 1996. Wetlands of Ohio. Poster. Columbus, OH.

Oklahoma Tourism and Recreation Department. 1987. Oklahoma Statewide Comprehensive Outdoor Recreation Plan. Oklahoma City, OK.

Osvald, H. 1955. The vegetation of two raised bogs in Northeastern Maine. *Srensk Botanisk Tidkrift,* 49: 110-118.

Paratley, R.D. and T.J. Fahey. 1986. Vegetation-environment relations in a conifer swamp in central New York. *Bull. Torrey Bot. Club,* 113: 357-371.

Patrick, W.H., Jr., G. Dissmyer, D.D. Hook, V.M. Lambou, H.M. Leitman, and C.H. Wharton. 1981. Characteristics of wetlands ecosystems of southeastern bottomland hardwood forests. In *Report on a Bottomland Hardwood Wetlands Workshop.* J.R. Clark and J. Benforado (Eds.). Lake Lanier, Georgia. June 1-6 1980, National Wetlands Technical Council, Washington, D.C. 64-89.

Patterson, G.G., G.K. Speiran, and B.H. Whetstone. 1985. Hydrology and Its Effects on Distribution of Vegetation in Congaree Swamp National Monument, South Carolina. Water-Resources Investigations Report 85-4256. U.S. Geological Survey, Reston, VA.

Penfound, W.T. 1952. Southern swamps and marshes. *Bot. Rev.,* 18: 413-446.

Penfound, W.T. and E.S. Hathaway. 1938. Plant communities in the marshlands of southeastern Louisiana. *Ecol. Monogr.,* 8: 1-56.

Peters, D.D. 1990. Wetlands and Deepwater Habitats in the State of Washington. U.S. Fish and Wildlife Service, National Wetlands Inventory, Region 1, Portland, OR.

Pfister, R.D., B.L. Kovalchik, S.F. Arno, and R.C. Presby. 1977. Forest Habitat Types of Montana. Gen. Tech. Rept. INT-34. USDA Forest Service, Intermountain Forest and Range Expt. Stat., Ogden, UT.

Phillips, R.C. 1984. The Ecology of Eelgrass Meadows in the Pacific Northwest: A Community Profile. FWS/OBS-84/24. U.S. Fish and Wildlife Service, Washington, D.C.

Pomeroy, L.R. and R.G. Wiegert (Eds.). 1981. *The Ecology of a Salt Marsh.* Springer-Verlag, New York.

Pool, D.J., S.C. Snedaker, and A.E. Lugo. 1977. Structure of mangrove forests in Florida, Puerto Rico, Mexico, and Central America. *Biotropica,* 9: 195-212.

Post, R.A. 1996. Functional Profile of Black Spruce Wetlands in Alaska. EPA 910/R-96-006. U.S. Environmental Protection Agency, Region 10, Seattle, WA.

Purer, E.A. 1942. Plant ecology of the coastal salt marshlands of San Diego County, California. *Ecol. Monogr.,* 12: 81-111.

Redfield, A.C. 1965. The ontogeny of a New England salt marsh. *Science,* 147: 50-55.

Redfield, A.C. 1972. Development of a New England salt marsh. *Ecol. Monogr.,* 42: 201-237.

Reimold, R.J. 1977. Mangals and salt marshes of eastern United States. In *Wet Coastal Ecosystems.* V.J. Chapman (Ed.). Elsevier Scientific Publishing Co., Amsterdam, The Netherlands. 157-166.

Reimold, R.J. and W.H. Queen (Eds.). 1974. *Ecology of Halophytes.* Academic Press, New York.

Reinartz, J.A. and E.L. Warne. 1993. Development of vegetation in small created wetlands in southeastern Wisconsin. *Wetlands,* 13: 153-164.

Reiners, W.A. 1972. Structure and energetics of three Minnesota forests. *Ecol. Monogr.,* 42: 71-94.

Rheinhardt, R.D. 1991. Vegetation Ecology of Tidal Swamps of the Lower Chesapeake Bay, USA. Ph. D. dissertation. Virginia Institute of Marine Science, College of William and Mary, Gloucester, VA.

Rheinhardt, R.D. 1992. A multivariate analysis of vegetation patterns in tidal swamps of lower Chesapeake Bay, USA. *Bull. Torrey Bot. Club,* 119: 193-208.

Rice, E.L. 1965. Bottomland forests of northcentral Oklahoma. *Ecology,* 46: 708-714.

Richardson, C.J. (Ed.). 1981. *Pocosin Wetlands. An Integrated Analysis of Coastal Plain Freshwater Bogs in North Carolina.* Hutchinson Ross Publishing Co., Stroudsburg, PA.

Rigg, G.B. 1937. Some raised bogs of southeastern Alaska with notes on flat bogs and muskegs. *Am. J. Bot.,* 24: 194-198.

Rowell, C.M., Jr. 1971. Vascular plants of the playa lakes of the Texas Panhandle and South Plains. *Southwest. Nat.,* 15: 407-418.

Rowell, C.M., Jr. 1981. The flora of playa lakes. In FWS/OBS-81/07. Proc. Playa Lakes Symposium. U.S. Fish and Wildlife Service, Washington, D.C. 21-29.

Schlesinger, W.H. 1978a. Community structure, dynamics, and nutrient ecology in the Okefenokee Cypress Swamp-Forest. *Ecol. Monogr.,* 48: 43-65.

Schlesinger, W.H. 1978b. On the relative dominance of shrubs in Okefenokee Swamp. *Am. Nat.,* 112: 949-954.

Schomer, N.S. and R.D. Drew. 1982. An Ecological Characterization of the Lower Everglades, Florida Bay, and the Florida Keys. FWS/OBS-82/58.1. U.S. Fish and Wildlife Service, Washington, D.C.

Seliskar, D.M. and J.L. Gallagher. 1983. The Ecology of Tidal Marshes of the Pacific Northwest Coast: A Community Profile. FWS/OBS-82/32. U.S. Fish and Wildlife Service, Washington, D.C.

Seyer, S.C. 1979. Vegetation Ecology of a Montane Mire, Crater Lake National Park, Oregon. Master's thesis. Oregon State University, Corvallis.

Shaler, N.S. 1885. Sea coast swamps of the Atlantic Coast. *Rept. U.S. Geol. Surv.,* 6: 353-398.

Shaler, N.S. 1886. Beaches and Tidal Marshes of the Atlantic Coast. *Nat. Geog. Mon.,* 1.

Shanks, R.E. 1966. An ecological survey of the vegetation of Monroe County, New York. *Rochester Acad. Sci.,* 11: 108-252.

Sharitz, R.R. and J.W. Gibbons. 1982. The Ecology of Southeastern Shrub Bogs (Pocosins) and Carolina Bays: A Community Profile. FWS/OBS 82-04. U.S. Fish and Wildlife Service, Office of Biological Services, Washington, D.C.

Shear, T., M. Young, and R. Kellison. 1997. An Old-Growth Definition for Red River Bottom Forests in the Eastern United States. Gen. Tech. Rep. SRS-10. USDA Forest Service, Southern Research Station, Asheville, NC.

Shelford, V.E. 1954. Some Lower Mississippi Valley floodplain biotic communities: their age and elevation. *Ecology,* 35: 126-142.

Silberhorn, G.M., G.M. Dawes, and T.A. Barnard, Jr. 1974. Coastal Wetlands of Virginia: Interim Report No. 3. Virginia Institute of Marine Science, Gloucester Point, VA. Special Report in Applied Marine Science and Ocean Engineering No. 46.

Simpson, R.L., R.E. Good, M.A. Leck, and D.F. Whigham. 1983. The ecology of freshwater tidal wetlands. *BioScience,* 33: 255-259.

Steele, R., R.D. Pfister, R.A. Ryker, and J.A. Kittams. 1981. Forest Habitat Types of Central Idaho. Gen. Tech. Rept. INT-114. USDA Forest Service, Intermountain Forest and Range Expt. Stat., Ogden, UT.

Steele, R., S.V. Cooper, D.M. Ondov, D.W. Roberts, and R.D. Pfister. 1983. Forest Habitat Types of Eastern Idaho-Western Wyoming. Gen. Tech. Rept. INT-144. USDA Forest Service, Intermountain Forest and Range Expt. Stat., Ogden, UT.

Steen, O. and R. Dix. 1974. A Preliminary Classification of Colorado Subalpine Forests. Mimeo report. Colorado State University, Dept. of Biology and Plant Pathology, Ft. Collins, CO.

Stephens, F.R. and R.F. Billings. 1967. Plant communities of a tide-influenced meadow on Chicagof Island, Alaska. *Northwest. Sci.,* 41: 178-183.

Stevenson, R.E. and K.O. Emery. 1958. Marshlands at Newport Bay, California. *Occas. Pa. Allan Hancock Found.,* 20: 1-109.

Steward, K.K. and W.H. Ornes. 1975. The autecology of sawgrass in the Florida Everglades. *Ecology,* 56: 162-171.

Stewart, R.E. and H.A. Kantrud. 1971. Classification of Natural Ponds and Lakes in the Glaciated Prairie Region. Resource Publication 92. U.S. Fish and Wildlife Service, Washington, D.C.

Stewart, R.E. and H.A. Kantrud. 1972. Vegetation of Prairie Potholes, North Dakota, in Relation to Quality of Water and Other Environmental Factors. USGS Prof. Paper 585-D. U.S. Geological Survey, Reston, VA.

Stinnett, D.P., R.W. Smith, and S.W. Conrady. 1987. Riparian Areas of Western Oklahoma. U.S. Fish and Wildlife Service and Oklahoma Department of Conservation, Oklahoma City, OK.

Stout, J.P. 1984. The Ecology of Irregularly Flooded Salt Marshes of the Northeastern Gulf of Mexico: A Community Profile. Biological Report 85(7.1). U.S. Fish and Wildlife Service, Washington, D.C.

Stuckey, R.L. 1989. Western Lake Erie aquatic and wetland vascular plant flora: its origin and change. In Lake Erie Estuarine Systems: Issues, Resources, Status, and Management. U.S. Department of Commerce, National Oceanic and Atmospheric Administration, Estuarine Programs Office, Washington, D.C. 205-256.

Suloway, L. and M. Hubbell. 1994. Wetland Resources of Illinois: An Analysis and Atlas. Spec. Pub. 15. Illinois Natural History Survey, Champaign, IL.

Swan, J.M.A. and A.M. Gill. 1970. The origins, spread, and consolidation of a floating bog in Harvard Pond, Petersham, MA. *Ecology,* 51: 829-840.

Swinehart, A.L. and G.D. Starks. 1994. A record of the natural history and anthropogenic senescence of an Indiana tamarck bog. *Proc. IN Acad. Sci.,* 103: 225-239.

Szaro, R.C. 1989. Riparian forest and shrubland community types of Arizona and New Mexico. Special Issue. *Desert Plants,* 9: 1-138.

Taylor, J.R. 1985. Community Structure and Primary Productivity of Forested Wetlands in Western Kentucky. Ph. D. dissertation. University of Louisville, Louisville.

Taylor, J.R., P.L. Hill, R.W. Bosserman, and W.J. Mitsch. 1982. Ecosystem analysis of selected wetlands in the western Kentucky coalfield. In Proc. Symposium on Wetlands in the Unglaciated Appalachian Region. B.R. McDonald (Ed.). West Virginia University, Morgantown; May 20-28, 1982. West Virginia Department of Natural Resources, Wildlife Resources Division, Elkins, WV. 75-85.

Taylor, N. 1938. A preliminary report on the salt marsh vegetation of Long Island, New York. *Bull. NY State Mus.,* 316: 21-84.

Teal, J. and M. Teal. 1969. *Life and Death of the Salt Marsh.* Audubon/Ballantine Books, New York.

Texas Parks and Wildlife Department. 1988. The Texas Wetland Plan — Addendum to the 1985 Texas Outdoor Recreation Plan. Austin, TX.

Thayer, G.W., W.J. Kenworthy, and M.S. Fonseca. 1984. The Ecology of Eelgrass Meadows of the Atlantic Coast: A Community Profile. FWS/OBS-84/02. U.S. Fish and Wildlife Service, Washington, D.C.

Tiner, R.W., Jr. 1977. An Inventory of South Carolina's Coastal Marshes. Tech. Rep. No. 23. South Carolina Marine Resources Center, Charleston, SC.

Tiner, R.W., Jr. 1985a. Wetlands of New Jersey. U.S. Fish and Wildlife Service, National Wetlands Inventory, Newton Corner, MA.

Tiner, R.W., Jr. 1985b. Wetlands of Delaware. U.S. Fish and Wildlife Service. National Wetlands Inventory, Newton Corner, MA and Delaware Department of Natural Resources and Environmental Control, Wetlands Section, Dover, DE. Cooperative Publication.

Tiner, R.W., Jr. 1987a. Preliminary National Wetlands Inventory Report on Vermont's Wetland Acreage. U.S. Fish and Wildlife Service, Newton Corner, MA.

Tiner, R.W., Jr. 1987b. *A Field Guide to Coastal Wetland Plants of the Northeastern United States.* University of Massachusetts Press, Amherst.

Tiner, R.W., Jr. 1988. Field Guide to Nontidal Wetland Identification. U.S. Fish and Wildlife Service. Region 5, Newton Corner, MA and Maryland Dept. of Natural Resources, Annapolis, MD. Cooperative Publication.

Tiner, R.W., Jr. 1989. Wetlands of Rhode Island, U.S. Fish and Wildlife Service, Region 5 Newton Corner, MA and U.S. Environmental Protection Agency, Region 1 Boston, MA. Cooperative Publication.

Tiner, R.W. 1990. Pennsylvania's Wetlands: Current Status and Recent Trends. U.S. Fish and Wildlife Service, Region 5, Hadley, MA.

Tiner, R.W. 1992. Preliminary National Wetlands Inventory Report on Massachusetts' Wetland Acreage. U.S. Fish and Wildlife Service, Hadley, MA.

Tiner, R.W. 1993. *Field Guide to Coastal Wetland Plants of the Southeastern United States.* University of Massachusetts Press, Amherst.

Tiner, R.W. 1996. Current Status of West Virginia's Wetlands: Results of the National Wetlands Inventory. U.S. Fish and Wildlife Service, Region 5, Hadley, MA.

Tiner, R.W. 1998. *In Search of Swampland. A Wetland Sourcebook and Field Guide.* Rutgers University Press, New Brunswick, NJ.

Tiner, R.W. and D.G. Burke. 1995. Wetlands of Maryland. U.S. Fish and Wildlife Service, Region 5, Hadley, MA and Maryland Department of Natural Resources, Annapolis, MD. Cooperative publication.

Tiner, R.W., Jr. and J.T. Finn. 1986. Status and Recent Trends of Wetlands in Five Mid-Atlantic States. U.S. Fish and Wildlife Service, Newton Corner, MA and U.S. Environmental Protection Agency, Philadelphia, PA. Cooperative Report.

Trettin, C.C., M.E. Jurgensen, M.R. Gale, D.F. Grigal, and J.K. Jeglum (Eds.). 1997. *Northern Forested Wetlands. Ecology and Management.* Lewis Publishers, Boca Raton, FL.

Twilley, R.R. 1998. Mangrove wetlands. Chapter 18. In *Southern Forested Wetlands Ecology and Management.* M.G. Messina and W.H. Conner (Eds.). Lewis Publishers, Boca Raton, FL. 445-473.

Tyndall, R.W., K.A. McCarthy, J.C. Ludwig, and A. Rome. 1990. Vegetation of six Carolina Bays in Maryland. *Castanea,* 55: 1-21.

Ungar, I.A. 1974. Inland halophytes of the United States. In *Ecology of Halophytes.* R.J. Reimold and W.H. Queen (Eds.). Academic Press, New York. 235-305.

U.S. Department of Agriculture. 1982. Natural Resources Inventory. Washington, D.C.

U.S. Fish and Wildlife Service. 1979. The Playas of the Western Texas Region. Region 2, Albuquerque, NM.

Utah Division of Parks and Recreation and Utah Department of Natural Resources. 1988. Utah's Wetlands: An Important Outdoor Recreation Resources. Salt Lake City, UT.

van der Valk, A.G. 1985. Vegetation dynamics of prairie glacial marshes. In *The Population Structure of Vegetation.* J. White (Ed.). Dr. W. Junk Publishers, Dordrecht, The Netherlands. 293-312.

van der Valk, A.G. (Ed.). 1989. *Northern Prairie Wetlands.* Iowa State University Press, Ames.

Veneman, P.L.M. and R.W. Tiner. 1990. Soil-Vegetation Correlations in the Connecticut River Floodplain of Western Massachusetts. Biol. Rep. 90(6). U.S. Fish and Wildlife Service, Washington, D.C.

Viereck, L.A., C.T. Dyrness, C.T. Batten, and K.J. Wenzlick. 1992. The Alaska Vegetation Classification. General Tech. Rep. PNW-GTR-286. USDA Forest Service, Pacific Northwest Research Station, Portland, OR.

Vince, S.W., S.R. Humphrey, and R.W. Simons. 1989. The Ecology of Hydric Hammocks: A Community Profile. Biol. Rep. 85 (7.26). U.S. Fish and Wildlife Service, Washington, D.C.

Vogl, R.J. 1966. Salt-marsh vegetation of Upper Newport Bay, California. *Ecology,* 47: 80-87.

Vogl, R.J. and J. Henrickson. 1971. Vegetation of an alpine bog on East Maui, Hawaii. *Pac. Sci.,* 25: 475-483.

Volland, L.A. 1976. Plant Communities of the Central Oregon Pumice Zone. R6-ECOL-104-1985. USDA Forest Service, Pacific Northwest Region, Portland, OR.

Walker, M.D., D.A. Walker, and K.R. Everett. 1989. Wetland Soils and Vegetation, Arctic Foothills, Alaska. Biol. Rep. 89(7). U.S. Fish and Wildlife Service, Washington, D.C.

Wass, M.L. and T.D. Wright. 1969. *Coastal Wetlands of Virginia.* Spec. Rep. Appl. Mar. Sci. and Ocean Eng. No. 10. Virginia Institute of Marine Sciences, Gloucester, VA.

Weinmann, F.C. and K. Kunz. 1995. A common error in assessing hydrophytic vegetation for wetland identification. In National Workshop on Wetlands, Technology Advances for Wetland Science. April 1995, New Orleans, LA. U.S. Army Engineer Waterways Expt. Station, Vicksburg, MS. 110-115.

Weller, M.W. 1981. *Freshwater Marshes. Ecology and Wildlife Management.* University of Minnesota Press, Minneapolis.

Wells, B.W. 1928. Plant communities of the coastal plain of North Carolina and their successional relations. *Ecology,* 9: 230-242.

Wharton, C.H. 1978. The Natural Environments of Georgia. Georgia Department of Natural Resources, Atlanta, GA.

Wharton, C.H., H.T. Odum, K. Ewel, M. Duever, A. Lugo, R. Boyt, J. Bartholomew, E. DeBellevue, W. Brown, M. Bown, and L. Duever. 1976. *Forested Wetlands of Florida — Their Management and Use, Center for Wetlands.* University of Florida, Gainesville.

Wharton, C.H., W.M. Kitchens, E.C. Pendleton, and T.M. Sipe. 1982. The Ecology of Bottomland Hardwood Swamps of the Southeast: A Community Profile. FWS/OBS-81/37. U.S. Fish and Wildlife Service, Washington, D.C.

Whigham, D.F., D. Dykyjová, and S. Hejný (Eds.). 1993. *Wetlands of the World: Inventory, Ecology and Management,* vol. I. Kluwer Academic Publishers, Dordrecht, The Netherlands.

White, W.A., T.A. Tremblay, E.G. Wermund, Jr., and L.R. Handley. 1993. Trends and Status of Wetland and Aquatic Habitats in the Galveston Bay System, Texas. Galveston Bay National Estuarine Program. Pub. No. GBNEP-31.

Whitehead, D.R. 1972. Development and environmental history of the Dismal Swamp. *Ecol. Monogr.,* 42: 301-315.

Widoff, L. 1988. Maine Wetlands Conservation Priority Plan. Maine Bureau of Parks and Recreation, Maine State Planning Office, and Wetlands Subcommittee, Land and Water Resources Council. Augusta, ME.

Wiegert, R.G. and B.J. Freeman. 1990. Tidal Salt Marshes of the Southeast Atlantic Coast: A Community Profile. Biol. Rep. 85 (7.29). U.S. Fish and Wildlife Service, Washington, D.C.

Williams, C.K. and T.R. Lillybridge. 1983. Forested Plant Associations of the Okanogan National Forest. R6-Ecol-132b-1983. USDA Forest Service, Pacific Northwest Region, Portland, OR.

Wilkinson, D.L., K. Schneller-McDonald, R.W. Olson, and G.T. Auble. 1987. Synopsis of Wetland Functions and Values: Bottomland Hardwoods with Special Emphasis on Eastern Texas and Oklahoma. Biol. Rep. 87(12). U.S. Fish and Wildlife Service, Washington, D.C.

Windell, J.T., B.E. Willard, D.J. Cooper, S.Q. Foster, C.F. Knud-Hanson, L.P. Rink, and G.N. Kiladis. 1986. An Ecological Characterization of Rocky Mountain Montane and Subalpine Wetlands. Biol. Rep. 86(11). U.S. Fish and Wildlife Service, Washington, D.C.

Winter, T.C. and M-K. Woo. 1990. Hydrology of lakes and wetlands. In *Surface Water Hydrology.* The Geological Society of America, vol. 0-1. 159-187.

Wisconsin Department of Natural Resources. 1992. Wisconsin Wetlands Priority Plan — An Addendum to Wisconsin's 1986-91 Statewide Comprehensive Outdoor Recreation Plan. Madison, WI.

Wistendahl, W.A. 1958. The flood plain of the Raritan River, New Jersey. *Ecol. Monogr.,* 28: 129-153.

Worley, I.A. 1981. Maine Peatlands. Their Abundance, Ecology, and Relevance to the Critical Areas Program of the State Planning Office. Planning Rept. No. 73. Maine Critical Areas Program, Augusta, ME.

Wright, A.H. and A.A. Wright. 1932. The habitats and composition of the vegetation of Okefenokee Swamp, Georgia. *Ecol. Monogr.,* 2: 109-232.

Wright, H.E., Jr., B.A. Coffin, and N.E. Aaseng (Eds.). 1992. *The Patterned Peatlands of Minnesota.* University of Minnesota Press, Minneapolis.

Wyoming Game and Fish Department. 1987. Wetlands Component — 1985 Wyoming State Comprehensive Outdoor Recreation Plan. Cheyenne, WY.

Youngblood, A.P. 1984. Coniferous Forest Habitat Types of Central and Southern Utah. USDA Forest Service, Intermountain Forest and Range Expt. Stat., Ogden, UT.

Youngblood, A.P. and W.F. Mueggler. 1981. Aspen Community Types on the Bridger-Teton National Forest in Western Wyoming. Res. Paper INT-272. USDA Forest Service, Intermountain Forest and Range Expt. Stat., Ogden, UT.

Youngblood, A.P., W.G. Padgett, and A.H. Winard. 1985. Riparian Community Type Classification for Eastern Idaho–Western Wyoming. R4-ECOL-85-01. USDA Forest Service, Intermountain Forest and Range Expt. Stat., Ogden, UT.

Zedler, J.B. 1982. The Ecology of Southern California Coastal Salt Marshes: A Community Profile. FWS/OBS-81/54, U.S. Fish and Wildlife Service, Washington, D.C. 110.

Zedler, P.H. 1987. The Ecology of Southern California Vernal Pools: A Community Profile. Biol. Rep. 85(7.11). U.S. Fish and Wildlife Service, Washington, D.C.

Zedler, J.B. and C.S. Nordby. 1986. The Ecology of Tijuana Estuary, California: An Estuarine Profile. Biol. Rep. 85(7.5). U.S. Fish and Wildlife Service, Washington, D.C.

Zieman, J.C. 1982. The Ecology of the Seagrasses of South Florida: A Community Profile. FWS/OBS-82/25. U.S. Fish and Wildlife Service, Washington, D.C.

Zieman, J.C. and R.T. Zieman. 1989. The Ecology of the Seagrass Meadows of the West Coast of Florida: A Community Profile. Biol. Rep. 85(7.25). U.S. Fish and Wildlife Service, Washington, D.C.

10 Wetland Mapping and Photointerpretation

INTRODUCTION

While much of this book has been devoted to field indicators of wetlands, signs of wetlands also may be observed on remotely sensed imagery (i.e., aerial photographs and satellite imagery). Aerial photographs have been used for decades to identify wetlands and other natural resources for site-specific assessments and for producing thematic maps for wetland inventories and land cover/use surveys. Satellite imagery is a more recent source for geospatial data.

Use of remotely sensed information also is important for wetland delineation. While positive identification of wetlands and accurate delineation of their boundaries requires field inspection, a review of existing information (maps and aerial photos) for a project site is recommended before conducting field investigations. Reviewing existing aerial photographs is particularly valuable as it permits an update or verification of features shown on various wetland maps or soil survey maps. Changes may have occurred since the maps were prepared and maps also have scale limitations that may prohibit small wetlands from being depicted. These wetlands may be observed on aerial photography.

This chapter offers an overview of wetland mapping and photointerpretation techniques as these topics are most useful for wetland identification and delineation, especially site-specific evaluations. Satellite imagery is not discussed since it requires technical skills and specialized equipment not available to most wetland delineators. The discussion of wetland mapping will focus on the U.S. Fish and Wildlife Service's National Wetlands Inventory (NWI) maps and the USDA Natural Resource Conservation Service's soil survey reports. These two products are the most widely available maps useful for wetland identification and delineation. For more in-depth treatment of remote sensing applications, consult textbooks such as *Manual of Photographic Interpretation* (Philipson, 1997) and *Interpretation of Airphotos and Remotely Sensed Imagery* (Arnold, 1997).

MAPS TO AID IN WETLAND IDENTIFICATION

Wetland maps are useful tools for wetland identification and classification. The availability of such maps in digital form (e.g., ARC/INFO) now facilitates geospatial analysis over broad geographic areas. This allows wetland data to be combined with other digital data for more sophisticated analyses of land and water resources.

The federal government has been producing fairly detailed wetland maps since the late 1970s and conducting soil surveys since the early 1900s. More recently, the USDA Natural Resources Conservation Service (NRCS) has been mapping wetlands for implementing the Swampbuster provision of the Food Security Act and for compiling statistics for its Natural Resources Inventory Program.* The National Marine Fisheries Service also has prepared maps showing the distribution of submerged aquatic vegetation in many of the nation's estuaries. The National Oceanic and Atmospheric Administration (NOAA) is monitoring coastal wetland changes through the use of satellite imagery (Coastal Change Analysis Program, C-CAP). The EPA has an environmental

---

* This mapping utilizes 35 mm color slides captured in summer to confirm crop status for compliance with agricultural subsidy programs (National Research Council, 1995). Unfortunately, this time of year is not conducive to wetland identification, hence numerous conventions need to be established to interpret wetlands (see Chapter 6, Table 6.2).

monitoring program (Environmental Monitoring and Assessment Program, E-MAP) that also uses satellite imagery to monitor the nation's wetlands and surface waters. E-MAP uses NWI maps to aid in processing satellite imagery. Kiraly et al. (1990) provide a review of federal programs used to monitor coastal resources.

Some states, counties, and municipalities have published detailed wetland maps for regulatory purposes. Many states on the Atlantic Coast have produced maps that actually designate official regulatory boundaries of coastal wetlands (e.g., Connecticut, New York, New Jersey, Maryland, Delaware, and Virginia). Some states have prepared maps of inland wetlands also that show regulatory boundaries (e.g., New York and New Jersey). The state of Vermont has adopted NWI maps as a regulatory tool to designate wetlands of significance. Massachusetts had a unique program that produced highly detailed wetland maps that were once used to place deed restrictions on private property containing wetlands — the Wetlands Conservancy Program (MacConnell et al., 1992); the maps are now used for resource planning and management purposes only. Other states have chosen to publish inland wetland maps for guidance purposes (e.g., Delaware, Maryland, and Wisconsin). For information on state mapping, contact the appropriate state wetland regulatory agency. Local governments in some affluent and growing suburban areas have produced wetland maps to aid in adminstering wetland ordinances or by-laws.

Map Limitations

Maps prepared by interpreting remotely sensed imagery (e.g., aerial photos or satellite imagery) have inherent limitations related to many factors, including the nature of the resources being mapped (e.g., their ease or difficulty of recognition), map scale (e.g., balancing minimum mapping units against map legibility), quality and scale of source imagery, environmental conditions present when imagery was captured (e.g., leaf-on, leaf-off, wet season, or dry season), the emulsion (for photos), the spectral bands analyzed (for satellite imagery), and the cartographic equipment used to prepare the maps, plus the skills of the photointerpreters, image processors, and cartographers. Even detailed site-specific maps prepared from on-the-ground surveys have limitations due to scale as well as some of the other factors listed above (e.g., nature of the resource mapped).

Wetlands pose special problems for accurate mapping due to their alternating wet–dry nature. While many wetlands are quite distinct due to observed wetness or unique vegetation, many others are not readily identified either on-the-ground or by interpretation of remotely sensed imagery. Wetland identification often requires analyzing subtle changes in vegetation patterns, soil properties, and signs of hydrology, especially for drier-end wetlands. The National Research Council (NRC, 1995) reported that "mapping wetlands in level landscapes, such as coastal or glaciolacustrine plains, is less precise because boundaries are not as evident." Problems associated with photo-interpreting forested wetlands have been reported (Tiner, 1990; Peters, 1994; NRC, 1995; see discussion later in this chapter). The point to remember is that the more difficult the wetland type is to identify on the ground, the more conservatively such types will be represented on maps produced by remote sensing techniques.

Maps produced by different agencies may reflect differences in the definition of wetlands (see Chapter 1). Also advances in the science of wetlands have dramatically improved our ability to recognize and delineate wetlands. Since 1989, the federal government has been using standardized procedures for delineating wetlands for regulatory purposes across the country. Prior to this time, wetland delineations were inconsistent nationally and probably inaccurate especially in areas of low relief due to the lack of science-based, standardized procedures. During the 1980s, these techniques were under development. Consequently, early (pre-mid-1980s) wetland maps may not reflect the current concept of wetland and may be based more on wetlands that are readily recognized by their vegetation and hydrology, while more recent mapping should include undrained hydric soils.

There also are other factors that affect wetland mapping. For example, the date and season of photos (leaf-on vs. leaf-off), the wetness conditions at the time of photography, the photo scale,

TABLE 10.1
Examples of Minimum Mapping Unit Sizes
for Different Scales of Soil Maps

Map Scale	Minimum Size Delineation (acres)	Map Scale	Minimum Size Delineation (acres)
1:500	0.0025	1:15840	2.5
1:2000	0.40	1:20000	4.0
1:5000	0.25	1:24000	5.7
1:7920	0.62	1:31680	10.0
1:10000	1.00	1:62500	39.0
1:12000	1.43	1:100000	100.0

Note: The minimum-sized delineation is based on a ¼ × ¼ in. square which is roughly the smallest area in which a symbol can be placed.

Source: Soil Survey Division Staff. 1993. Soil Survey Manual. U.S. Department of Agriculture, Washington, D.C.

the quality of the photo processing, and skill of the photointerpreter affect the ability to produce a high-quality wetland map (see discussion on wetland photointerpretation later in this chapter).

Map scale is an important issue. All maps have a minimum mapping unit (mmu) whether specified or not. It is often related to the scale of the map given considerations of map legibility (Table 10.1). These mmu values are for land use/cover mapping or soil mapping where the entire map is labelled with fairly specific classifications. In contrast, when a map is emphasizing one or two features (e.g., wetlands and deepwater habitats), the minimum mapping unit can be smaller than the ones listed in the table. Such units may be called target mapping units (tmu). For a wetland map, a tmu is an estimate of the minimum sized wetland that will be consistently mapped. It is not the smallest wetland that appears on the map, but it is the size class of the smallest group of wetlands that are consistently shown. While knowledge of the tmu may be important to the users, accurately determining the tmu is another matter. For wetlands, some types are conspicuous (e.g., ponds, flooded basin wetlands, and bogs), allowing smaller ones to be mapped. Other types (e.g., evergreen forested wetlands, drier-end wetlands, and significantly drained wetlands) may be more difficult to photointerpret and larger ones may be missed. This is inherent in the use of remote sensing to map wetlands.

Maps produced by photointerpretation will never be as accurate as a detailed on-the-ground delineation, except perhaps where topographic differences are abrupt and hydrologic differences obvious. Minutes of photointerpretation time cannot hope to improve upon hours of field work examining plants, soils, and signs of hydrology and flagging the often undulating boundaries of wetlands. This is not to say that photointerpretation cannot produce relatively accurate boundaries at a fraction of the cost of doing on-the-ground delineation. For some types in certain landscapes (e.g., marshes, fens, bogs, and wetter swamps in depressional areas surrounded by upland or in open water), photointerpretation works well for locating the boundary of these types. For other types and in drier situations (e.g., evergreen forested wetlands, drier-end wetlands in relatively flat landscapes, and significantly drained wetlands), photointerpretation will only produce generalized boundaries that may vary considerably in the field. One must also remember that for regulatory purposes, a site inspection is required to accurately delineate the wetland boundaries of all wetlands at a project site, whether or not they appear on various wetland maps. In cases where the maps designate official regulatory boundaries, field work is needed to verify the accuracy of such mapping.

Finally, there are two general ways to approach wetland mapping, recognizing the difficulties of mapping this resource. The first is driven by a desire to map wetlands that are more or less readily photointerpreted. Following this approach means that if an area is mapped as a wetland, it should be correct or have a very high probability of being a wetland. This approach leads to more Type I errors (errors of omission), as emphasis is placed on mapping photointerpretable wetlands, so wetlands that are not, are missed. This approach is typically used in making National Wetlands Inventory maps (see below). The other approach is based on the goal of showing all possible wetlands and accepting misclassifications in the process. This type of mapping will likely lead to more Type II errors (errors of commission), where some to many designated wetlands are actually uplands. Each approach has its merits, and it may actually be most desirable to have a map showing both the photointerpretable wetlands and other "possible" wetlands based on other considerations (e.g., landscape position, landform contiguous to interpretable wetlands, etc.).

NATIONAL WETLANDS INVENTORY MAPS

The U.S. Fish and Wildlife Service (FWS), through its National Wetlands Inventory (NWI) Project, is producing a series of large-scale (1:24,000; 1:63,360 for Alaska) maps that show the location, size, and type of wetlands within defined geographical areas for the entire country. The maps are produced to aid resource managers and planners in making wise decisions regarding the fate of wetlands. They serve to advise the public on where wetlands are located, e.g., to help assess the development potential of real estate. The NWI maps also may serve as a base to monitor future changes in the nation's wetland resource.

Mapping Procedure

The NWI maps are prepared through conventional photointerpretation techniques using mid- to high-altitude aerial photographs. Most of the photography is acquired by the federal government's aerial photography programs, typically 1:58,000 color infrared (early to mid-1980s) and 1:40,000 color infrared or black and white (late 1980s to present). The earliest NWI maps used 1:80,000 black and white photography which was the best national high-altitude photography available in the late 1970s. For wetland mapping, leaf-off color infrared photography is best (see discussion at end of chapter). Consequently, the source imagery imposes certain limitations on the accuracy of the NWI maps.

Once aerial photos have been received and data-preped (e.g., work areas drawn and overlays attached), the following steps are taken:

1. Review aerial photographs to identify obvious wetland types and problematical areas (i.e., wetland vs. upland and classification questions — cover types, water regimes, etc.).
2. Select sites and a route for field checking problematic areas and obvious wetland types, emphasizing the former.
3. Conduct field checks in the study area (usually one or two 1:100,000 scale map sheets per week of field work, depending on wetland density and complexity) and collect site-specific data to resolve photointerpretation questions.
4. After the field trip, review field sites on stereoscopic aerial photographs to become more familiar with the appearance of the diversity of wetlands in the study area; prepare a field trip report.
5. Perform stereoscopic photointerpretation of the study area following NWI standard conventions (U.S. Fish and Wildlife Service, 1995), delineate wetland boundaries on photographic overlays, classify each wetland polygon according to the official FWS wetland classification system (Cowardin et al., 1979), and consult existing collateral information (e.g., soil survey reports) as needed.
6. Conduct a followup field trip, if necessary, to resolve new problems that arose during photointerpretation before finalizing overlays.

7. Perform quality control of interpretations at regional and national levels. (Regional quality control involves reviewing every interpreted photograph for possible additions, deletions, and misclassifications; this is done at the FWS Regional Offices. National consistency quality control includes random checking of interpreted photographs to ensure compliance with standards for classification and delineation; this work is performed at the NWI national mapping office.)

8. Prepare draft large-scale (1:24,000) wetland maps following NWI cartographic conventions and using the U.S. Geological Survey topographic maps as bases (U.S. Fish and Wildlife Service, 1994).

9. Coordinate interagency (federal and state) review of draft maps and, if necessary, conduct field checking.

10. Prepare "edited" draft map for final map production.

11. Produce final NWI maps (Figure 10.1) and digitize maps where funding is available. (NWI maps can be ordered by calling 1-800-USA-MAPS; maps also are available from NWI map distribution centers in many states. Maps are available in digital form for many areas through the Internet: *www.nwi.fws.gov.*)

Features Mapped

The NWI maps show: (1) the location and shape of wetlands and deepwater habitats; (2) the type of wetland based on vegetation (or substrate, where vegetation is absent), water regime, salinity (for tidal areas), and other characteristics; and (3) the type of deepwater habitat based on ecological system, hydrology (tidal/nontidal), and other features (e.g., impounded). While the NWI maps depict the location of a large number of wetlands and probably the ones most important to wetland-dependent fish and wildlife resources and flood storage, not all wetlands are shown.

Wetlands and deepwater habitats are classified according to the FWS's official wetland classification system (Cowardin et al., 1979; see Chapter 8 for overview). NWI maps use an alpha-numeric code to designate the wetland type according to this system. For example, the code PFO1C, representing a common nontidal wetland type, can be broken down as follows: P–Palustrine (system); FO–Forested Wetland (class); 1–Broad-Leaved Deciduous (subclass); and C–Seasonally Flooded (water regime). This type includes red maple swamps, ash swamps, bottomland hardwood swamps, and cottonwood floodplain wetlands. The code PEM1F indicates a Palustrine (P) Emergent Wetland (EM), Persistent (1), and Semipermanently Flooded (F). Two common examples of this type are cattail and tule (bulrush) marshes. Leatherleaf-dominated shrub bogs are designated as PSS3Ba — Palustrine (P) Scrub-Shrub Wetland (SS), Broad-leaved Evergreen (3), Saturated (B), and acidic (a). Tidal salt marshes are shown as E2EM1P or E2EM1N — Estuarine (E) Emergent Wetland (EM) Persistent (1) and Irregularly Flooded (P) or Regularly Flooded (N). A legend at the bottom of each NWI map explains the alpha-numeric code symbology (Figures 10.2a/b).

All wetlands are not shown on the maps as they are derived from photointerpretation of mid-level to high-altitude aerial photographs with selective field checking. The effective date of the map is based on the aerial photos used. In many regions, the NWI maps are now 20 years old and are, for most practical purposes, obsolete, especially in areas of high- to moderate-growth, agricultural expansion, and growing beaver populations. The photo data used to prepare each map are recorded in the legend. Another important point to know is the target mapping unit (tmu; i.e., the smallest area consistently mapped). The tmu varies with the scale of the photography used in the mapping effort (see Table 10.2 for regional targets). The tmu is generally the minimum-sized wetland or class size of wetlands consistently mapped within the limits of the aerial photography's ability to reveal various wetland types.* For conspicuous types like ponds, wetlands even smaller than the

---

* A tmu of 1 to 3 acres means that nearly all of wetlands larger than this size class should be mapped (within the limits of photointerpretation), while more than half of the wetlands in this range also should be mapped, along with some conspicuous smaller ones (e.g., ponds).

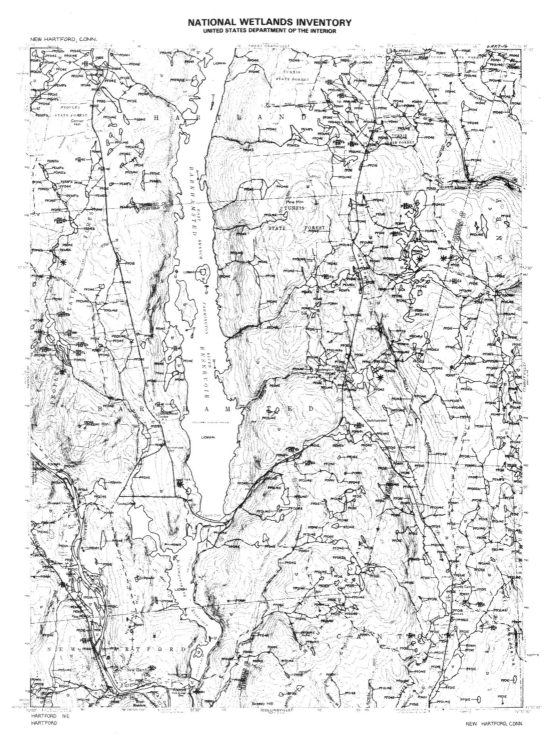

FIGURE 10.1   Example of a National Wetlands Inventory (NWI) map (reduced, original scale is 1: 24,000).

target are observed on the photos and are frequently mapped. Yet, other types such as forested wetlands, especially drier ones and evergreen ones, are more difficult to photointerpret regardless of photo scale and seasonality. Such types will be more conservatively mapped. For grassland

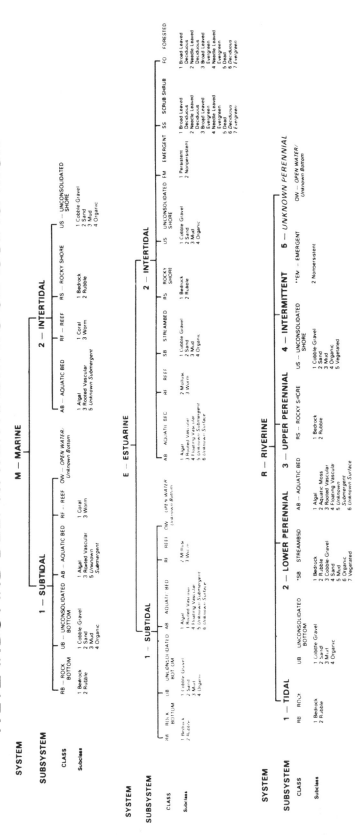

FIGURE 10.2 Legend for National Wetlands Inventory (NWI) map.

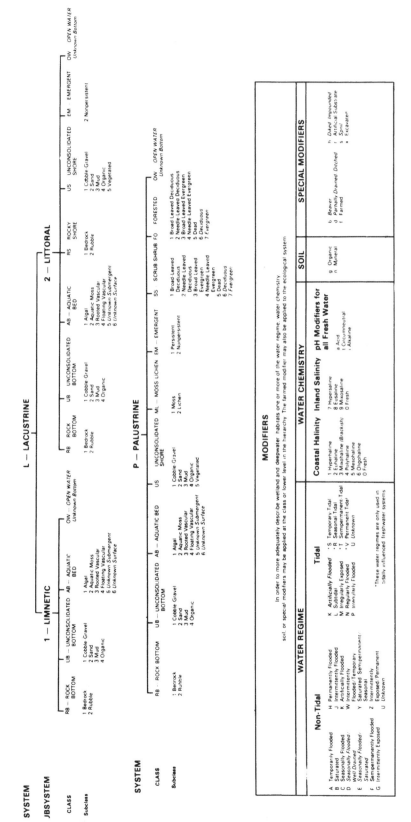

FIGURE 10.2 (continued)

TABLE 10.2
Examples of Target Mapping Units for NWI Maps Across the Country

Region	Target Mapping Unit
Northeast	Varies with photo scale:
	1:80,000 = 3–5 acres
	1:58,000 = 1–3 acres
	1:40,000 = 1 acre
	(Must recognize limitations for mapping temporarily flooded and seasonally saturated types)
Southeast	Same as for Northeast
Midwest	1–3 acres (in general)
	1/4–1 acre in Pothole Region and other agricultural areas
Interior West	No minimum established due to occurrence of most wetlands in open grasslands or agricultural fields or in conspicuous narrow drainages in mountains
Northwest	Same as for the Northeast
Southwest	1–3 acres
Alaska	2–5 acres

Source: Tiner, R.W. 1997a. NWI Maps: What they tell us. *National Wetlands Newsletter,* 19(2): 7-12.

regions like the Midwest prairies, very small pothole wetlands (less than 0.25 acres) are readily identified and usually mapped, while in forested regions like the eastern U.S., mapping of forested wetlands will be more conservative. Wetlands must be photointerpretable in order to be mapped, unless detected during field investigations.

NWI Map Strengths and Weaknesses

In many areas of the country, NWI maps are the only wetland maps available. Since they are specifically designed to show wetlands and deepwater habitats, they are more comprehensive than the U.S. Geological Survey maps. The wetland data also are more current than the U.S.G.S. base map information (e.g., swamp and marsh symbols).

In evaluating various remote sensors for wetland mapping, the Wetland Subcommittee of the Federal Geographic Data Committee concluded that "the best technique for initial wetland habitat mapping and inventory is the technique currently used by the FWS's NWI project…" (Federal Geographic Data Committee, 1992). While admittedly, the use of larger scale aerial photos and performing more field checking would improve the resolution and accuracy of NWI maps, the maps as currently produced provide an acceptable level of spatial and classification information for general planning purposes and for tracking broad wetland trends.

By current design, NWI maps tend to err more by omission (Type I error) than by commission (Type II error). This means that if an NWI map indicates the presence of wetland in a given area, it is highly likely that a wetland is there. This is supported by several studies (e.g., Swartwout et al., 1982; Nichols, 1994; Stolt and Baker, 1995). Conversely, if an NWI map does not indicate a wetland, one is usually not there, but users should not be surprised to find unmapped wetlands (especially drier-end wetlands and difficult to photointerpret wetlands, e.g., certain evergreen forested wetlands, farmed wetlands, mowed wetlands, and significantly drained wetlands) on the ground, particularly in favorable landscape positions such as along streams in narrow valleys or in depressions.

In most cases, the larger and wetter wetlands plus most open waterbodies are depicted on the NWI maps. An exception is NWI mapping in the Prairie Pothole region and other grassland areas where wetlands are highly photointerpretable, allowing very small wetlands to be mapped (Table 10.2).

The NWI maps and published state wetland reports have provided the public with better information on the distribution of wetlands within states than previously available. NWI maps can form the base for more detailed local inventories, such as was done in Puget Sound, Washington (Granger, 1995).

Another strength of NWI maps is that they attempt to show all types of wetland regardless of whether they are regulated or not. This allows the maps to be used by federal, state, and local governments and others interested in wetland resources as well as by researchers and natural resource managers. In some areas, such as the Atlantic–Gulf coastal plain, many designated wetlands are not regulated by the Corps because they fail the Corps manual's 3-parameter test. Many NWI wetlands: have plant communities dominated by FAC– or FACU species, possess wet soils that lack typical hydric soil properties, lack currently accepted wetland hydrology indicators, or are not wet enough during the defined growing season to qualify as a regulated wetland, despite significant wetness during the year (Tiner, 1993 a;b; National Research Council, 1995). While regulatory requirements may change, the NWI maps attempt to show scientifically accepted wetlands. As a result, the FWS's wetland trends studies show how the nation's wetland resource (at least that which is photointerpretable) is faring and, therefore, provide a consistent means of assessing wetland trends.

Table 10.3 lists some major limitations of NWI mapping due to reliance on photointerpretation. Despite these limitations, the NWI maps were "generally found to be very accurate" in a multi-agency field evaluation of NWI maps and satellite mapping (produced by NOAA) in Maryland (Maryland Department of Natural Resources, 1991). Several studies have reported high accuracies of NWI maps (Swartwout et al., 1982, Crowley et al., 1988, U.S. Fish and Wildlife Service, 1992, and Nichols, 1994) in Massachusetts, Vermont, New Jersey, and Maine, respectively. The National Research Council (1995) noted that "wetland delineation on NWI maps is generally accurate [in] areas where there is an abrupt change in hydrology, soil, or vegetation at the wetland boundary." Other studies have found significant omissions of wetlands from NWI maps when compared to regulatory boundaries (Lukin and Mauger 1983; Thurston Regional Planning Council, 1991; Moorhead and Cook, 1992; Stolt and Baker, 1995; McMullen and Meacham, 1996). The Thurston Regional Planning Council study reported significant omissions on NWI maps that were later found to be the result of a digitizing error by the researchers. Forested wetlands, small wetlands, and narrow (linear) wetlands tend to be the major sources of omissions. Drier-end hydric mineral soils also pose problems for wetland mapping as they do for onsite delineation. Also, the fact that NWI maps intentionally do not show farmed wetlands in many parts of the country also leads to a significant underestimate of the amount of wetland.

Uses of NWI Maps

NWI maps have been used for a multitude of purposes. The most frequent usage is by wetland regulators, the regulated public, and environmental consultants for preliminary site assessment (i.e., identification of potentially regulated wetlands). Other map uses include:

- Refuge planning and acquisition
- Park management
- Watershed planning
- Habitat protection planning
- Environmental impact statements
- Preliminary facility and transportation/utility corridor siting impact analysis
- Oil spill contingency planning
- Potential wetland restoration site identification
- Natural resource inventories
- Wildlife surveys (identification of important habitats)
- Preliminary assessment of damaged resources at Superfund sites
- Land appraisals (e.g., payments in lieu of taxes and to determine appropriate taxes)

TABLE 10.3
Examples of Major NWI Map Limitations

1. **Target Mapping Unit (tmu).** A tmu is an estimate of the minimum sized wetland that the NWI is attempting to consistently map. It is not the smallest wetland shown on the maps. The tmu for wetlands generally varies with the scale of the aerial photography used, wetland type, project design, and funding.

2. **Spring Photography.** Where spring photography was used, aquatic beds and nonpersistent emergent wetlands may be undermapped; these areas were classified as open water, unless vegetation was observed during field investigations. In a few cases, scrub-shrub wetlands such as buttonbush swamps were submerged, avoiding photo-detection; they too would be included within mapped open waterbodies. In some instances, flooded emergents may have been misclassified as scrub-shrub wetlands due to misinterpretation of photo-signature.

3. **Leaf-on Photography.** The canopy of deciduous trees makes it extremely difficult to separate all but the wettest forested wetlands from upland forests. In some areas, such as the Pacific Northwest, spring photography is difficult to acquire due to cloud cover, so leaf-on photography was used for wetland mapping. In Alaska, most of the aerial photography is acquired in mid-summer which results in conservative mapping of forested wetlands. While posing problems for forested wetland interpretation, growing season photography usually improves interpretation of aquatic beds and nonpersistent emergent wetlands.

4. **Forested Wetlands.** These are among the more difficult types to photointerpret; these types are conservatively mapped.
   - Forested wetlands on glacial till are difficult to photointerpret, so many of these wetlands do not appear on NWI maps.
   - The location of temporarily flooded or seasonally saturated forested wetlands are among the most difficult to identify on the ground as well as through photointerpretation, so many of these wetlands do not appear on the NWI maps. This problem is particulary evident on mapping done prior to 1989. This limitation is common along the coastal plain and in glaciolacustrine plains such as the Lake Ontario Plain (New York).
   - In areas where 1:80,000 black and white photography was used, many forested wetlands were not photointerpretable.
   - In Southeast Alaska, extremely subtle differences in the photo-signatures of evergreen forested wetlands and evergreen forested uplands result in both errors of omission and commission.

5. **Upland Inclusions.** Small upland areas may occur within delineated wetlands due to minimum mapping size and on the coastal plain due to the complexity of wetland-upland interspersion. Field inspections and/or use of larger-scale photography may be used to refine wetland boundaries when necessary.

6. **Estuarine and Tidal Waters.** Delineation of the break between the estuarine and riverine (tidal) systems and the oligohaline (slightly brackish) segment of estuaries should be considered approximate based on available reports and/or limited field checking. In Maine, the irregular rocky coastline complicated delineation of the boundary between the marine and estuarine systems.

7. **Intertidal Flats.** Since the aerial photos were not always captured at low tide, all intertidal flats were not visible. Boundaries of these nonvegetated wetlands were approximated from coastal and geodetic survey maps and topographic maps where necessary. Flats in sandy areas tend to shift over time, so boundaries may be different than depicted on the NWI maps.

8. **Coastal Wetlands.** Identification of high marsh (irregularly flooded zone) vs. low marsh (regularly flooded zone) in estuarine wetlands is conservative; the photo-signatures of these zones are not distinctive in many instances.

9. **Water Regimes.** Water regimes designated were based on photosignatures coupled with limited field verification; they should be considered approximate. Long-term hydrologic studies are required to accurately describe the hydrology of any particular wetland. Recent NWI maps in the coastal zone indicating a saturated water regime (B) reflect seasonal saturation related to high water tables in winter and early spring.

10. **Linear wetlands (long, narrow)** that follow drainageways and stream corridors may or may not be mapped depending on project objectives. Most NWI maps identify at least some of these features using a dashed pattern. In most cases, no attempt was made to map all linear wetlands. Users can infer the possible occurrence of these wetlands by looking for pertinent topographic features on the NWI maps.

11. **In general, only five types of farmed wetlands** are shown on NWI maps: cranberry bogs, prairie potholes, pothhole-like depressions, playa lakes, and California's seasonally flooded diked former tidelands. This is based on technical considerations and an interagency agreement between the U.S. Fish and Wildlife Service and the USDA. Natural Resource Conservation Service developed back in the 1970s.

12. **Partly drained wetlands** were mapped based on recognizable photo-signatures; many of these wetlands may not appear on the NWI maps.

TABLE 10.3 (continued)
Examples of Major NWI Map Limitations

13. **Tundra.** Moist tundra (usually wetland) is often difficult to separate from dry upland tundra due to photo-signatures. This is especially true where wide transition zones exist between the two types.

14. **NWI map data** are dependent on the date of the photography. The activities of humans (e.g., filling and drainage) or beavers may have caused changes in wetlands since the aerial photos were taken. Maps do not show losses or gains in wetlands since that date. The date of the photography used to prepare the maps is shown on the legend.

15. **The aerial photographs** reflect wetness conditions during the specific year and season when they were taken — if taken during a dry season or a dry year, many wetlands will be missed and not mapped.

16. **Drier-end wetlands** (e.g., seasonally saturated and temporarily flooded) are conservatively mapped. (Note: These wetlands are the most difficult to identify on-the-ground as well as through remote sensing.)

17. **The mapped boundaries** may be somewhat different than if based on detailed field observations, especially in areas with subtle changes in topography (e.g., pitch pine lowlands, coastal plain flatwoods, or glaciolacustrine plains).

18. **Knowledge of wetland identification** has improved since the NWI commenced in the late 1970s, so earlier mapping may have missed many drier-end wetlands simply due to a lack of recognition in the field. Maps produced since the late-1980s should better reflect wetlands due to improved knowledge of hydric soils and their relationship to plants and hydrology.

Source: Adapted from Tiner, R.W. 1997a. NWI Maps: What they tell us. *National Wetlands Newsletter,* 19(2): 7-12.

Scientists studying the application of satellite technology for wetland mapping have used NWI maps as training sites, while sportsmen have used NWI maps to locate areas for hunting and fishing.

A recent FWS report (1996) presented examples of the agency's use of NWI maps including refuge management, wetland permitting, and wetland restoration planning. States also are using NWI maps in a variety of ways. For example, Vermont uses NWI maps to identify "Class 2 wetlands" (wetlands so significant that they merit protection under Vermont Wetland Rules). NWI wetlands and any unmapped wetlands contiguous to them are given this special status. Massachusetts used NWI maps and digital map data as a base to prepare an atlas of tidally restricted marshes (Massachusetts Wetlands Restoration & Banking Program and Natural Resources Assessment Group, 1996). Indiana and Illinois use NWI maps to assess property taxes with wetland acreage for reducing tax bills. The Internal Revenue Service used NWI maps in Massachusetts to estimate land value for inheritance tax purposes.

Besides being aware of the limitations of NWI maps, users should learn to recognize the landforms that favor the establishment of wetlands (e.g., depressions, broad toes of slopes, saddles between mountains, natural drainageways between rolling hills, and margins of streams and other waterbodies). These features can be interpreted from the NWI map by examining the topography and hydrography from the base map. NWI maps may be used to establish the presence and general configuration (i.e., general boundary) of a wetland in a given location. NWI wetland boundaries may be quite accurate in wetlands abrupt boundaries (e.g., distinct depressions vs. surrounding upland) or where vegetation pattern is highly contrasting (e.g., red maple swamp vs. upland pine forest). Yet, in broad flat topographic regions like coastal plains or glaciolacustrine plains where wetland boundaries are more diffuse, the NWI wetland boundaries are quite general and definitely require field work to locate the exact boundary when needed. When used in combination with county soil survey maps, a more complete universe of potential wetland sites are portrayed. Such analysis coupled with review of current aerial photos also may reveal potential sites for wetland restoration.

COUNTY SOIL SURVEYS

The USDA Natural Resources Conservation Service (formerly the Soil Conservation Service) has been conducting soil surveys throughout the country since 1896 (Soil Survey Division Staff, 1993).

Originally conceived to help farmers locate suitable soils and to aid in their management, soil surveys now provide valuable information for engineers, wetland scientists, and natural resource planners. When the survey is completed for a given county or a major portion of a county, a soil survey report is published. The report contains invaluable information about the county and its soils (e.g., climate, soil series and map unit descriptions, use and management of soils, and formation and classification of soils), plus large-scale, photo-based maps showing the location and configuration of individual soil map units.

Mapping Procedures

Guidance for conducting soil surveys is presented in *Soil Survey Manual* (Soil Survey Division Staff, 1993). Soil surveys may be done at different levels of detail. A first order survey has a minimum delineation of ≤2.5 acres, a second order –1.5 to 10 acres, a third order –4 to 40 acres, a fourth order –40 to 640, and a fifth order –640 to 10,000 acres. Most soil survey reports summarize the findings of second order surveys. A county soil survey may include different order surveys within the county based on soil management considerations. For example, an irrigated section of the county may be mapped at the second order level, while the rangeland section may be surveyed at the third order level. The order of the survey should appear in the published soil survey with a locator map.

In preparing for a survey, soil scientists review aerial photos and other materials to become familiar with the area's land, water, and cultural resources. In the field, soil scientists dig hundreds of soil pits and auger holes, examining changes in soil properties in different plant communities, landscape positions, and soil parent materials. Soils are described following national conventions including *Keys to Soil Taxonomy* (Soil Survey Staff, 1998). When working in a given area, soil scientists tend to form a model of the soils landscape based on the relationships observed during field work. For example, a certain sequence of series occurs on a given parent material ranging from excessively drained sites to very poorly drained sites. As they are conducting extensive field work (i.e., months to years), soil scientists delineate the boundaries of individual soil types or groups of soils on aerial photographs. At the conclusion of the soil survey, soil maps are prepared using the aerial photos as the base maps. The maps vary in scale from 1:15,480 to 1:25,000. These maps are usually included at the back of the soil survey report, while in other cases, the maps may be filed in a separate folder. An example of a soil survey map is presented in Figure 10.3.

Features Mapped

Polygonal soil map features called *soil map units* represent areas of various soil types that are large enough to show on the maps. The name of these units usually corresponds with the predominant soil series and phase* (e.g., Freetown muck, Freetown muck ponded, and Ridgebury fine sandy loam, 3 to 8% slope). Some units are mapped as land types or miscellaneous areas with common names such as tidal marsh, swamp, bog, or muck, beaches, alluvial land, or urban land. When one soil represents most of the unit, the unit is named after that series. This type of soil map unit is called a *consociation*. In other cases, where two or more dissimilar soils are dominant, a soil *association* or *complex* is mapped. The distinction between an association and a complex is based on mapping detail. The major soils of the former can be separated at a scale of 1:24,000, while the component soils of the latter cannot (Soil Survey Division Staff, 1993). These units have the names of all the dominant series or land types. The proportions of each soil will vary in complex or association map units but the soils should be present in all units. Other map units with more than one soil name are called *undifferentiated groups* where two or more soil taxa that do not regularly occur together on the landscape are delineated as a unit. All mapping units are designated

---

* Phases are designated for such features as surface texture, depth, slope, accelerated erosion, stoniness, ponding, and flooding.

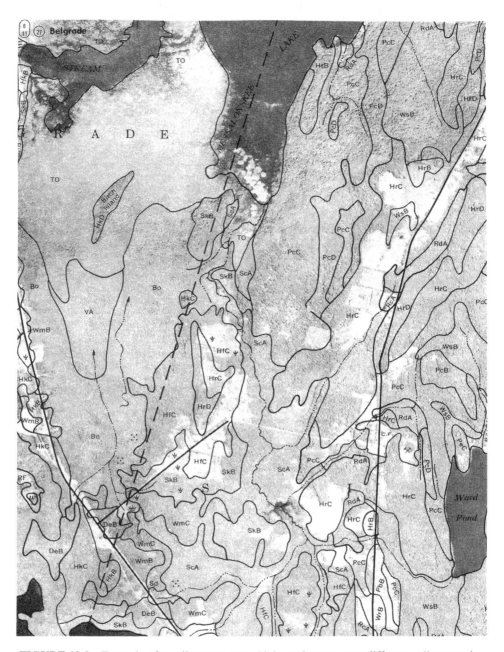

FIGURE 10.3 Example of a soil survey map. Alpha codes represent different soil map units.

by either an alpha-code or a number-code on the maps. A legend on a separate sheet correlates the codes with their corresponding names.

Of interest to wetland delineators are additional features that may be shown on the maps. *Miscellaneous areas* (essentially lacking soil and supporting little or no vegetation) include beaches, glaciers, oil-waste lands (slush pits), playas, riverwash, salt flats, and water (Soil Survey Division Staff, 1993). The location of small wet spots, springs, and depressions that contain water most of the time are indicated by symbols. Linear features like narrow streams and alluvial fans also are designated.

Soil Survey Strengths and Weaknesses

The soil survey reports provide useful information on the area's landforms, soil properties, and uses of soils. The maps display the relationships between soils, vegetation, and landscape position. Given the extensive field work performed in producing the soil surveys, the maps are often quite accurate. There are local variations in the quality of the surveys as with any mapping effort due to a combination of the complexity of soil patterns in the area, the skill level of the soil mappers, and other factors. The date of the soil survey also is important as new series will not be shown in older survey (older taxonomy will be used; the concept of some series have changed) and older maps tend to be less detailed than newer surveys especially in urbanizing areas, forested regions, and other areas that were not immediately useful for agriculture. More soil phase mapping may be reflected in recent surveys.

When used in conjunction with a list of hydric soil map units (available for each county from the Natural Resources Conservation Service's county offices), the soil survey maps can help identify the likely presence of wetland in a given area. Mapping units of the wetter hydric soils (e.g., Histosols and Histic subgroups) should indicate obvious wetlands. Units of drier or more transitional hydric soils (e.g., Aeric subgroups) should identify areas where the delineator will have to do considerable digging to identify wetlands and their boundaries. Units of Typic subgroups of Aquic suborders indicate areas of intermediate difficulty in delineating regulatory wetlands, as wetland hydrology indicators may be lacking especially during the summer or dry season.

As with any map, soil maps have their limitations. In general, the minimum map unit of soil types ranges from 1.5 to 10 acres, depending on landscape diversity, survey objectives, and final map scale. When a 1:24,000-scale map is desired, a minimum map unit of 5.7 acres is the general rule (Soil Survey Division Staff, 1993). Yet most map units are much larger than this especially in forested areas. For example, the Umatilla County area, Oregon soil survey identifies a 5-acre minimum for strongly contrasting soils, a 40-acre minimum for small grain-fallow and annual cropping areas, and a 100-acre minimum for rangeland and woodland (Johnson and Makinson, 1988). This can result in large units of mixed soil types, since soil map units often contain more than one type of soil. For some hydric soil map units, hydric soils comprise 60% of the unit and nonhydric soils as much as 40% although inclusions of other soils may usually represent less than 20%. Thus, only 60 to 80% of any hydric soil map unit may actually contain hydric soils and have a potential for being a wetland or regulated wetland. Soil map units often have inclusions of other soils, so that a large area mapped as nonhydric (upland) soil may, in fact, have substantial acres of hydric soil (wetland) within its borders, and vice versa. As much as 40% of a map unit may contain other soils. The names of these inclusions are given in the description of the soil map unit in the survey report.

Soil maps usually do not differentiate between undrained hydric soils and drained or recently filled hydric soils, thus making it virtually impossible to separate current wetlands from historic wetlands, except where a flooded phase or undrained phase or made-land (filled) is mapped.

Finally, the series level of soil classification was never intended to separate hydric soils from nonhydric soils. Some series are so broadly characterized that they include both hydric and nonhydric members. These "facultative hydric soils" may or may not be hydric depending largely on landscape position (those in depressions are usually hydric, but those on slopes are often not); soil maps do not distinguish between them. This situation applies to most, if not all, of the soils with an Aeric subgroup. The designation of drainage class for these soils may differ from one state to another. For example, one state might consider the series to be somewhat poorly drained (nonhydric), while another state considers it poorly drained (hydric). These types of series need to be subdivided into two series, one that has hydric soil morphology and another that does not. In the meantime, considering mapping units with such series as wetlands, in all likelihood, will overestimate wetland acreage.

Uses of Soil Surveys

Soil surveys provide much valuable information for wetland delineators. First, the maps will give the wetland delineator a good sense of what to expect in the field. Second, the survey reports contain descriptions of the soil series and soil mapping units which wetland delineators can use to determine the likely series of a soil observed in the field. The mapping unit descriptions point out the types of inclusions that occur in the unit. This coupled with the descriptions of individual series will help verify the series of the soil in question by matching the observed soil properties with the features described.

Some tables are particularly important to wetland delineators. The freeze dates table (usually Table 2 or 3 in modern surveys) aids in determining the beginning and end of the "growing season" for use in assessing wetland hydrology (Chapters 2 and 6). The soil and water features table is useful for documenting potential hydrologic conditions associated with mapped series (Table 10.4). The soil classification table (usually the last table in the soil survey report) contains the taxonomy of each soil series. In the absence of a list of hydric soils, one can interpret likely hydric soils from nonhydric soils from possible hydric soils by making interpretations from their taxonomic classification (see end of Chapter 5 for guidance).

Some misuses of soil surveys have occurred. For example, some regulators have claimed that any hydric soil mapping unit is a potentially regulated area. While the potential may be there, many hydric soils have been drained, filled, or otherwise developed to the point that they are no longer wetlands. Afterall, much wetland has been converted to agricultural land during the past 200 years. This misuse of the soil survey data also has been applied to estimate the acreage of land subject to wetland regulations, resulting in grossly exaggerated reports on acres of wetlands subject to government regulations across the country. The soil surveys and acreage of hydric soil map units should not, therefore, be the sole sources used to identify the presence of wetlands or to determine the acreage of land subject to regulation. Many hydric soil mapping units no longer meet the federal government's three-parameter test for regulation. Soil maps should be used in conjunction with the NWI maps, and other information (e.g., current aerial photos) to identify the likely presence of wetland in a given area and soils that may be encountered during field inspections. After reviewing these materials, on-site inspections should be performed to verify the presence of wetland and determine its boundaries.

COMPARING NWI MAPS WITH SOIL SURVEY MAPS

The availability of both data sets in digital form makes it easy for researchers to compare and contrast NWI maps vs. soil survey data. Using GIS technology, some researchers have reported on such comparisons (Thurston Regional Planning Council, 1991; Klemow and Mohseni, 1996). These studies did not establish or assess the accuracy of either source in wetland identification since field verification was not performed. Consequently, they could not make valid claims about which source more accurately identified the wetland resources. One study introduced significant digitizing errors that affected the study results, so accurate digitizing of data sources must be assured in future studies (Morrison, 1993). Also, when used in the GIS environment, wetland boundaries of NWI maps (which are about 50 ft wide), for example lose this dimension and are converted to a line at whatever the map scale is generated by the computer. The line width should be maintained in all GIS applications for maps of all kinds, so at larger scales, the line should be designated as a band of the appropriate width.

It should not be surprising to find significant discrepancies between wetlands identified on NWI maps and "hydric soil map units" of soil surveys. The soil survey focuses on management (uses of soils), while NWI is concerned strictly with wetland identification. Soil map units often contain both hydric and nonhydric soils if the management of those soils is similar, leading to more Type II errors (commissions). Since NWI maps tend to make more Type I errors (omissions), this difference

TABLE 10.4
Example of Soil–Water Features Table from the Hampshire County, MA Soil Survey

Soil Name and Map Symbol	Hydrologic Group	Flooding Frequency	Duration	Months	High Water Table Depth Ft	Kind	Months	Bedrock Depth In	Hardness	Potential Frost Action	Risk of Corrosion Uncoated Steel	Concrete
Fm------- Freetown	D	None------	---	---	0–1.0	Apparent	Jan–Dec	>60	---	High------	High------	High.
GfB, GfC, GhB, GhC, GxB, GxC, GxD------ Gloucester	A	None------	---	---	>6.0	---	---	>60	---	Low------	Low------	High.
Ha------- Hadley Hd:	B	Occasional	Brief-----	Feb–Apr	4.0–6.0	Apparent	Nov–Apr	>60	---	High------	Low------	Moderate.
Hadley---------	B	Occasional	Brief-----	Feb–Apr	4.0–6.0	Apparent	Nov–Apr	>60	---	High------	Low------	Moderate.
Winooski------- Urban land.	B	Occasional	Brief-----	Feb–Apr	1.5–3.0	Apparent	Nov–Apr	>60	---	High------	Moderate	Moderate.
HfB, HfC-------- Haven	B	None------	---	---	>6.0	---	---	>60	---	Moderate	Low------	High.
HgA,HgB,HgC,HgD,HgE------ Hinckley Hu:	A	None------	---	---	>6.0	---	---	>60	---	Low------	Low------	High.
Hinckley-------	A	None------	---	---	>6.0	---	---	>60	---	Low------	Low------	High.
Merrimac------- Urban land.	A	None------	---	---	>6.0	---	---	>60	---	Low------	Low------	High.
HvC-------- Holyoke	C/D	None------	---	---	>6.0	---	---	10–20	Hard	Moderate	Low------	High.
Lk-------- Limerick	C	Frequent---	Brief-----	Jan–Jun	0.5–1.5	Apparent	Nov–Jun	>60	---	High------	High------	Moderate.
Ma-------- Maybid	D	None------	---	---	+1–0.5	Apparent	Oct–Aug	>60	---	High------	High------	Moderate.

TABLE 10.4 (continued)
Example of Soil–Water Features Table from the Hampshire County, MA Soil Survey

Soil Name and Map Symbol	Hydrologic Group	Flooding			High Water Table			Bedrock		Potential Frost Action	Risk of Corrosion	
		Frequency	Duration	Months	Depth Ft	Kind	Months	Depth In	Hardness		Uncoated Steel	Concrete
MeA, MeB, MeC, MeD----- Merrimac	A	None------	---	---	>6.0	---	---	>60	---	Low------	Low------	High.
MoB, MoC, MsC, MxB, MxC, MxD----- Montauk NaC, NaD:	C	None------	---	---	2.0–2.5	Perched	Feb–May	>60	---	Moderate	Low------	High.
Narragansett---	B	None------	---	---	>6.0	---	---	>60	---	Moderate	Low------	Moderate.
Holyoke--------	C/D	None-------	---	---	>6.0	---	---	10–20	Hard	Moderate	Moderate	High.

Note: The Corps recommends using this table to help verify wetland hydrology for a soil classified as one of the listed series by March 1992 guidance (Williams, 1992). Flooding or saturation must fall within 5% of the growing season. In this table, the following soils are hydric: Freetown (typic medisaprist), Limerick (typic fluvaquent), and Maybid (typic humaquept). Note that although Maybid is not "flooded," it does have a water table ranging from +1 to 0.5 ft from October to August, meaning that it is ponded with 1-ft of water for some time during that period and when not, the water table is usually within 0.5 ft from the surface.

Source: Swenson, E.I. 1981. Soil Survey of Hampshire County, MA, Central Part, USDA Soil Conservation Service, Amherst, MA.

in survey or inventory objectives can lead to enormous differences in the extent of potential wetlands. A comprehensive investigation of wetland maps produced by different methods in eastern Maryland verified these observations (Shapiro, 1995). Kalla and Couch (1996) compared NWI maps and soil survey maps in Georgia flatwoods in the field and found that NWI maps were a better predictor of federally regulated (jurisdictional) wetlands than the hydric soil map units on the soil survey, yet cautioned that NWI maps should not be used alone for this purpose.

USE OF MAPS FOR REGULATORY DETERMINATIONS

Where states have produced maps showing the official boundaries of regulated wetlands, such maps can be relied upon for determining the general extent of regulated areas. In most cases, field verification is required to ensure the accuracy of the mapping as the states usually have provisions to adjust the boundary as needed based on field conditions.

For federal regulatory purposes, however, such maps do not exist. The following guidance is offered to help use existing maps as tools to aid in identifying potentially regulated wetlands.

1. Consult both NWI maps and soil surveys (plus any available state or local wetland maps) to get an idea of where wetlands may be located on a particular parcel of land. Given that NWI maps often underestimate the extent of wetlands (especially in forested regions) and soil surveys (hydric soil map units) usually overestimate this, the real extent of wetlands is probably somewhere inbetween. Also recognizing the landscapes where wetlands tend to form (e.g., floodplains, drainageways, toes of slopes, flats, depressions, and saddles between mountains) and considering these sites as potential wetland areas also is a good practice. These landscapes can be seen on the NWI maps by interpreting topographic contours.
2. Review recent aerial photos of the site to update published map information and to locate smaller wetlands and possible landscape positions that will likely support wetlands. Different plant communities also can be interpreted. Stereoscopic interpretation will reveal topographic differences important in wetland recognition.
3. To answer the question of the existence of a regulated wetland on one's property and if present, where are its boundaries? Field assessments of vegetation, soils, and hydrology are required.

FUTURE MAPS

Since field assessment of wetlands is practical only for small areas, the production of a set of "perfect" wetland maps for the nation or individual states is virtually impossibility given the transitional nature of the resource and the limitations of photointerpretation or other remote sensors. A wetland map could be designed to favor Type II errors (commission) over Type I errors (omission). This would mean that in addition to the photointerpretable wetlands (e.g., as currently mapped by the NWI program), certain landscapes that favor wetland formation would be mapped. Perhaps digital hydric soil data could be combined with digital NWI data to generate such a map. Additional photointerpretation would be advisable to eliminate hydric soil areas that are clearly no longer wetlands, to add smaller photointerpretable wetlands or other landscape positions that may support wetlands, and to update the digital database. The questionable "wetland" areas could be labelled with a unique code to separate them from the photointerpretable wetlands. These additions could be called "areas potentially supporting some wetlands," for example. Some of these areas may be entirely or partially wetland, while others may not. In some areas, NWI maps already identify mixed wetland–upland complexes where mapping discrete wetlands was not possible (e.g., Hawaii, Oregon, and California). Tiner (1996) recommended that the government consider producing regulatory guidance maps to help inform the public on the likely presence of regulated wetlands. He proposed separating land into three categories: wetland, upland, and land requiring field inspection (to identify the presence or absence of wetlands). The publication of these maps would not eliminate the need for field investigations in most areas with obvious wetlands as the upper boundaries would still need to be established for specific permit applications.

A map showing wetlands and potential wetland areas may be more valuable than the current NWI maps, especially to regulatory agencies and the regulated community. This type of wetland map product could serve more uses and would duly inform landowners and developers to where regulated wetlands may exist. These maps would continue to be an asset to current regulatory programs, at least in part by helping guide development away from mapped wetlands to more suitable locations.

Another future option involves assembling information from field delineations to update and enhance the national digital map database. If regulatory agencies required applicants to approximate verified wetland boundaries on a U.S.G.S. topographic map or on an NWI map and provided such information to the FWS, future updates of NWI maps, for example, could incorporate this valuable site-specific information. This would make good use of the work of professional wetland delineators. Moreover, in this way, maximum benefits are derived from these intensive field surveys as the work would contribute to development of a better national wetland map and spatial database.

WETLAND PHOTOINTERPRETATION

Interpretation of aerial photographs has been a widely used method for inventorying wetlands. In the U.S., several states have produced maps of coastal wetlands using large-scale aerial photographs, including New Jersey, New York, Delaware, and South Carolina (Klemas et al., 1973; Tiner, 1977; Brown, 1978). Coastal wetlands and some types of inland wetlands have been inventoried through air photointerpretation for land-cover and land-use mapping projects in Massachusetts and Rhode Island (MacConnell, 1974; 1975). A few other states have mapped inland wetlands through use of large-scale aerial photographs, including Maine, New York, and Wisconsin (McCall, 1972; Cole and Fried, 1981; Wisconsin Coastal Management Program, 1995). Table 10.5 identifies photography used for some significant wetland mapping projects in the U.S. Local studies involving wetland photointerpretation include Anderson and Wobber (1972), Cowardin and Myers (1974), Gammon and Carter (1979), Kennard et al. (1980), Lovvorn and Kirkpatrick (1982), Miller et al. (1976), Olson (1964), Reimold et al. (1973), Roller (1977), Seher and Tueller (1973), Shima et al. (1976), and Stroud (1969). In addition to these efforts, the Fish and Wildlife Service (FWS) of the Department of the Interior established its National Wetlands Inventory Project (NWI) in 1975 to conduct an inventory of U.S. wetlands by interpreting high-altitude aerial photographs. Lee (1991) summarized the uses of various types of remote sensors for wetland detection. Much of the following discussion comes from Tiner (1990, 1997b), focusing on experiences of the NWI Program.

PHOTOINTERPRETATION CONCERNS

Wetland photointerpretation is not a simple task. Wetlands occur along a soil moisture continuum between permanently flooded, deepwater habitats and drier habitats that are not wet long enough to develop anaerobic soil conditions. This makes many wetlands, especially those subject to only brief flooding and seasonal saturation, particularly difficult to identify on the ground, let alone on aerial photographs. Moreover, wetlands do not only form in depressions, but occur on sloping areas as well. This further complicates their interpretation. In addition, wetlands vary widely from one region to another.

In any photointerpretation project, the quality and timing of the photography are prerequisites for accuracy. Overexposed or underexposed film is of little value as are photographs with considerable cloud or snow cover. Stereoscopic coverage with sufficient overlap is essential to assess topographic relief and to separate trees from shrubs based on height. Seasonality of the coverage (e.g., spring, summer, and fall) is another important consideration since the predominant vegetation

TABLE 10.5
Examples of Photography Used for Some Major Wetland Mapping Projects in the U.S.

State or Region (Type of Wetland)	Photography Emulsion	Scale	Source
Entire Country	CIR	1:58,000	FWS NWI
	CIR	1:40,000	FWS NWI
(All)	BW	1:80,000	FWS NWI
	BW	1:40,000	FWS NWI
Delaware (Coastal)	CIR, TC, & BW	1:12,000	Klemas et al. (1973)
Hawaii (All)	TC & CIR	1:130,000	Chime et al. (1978)
(>5 acres)		1:65,000	
		1:32,500	
Maine (All)	BW	1:1320	McCall (1972)
(>10 acres)			
Maryland (Coastal)	TC (mostly)	1:12,000	McCormick and Somes (1982)
(>0.25 acres)			
Massachusetts	BW	1:20,000	MacConnell (1975)
(All, except forested)			
New Jersey	CIR	1:12,000	Brown (1978)
(Coastal)			
New York	BW	1:24,000	Cole and Fried (1981)
(Inland)			
New York	CIR	1:12,000	Brown (1978)
(Coastal)			
Rhode Island	BW	1:12,000	MacConnell (1974)
(All, except forested)			
South Carolina	BW	1:40,000	
(Coastal)	CIR	1:12,000	Tiner (1977)
		1:6,000	

Note: CIR – color infrared; BW – black and white panchromatic; TC – true color.

Source: Tiner, R.W. 1997b. In *Manual of Photographic Interpretation,* 2nd ed. W.R. Philipson (Ed.). American Society for Photogrammetry and Remote Sensing, Bethesda, MD.

and the hydrologic characteristics (i.e., water regime) largely determine the relative ease of difficulty with which wetlands can be interpreted. The wettest wetlands are usually easiest to interpret, while the drier ones are most problematical. Antecedent weather conditions (prior to photo acquisition overflights) are important considerations. Extreme flooding conditions as well as extreme droughts create problems for accurate wetland photointerpretation.

Photographic scale establishes limits on what can be interpreted (e.g., minimum mapping unit, degree of resolution between different wetland types, and the detail and width of wetland boundaries). Large-scale photography (generally 1:24,000 and larger) is best for local wetland mapping efforts where precise boundaries of wetlands and identification of small wetlands are required. Several states, including New Jersey, Delaware, and Maryland have produced regulatory maps for tidal wetlands based on wetland interpretation of large-scale photographs (e.g., 1:12,000); the maps depict the official wetland boundaries. Large-scale photographs also facilitate identification of discrete plant communities. In contrast, smaller-scale photographs, such as high-altitude aerial photographs (1:58,000), are more useful for national or regional inventories where less detail is required. With this type of photography, general wetland boundaries can be delineated for wetlands

TABLE 10.6
Pen Line Width Relative to Pen Point Size and Photo Scale

Photo Scale	Pen Line Width on the Ground		
	3 × 0 (.25mm)	4 × 0 (.18mm)	6 × 0 (.13mm)
1:4,800	4.0 ft	2.8 ft	2.0 ft
1:7,200	6.0	4.2	3.0
1:12,000	10.0	7.0	5.0
1:24,000	20.0	14.0	10.0
1:40,000	33.3	23.3	17.0
1:58,000	48.3	33.8	24.7

Source: MacConnell, W. et al. 1992. Recording Wetland Delineation on Property Records. Department of Forestry and Wildlife Management. University of Massachusetts, Amherst.

larger than one acre in size and for even smaller conspicuous wetlands (e.g., ponds or pothole wetlands in agricultural lands).

An intermediate scale 1:40,000 may be the best compromise as considerable detail can be captured in less time and for lower costs. This is probably the best scale for producing wetland guidance maps. In studying the differences in results of interpreting several scales of color infrared photography for mapping wetlands in the "pothole" region of Maryland, Tiner and Smith (1992) recommended the following minimum mapping units: (1) one acre for 1:58,000, (2) 0.5 to 1.0 acre for 1:36,000, (3) 0.25 to 0.5 for 1:24,000, and (4) 0.1 to 0.25 acre at 1:12,000; conspicuous smaller wetlands could be mapped as dots on these photos, if necessary. They found that differences in total wetland and deepwater habitat acreage derived from interpreting at the three larger scales were rather small: a 7% increase from 1:36,000 to 1:12,000, a 3% increase from 1:24,000 to 1:12,000, and a 1% decrease from 1:36,000 to 1:24,000. The latter decrease was attributed to more generalized mapping at 1:36,000. The level of effort, however, to achieve the gains was not insignificant. For every increase in scale, there was nearly a doubling of effort in the study area. The study area had a high density of small pothole-like forested wetlands and extensive floodplain forested wetlands. For simpler landscapes, the increased level of effort may not be as large. The width of the wetland boundary delineated on aerial photos also is dependent on photo scale (Table 10.6).

Emulsion is another vital characteristic of aerial photographs. Black and white (panchromatic) photography yields shades of gray and textural differences for interpretation (Figures 10.4–10.6). True color and color infrared photos produce an array of colors (plus shades) and textural patterns for recognition (Plates 50-55). Color infrared is generally the preferred imagery for wetland mapping and vegetation mapping in general because this film records a wider range of colors and tones than true color (Arnold, 1997). There are exceptions: submerged aquatic vegetation (SAV) mapping and identification of purple loosestrife (*Lythrum salicaria*) stands are both better observed on true color, and on black and white (panchromatic) film for the former. These films have better water depth penetration than color infrared film, necessary for SAV detection. Purple loosestrife is easily interpreted from true color photos captured when its purplish pink flowers are in full bloom (e.g., around first week in August in Massachusetts). Frazier and Moore (1993) tested different film types for identifying small stands of purple loosestrife. They found that Fuji Velvia 50 ASA was the easiest to interpret then Ektachroma 100 ASA, followed by Kodachrome 64 ASA.

Finally, the skills of the photointerpreter also are a significant factor in the quality of the interpretation. Photointerpreters must have certain physical skills (e.g., ability to see in stereo, to distinguish colors, and to draw neatly) and cognitive skills (e.g., knowledge of landscapes and

FIGURE 10.4   Aerial photograph (reduced 60%; original scale 1:12,000) flown in March at low tide, showing salt marshes and associated intertidal flats along the waterway. In upper left corner is a diked salt marsh (dark-colored area behind light-colored dike with dark excavated channel in front).

wetland ecology) (MacConnell et al., 1992). They also must be able to identify wetlands and their boundaries in the field during ground truthing exercises.

Interpreting Estuarine Wetlands

Coastal wetlands occur in estuaries where tidally influenced waters are salty or brackish due to ocean-derived salts. Since tidal flooding gradually moves upstream in the estuaries with the incoming rising tide, marshes at different locations within a large estuarine system are in various stages of flooding at a single point in time. For example, when marshes closest to the ocean are flooded, similar marshes farthest upstream are usually not. This presents a basic problem for acquisition of low-tide photographs for vast estuarine systems. Moreover, the tidal cycle requires 24 h and 50 min. to make a complete cycle (e.g., two high tides and two low tides on the Atlantic Coast), so low tide does not always fall within the optimal time of day for taking aerial photographs.

To capture low-tide photographs for such systems requires considerable planning and expense as well as good fortune because local weather conditions can pose additional problems (Figure 10.4). Consequently, aerial photographic missions that cover large blocks of land, such as the National High-Altitude Photography Program (NHAPP) and National Aerial Photography Program (NAPP), are not synchronized with low tide throughout the estuary. If flooding is particularly deep, vegetation may be obscured making the true waterward boundary difficult to define. In some coastal areas with large tidal ranges, such as Maine and Alaska, there are extensive areas of nonvegetated tidal flats. When covered by deep water, these areas cannot be accurately detected.

When tidal flooding is not a problem, coastal wetlands are among the simplest to interpret for several reasons:

1. They are mostly large, expansive systems lying between conspicuous uplands and deep-water areas and often behind coastal barriers (e.g., between the mainland and the Atlantic Ocean or Gulf of Mexico).
2. They are open systems dominated by herbaceous vegetation (except in the tropics) and in many cases are bordered by trees intolerant of salt and brackish water flooding.
3. In many coastal wetlands, vegetative communities are relatively monospecific, dominated by conspicuous species such as smooth cordgrass (*Spartina alterniflora*), salt hay grass

(*S. patens*), tall cordgrass (*S. cynosuroides*), marsh spikegrass (*Distichlis spicata*), black needlerush (*Juncus roemerianus*), black mangrove (*Avicennia germinans*), and red mangrove (*Rhizophora mangle*) that give these wetlands a characteristic photo appearance.

4. Many of the dominant plants only grow in these habitats and do not grow in adjacent uplands.
5. They can be readily interpreted during virtually any season.

While the waterward and landward limits of coastal wetlands usually can be readily established, it is often more difficult to separate low marsh from high marsh on the high-altitude photographs because smooth cordgrass occurs in both zones. These zones may be separable on low-altitude summer photography at the peak of the growing season due to differences in plant height and vigor.

Black needlerush is a persistent emergent plant that has a characteristic appearance on color infrared (CIR) and true color photographs during any season. Dense stands of common reed (*Phragmites australis*) or cattails (*Typha* spp.) can be interpreted, especially on large-scale photographs. Salt flats within the high marsh can be observed as barren sandy areas or as areas vegetated by glassworts or pickleweed (*Salicornia* spp.) or saltwort (*Batis maritina*), depending on plant density. High reflectance of sand may wash out reflectance of vegetation, thereby obscuring vegetative cover.

Florida's mangrove swamps can be interpreted with either high-altitude or low-altitude aerial photographs. Black mangrove can be separated from red mangrove on large-scale photographs due to the observed morphological differences. On high-altitude photographs (1:80,000), these differences are not apparent and it is not possible to separate the two species. Patterson (1986) found that 1:2,000 to 1:12,000 color infrared photography was best for separating fringing, mixed, and black mangrove communities. With smaller photo scales, it becomes increasingly more difficult to distinguish these communities, yet even at 1:60,000 using magnification of stereoscopes, he recognized all communities except the thin fringing community.

Vegetated rocky shore covered with rockweeds (*Fucus* spp. and *Ascophyllum nodosum*) are best interpreted with color infrared photography captured at low- to mid-tide. Dense beds stand out as bright orange or reddish patches (depending on the emulsion).

Losses and gains in coastal wetlands and other types also can be detected on aerial photographs (Figures 10.5a/b). With increasing scale of the photographs, smaller wetland changes can be detected and delineated more accurately. In a trend analysis study of tidal wetlands within Shinnecock Bay on Long Island, New York, Tiner (1987) identified wetland changes smaller than 0.1 acre on

FIGURE 10.5 Aerial photographs revealing wetland trends in South Bethany, DE: (a) 1937 photo and (b) 1973 photo. (Courtesy of Delaware Department of Natural Resources and Environmental Control.)

FIGURE 10.6   Black and white aerial photograph (reduced 60%; original scale 1:24,000) showing freshwater swamps, bogs, and marshes. Large emergent-shrub wetland (lower left), shrub bogs (top middle) with small whitish patches of emergent wetlands, and forested wetland in distinct isolated depression (middle right).

1:12,000 CIR photographs. Conspicuous changes of this size were detected on 1:36,000 CIR photographs, but these changes were difficult to delineate on a photographic overlay. Moreover, many of these small changes might have been overlooked if the 1:12,000 photographs were not examined.

INTERPRETING PALUSTRINE WETLANDS

Inland wetlands may be divided into four types based on predominant vegetation: aquatic bed, emergent (dominated by herbaceous plants), scrub-shrub (dominated by woody plants less than 20 ft high), and forested (dominated by woody plants 20 ft or taller) (see Chapters 8 and 9).

Although differences in predominant vegetation are a prime factor in photointerpretation, in general, the wetter the wetland, the easier it is to identify regardless of vegetative cover, provided the photographs are acquired at an optimal time period (Figure 10.6). The optimal time period, however, varies between wetland types. For example, aquatic beds are best identified on photographs acquired at the peak of the growing season, when vegetation can be observed floating on the surface or submerged (in clear waters). This is also true for many emergent wetlands, although not all. The boundaries of prairie pothole marshes are best delineated when the basins are filled with water in early spring. Deciduous forested wetlands are best observed on leaf-off photographs taken during the spring when water tables are usually highest and saturated or flooded soils can be observed beneath the canopy (Plates 50-51). For dense evergreen wetland forests, there is no optimal time, since leaves are not shed annually.

In practice, when conducting a comprehensive inventory of all wetland types, if budgetary constraints do not permit acquisition of multiseason photographs, the available photographs predetermine many of the problems to be encountered. If possible, photographs that facilitate identification of the most abundant type should be selected.

Aquatic beds are usually missed when spring photographs are used. They are included, however, within the boundaries of their associated open water bodies. On occasion through field studies, aquatic beds are observed and their boundaries approximated on the photographic overlays. In this way, at least, some of the beds are mapped. When summer or fall photographs are interpreted, aquatic beds are easily identified.

Emergent wetlands include marshes, wet meadows, herbaceous fens, prairie potholes (in the Upper Midwest), and certain playas (in the Southwest). Inland marshes are among the most easily recognized. Topographic position, the smooth texture of the vegetation, and the close relationship to water (i.e., many marshes are associated with a permanent water body) are the major interpretive elements. Beaver-influenced marshes are easily recognized by the presense of beaver dams/lodges and dead trees. Marshes affected by muskrats also can be interpreted; their lodges are quite apparent on 1:12,000 photographs. Frazier and Moore (1993) found that individual flowering plants of purple loosestrife could be detected on 1:5,000 or larger color slides.

Wet meadows occur in agricultural areas and may be found in isolated depressions, on gentle groundwater seepage slopes, or along narrow streams. They are best observed on spring photographs that show saturated soils due to the seasonal high water table. Human impacts, such as grazing, mowing and irrigation, make identification of wet meadows difficult. For example, distinguishing wet pastures (wetlands) from moist pastures (uplands) in irrigated regions, such as Nevada and Utah, can be problematic. The existence of drainage structures (e.g, ditches and tile drains) adds confusion to wetland photointerpretation as it does for on-the-ground delineation as well.

Other difficult-to-identify emergent wetlands include pitcher plant bogs of the southeastern U.S. and temporarily flooded floodplain meadows or marshes. The bogs occur at varied positions on the landscape ranging from depressions to adjacent hillsides (crossing 20-ft contours). Moreover, these wetlands lack standing water and consequently, do not have a distinctive appearance, i.e., they look similar to adjacent uplands on aerial photos. Herbaceous fens occur in boreal regions such as Minnesota and Maine, and are often observed as emergent wetland patches within larger shrub-bog wetland complexes. In Alaska, distinguishing between moist tundra (wetland) and alpine tundra (nonwetland) can be very difficult, yet subtle tonal differences along the topographic gradient can be detected by a skilled interpreter.

Prairie potholes are glacially formed depressional wetlands characteristic of the Upper Midwest (e.g., the Dakotas; see Figure 7.6). This region is known for its wide-ranging rainfall patterns that cause a marked change in plant species composition over a relatively short time period (e.g., 20 years; see Chapter 7). In dry years and seasons, the limits of the pothole basins are difficult to identify, since the drier portions of these wetlands are often tilled and cultivated at these times. Consequently, delineation of the basins is best performed on photographs acquired when the basins are filled with water. Prior to mapping prairie potholes for the NWI Project, special photographic missions were flown when the basins were filled (Plate 52). It took several years to acquire these photographs, but they proved most useful for wetlands mapping. In the larger potholes, there is a distinct vegetation pattern associated with degrees of wetness or water regimes. In the center, the deeper potholes have a permanently flooded zone where aquatic beds may predominate. This zone is fringed by emergent vegetation and aquatic beds in a semipermanently flooded zone. Above this zone is a seasonally flooded zone followed by a temporarily flooded zone. These concentric bands of vegetation are quite apparent during the growing season.

Scrub-shrub wetlands are comprised of bogs and pocosins dominated by ericaceous shrubs, and other wet areas dominated by true shrubs or tree saplings. Some of these wetlands have characteristic plant species that may be easily interpreted, such as leatherleaf (*Chamadaphne calyculata*), alders (*Alnus* spp.), buttonbush (*Cephalanthus occidentalis*), willows (*Salix* spp.), swamp cyrilla (*Cyrilla racemiflora*), fetterbush (*Lyonia lucida*), St. John's-wort (*Hypericum* spp.), cranberry (*Vaccinium macrocarpon*), and meadowsweets (*Spiraea* spp.) especially on CIR photos. Leatherleaf bogs are particularly evident on spring CIR photographs, appearing as a smooth orange signature along the water's edge or in upland depressions. Cultivated cranberry bogs produce similar colors, but the pattern is different due to the presence of ditches and dikes (Plate 51). The evergreen shrubs of the pocosins (coastal North Carolina) are smooth red on fall, winter, or early spring photographs. Mixed shrub communities in temporarily flooded wetlands are difficult to interpret as are wet alder thickets in more northern areas. Alder is more restricted to wetlands in the southern part of its range which facilitates identification of this wetland type; however, farther north (e.g., in northern Maine) alder becomes less wetland-specific and can be found in large numbers in uplands, creating difficulty for wetland photointerpretation. Here topography and existing soils information are often considered in separating alder wetlands from dry alder areas. In the Southwest, honey mesquite (*Prosopis juliflora*) dominates certain riparian habitats. It grows from the water's edge (on sand bars) to floodplain terraces and adjacent hillsides. To separate the wet sites from the dry sites, plant size and density and topography must be considered. The wetter areas are usually covered by a dense growth of taller individuals.

Forested wetlands are dominated by deciduous and/or evergreen species. As a general rule, evergreen forested wetlands are more difficult to interpret since they retain their canopy, but deciduous forested wetlands also are difficult to identify on summer photographs. As with other wetland types, the drier the wetland, the more difficult it is to interpret because soils are usually not saturated or flooded at the time of photography.

For most of the conterminous U.S., early spring photographs are best for identifying the limits of forested wetlands. At this time, the water table should be closest to the surface, new leaves are not yet developed on deciduous trees, and saturated or inundated soils are visible through the forest canopy (Figure 10.6). Late fall photography also has deciduous trees in leaf-off condition, but it is normally not as useful for forested wetland detection because the soil is usually not saturated at or near the surface at this time. Despite this disadvantage for delineating wetland boundaries, fall photography may facilitate identification of certain forested wetland communities, such as larch (*Larix laricina*) wetlands in the boreal regions and bald cypress (*Taxodium distichum*) in the southeastern U.S. The leaves of larch, for example, turn yellow in the fall, so leaf color in combination with the tree's pyramid shape make it easy to identify on panchromatic as well as CIR photographs.

When leaf-off photographs are not available, even the wetter deciduous forested wetlands may be difficult to identify. This is especially a problem in regions where weather or sun angle favor the acquisition of summer coverage, such as the northwestern states and Alaska. The weather conditions (much fog and rain) in the Northwest generally preclude acquiring leaf-off photographs for large blocks of land, such as required by the NHAPP and the Alaska High-Altitude Photography Program (AHAPP). For smaller scale projects, it is possible, however, to capture leaf-off photographs. For Alaska, the low sun angle in spring and fall creates extensive shading and shadows that make acquisition of useable leaf-off photographs virtually impossible to obtain for a large geographic area.

When using leaf-on photographs, one must usually rely on features other than saturated soils to separate deciduous forested wetlands from deciduous forested uplands. Topography plays a particularly important role in these cases. Forested areas at low positions on floodplains may be identified as forested wetlands, especially if collateral information, such as soil survey reports indicate hydric soils in the area. In all cases, field work must be conducted to identify a general contour at which wetland ends and upland begins within a representative sample of drainage basins in the project area.

Recognizing that wetlands also occur on slopes with hillside seepage, one must consider whether the forested slope is wetland. For example, along the Pacific Coast west of the Cascade Mountains, forested wetlands of red alder (*Alnus rubra*) commonly occur on seepage slopes. In the Pacific Northwest, red alder dominates both wetland and upland slopes, thus separating the wet alder from the dry alder on leaf-on photographs is most difficult. In some cases, there are canopy openings in which wetland understory vegetation (e.g., sedges and willows) or beaver dams may be observed. In the absence of these openings, distinguishing the red alder wetlands from the red alder uplands remains problematic. As one moves south into southern Oregon and northern California (where the climate is much drier), red alder is restricted to wetlands, including hillside seeps, thus red alder communities are only wetland communities here and can be photointerpreted as such. This situation is similar to the alder shrub wetlands on the East Coast mentioned earlier.

Temporarily flooded forested wetlands are the most difficult type of deciduous forested wetland to recognize, regardless of the photographs used. In fact, wetlands of this type are often difficult to identify on the ground, since many associated plants also are found in uplands. These wetlands commonly occur on floodplains, where they are flooded for brief periods (e.g., usually less than 2 weeks) during the growing season, or in isolated depressions subject to saturation from a seasonally high water table or to ponding from precipitation (Plate 53). Since saturated soils are not usually observed unless leaf-off photographs are acquired at the time of saturation or inundation, one must rely more on topographic position (e.g., low positions on the floodplain), drainage patterns, soil survey reports, and field work. Field work is necessary to demarcate wetland–nonwetland boundaries within floodplains and other positions on the landscape; this is vital to any photo-interpretation project.

In their natural state, temporarily flooded forested wetlands are difficult enough to identify, but to make matters worse, the hydrology of many of these wetlands has been modified to varying degrees throughout the country. This situation is particularly widespread in the Southeast and the West where dams, levees, channelization projects, and stream diversions often prevent seasonal flooding. In these cases, the interpreter must decide whether the hydrology has been altered to the extent that the area is no longer wetland. In the western U.S., cottonwoods (*Populus* spp.) are the dominant trees in these situations. The wetland community often has a denser canopy and a thicker understory, while the nonwetland (relict) cottonwood community has a more open canopy and thinner understory. The extremes are relatively diagnostic, but there are many intermediate situations that remain problematical. Moreover, there is considerable debate over how much of the riparian habitat along western rivers and streams is wetland, further complicating the photointerpretation of forested wetlands in this region. For practical and natural resource managment purposes, riparian habitats should be mapped as a natural system given its ecological significant (Plate 54). The NWI has initiated riparian habitat mapping in the West.

Evergreen forested wetlands are among the most difficult wetlands to interpret. Since evergreens do not lose their leaves each year, the foliage often prevents observation of saturated soils, at least through dense stands which are extremely common. Where the canopy is more open, however, wetland detection is aided by observing saturated soils or characteristic understory vegetation. For example, pocosin forested wetlands along the Coastal Plain of North Carolina dominated by pond pine (*Pinus serotina*) may be recognized by their characteristic smooth-textured shrub understory. In the Pacific Northwest, lodgepole pine (*Pinus contorta*) wetlands have a dense understory of willows and sedges in contrast to the thinner understory of lodgepole pine uplands. In the Northeast, black spruce (*Picea mariana*) wetlands may have openings of leatherleaf (an easily interpreted evergreen shrub) or may occur contiguous with and at about the same elevation as leatherleaf bogs and other wetlands.

The biggest problem in interpreting evergreen forested wetlands involves dense evergreen stands that occur both in wetlands and adjacent uplands. Sometimes, the height of the canopy may reflect a difference in wetness. In Alaska, for example, wetland evergreens are somewhat shorter than upland evergreens (about 60 vs. 100 ft tall) in certain areas. Wet evergreens also may show signs of water stress, as evidenced by the yellowing of some of their leaves (chlorosis). In other cases, evergreen forested wetlands may be dominated by a species that actually looks different from evergreens on the adjacent upland, such as Atlantic white cedar (*Chamaecyparis thyoides*; only in wetlands) vs. pitch pine (*Pinus rigida*; in wetlands and uplands) in the Pine Barrens of southern New Jersey. Here, it is relatively easy to separate the white cedar stands from the pitch pine stands, but separating pitch pine wetlands from pitch pine uplands is difficult (Plate 55). The saturated soils of the wetter pines may be evident in canopy openings, but many pitch pine wetlands are only temporarily flooded, and their soils are usually not saturated at the surface. Again, landscape and topographic position and available soils information must be considered and field studies conducted to separate seasonally saturated pitch pine wetlands from pitch pine upland forests. A similar problem is encountered where the life form of the upland forest dominant resembles the wetland forest dominant, such as white spruce (*Picea glauca*) and black spruce, respectively, in the Northeast and Alaska.

In the southern U.S., pine plantations are established in many wetlands (wet flatwoods). Some of these wetlands are drained, while others are not. These planted pines look much like upland pines in that they form dense stands. Consequently, they present problems for wetland photo-interpretation and one must again consider topographic position, consult soils information, if available, and perform ground-truthing. This approach provides a useful, generalized wetland boundary for planning purposes, recognizing its limitations, e.g., the boundaries are not exact. Also, some types of evergreen forested wetlands are difficult to identify even in the field and require extensive soil sampling to determine the limits of wetland.

In some parts of the U.S. (e.g, southeast Alaska and Hawaii), rainforests make photointerpre-tation of forested wetlands extremely difficult. In Southeast Alaska, topographic features (such as drainage, depressions, and level terrain) may be used to assist in identifying evergreen forested wetlands. In Hawaii, leaf-on, panchromatic photographs were used by the NWI Project and, at the higher elevations, areas were identified as mixes of forested wetland and forested upland, depending on the recognition of a saturated emergent wetland within the forests.

INTERPRETING SUBMERGED AQUATIC VEGETATION

Underwater aquatic beds are important fish spawning and nursery grounds and have been the target of numerous studies by federal and state agencies. NOAA recently worked with several researchers to develop guidance for monitoring submerged rooted vascular plants (SRV) for its Coastal Change Analysis Program (C-CAP) (Dobson et al., 1995). The recommended film is Aerocolor 2445 color-negative film, with Aerochrome 2448 color-reversal and Aerographic 2405 black and white negative film also satisfactory. The photo scale depends on water clarity, with 1:12,000 to 1:24,000 being the typical scale, but 1:12,000 suggested for more turbid waters and 1:48,000 possibly for extreme clear waters such as Florida Keys. Timing of photography also is critical.

- During maximum biomass (e.g., June in the Northwest and Northeast; April to May for eelgrass in North Carolina; September for other species in the East)
- No clouds and minimum haze
- Low turbidity (e.g., not during phytoplankton blooms or persistent winds or immediately after heavy rains)
- Within 2 hours of lowest tide during the early morning (low sun angle of 15 to 30%)
- Minimal wind and wave action (winds <10 mph; no breaking waves)

At 1:24,000, the smallest detectable stand of SRV is about 1 m. C-CAP's minimum mapping unit is set at 0.03 ha (approximately a 20 m-diameter stand). One must recognize that like wetland mapping, SRV mapping tends to be conservative, being dependent on the quality of the aerial photography, the ability of the photointerpreter, and the nature of the beds (i.e., their ease or difficulty of photointerpretation). The aerial photograph in Figure 10.7 shows the ease of interpreting beds of SRV with large-scale imagery. For change detection (monitoring), same season photography from different years is recommended, noting the limitations of historical analysis using available images. Some examples of SRV or submerged aquatic vegetation (SAV) mapping are found in Ferguson and Wood (1990, 1994), Haddad and Harris (1985), Jensen et al. (1980), Orth and Moore (1983), and Orth et al. (1990, 1991).

Studies of submerged aquatic vegetation are not limited to estuarine systems. Lee (1991) described the results of a study where 1:6,000 color aerial photography was successfully used to identify SAV in the Lake St. Clair–Detroit River system in Michigan. With this photography, five genera were distinguishable with an estimated accuracy of 68%.

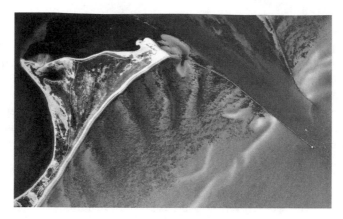

FIGURE 10.7   Black and white copy of true color aerial photograph (reduced 60%; original scale 1:24,000) showing submerged rooted vegetation in shallow water off Nantucket Island, MA. (Courtesy of the Massachusetts Dept. of Environmental Protection.)

REFERENCES

Anderson, R.R. and F.J. Wobber. 1972. Wetlands mapping in New Jersey. *Photogram. Eng.,* 39: 353-358.

Arnold, R.H. 1997. *Interpretation of Airphotos and Remotely Sensed Imagery.* Prentice Hall, Upper Saddle River, NJ.

Brown, W.W. 1978. Wetland mapping in New Jersey and New York. *Photogram. Eng. Remote Sens.,* 44(3): 303-314.

Chime, L.R., G.E. Gnauck, and J. Maragos. 1978. Hawaii wetlands mapping. In *Coastal Zone 1978.*

Cole, N.B. and E. Fried. 1981. Technical Manual. Freshwater Wetlands Inventory. NY Department of Environmental Conservation, Division of Fish and Wildlife, Albany.

Cowardin, L.M., V. Carter, F.C. Golet, and E.T. LaRoe. 1979. Classification of Wetlands and Deepwater Habitats of the United States. FWS/OBS-79/31. U.S. Fish and Wildlife Service, Washington, D.C.

Cowardin, L.M. and V.I. Myers. 1974. Remote sensing for identification and classification of wetland vegetation. *J. Wildl. Manage.,* 38(2): 308-314.

Crowley, S., C. O'Brien, and S. Shea. 1988. Results of the Wetland Study on the 1988 Draft Wetland Rules. Agency of Natural Resources, Division of Water Quality, State of Vermont, Waterbury.

Dobson, J.E., E.A. Bright, R.L. Ferguson, D.W. Field, L.L. Wood, K.D. Haddad, H. Iredale III, J.R. Jensen, V.V. Klemas, R.J. Orth, and J.P. Thomas. 1995. NOAA Coastal Change Analysis Program (C-CAP): Guidance for Regional Implementation. NOAA Tech. Rep. NMFS 123. U.S. Department of Commerce, Seattle, WA.

Federal Geographic Data Committee, Wetlands Subcommitttee. 1992. Application of Satellite Data for Mapping and Monitoring Wetlands. Technical Report 1. Fact Finding Report. Reston, VA.

Ferguson, R.L. and L.L. Wood. 1990. Mapping submerged aquatic vegetation in North Carolina with conventional aerial photography. In Federal Coastal Wetland Mapping Programs — A Report by the National Ocean Pollution Policy Board's Habitat Loss and Modification Work Group. S.J. Kiraly, F.A. Cross, and J.D. Buffington (Eds.). Biol. Rep. 90 (18): 125-133. U.S. Department of the Interior, Fish and Wildlife Service, Washington, D.C.

Ferguson, R.L. and L.L. Wood. 1994. Rooted Vascular Beds in the Albemarle-Pamlico Estuarine System. Rep. No. 94-02. North Carolina Department of Environment, Health, and Natural Resources, Raleigh, NC.

Frazier, B.E. and B.C. Moore. 1993. Some tests of film types for remote sensing of purple loosestrife, Lythrum salicaria, at low densities. *Wetlands,* 13: 145-152.

Gammon, P.T. and V. Carter. 1979. Vegetation mapping with seasonal color infrared photographs. *Photogram. Eng. Remote Sens.,* 45(1): 87-97.

Gilbert, M.C., M.W. Freel, and A. J. Bieber. 1980. Remote Sensing and Field Evaluation of Wetlands in the Sandhills of Nebraska. U.S. Army Corps of Engineers, Omaha District.

Granger, T. 1995. Memorandum to E. Summerfield. Subject: *Comparison of Wetland Acres Identified on NWI Maps and on River Basin Team Composite Wetland Maps for Dungeness, Port Angeles, and Liberty Bay/Miller Bay Watersheds and Bainbridge Island.* Pubet Sound Cooperative River Basin Team, Olympia, Washington.

Haddad, K.D. and B.A. Harris. 1985. Use of remote sensing to assess estuarine habitats. In *Coastal Zone '85.* O. Magoon, H. Converse, D. Miner, D. Clark, and L. Tobin (Eds.). American Society of Civil Engineers, New York.

Jensen, J.R., L. Tinney, and J. Estes. 1980. Remote sensing techniques for kelp surveys. *Photogramm. Eng. Remote Sens.,* 46: 743-755.

Johnson, D.R. and A.J. Makinson. 1988. Soil Survey of Umatilla County Area, Oregon. USDA Soil Conservation Service.

Kalla, P. and J. Couch. 1996. Flatwoods wetland determination by GIS. In Wetland Data and Data Acquisition Workshop. Wetland, Floodplain, and River Online Services and GIS Applications. B. Wilen (Mod.). 12th Annual Meeting (July 9-12, 1996; Arlington, VA). The Association of State Wetland Managers, New Berne, NY. 1-2.

Kennard, W.C., M.W. Lefor, and D.L. Circo. 1980. Identification of Inland Wetlands for Transportation Planning Using Color Infrared Aerial Photography. Final Report JHR 80-132. University of Connecticut, Natural Resources Department and Biological Sciences Group, Storrs.

Kiraly, S.J., F.A. Cross, and J.D. Buffington (Eds.). 1990. Federal Coastal Wetland Mapping Programs — A Report by the National Ocean Pollution Policy Board's Habitat Loss and Modification Work Group. Biol. Rep. 90 (18). U.S. Department of the Interior, Fish and Wildlife Service, Washington, D.C.

Klemas, V., F.D. Daiber, D.S. Bartlett, O.W. Crichton, and A.O. Fornes. 1973. Coastal Vegetation of Delaware. University of Delaware, Newark.

Klemow, K.M. and M. Mohseni. 1996. How accurate are National Wetland Inventory Maps? An analysis from northeastern Pennsylvania. Presentation at the 1996 Ecological Society of America meeting, Providence, RI, July 1996.

Lee, K.H. 1991. Wetlands Detection Methods Investigation. Report No. 600/4-91/014. EPA–Environmental Monitoring Systems Labortory, Las Vegas, NV.

Lovvorn, J.R. and C. M. Kirkpatrick. 1982. Analysis of freshwater wetland vegetation with large-scale color infrared aerial photography. *J. Wildl. Manage.,* 46(1): 61-70.

Lukin, C.G. and L.L. Mauger. 1983. Environmental Geological Atlas of the Coastal Zone of North Carolina: Dare, Hyde, Tyrrell, and Washington Counties. Coastal Energy Impact Program Report No. 32. North Carolina Department of Health, Environment and Natural Resources, Raleigh.

MacConnell, W.P. 1974. Remote Sensing Land Use and Vegetative Cover in Rhode Island. Bulletin No. 200. Cooperative Extension Service, University of Rhode Island, Kingston.

MacConnell, W.P. 1975. *Remote Sensing 20 Years of Change in Massachusetts 1951/52–1971/72.* Bulletin No. 630. Massachusetts Agricultural Experiment Station, University of Massachusetts, Amherst.

MacConnell, W., J. Stone, D. Goodwin, D. Swartwout, and C. Costello. 1992. Recording Wetland Delineations on Property Records. The Massachusetts DEP Experience 1972-1992. Department of Forestry and Wildlife Management, University of Massachusetts, Amherst.

Maryland Department of Natural Resources. 1991. Results of a Field Reconnaissance of Remotely Sensed Land Cover Data (July 16-18, 1991; Salisbury, MD). Sponsored by Salisbury State University, Maryland Department of Natural Resources, NOAA–Coastal Ocean Program, and U.S. Fish and Wildlife Service–National Wetlands Inventory.

Massachusetts Wetlands Restoration & Banking Program and Natural Resources Assessment Group. 1996. Atlas of Tidally Restricted Marshes: North Shore of Massachusetts. Executive Office of Environmental Affairs, Boston, MA and Department of Plant and Soil Sciences, University of Massachusetts, Amherst.

McCall, C.A. 1972. Manual for Maine Wetlands Inventory. Maine Dept. of Inland Fisheries and Game, Augusta.

McCormack, J. and H.A. Somes, Jr. 1982. The Coastal Wetlands of Maryland. Maryland Dept. of Natural Resources, Annapolis.

McMullen, J.M. and P.A. Meacham. 1996. A comparison of wetland boundaries delineated in the field to those boundaries on existing state and federal wetland maps in central New York State. In *Wetlands: Environmental Gradients, Boundaries, and Buffers.* G. Mulamoottil, B.G. Warner, and E.A. McBean (Eds.). CRC Press, Lewis Publishers, Boca Raton, FL. 193-205.

Miller, T.B., R.C. Heller, J.J. Ulliman, and F.D. Johnson, 1976. *Evaluating Riparian Habitats from Aerial Color Photography.* Bulletin No. 11. University of Idaho Forest, Wildlife and Range Experiment Station, Moscow.

Moorhead, K.K. and A.E. Cook. 1992. A comparison of hydric soils, wetlands, and land use in coastal North Carolina. *Wetlands,* 12: 99-105.

Morrison, S.W. 1993. June 14 letter to Dennis Peters. Thurston Regional Planning Council, Olympia, WA.

National Research Council. 1995. *Wetlands Characteristics and Boundaries.* National Academy Press, Washington, D.C.

Nichols, C. 1994. Map Accuracy of National Wetlands Inventory Maps for Areas Subject to Land Use Regulation Commission Jurisdiction. U.S. Fish and Wildlife Service, Hadley, MA.

Olson, D.P. 1964. The Use of Aerial Photographs in Studies of Marsh Vegetation. Bulletin 13. Maine Agricultural Experiment Station, Orono.

Orth, R.J. and K.A. Moore. 1983. Chesapeake Bay: an unprecedented decline in submerged aquatic vegetation. *Science,* 222: 51-53.

Orth, R.J., K.A. Moore, and J.F. Nowak. 1990. Monitoring seagrass distribution and abundance patterns: a case study from the Chesapeake Bay. In Federal Coastal Wetland Mapping Programs — A Report by the National Ocean Pollution Policy Board's Habitat Loss and Modification Work Group. S.J. Kiraly, F.A. Cross, and J.D. Buffington (Eds.). Biol. Rep. 90 (18): 111-123. U.S. Department of the Interior, Fish and Wildlife Service, Washington, D.C.

Orth, R.J., J.F. Nowak, A.A. Frisch, K.P. Kiley, and J.R. Whiting. 1991. Distribution of Submerged Aquatic Vegetation in Chesapeake Bay. U.S. Environmental Protection Agency, Chesapeake Bay Program, Annapolis, MD.

Patterson, S.G. 1986. Mangrove Community Boundary Interpretation and Detection of Areal Changes on Marco Island, FL: Application of Digital Image Processing and Remote Sensing Techniques. Biol. Rep. 86(10). U.S. Fish and Wildlife Service, Washington, D.C.

Peters, D.D. 1994. Use of aerial photography for mapping wetlands in the United States: National Wetlands Inventory. Proc. 1st Intl. Airborne Remote Sensing Conference and Exhibition, Strasbourg, France. Volume III: 165-173.

Philipson, W.R. (Editor-in-Chief). 1997. *Manual of Photographic Interpretation, 2nd ed.* American Society for Photogrammetry and Remote Sensing, Bethesda, MD.

Reimold, R.J., J.L. Gallagher, and D.E. Thompson. 1973. Remote sensing of tidal marsh. *Photogram. Eng. Remote Sens.,* 19(5): 477-488.

Roller, N.E.G. 1977. *Remote Sensing of Wetlands.* Rept. No. 193400-14-T. Environmental Research Institute of Michigan, Ann Arbor.

Seher, J.S. and P.T. Tueller. 1973. Color aerial photos for marshland. *Photogram. Eng.,* 19(5): 489-499.

Shapiro, C. 1995. Coordination and Integration of Wetland Data for Status and Trends and Inventory Estimates. Tech. Rept. No. 2. Federal Geographic Data Committee, Reston, VA.

Shima, L.J., R.R. Anderson, and V.P. Carter. 1976. The use of aerial color infrared photography in mapping the vegetation of a freshwater marsh. *Chesap. Sci.,* 19(2): 74-85.

Soil Survey Division Staff. 1993. Soil Survey Manual. Agriculture Handbook No. 18 Revised. U.S. Department of Agriculture, Washington, D.C.

Soil Survey Staff. 1998. Keys to Soil Taxonomy, 8th ed. U.S. Department of Agriculture, Natural Resources Conservation Service, Washington, D.C.

Stolt, M.H. and J.C. Baker. 1995. Evaluation of National Wetland Inventory maps to inventory wetlands in the southern Blue Ridge of Virginia. *Wetlands,* 15(4): 346-353.

Stroud, L.M. 1969. Color-infrared Aerial Photographic Interpretation and New Primary Productivity of a Regularly-Flooded North Carolina Salt Marsh. Master's Thesis. North Carolina State University, Raleigh.

Swartwout, D.J., W.P. MacConnell, and J.T. Finn. 1982. An evaluation of the National Wetlands Inventory in Massachusetts. Proc. of In-Place Resource Inventories Workshop (University of Maine, Orono, ME; August 9-14, 1981). 685-691.

Swenson, E.I. 1981. Soil Survey of Hampshire County, MA, Central Part. USDA Soil Conservation Service, Amherst, MA.

Thurston Regional Planning Council. 1991. *Thurston Regional Wetland Pilot Project.* Olympia, WA.

Tiner, R. W., Jr. 1977. *An Inventory of South Carolina's Coastal Marshes.* Tech. Rep. 23. South Carolina Marine Resources Center, Charleston, SC.

Tiner, R.W., Jr. 1987. New York Tidal Wetlands Trends: Pilot Study in Shinnecock Estuary and Recommendations for Statewide Analysis. U.S. Fish and Wildlife Service, Fish and Wildlife Enhancement, Newton Corner, MA.

Tiner, R.W. Jr. 1990. Use of high-altitude aerial photography for inventorying forested wetlands in the United States. *For. Ecol. Manage.*, 33/34: 593-604.

Tiner, R.W. 1996. Practical considerations for wetland identification and boundary delineation. In *Wetlands: Environmental Gradients, Boundaries, and Buffers.* G. Mulamoottil, B.G. Warner, and E.A. McBean (Eds.). CRC Press, Lewis Publishers, Boca Raton, FL. 113-137.

Tiner, R.W. 1997a. NWI Maps: what they tell us. *National Wetlands Newsletter,* 19(2): 7-12.

Tiner, R.W. 1997b. Wetlands, chap. 13. In *Manual of Photographic Interpretation*, 2nd ed. W.R. Philipson (Ed.). American Society for Photogrammetry and Remote Sensing, Bethesda, MD. 475-494.

Tiner, R.W. and G.S. Smith. 1992. Comparison of Four Scales of Color Infrared Photography for Wetland Mapping in Maryland. National Wetlands Inventory Report R5-92/03. U.S. Fish and Wildlife Service, Hadley, MA.

Tiner, R.W. and P.L.M. Veneman. 1989. *Hydric Soils of New England.* Bulletin C-183R. University of Massachusetts Cooperative Extension, Amherst.

U.S. Fish and Wildlife Service. 1992. An Investigation and Verification of Draft NWI Maps for Cape May County, New Jersey. New Jersey Field Office, Pleasantville.

U.S. Fish and Wildlife Service. 1994. Cartographic Conventions for the National Wetlands Inventory. NWI Project, St. Petersburg, FL.

U.S. Fish and Wildlife Service. 1995. Photo Interpretation Conventions for the National Wetlands Inventory. NWI Project, St. Petersburg, FL.

U.S. Fish and Wildlife Service. 1996. Some Uses of National Wetlands Inventory Maps and Digital Map Data in the Northeast. Ecological Services, Region 5, Hadley, MA.

Williams, Major General A.E. 1992. U.S. Army Corps of Engineers memorandum on clarification and interpretation of the 1987 manual. March 6, 1992.

Wisconsin Coastal Management Program. 1995. Basic Guide to Wisconsin's Wetlands and their Boundaries. Wisconsin Department of Administration, Madison.

Index

Note: *Page numbers in this index for plant species refer to all text and tables, except for nondominant species listed in Tables 9.3 through 9.10 which provide extensive listings of species for representative wetland plant communities in the U.S.*

A

Abies, 25-26, 80, 86, 222, 301, 303-304, 322-325, 328-329
Abundance classes, 110
Acer, 22, 24, 26, 57-58, 67, 71, 73, 79-80, 82, 302-304, 316-321, 326
Achillea, 223
Acidophiles, 85-86
Acid soils, 52
Acorus, 82, 84, 87, 226, 263, 298
Adventitious roots, 19, 55, 60-62, 79
α, α'-dipyridyl, 175-176, 230, 240
Aerenchyma, 19, 55, 57, 60, 62-65, 67
Aerobic conditions, 30-31, 52-53, 147, 238-239
Agropyron, 82, 223, 325
Agrostis, 74, 298
Alabama, 23, 171
Alaska, 23, 77, 133, 229, 234, 240, 291-297, 299-304, 306-308, 372
　forested wetland communities, 331
　shrub wetland communities, 314-315
Alder swamps, 310, 312, 314-315, 322, 326-327, 330-331
Alfisols, 153, 178, 241
Algae-covered rocky shores, 31, 38, 309
Algal flats, 309
Algal mats, 174, 201, 228
Alisma, 68, 72, 82, 87, 263, 298
Allenrolfea, 237, 300
Alliaria, 82, 223
Allium, 223, 225
Allocarya, 227
Alluvial soils, 138, 140, 167, 170-172, 175-176, 233-234, 237
Alnus, 19, 25-26, 60-61, 67, 87, 263, 303-304, 310, 312, 314-316, 322-323, 326-327, 329-331, 372-373
Alopecurus, 82, 85, 87, 298-299
Alternanthera, 82, 298
Amaranthus, 82, 84, 226
Ambrosia, 87, 300
Amelanchier, 87
Ammania, 225
Amorpha, 60
Amphicarpa, 87
Anaerobiosis, 30-31, 51-53, 60, 133-134
Andic soils (Andisols), 178, 229-230, 240
Andromeda, 72, 82, 301, 310-311, 315
Andropogon, 73, 297-298

Anemone, 225
Angelica, 87, 298
Angle-gauge, 107, 109
Anthoxanthum, 223
Anthraquic soils, 241-242
Anthric saturation, 34, 133
Apocynum, 223
Aqualfs, 31
Aquatic beds, 375-376
Aquatic invertebrates, 174, 200-201, 203
Aquatic mosses, 31, 201
Aquatic plants, 54, 371, 375-376
Aquic conditions, 133, 151-152
Aquic moisture regime, 133, 151-152, 239
Aquic suborders, 152-154, 176-178, 202
Aralia, 224
Arctophila, 82, 299
Areal cover, 104-106
Aridisols, 178, 237-238
Arid wetlands, 20, 42-43, 137, 170, 227-229, 235-238, 248-249, 282-285
Arisaema, 87
Aristida, 297-298
Arizona
　forested wetland communities, 326-327
　shrub wetland communities, 313
Arkansas, 23
Aronia, 85, 311
Artemisia, 85
Asclepias, 82, 87
Ascophyllum, 38, 82, 309, 370
Ash wetlands, 317, 319, 326, 329-330
Asimina, 82, 224
Aspen wetlands, 322-325, 329
Aster, 25, 67, 82, 84, 86-87, 225, 297-298, 300, 307
Athyrium, 322
Atriplex, 26, 237, 300, 307
Australia, 20, 42-43, 140
　wetland classification systems, 282-285
　wetland definitions, 8-9
Avicennia, 62-63, 67, 308, 370
Azolla, 72

B

Baccharis, 307, 309
Balsam poplar wetland, 325

Basal area
    actual, 106-109, 212
    basal area factor prisms, 107, 109
    calculating, 107, 207
    conversion charts, 108-109
    estimated, 107
    relative, 107, 110, 209
Batis, 82, 84, 309, 370
Baumea, 282
Bayhead swamp, 320
Beaver, 40-41, 147, 241, 372
Beckmannia, 82, 298
Belgium, 136
Belt transect method, 121
Betula, 26, 69, 72, 81-82, 86-87, 263, 302, 304, 310, 314-
    315, 317, 331
Berberis, 223
Bidens, 25, 82, 84, 87, 226, 298, 308
Biologic zero, 22, 24, 133-134, 240
Bitterlich variable plot method, 121, 209
Blackbirds, red-winged, 28
Boehmeria, 87
Bog communities, 210, 313-316
Bogs, 28, 33, 38, 40-41, 52, 68, 259-261, 277, 280-281,
    301, 310, 372
Boltonia, 84, 298
Borrichia, 82, 84, 237, 309
Bottomland hardwood wetlands, 23, 52, 274, 276, 317, 319
Brasenia, 72, 82
Buttonbush swamp, 310
Buttressing, 56-58, 171, 174

C

Cabbage, 246
Calamagrostis, 82, 87, 297-299, 322-325
Calciphiles, 85-86, 297
California, 23, 77, 170, 236-237, 240, 250, 276, 306-308,
    373
California forested wetlands, 326-327
Calla, 72, 263, 298
Callitriche, 86, 226, 298-299
Caltha, 24, 72, 82, 87, 298, 324-325
Calystegia, 98
Campanula, 85, 87
Canada, 137
    wetland classification, 276-279
    wetland definition, 7-8
Capillary fringe, 129, 131, 137, 147, 238
Cardamine, 26, 323-324
Carex, 25, 60, 67, 72, 82-87, 263, 297-299, 307-308, 312-
    314, 328, 330
Caribbean, 77
Carolina bay wetland, 83
Carpinus, 79, 86-87, 319, 321
Carrots, 246
Carya, 72, 82, 87, 304, 319
Cedar wetlands, 28
    Atlantic white, 321
    northern white, 316

    photointerpretation, 374
    dalt, 327
    western red, 322, 325
Celtis, 58, 73, 87, 304, 319
Central plains, 77
Cephalanthus, 60, 67-68, 72, 82, 84, 87, 301, 310, 372
Ceratophyllum, 82
Chamaecyparis, 25, 72, 82, 259, 301, 303, 321, 374
Chamaedaphne, 24, 87, 301, 310-311, 372
Chelone, 82, 87
Chenopodium, 223
Chesapeake Bay, 138
Chrysothamus, 26
Cicuta, 72, 298-300, 307
Cinna, 87
Circaea, 87
Circium, 223
Cladium, 82, 84-85, 263, 297-298
Cladonia, 279
Clayey soils, 127-129, 136, 236-237
Claytonia, 24, 87, 103, 224
Clean Water Act, see Federal Clean Water Act
Clethra, 301
Climatic effects, 39, 53, 242
Coastal plain, 127, 135, 147, 160
Coastal plain swamp, 317
Cochlearia, 307
Coefficient of conservation, 87
Colorado, 23, 170
    forested wetlands, 322-325
Connecticut, 138, 189, 238, 348
    forested wetland communities, 317-318
    shrub wetland communities, 310-311
    wetland definition, 12
    wetland delineation method, 200
Conocarpus, 82, 308
Converted wetland, 196
Coptis, 83
Corn, 60, 68-69, 246
Cornus, 26, 81, 301, 310
Corps wetland delineation manual
    approach, 192-194, 221-223, 225
    hydric soil indicators, 158-159
    hydrophytic vegetation indicators, 79
    methods, 205-208
Corylus, 26
Cotton, 246
Cottonwood wetland communities, 313, 319, 322-323, 325-
    327, 374
Cotula, 84
Cover
    actual, 105
    areal, 104-106
    classes, 106, 212
    relative, 105
Cranberry bogs, 241-242, 372
Crassula, 82, 227
Crayfish burrows, 173-174, 247
Crayfish, pitcher-plant, 28
Cressa, 82, 307
Crinum, 298

Cryic soils, 160, 240
Cryptotaenia, 82
Cuscuta, 82, 227, 307
Cyclical wetlands, 42-43, 226-229
Cyperus, 87, 225, 282, 298
Cypress swamps, 276, 320-321
Cypripedium, 85, 87, 297
Cyrilla, 84, 301, 311, 372
Cystopteris, 223

D

Dactylis, 223
Dark-colored soils, 168, 239-240; see also Mollisols and
          Vertisols
Daucus, 223
Decodon, 57, 72, 82, 298
Delaware, 75, 189, 348, 370
      forested wetland, 318
Denitrification, 29, 66, 134
Density, 110
Depleted matrix, 135, 169-170
Depressional indicators of hydric soils, 170-174
Deschampsia, 84, 87, 227, 299-300, 307
Dianthus, 68
Dicranum, 85
Diospyros, 75
Distichlis, 82-84, 227, 237, 259, 299-300, 370
Dodecantheon, 299
Dominance
      calculation, 112
      definition, 111
      examples, 122-124
      types, 269
Dowingia, 227
Drainage, 19, 39-41
      assessment, 242-245
      effects on soil properties, 140, 146-147
      effects on wetland determinations, 198, 216-217
Drained hydric soils, 146, 243-245
Drepanocladus, 279
Drosera, 72, 82, 85-87, 300
Dryopteris, 87
Duchesnea, 223, 225
Dulichium, 72, 263, 298-299
Dupontia, 300

E

Echinochloa, 82, 87, 298, 300
Ecotones
      definition, 34-35
      riparian, 36
      wetland, 35-39
Ecotypes
      definition, 71
      wetland, 54, 71-76
Eichhornia, 282
Elatine, 227

Eleocharis, 72, 82, 87, 223, 227, 263, 298-300, 313, 328
Elodea, 72
Elymus, 83, 87
Emergent wetlands
      in the U.S., 296-301, 306-308
      photointerpretation, 369-372
Empetrum, 82, 315
Endosaturation, 34, 133
Enteromorpha, 309
Entisols, 153, 178
EPA wetland delineation manual
      approach, 194-195
      hydrophytic vegetation indicators, 79
      methods, 208-209
Epilobium, 87, 263
Episaturation, 34, 133
Equisetum, 82, 87, 223, 299-300, 322-325, 328
Eremocarpus, 227
Erianthus, 84, 298
Erigeron, 223
Eriocaulon, 300
Eriophorum, 60, 72, 82, 85-86, 297-301, 314
Erythronium, 25, 27
Estonian mires, 281
Estuarine wetlands, 268-270, 272-273
      emergent, 306-308
      intertidal shores, 309
      photointerpretation, 369-371
      scrub-shrub, 308-309
Eucalyptus, 74
Eulalia, 86
Eupatoriadelphus, 297-298
Eupatorium, 82, 84, 297-298
Europe, 174
Euthamia, 25, 297
Everglades, 297

F

50% rule, 189
50/20 rule, 111-112
FAC neutral rule, 78-79, 247
FACU-dominated wetlands, 80, 222-225
Faculative wetland plants, 36
Fagus, 73, 80, 86, 304
Fall flowering, 25
Fauria, 300
Federal Clean Water Act, 156, 189-191, 229
      wetland definition, 9-10
      wetland delineation, 189-195, 208; see also Corps
                  manual and EPA manual
      wetland hydrology, 21
Federal interagency wetland delineation manual, 190-191
      approach, 196-197, 221, 225
      hydrophytic vegetation indicators, 78-79
      methods, 209-213
Fens, 33, 38, 277, 280, 297, 311, 316
Ferrihydrites, 139
Ferrous iron test, 175
Festuca, 69, 74

Fibrists, 151, 165
Fimbristylis, 300
Finnish peatlands, 281
Fire, 40, 43, 139, 229, 240
Fir wetlands
    silver, 329
    subalpine, 322-325, 328
Fish, 28
Flarks, 280-281
Flatwoods, 19, 30, 32-33, 68, 80, 231, 248, 273-274, 296
Flatwood wetland communities, 319-320
Flooding
    duration, 34
    effects on plants, 52-69, 71, 74
    effects on soils, 51-54
    environmental changes, 51-53
    frequency, 32
    indicators, 171-174
    plant adaptations, 53-69
    plant tolerance, 53-54, 64
Floodplain soils, see Alluvial soils
Floodplain wetlands, 22, 28, 33, 53-54, 248, 274-276, 372-374; see also Alluvial soils
Florida, 23, 37, 53, 74-75, 156, 171, 231-232, 234, 250, 295-297, 309
    field indicators of hydrophytic vegetation, 79
    forested wetland communities, 319-321
    wetland definition, 12, 200
    wetland delineation manual, 200-202
Flowering, 24-27
Fluted trunks, 57-59, 174
Folists, 151, 229-230
Food Security Act, 149, 190
    converted wetland, 196
    farmed wetland, 21, 195-196
    manual, 195-196
    normal circumstances, 191
    prior converted wetland, 195-196
    wetland definition, 10-11
    wetland delineation approach, 195-196
    wetland hydrology, 21
    wetland mapping conventions, 195-196
Forestiera, 67, 72
Forested wetlands, 303-305
    photointerpretation, 357, 373-375
    plant communities, 316-331
Frankensia, 82, 307
Fraxinus, 25, 54, 60-61, 65, 67-68, 72, 86, 224, 303-304, 316, 319, 326, 329-330
Frequency
    analysis of vegetation, 118-121, 212-213
    definition, 110-111
Frogs, 28
Fucus, 38, 82, 84, 309, 370

G

Galium, 86, 223, 299
Gaultheria, 85, 223-224
Gaylussacia, 224, 300

Gelisols, 178
Georgia, 62, 171, 234
Geologic processes, 243
Geum, 82, 85, 225
Gilgai topography, 236, 249-250
Glacial rebound, 41, 242
Gleditsia, 72
Gleization, 135-136
Gleyed matrix, 135, 168-169, 241
Gley soils, 141, 168
Glyceria, 72, 82, 85, 298-300
Goethite, 139
Gordonia, 303, 311, 320
Grady ponds, 170
Gratiola, 86
Great Lakes, 295
    coastal marshes, 36, 39, 81, 85
    soils, 238
Great Salt Lake, 300
Greenhouse effect, 39
Greenrust, 139
Grindelia, 84
Groundwater gley soils, 141, 175
Growing season
    examples across U.S., 23
    federal regulatory definition, 21-22, 194
    microbial activity, 22-24, 29
    relative to plant activity, 21-22, 24-28
    relevance to wetland animals, 28, 30
    relevance to wetland functions, 29-30
    significance to wetland hydrology, 24, 30
Gulf-Atlantic coastal plain, 127, 135, 147, 160, 174, 291, 294, 297
Gum swamp, 320
Gypsum formation, 137

H

Habenaria, 300
Halomorphic soils, 140-141
Halophytes, 83-85, 188, 237, 300, 306-310
Hamamelis, 25
Hardwood forested wetlands, 317, 319-320, 329-330
Hawaii, 77, 296
Height classes, 110
Helianthus, 60
Helonias, 86
Helophytes, 70
Hematite, 139, 175, 230
Hemists, 151
Hemlock wetlands
    eastern, 318
    mountain, 331
    western, 329, 331
Heteranthera, 82
Hibiscus, 82, 84, 298, 308
Hilaria, 26
Hippurus, 82, 84, 300, 308
Histic epipedon, 40, 161, 167, 231, 235, 240
Histic materials, 160-168, 231

Histosols, 178; see also Organic soils
Holcus, 223
Honckenya, 307
Hordeum, 82, 84, 299-300
Human influences, see also Drained hydric soils
    on plant distribution, 80-81, 86-87, 223, 225-226
    on soils, 140, 146-147, 233
Hydrarch succession, 39-40
Hydric soils
    aerobic conditions, 30-31, 147
    andisols, 178, 229-230
    anthraquic, 241-242
    aquic suborders, 152-154
    aridisols, 178, 237-238
    buried horizons, 138
    dark-colored parent material, 239-240
    definition and criteria, 141-146
    drained, 146, 243-245
    entisols, 230, 232-234
    farmed, 246-247
    fidelity categories, 145, 148-149
    formation, 20
    general concept, 1, 146-147
    indicators
        development of U.S. lists, 146, 152, 156-157
        in other countries, 174-175
        U.S. list, 154, 156-174, 193, 199
    lists, 142, 146, 148-150
    mineral, 151-152
    mollisols, 31, 230, 235-236; see also Mollisols
    newly formed, 240-241
    organic, see Organic soils
    partly drained, 243-245
    problematic, 159, 230-247
    red parent material, 233
    relict, 136, 242-243
    sandy, 231-233
    saturated but lacking hydric properties, 238-239
    soil orders, 178
    spodosols, 139, 230, 234-235
    tropical, 230
    vertisols, 230, 236-237
Hydric substrates, 1
Hydrodynamics, 260
Hydrogen ion concentration (pH), 85-86, 271
Hydrogen sulfide, 66, 134, 170, 199, 204, 231
Hydrologic assessment, 242-245
Hydrologic models, 245
Hydrology, 18-19; see also Wetland hydrology
Hydromorphic soils, 140-141, 174
Hydroperiods, 33
Hydrophytes, 1
    common herbaceous species, 298-300
    common obligates, 72
    definition, 69-71
    fall-winter growth, 25, 27-28
    problems interpreting, 71-76, 78-81, 222-230
    specificity for wetlands, 71-76
    spring bloomers, 22, 24-27
    wetland ecotypes, 71-76
    winter wetness effects on, 27-28, 74

Hydrophytic vegetation
    exercises to determine, 122-124
    field indicators, 78-80, 193, 199, 203-204
    problematic communities, 222-230
    representative forested communities, 316-331
    representative shrub communities, 310-315
Hydroponics, 52
Hylocomium, 279
Hypericum, 299, 302, 372
Hypertrophied
    lenticels, 19-20, 55-57, 65-66, 79, 174, 222
    stems, 20, 55-58, 79
Hyporheic zone, 42

I

Idaho, 23, 240
    forested wetland communities, 322, 324-325
    shrub wetland communities, 313
Ilex, 82, 224, 300-301, 311, 319
Illinois, 23, 87, 236
Impatiens, 84, 86, 226
Importance value, 110-111
Inceptisols, 153, 178
Indiana, 233
Interfluve, 127, 275
Intermountain region, 77
International
    wetland classification systems, 276-285
    wetland definitions, 7-9
Intrazonal soils, 141
Iowa, 139
Iris, 72, 263, 298
Iron oxides, 66, 134-137, 171, 233
Irrigated lands, 250, 372
Isoetes, 72, 82, 226
Itea, 82
Iva, 82, 84, 307, 309

J

Jaumea, 82, 84, 299, 307
Juglans, 224, 326
Juncus, 24, 72, 74, 82, 84, 297-300, 307, 309, 370
Juniperus, 25

K

Kalmia, 60, 82, 85, 224, 301, 310, 315
Kansas, 26, 170, 223-225, 238
Kentucky forested wetland, 319
Kochia, 223, 300
Kosteletzkya, 82

L

Lactuca, 68
Laguncularia, 63, 308

Lake Ontario, 81
Lamium, 26
Land resource regions, 166
Landscape changes, 243
Laportea, 82
Larch swamps, 316, 331
Larix, 26, 80, 301, 303-304, 316, 331, 373
Lasthenia, 227
Lathyrus, 300, 307
Ledum, 71, 82, 85, 301, 310, 315, 325
Leersia, 72, 82, 226, 282, 298
Lemna, 72
Lepidium, 27, 227
Lepidocrocite, 139
Lepironia, 282
Leptochloa, 300
Lettuce, 246
Leucothoe, 311
Lichen lines, 61, 171, 201
Ligustrum, 223
Lilaea, 226
Lilaeopsis, 84
Limnogenous mire, 281
Limonium, 83-84, 307
Limosella, 82
Lindera, 26
Lindernia, 72
Line intercept method, 119, 121
Linum, 27
Liquidambar, 25, 53, 62, 68, 79, 303-304, 317-318, 321
Liriodendron, 54, 60, 224, 304, 317
Littoral ecotones, 36
Lobelia, 65, 85
Lolium, 69, 223
Lonicera, 72, 85, 223
Louisiana, 23, 27, 67, 80, 138, 145, 242
    forested wetlands, 319
Ludwigia, 57, 60, 84, 86, 299-300
Lycium, 84
Lycopersicon, 60
Lycopus, 298
Lyonia, 301, 311, 372
Lysichitum, 72, 223, 299, 329-330
Lysimachia, 85, 263
Lythrum, 60, 82, 263, 298-299, 368, 372

M

Maghemite, 139
Magnolia, 82, 303, 320
Maine, 23, 71, 156, 190-191, 242, 372
    forested wetland communities, 316
    shrub wetland communities, 310-311
Malvella, 300
Manganese oxides, 134, 139, 171, 233
Mangrove swamps
    in the U.S., 308-309
    photointerpretation, 370
Maple wetlands
    box elder, 326

red, 317, 319, 321
silver, 317, 319
Maps, 347-348
    future wetland maps, 365-366
    limitations, 348-350
    soil maps, 358-365
    use for regulatory purposes, 365
    wetland maps, 350-358
Marine
    water, 262
    wetlands, 269
Marl, 171
Maryland, 12, 23, 36, 75, 149, 169, 189-191, 238, 263, 348
    forested wetland communities, 317
    shrub wetland communities, 310-311
Marshes, 33, 39-40, 266-267, 274, 278; see also Palustrine
                emergent wetlands
    photointerpretation, 369-372
    U.S., 297-300, 306-308
Marsilea, 225-226
Massachusetts, 22, 26, 30, 189, 191, 231, 233, 348, 376
    field indicators of hydrophytic vegetation, 79, 202-203
    wetland definition, 12
    wetland delineation manual, 202-203
Melaleuca, 60, 282
Mentha, 299
Menyanthes, 67, 263, 297, 299-300
Menziesia, 223
Mertensia, 307, 323-324
Meteoric water, 262
Methane, 134, 174
Michigan, 39, 80, 234, 375
Microbial activity, 22-24, 29, 51, 133, 239
Midwest, 291, 372
Mikania, 82, 86
Mimulus, 75, 299
Minerotrophic wetland, 261, 281
Mints, 241
Minimum area for vegetation sampling, 114-116
Minimum map unit sizes, 349, 355, 359
Minnesota, 23-24, 26, 31, 86, 133, 372
    forested wetland communities, 316
    shrub wetland communities, 310-311
Mires, 281-282
Mississippi, 28, 138, 244, 291, 294
Missouri, 23, 25
Mitchella, 224
Mitella, 299
Molinia, 67
Mollic epipedon, 168, 236
Mollisols, 31, 139-140, 154, 168, 178, 235-236
Monanthochloe, 82-84, 307
Montana, 236-237
    forested wetland communities, 322-325
    shrub wetland communities, 312
Mottled depleted matrix, 170
Mottles, 149, 152, 169; see also Redoximorphic features
Muck, see Saprists
Mucklands, 246
Mucky mineral layer, 161, 167, 234
Mucky sand, 155, 157, 234

Muhlenbergia, 85, 297-299
Mycorrhizae, 66
Myrica, 67, 72, 84, 300-301, 315
Myriophyllum, 72, 82, 300

N

Najas, 72, 82
Narthecium, 67
National Research Council
    recommendations for wetland delineation, 21, 29-30,
        149, 194, 198, 214-215, 229, 247, 250
    wetland definition, 6-7
National Wetlands Inventory maps, 148, 196, 350-351, 372
    features mapped, 351-355
    strengths and weaknesses, 355-358
    uses, 356-358
Nativeness, 87
Natural area index, 87
Nebraska, 20, 23,137, 170, 296
Nemopanthus, 71, 311
Neostapfia, 227
Netherlands, 27
Nevada, 26, 372
New England, 24, 28, 38, 80, 133, 155-156, 167, 169, 238,
    295
New Hampshire, 30, 127, 191, 234-235, 242, 246
New Jersey, 13, 54, 75, 86, 148, 189-190, 231, 235, 238,
    348
    forested wetland communities, 317-318
    shrub wetland communities, 310
New Mexico, 23
    forested wetland communities, 326-327
    shrub wetland communities, 310
New York, 23, 25, 75, 80, 86, 101, 171, 189, 191, 198, 348
    field indicators of hydrophytic vegetation, 79
    wetland definition, 13
    wetland delineation manual, 203-204
Nipa, 263
Nonsoil, 129
Nontidal wetlands, 32; see also Palustrine wetlands
Normal circumstances
    Clean Water Act, 191
    Food Security Act, 191
North Carolina, 23, 26-27, 127, 135, 140, 147, 233-234,
    372, 375
    forested wetland communities, 320-321
    shrub wetland communities, 311
North Central Region, 77
North Dakota, 22-23, 137-138, 233, 250
Northeast, 25, 28, 30, 71, 77, 80, 83-84, 103, 140, 155, 168,
    223-225, 238, 274, 291, 306-308, 375
North Plains region, 77
Northwest, 77, 375; see also Pacific Northwest
Nuphar, 65, 72, 82, 84, 226, 298-300
Nymphaea, 72, 82, 263, 298-300
Nyssa, 52, 54, 57-58, 60, 62, 67-69, 73-74, 79, 81-82, 303,
    319-321

O

Oak swamps, 317, 319, 321
Obligate hydrophytes, 72, 147
Oemlaria, 223
Ohio, 23
Oklahoma, 23
Oligotrophic wetland, 281
Ombrotrophic wetland, 261
Onions, 246
Ontario, 170
Oplopanax, 322-325, 329
Oregon, 22-23, 31, 69, 145, 240, 250, 373
    forested wetland communities, 328-329
    shrub wetland communities, 312-313
Organic bodies, 161, 167, 231
Organic matter, 129, 134-135, 139, 140, 151, 161-168, 231-
    232
Organic soils, 129
    definition, 149-151
    indicators, 160-161, 165
Orontium, 25, 82, 86
Oryza sativa, 20, 60, 64, 67-68, 74, 282
Oryzopsis, 26
Osmunda, 72
Oxidation-reduction reactions, 134
Oxidized rhizospheres
    development, 66-67, 137, 241
    use in wetland identification/delineation, 67, 193, 201,
        203-204, 235, 241
Oxisols, 178, 230, 238

P

Pacific Northwest, 25, 27, 138, 140, 222-223, 307-308, 373
Paludification, 1, 40-41
Palustrine wetlands, 295
    emergent, 296-301
    forested, 302-305, 315-331
    photointerpretation, 371-375
    scrub-shrub, 301-302, 310-315
Panicum, 38, 73-74, 84, 297-298, 307-308
Parnassia, 85, 299
Parthenocissus, 225
Paspalum, 282, 298, 300
Peat, see Fibrists
Peat accumulation, 134-135
Peatland classification, 259, 280-282
Peatlands, 39, 85-86
Peat moss, 79, 174, 199, 202, 204, 296
Pedicularis, 82, 299
Peltandra, 68, 72, 82, 84, 226, 298
Pennsylvania, 23, 81, 190-191, 238, 273-274, 296
Pennsylvania forested wetland, 316
Perched layers, 133
Permafrost wetlands, 41, 133, 229, 295-296
Persea, 303, 320
Petasites, 299-300
Petunia, 68
pH, 85-86, 271

Phalaris, 85, 263, 297-299
Philoxerus, 86
Phleum, 68, 83, 223
Photointerpretation, 366-375
  estuarine wetlands, 369-371
  palustrine wetlands, 371-375
  purple loosestrife, 368, 372
  rainforests, 375
  riparian habitat, 374
  submerged aquatic vegetation, 375-376
Photosynthesis, 27
Phyla, 86
Phragmites, 60, 67, 84, 259, 263, 298-299, 307-308, 370
Picea, 60, 65, 67, 79-80, 82, 85, 222, 224, 263, 303-304, 314, 316, 322-325, 328, 331, 374-375
Piedmont, 238
Pilularia, 226
Pine flatwoods, 318-320
Pine wetlands
  eastern white, 317-318
  loblolly, 318
  lodgepole, 328, 331
  longleaf, 319
  pitch, 318
  photointerpretation, 374
  pond, 320-321
  slash, 320-321
Pinus, 25, 263, 374
  banksiana, 65, 80, 222
  contorta, 60, 67, 69, 304, 314, 328-329, 331, 374
  echinata, 52
  elliottii, 27, 60, 64, 67, 300, 303, 320-321
  monticola, 74
  palustris, 222, 300, 319
  ponderosa, 69
  rigida, 65, 75, 80, 222, 224, 318, 374
  serotina, 19, 52, 61, 301, 311, 320-321, 374
  strobus, 27, 65, 80, 222, 224, 317-318
  taeda, 19-20, 25, 27, 52, 67, 69, 74, 79, 80, 304, 318
Pit and mound relief, 226-227, 249-250
Planera, 72
Plant adaptations
  morphological, 19-20, 53-69
  physiological, 54-55
  to flooding and waterlogging, 19-20, 51-55
  to salt stress, 52, 56, 74
Plantago, 307
Plant indicators
  of disturbance, 80-81, 86-87, 223, 225
  of hydrology, 81-83
  of nativeness, 87
  of soil type, 86
  of water chemistry, 83-86
  of water sources, 81, 83
  of wetlands, 69-81
Plant specificity for wetlands, 71-76
Plant tolerance to flooding, 53-54, 64
Platanus, 53, 65, 68, 82, 304, 326
Playas, 38, 170, 195, 223, 225, 228-229, 273-274, 276, 300
Pleurozium, 85, 279

Plotless sampling methods, 107, 121-122, 212
Plot sampling methods, 114-118, 207, 209, 211-212
Plot
  size, 114-116
  number, 116-117
  shape, 117-118
Pnematophores, 55, 62-63
Poa, 82, 223
Pocosin, 195, 273, 276, 372
Pocosin communities, 311, 321
Podzolization, 234
Pogogyne, 227
Polychromatic matrix, 232
Polygala, 300
Polygonum, 72, 82, 84, 86, 223, 226, 297-300, 308
Point intercept method, 118-121, 212-213
Ponding, 133, 142-147, 246
  duration, 34
  frequency, 32
  indicators, 171-174
Pontederia, 72, 82, 84, 263, 298, 308
Poorly drained soils, 141, 146; see also Hydric soils, definition and criteria
Populus, 25-26, 54, 60, 65, 68, 79, 82, 304, 313, 319, 322-327, 329, 374
Potamogeton, 65, 67, 72, 82, 300
Potentilla, 83-85, 297, 299-300
Prairie pothole wetlands, 28, 38-39, 42, 84-85, 137, 195, 226, 236, 297, 372
Prevalence index, 113, 119-121
Primary indicators method for wetland delineation, 78-79, 191, 198-200
Problematic wetland delineation, 221-251
  field situations, 209, 249-250
  hydrologically difficult, 243-249
  soils, 159, 230-247
  wetland types, 208, 213-230
Professional judgment, 193, 218
Proserpinaca, 72
Prosopis, 373
Prunella, 223
Prunus, 81, 223-224, 245
Pseudo-gley soils, 141, 168-169, 175
Pseudotsuga, 27
Pteridium, 223
Ptilium, 279
Puccinellia, 84, 299, 306-307
Puerto Rico, 308

Q

Quebec wetland delineation method, 205
Quercus, 25-26, 53, 57-59, 65, 67, 72-73, 86, 224, 303-304, 317, 319, 321

R

Radial oxygen loss, 66-67

Ramar
   wetland classification, 279-280
   wetland definition, 8
Ranunculus, 299, 324-325
Red parent material soils, 147, 233, 238
Redox dark surface, 168
Redox potential, 30, 51, 53, 133-134, 175-176
Redox potential measurement, 175-176
Redoximorphic features, 135-137
   redox concentrations, 137, 171, 241
   redox depletions, 135-137, 241
Reduction, 30-31, 134, 139, 147, 175-176, 239
Regulatory wetland definitions, 9-13
Relative cover, 105
Releve method, 101
Rhamnus, 85, 223, 263, 297
Rhexia, 300
Rhizophora, 62, 67, 72, 308, 370
Rhode Island, 29, 191, 198, 239-240
   field indicators of hydrophytic vegetation, 79
   forested wetland communities, 317-318
   shrub wetland communities, 311
   wetland definition, 13
   wetland delineation method, 204
Rhododendron, 81-82, 84, 300, 311
Rhynchospora, 82, 84, 282, 298-300
Rice, 20; see also *Oryza sativa*
Rice paddy soils, 241-242
Riparian forest, 326
Riparian wetlands, 20, 248-249, 373-374; see also
      Palustrine wetlands
River swamps, 320-321
Roots
   adventitious, 19, 55, 60-61
   growth, 20, 25, 27, 59-60, 62
   pneumatophores, 55, 62
   shallow, 59-60
   soil water, 62, 79
Rorippa, 299
Rosa, 72, 82-84, 223-224
Rubus, 83, 85, 223
Rumex, 54, 60, 62, 64-65, 69, 84, 300
Ruppia, 72

S

Sabal, 303, 321-321
Sagittaria, 60, 72, 82, 84, 263, 298-300, 308
Salamanders, 28
Salicornia, 72, 82-84, 300, 306-307, 309, 370
Salinity regimes, 83-85
Salix, 24, 26, 60, 65, 67-68, 72, 82, 85, 263, 297, 302-304,
      312-315, 319, 323-325, 327
Salt cedar wetland, 326
Salt marshes, 38-40, 53, 74, 300; see also Estuarine
      emergent wetlands
Salt stress on plants, 52, 74
Salty soils, 31, 52, 137, 140-141, 228, 237-238
San Francisco Bay, 138
Sandy hydric soils, 136, 155-157, 159-164, 167, 231-233

Saprists, 151, 165; see also Organic soils
Sarcobatus, 237
Sarracenia, 82, 300
Saturated, but not reduced soils, 238-239
Saturation, 34, 144; see also Waterlogging
Saururus, 82, 84
Savannahs, 297-298
Saxifraga, 299
Schoenus, 282
Scirpus, 60, 72, 82, 84-86, 226, 263, 297-300, 307-308
Scleria, 282, 300
Scolochloa, 82, 85, 299
Scorpidium, 279
Scrub-shrub wetlands, see Shrub wetlands
Sea level rise, 39-41
Sedge meadows, 297
Sediment accretion in wetlands, 138, 172-173
Seed germination, 27, 68-69
Seepage wetlands, 276
Semi-arid wetlands, see Wetlands, semiarid
Semi-terrestrial soils, 141
Senecio, 82, 299-300, 324-325
Sequoia, 60
Serenoa, 68
Sesbania, 57
Sesuvium, 86
Setaria, 227
Shrub wetlands
   photointerpretation, 370, 373-373
   plant communities, 301-303, 308-315
Sinapis, 27
Sinkholes, 276, 295-296
Silver fir forest, 329
Silver maple floodplain swamp, 317
Sisyrinchium, 299
Sium, 72, 82, 84, 263, 298-299
Smilacina, 85
Smilax, 225, 311
Soil color
   describing, 130-133
   factors affecting, 135-137, 139-140
Soil composition, 129
Soil definition, 128
Soil development, 127-128
Soil drainage classes, 141
Soil survey maps, 195, 358
   comparison with NWI maps, 362-365
   features mapped, 359-364
   mapping procedures, 359
   strengths and weaknesses, 361
Soil taxonomy interpretation, 151-152, 176-177
Soil temperature, 52
Soil texture, 129-130
Soil wetness classes, 136
Solidago, 25, 72, 74, 82, 223, 297
Soligenous mire, 281
Somewhat poorly drained soils, 141, 145; see also Hydric
      soils, definition and criteria
South Carolina, 23, 26, 171
South Dakota, 237

Southeast, 30, 77, 80, 168, 233, 238, 295, 297-298, 300, 303-304, 306-309
South Plains region, 77
Southwest, 42, 77, 236-237, 300, 373
Soybeans, 246
Sparganium, 65, 72, 82, 84, 263, 299
Spartina, 38, 64-65, 72, 74, 80, 82-85, 237, 259, 263, 299, 306-309, 369
Species-area curve, 101-102, 116-118
Spergularia, 84, 86, 307
Sphagnum, 223, 279, 282, 296, 301, 310
Spiraea, 262, 311-312, 328, 372
Spodic horizon, 135, 140, 234-235
Spodosols, 139-140, 154, 175, 178, 234-235
Sporobolus, 74, 237, 300
Spring bloomers, 22, 24-27
Spruce wetlands
    black, 314, 316, 331
    blue, 322, 324
    Engelmann, 322-325, 328
    photointerpretation, 374-375
    white, 331, 375
Stagnant water conditions, 52, 60
State distribution of wetlands, 292, 295
Stellaria, 26
Stems
    hollow, 58-59
    buttressed, 56-58
Streptopus, 325
Strings, 280
Stripped matrix, 161, 232
Suaeda, 84, 307
Subalpine fir wetlands, 322-325, 328
Subhumid wetlands, see Wetlands, semiarid to subhumid
Subhydric soils, 141, 174
Submerged aquatic vegetation, photointerpretation, 375-376
Sugar cane, 241
Sulfate-reducing bacteria, 170
Sulfidic materials, 170; see also Hydrogen sulfide
Surface-water gley soils, 141
Surface water soils, 141
Swamps, 1, 266-267, 279-280, 302-304; see also Palustrine wetlands
Symphoriocarpos, 330
Symplocarpus, 24, 26, 72, 82-83, 103, 263

T

Tamarix, 327
Taraxacum, 26, 83
Taxodium, 52, 57-58, 60-63, 65, 67-69, 72, 75, 84, 303, 320-321, 373
Telluric water, 262
Tennessee, 23, 295
    forested wetland, 319
Tensiometers, 136
Terrestrialization, 35, 39
Texas, 23, 147, 168, 175, 236-237, 250
    forested wetlands, 319

Thalassia, 72
Thalia, 298
Thelypteris, 82-83, 87, 298
Thuja, 26, 60, 79-80, 85, 297, 303-304, 316, 322, 325, 331
Tidal wetlands, 31-32, 38-40, 66, 83, 140, 170, 188, 226, 266-267; see also Estuarine wetlands
Toxicodendron, 83
Transitional
    communities, 35-37
    mire, 281
Triadenum, 81
Trichostema, 227
Trifolium, 223
Triglochin, 72, 83-85, 299, 306-307
Trollius, 299
Tropical wetlands, 229-230
Tsuga, 59-60, 65, 80-81, 86, 222-224, 303-304, 314, 318, 329, 331
Tuctoria, 227
Tundra, 358, 372
Typha, 25, 38, 60, 64, 67-68, 72, 82, 84, 263, 298-300, 307, 370

U

Ulmus, 25-26, 58-60, 65, 68, 75, 86, 304, 319-320
Ultisoils, 154, 178, 238
Ulva, 82, 309
Umbric epipedon, 168
U.S. Army Corps of Engineers, 189-191
    wetland definition, 9-10
    wetland delineation manual, 192-194, 205-208
U.S.D.A. Natural Resources Conservation Service, 190
    hydric soils lists, 142, 146, 148-150
    soil surveys, 358-365
    wetland definition, 10-11, 265
    wetland delineation, 195-196
    wetland mapping conventions, 195-196
U.S.D.A. Soil Conservation Service, see U.S.D.A. Natural Resources Conservation Service
U.S. Environmental Protection Agency
    wetland definition, 9-10
    wetland delineation, 189-191, 193, 195
    wetland delineation manual, 193-195, 208-209
U.S. Fish and Wildlife Service
    wetland classification system, 198, 265-272, 274-276
    wetland definitions, 4-6
    wetland mapping, 169, 189, 198, 350-358, 362, 365-366
U.S. wetlands
    distribution, 291-294
    estuarine, 306-309
    palustrine, 294-305, 310-331
Upper Midwest, 28, 42
Urtica, 83
Utah forested wetlands, 322-325
Utricularia, 72, 85-86

V

Vaccinium, 25, 27, 72, 83, 85, 223-224, 300-301, 310-313, 315, 325, 328, 372
Vallisneria, 72
Variable plot method, 107
Vegetation sampling, 101-122
    exercises, 122-124
    line intercept method, 119, 121
    locating sampling areas, 103
    plotless methods, 107, 121-122
    plot (or quadrat) methods, 114-118, 207, 209, 211-212
    point intercept method, 118-121
    releve method, 101
    sampling variables, 104-114
    species-area curve, 101-102, 116-118
    stand identification, 102
    strata, 103-104
    timing of sampling, 103
Venezuela, 237
Veratrum, 322
Vermont, 13, 23, 190
Vernal pools, 38, 170, 226-227, 273, 276
Veronica, 298-299
Vertisols, 168, 178, 230, 236-237, 241
Very poorly drained soils, 141, 146; see also Hydric soils, definition and criteria
Viburnum, 224, 263, 300, 311
Vinca, 26
Virginia, 25, 54, 75, 189, 348
    forested wetland, 320
Virgin Islands, 308

W

Washington
    District of Columbia, 26
    forested wetland communities, 328-330
    state, 223, 240
Washingtonia, 327
Waterbirds, 20
Water chemistry, 83-86, 137, 260-261
Water classes, 278
Waterlogging
    effects on soils, 30-31, 51-54, 133-138, 140, 146-147, 151
    effects on vegetation, 20, 52-55, 57, 60, 69, 74
    environmental changes, 51-53
Water regimes, 269, 272
Water-stained leaves, 171, 174
Weighted averages, 111-114
West coast, 75, 226-227
West Virginia, 238
Wetland accretion rates
    peat accumulation, 134-135
    sedimentation, 138, 172-173
Wetland animals, 20, 28, 30
Wetland alterations, 39-42; see also Drainage
Wetland classification
    Canada, 276-279

    early ecological systems, 263-264
    ecosystems, 267-270
    ecosystem form/energy source, 263
    disturbance criteria, 274
    hierarchial approach, 257-258, 267-279
    horizontal approach, 257-258, 265-267, 279-280
    hydrogeomorphology, 272-276
    hydrology, 260, 262, 269, 272, 274-275, 278, 282-285
    landscape position and landform, 262, 274-276, 281-282, 285
    nutrient status, 281
    origin, 262
    other countries, 279-285
    peatland, 280-282
    shape, 285
    size, 262
    soil or substrate type, 262, 271
    United States, 264-276
    vegetation, 259, 269, 271, 277-279, 281
    water chemistry, 260-261, 270-271, 273
    watershed areas, 281-282
Wetland definitions, 1-14
    Austrialia, 8-9
    field applications, 13-14
    Canada, 7-8
    Fish and Wildlife Service, 4-6
    National Research Council, 6-7
    Ramsar, 8
    Shaler's definition (1890), 3
    United States
        federal regulatory, 9-11
        scientific, 3-7
        state regulatory, 11-13
    Zambia, 9
Wetland delineation, 187
    complex landscapes, 249-250
    evolution of methods, 188-191
    hydrologically problematic wetlands, 242-245, 247-249
    key concepts, 17-43
    National Research Council findings, 214-215, 229
    problematic plant communities, 222 230
    problematic soils, 299-247
    rocky landscapes, 250
    tips, 215-218
    using various manuals, 192-205
Wetland determination, 187, 195-196
Wetland ecotones, 34-39
Wetland formation, 1-2, 127, 294-296
Wetland functions, 20, 24, 28-30, 259-260
Wetland hydrology, 1, 18-30; see also Water regimes
    common types, 31-34, 260-261
    critical depth of saturation, 20, 231
    definition, 19
    duration of flooding or waterlogging, 19-20
    ecological considerations, 20, 24-30
    environmental effects, 51-53
    frequency, 20, 32
    hydroperiods, 33
    indicators, 192, 193, 199-204
    inflow, 32, 260
    influence of vegetation, 19

minimum threshold, 19, 247
modifications, 39-41, 243, 246
monitoring, 231, 244-245, 251
nontidal, 31-32
outflow, 32, 260
problems for identification, 242-245, 247-249
throughflow, 32, 260
tidal, 31-32
timing of wetness, 20
Wetland identification, 18, 187
evolution and use of methods, 188-191
National Research Council recommendations, 214-215, 229
using plants, 18-19, 76, 78-81, 189, 199, 201-205
using soils, 18-19, 37, 189, 199-200
using various manuals, 192-205
Wetland landforms and landscapes, 1-2, 127, 216, 222, 230, 235-237, 239, 262, 274-276, 281-282, 285, 294-296
Wetland manuals, 187-213
federal, 192-197, 205-213
primary indicators method, 78-79, 191, 198-200
Quebec, 205
states, 200-205
Wetland mapping, 187, 347-348, 365-366
Food Security Act, 195-196
limitations, 348-350
National Wetlands Inventory maps, 350-358
soil survey maps, 358-365
Wetland photointerpretation, 357-358, 366-369
aquatic beds, 371
estuarine wetlands, 369-371
palustrine wetlands, 371-375
submerged aquatic vegetation, 375-376
Wetland plant communities
emergent, 296-301, 306-308
intertidal shores, 309
forested, 303-305, 316-331
scrub-shrub, 301-303, 308-315
Wetland plant field guides, 87-89
Wetland plant lists, 76-78
Wetland plant species, 72, 298-300, 310-331; see also Individual entries
Wetland soils, see Hydric soils
Wetlands, see also Wetland types
aerobic conditions, 30-31, 52-53, 147, 238-239
arid and subhumid regions, 42-43
cyclical, 42-43, 226-229
depressional, 7, 273-275, 285
drier-end, 36-37, 80, 232, 248-249
dominated by dry-site species, 80
ecotones, 34-39
episodic, 42
estuarine, 268-270, 273-274
farmed, 34, 195-196, 223, 225-226, 241-242, 246-247
flats, 273-274, 285

floodplain, 273
groundwater-driven, 247-248
interdunal, 271, 296
intermittent, 42
marine, 268-269, 275
lacustrine, 268-270, 273-274, 283
lentic, 275, 284-285
lotic, 275, 284-285
overflow, 7
partly drained, 243-245
prior converted, 195
riparian, 20, 273, 304
riverine, 273
seasonally saturated, 32, 36-37, 248, 260
seasonally variable, 225-226
seepage, 7
semiarid to subhumid, 20, 31
slope, 273
terrene, 275
vegetation dynamics, 39-40, 222-229
Wetland trends detection, 370-371
Wetland types in the U.S., 266-267
estuarine, 268-269, 275, 306-309
palustrine, 294-305, 310-331
Wet meadows, 20, 33, 38, 67, 110, 140, 223, 297-300
Willow-dominated wetlands, 31, 34, 312-315, 319, 322-325, 327
Winter effects on plants, 27-28
Wisconsin, 25, 27, 69, 80, 136, 260
forested wetland communities, 316-317
glacial effects, 242, 294
shrub wetland communities, 310
wetland definition, 11, 13
wetland delineation method, 204-205
Wood frogs, 28
Woodwardia, 72, 82
Wyoming, 22, 170
forested wetlands, 322-325

X

Xyris, 72

Z

Zambian wetland definition, 9
Zannichellia, 72
Zea, 60, 68-69, 246
Zenobia, 301, 311
Zizania, 65, 72, 82, 84, 226, 263, 298, 308
Zizaniopsis, 84, 298
Zonal soils, 141
Zostera, 72